# Statistical Extremes and Applications

# NATO ASI Series

## Advanced Science Institutes Series

*A series presenting the results of activities sponsored by the NATO Science Committee, which aims at the dissemination of advanced scientific and technological knowledge, with a view to strengthening links between scientific communities.*

The series is published by an international board of publishers in conjunction with the NATO Scientific Affairs Division

| | | |
|---|---|---|
| A | Life Sciences | Plenum Publishing Corporation |
| B | Physics | London and New York |
| C | Mathematical and Physical Sciences | D. Reidel Publishing Company Dordrecht, Boston and Lancaster |
| D | Behavioural and Social Sciences | Martinus Nijhoff Publishers |
| E | Engineering and Materials Sciences | The Hague, Boston and Lancaster |
| F | Computer and Systems Sciences | Springer-Verlag |
| G | Ecological Sciences | Berlin, Heidelberg, New York and Tokyo |

Series C: Mathematical and Physical Sciences Vol. 131

# Statistical Extremes and Applications

edited by

J. Tiago de Oliveira

Academy of Sciences of Lisbon
Faculty of Sciences of Lisbon
Center for Statistics and Applications (I.N.I.C.), Lisbon, Portugal

Springer-Science+Business Media, B.V.

Proceedings of the NATO Advanced Study Institute on
Statistical Extremes and Applications
Vimeiro, Portugal
31 August-14 September 1983

Library of Congress Cataloging in Publication Data

NATO Advanced Study Institute on Statistical Extremes and Applications (1983: Vimeiro, Lisbon, Portugal)
Statistical extremes and applications.

(NATO ASI series. Series C, Mathematical and physical sciences; vol. 131)
"Proceedings of the NATO Advanced Study Institute on Statistical Extremes and Applications, Vimeiro, Portugal, 31 August-14 September, 1983"—T.p. verso.
"Published in cooperation with NATO Scientific Affairs Division."
Includes index.
1. Mathematical statistics—Congresses. 2. Extreme value theory—Congresses.
I. Tiago de Oliveira, J. II. Title. III. Series: NATO ASI series. Series C, Mathematical and physical sciences; vol. 131.
QA276.A1N38 1983 519.5 84-15144
DOI 10.1007/978-94-017-3069-3


Originally published by D. Reidel Publishing Company, Dordrecht, Holland in 1984
MyCopy version of the original edition 1984


www.springer.com/mycopy

To the memory of

E. J. Gumbel

TABLE OF CONTENTS

SECTION IV: EXTREMES AND STOCHASTIC PROCESSES



SECTION V: APPLICATIONS



B - SPECIFIC LECTURES



C - WORKSHOPS

SECTION I: RATES OF CONVERGENCE



SECTION II: CONCOMITANTS



SECTION III: COMPUTING

# PREFACE

The first references to statistical extremes may perhaps be found in the Genesis (The Bible, vol.I): the largest age of Methu'selah and the concrete applications faced by Noah — the long rain, the large flood, the structural safety of the ark —. But as the pre-history of the area can be considered to last to the first quarter of our century, we can say that Statistical Extremes emerged in the last half-century. It began with the paper by Dodd in 1923, followed quickly by the papers of Frechet in 1927 and Fisher and Tippett in 1928, after by the papers by de Finetti in 1932, by Gumbel in 1935 and by von Mises in 1936, to cite the more relevant; the first complete frame in what regards probabilistic problems is due to Gnedenko in 1943. And by that time Extremes begin to explode not only in what regards applications (floods, breaking strength of materials, gusts of wind, etc.) but also in areas going from Probability to Stochastic Processes, from Multivariate Structures to Statistical Decision. The history, after the first essential steps, can't be written in few pages: the narrow and shallow stream gained momentum and is now a huge river, enlarging at every moment and flooding the margins. Statistical Extremes is, thus, a clear-cut field of Probability and Statistics and a new exploding area for research.

The purpose of the NATO-ASI "Statistical Extremes and Applications" was to obtain a complete perspective of the field, also with a series of promising directions of research and some recent results.

Thus the program of the book:
— the state of art of Statistical Extremes and Applications through the sequence of Probabilistic Aspects (univariate problems), Statistical Decision (univariate data), Multivariate Structures (Probability and Statistics), Stochastic Processes and, finally, some applications (to Hydrology, to Structural Safety, to Meteorology and to Insurance) followed by some lectures on specific topics;
— a set of workshops dealing with Rates of Convergence, Concomitants, Computing Methods and Seismic Problems;

— a set of contributed papers widely dispersed, with new results and suggestions for research.

The best thanks must be given to NATO (Scientific Affairs Division) for the financial support given to the meeting and also to the portuguese sponsors — Secretary of State for Universities, National Institute for Scientific Research, National Board for Scientific and Technological Research, General Post Office and Calouste Gulbenkian Foundation; without their help it would be impossible to realize this gathering.

The cooperation of all participants was very helpful for the exchange of ideas and to the good and friendly atmosphere throughout the meeting; my best thanks to all of them.

At last, but surely not the least, I want to thank the help and cooperation of the members of the Department of Statistics of the University of Lisbon, with special relevance to the co-directors M. Ivette Gomes and K.F. Turkman; it is a very good group that crossed the desert during the organization time and continues to work on...

J. Tiago de Oliveira

# STATISTICAL EXTREMES: THEORY AND APPLICATIONS, MOTIVATION AND PERSPECTIVES

Benjamin Epstein

Technion, Israel Institute of Technology

In this opening lecture we try to put in perspective some aspects of the theory of statistical extremes and emphasize the interplay between theory and its applications.

We begin by sketching the development of the classical theory of extremes, which is concerned primarily with the asymptotic distribution of the maximum and minimum of independent, identically distributed random variables. Particular emphasis is placed on the important roles played by E.L. Dodd and M. Fréchet in providing the conceptual framework and change in direction of thought needed to develop the asymptotic theory of extremes. It was Dodd in 1923 who was the first to relate the asymptotic growth of the maximum of n i.i.d. random variables to the rate at which the right tail of the underlying p.d.f. falls off to zero and Fréchet in 1927, who introduced the key idea that if there exists a limit law for maxima (minima), then it must be stable in the sense that the distribution of the maximum (minimum) of n i.i.d. observations drawn from it must be of the same type, except for a linear transformation. In retrospect, we realize that the ideas of Dodd and Fréchet were crucial to the development of the asymptotic theory of extremes.

We then emphasize the central role played by the exponential distribution in the asymptotic theory of extremes and, in particular, the relationship between extreme value distributions and the exponential distribution. This is followed by a discussion of the unique position and special contributions of E.J. Gumbel in the field of extremes.

An account is given of Griffith's flaw theory and the "weakest link"



*J. Tiago de Oliveira (ed.), Statistical Extremes and Applications, 1–2.*

principle which provide the empirical justification for applying extreme value theory to problems of material strength and the so-called size effect phenomenon.

We conclude with a number of examples, taken from a variety of fields, which illustrate the wide applicability of the statistical theory of extremes.

INTRODUCTION, ORDER STATISTICS, EXCEEDANCES.
LAWS OF LARGE NUMBERS.

Janos Galambos

Temple University, Philadelphia, Pa.

ABSTRACT

The paper is to discuss the distribution theory of order statistics for finite sample sizes. Although the detailed exposition is on the independent and identically distributed case, some results, mainly in terms of inequalities, are also presented for dependent structures.

Growth properties of order statistics are discussed from two different points of view: an order statistic exceeding a certain level in terms of the sample size, and the distributional properties of future observations exceeding previously observed order statistics.

In the form of asymptotic results, laws of large numbers are discussed.

## 1. INTRODUCTION

Let X be a random variable with distribution function F(x). Let $X_1, X_2, \ldots, X_n$ be independent observations on X. Then an arrangement of the $X_j$ in a nondecreasing sequence $X_{1:n} \leq X_{2:n} \leq \cdots \leq X_{n:n}$ is called the sequence of order statistics (of the $X_j$). Note that, while the identification of the subscript j in $X_j = X_{r:n}$ is ambiguous in the above definition, the value of $X_{r:n}$ is uniquely defined. If the population distribution F(x) is continu-



J. Tiago de Oliveira (ed.), Statistical Extremes and Applications, 3–17.

ous, then equalities among the $X_j$ occur only with probability zero, but for discontinuous F(x) identical observations are possible with positive probability. This distinction, however, is theoretical only; all observations come in discrete form (since one can record data up to a certain number of decimal digits of units of measurements), and one can therefore argue whether these discrete observations are good approximations to X with a continuous F(x), or whether (X,F(x)) is a sufficiently good approximation to (model for) the discrete observations $X_1, X_2, \ldots, X_n$. While at this early stage the posed question sounds merely philosophical, our aim of producing an asymptotic theory of extremes forces us to take side; namely, as it will turn out, for most discrete distributions, (linearly) normalized extremes do not have an asymptotic law, but for the most widely applied continuous populations a nice theory exists. We therefore adopt the view that the abstract model (X,F(x)) is well determined (whether we know F(x) or not), and the observations $X_j$ are approximations to X; in fact, we assume that this approximation is good enough for enabling us to freely interchange $X_j$ and X in theoretical discussion (we shall see the huge price we have to pay if we make a small mistake in this).

In addition to guaranteed distinct values of observations, the advantage of continuous population distribution F(x) is that one can transform the observations $X_j$ to

$$U_j = F(X_j),\ \text{or}\ U_j^* = 1 - F(X_j),\ 1 \le j \le n, \qquad (1)$$

leading to uniformly distributed variables on the interval (0,1), or to (log z is the natural logarithm)

$$Y_j = -\log F(X_j),\ \text{or}\ Y_j^* = -\log(1-F(X_j)),\ 1 \le j \le n, \qquad (2)$$

whose distribution is unit exponential (that is, their distribution is $1-e^{-x}$, $x \ge 0$), and, for $1 \le r \le n$,

$$U_{r:n} = F(X_{r:n}),\ U_{n-r+1:n}^* = 1-F(X_{r:n}), \qquad (3)$$

and

$$Y_{n-r+1:n} = -\log F(X_{r:n}),\ Y_{r:n}^* = -\log(1-F(X_{r:n})). \qquad (4)$$

Hence, for smooth enough function F(x), $X_{r:n}$ can be studied through the corresponding order statistics of uniform or exponential observations.

The theory of order statistics comes up frequently both as tools in statistical inference and in stochastic model building.

H.A. David (1981) describes a large variety of statistical applications, while the books Gumbel (1958), de Haan (1970) and Galambos (1978) are devoted to the theory and applications of extremes (the last one developing a theory without the assumption of independence of the $X_j$ as well as covering the multivariate case).

The following few examples are selected here both to show the flavor of applications of order statistics and to introduce the classification of order statistics as extremes and central terms.

Example 1. Let X be the random life of an electric bulb produced at a specific factory. Then, selecting at random 500 bulbs, one can mark the bulbs and record their actual lives $X_j$, $1 \le j \le 500$. However, a more natural way of recording these random values is in the order as they burn out; i.e., the order statistics $X_{r:n}$, $r \ge 1$, are naturally recorded by the experiment. One can, of course, make conclusions about the quality of the lot after some have burned out, thus saving a considerable amount of time in "completing" the experiment. This example is also suitable to observe that even in such a controlled experiment, one cannot always guarantee the independence of the observations $X_j$ because the selection of the 500 bulbs must be without replacement (a bulb selected is destroyed by the nature of the experiment), even "almost independence" is not guaranteed except if 500 is a very small proportion of the total lot.

If the experiment of the preceding example is terminated after a very few bulbs have burned out, and then a decision is made on the quality of the lot, we say that the decision was made on the base of extreme observations. We actually call $X_{1:n}$ and $X_{n:n}$ extremes, regardless of the value of n. Other order statistics are also called extremes, but only in a limiting sense: if n tends to infinity and if k is a fixed value (does not depend on n), then the order statistics $X_{k:n}$ and $X_{n-k:n}$ are called extremes; for emphasis, the former are also called lower extremes, and the latter upper extremes. The difference $R_n = X_{n:n}-X_{1:n}$ is the range of the sample, and it is widely applied as a quick estimator of the standard deviation of the population distribution. Another statistical application of extremes is in a negative sense through the so called trimmed mean: statisticians like to believe that the extremes are not reliable and thus their hope is to get a better estimator of the expected value if the sample mean is calculated without the extremes. The present author's view, however, is just the opposite: extremes are the best signals when a model is wrong. Although outliers do exist due to the sensitivity of extremes, in many instances not the outlier but the whole experiment ought to be rejected and repeated, when an accurate decision is to be made.

Example 2. If $X_j$ is the water level of a river at a given location on day j (starting point is immaterial), then $X_{n:n}$ is the only value needed for the analysis of floods, and $X_{1:n}$ for droughts.

Example 3. If a system of n components fails to function whenever less than k of its components function (so called k-out-of-n system), then, denoting by $X_j$ the random life of component j, the life of the system is $X_{n-k+1:n}$.

Example 4. When a sheet of metal is (hypothetically) divided into n pieces, and if $X_j$ is the strength of the j-th piece, the strength of the sheet, by the weakest link principle, is $X_{1:n}$. However, the strength of a bundle of threads is certainly neither the minimum nor the maximum of the strengths of the individual threads. Namely, if $X_j$ is the strength of the j-th thread, then the bundle will not break under a load S if there are at least k threads in the bundle, each of which can withstand a load S/k. Hence the strength $S_n$ of a bundle of n threads satisfies

$$S_n = \max\{(n-k+1)X_{k:n}: 1 \leq k \leq n\}.$$

Although $S_n$ is represented as a maximum, after some calculation, it turns out that its distribution is related to a central order statistic $X_{r:n}$: with some $0 < t < 1$, r is the integer part of nt. Such central terms are also called sample quantiles. For their relevance in statistical inference, see David (1981) (including all central terms such that $r/n \to t$ as $n \to +\infty$).

Although the classification of order statistics into extremes and central terms (and those in between) is accurate when $n \to +\infty$, for a fixed value of n, their meaning is very ambiguous. For example, if n = 200, $X_{190:200}$ can be viewed as an extreme (190 = n-10), as a quantile (190 = 0.95n), or as something in between ($190 = n - 1/2\ (2n)^{1/2}$). Therefore, their asymptotic distribution theory is meaningless without a good estimate of error terms.

## 2. DISTRIBUTION OF ORDER STATISTICS IN THE I.I.D. CASE

Although the examples of the preceding paragraph clearly illustrate that the basic observations $X_j$, $1 \leq j \leq n$, are not always independent and/or identically distributed, as a first approximation to the theory of order statistics, we assume that the $X_j$ are i.i.d. (independent and identically distributed).

The basic distribution theory of order statistics can easily be developed by a reference to the binomial distribution. Namely, let

$$I_j(x) = \begin{cases} 1 & \text{if } X_j < x, \\ 0 & \text{otherwise.} \end{cases}$$

Then the indicator variables $I_j(x)$ are independent, $P(I_j(x) = 1) = F(x)$, and

$$F_{r:n}(x) = P(X_{r:n} < x) = P(\sum_{j=1}^{n} I_j(x) \geq r) \tag{5}$$

$$= \sum_{j=r}^{n} \binom{n}{j} F^j(x)[1-F(x)]^{n-j}.$$

In particular,

$$F_{1:n}(x) = P(X_{1:n} < x) = 1-(1-F(x))^n, \tag{6}$$

and

$$F_{n:n}(x) = P(X_{n:n} < x) = F^n(x), \tag{7}$$

which formulas can, of course, be deduced directly without any reference to (5).

One can also write the binomial form (5) in the familiar integral form of $F_{r:n}(x)$:

$$F_{r:n}(x) = \frac{1}{B(r,n-r+1)} \int_0^{F(x)} t^{r-1}(1-t)^{n-r} dt, \tag{8}$$

where $B(r,n-r+1)$ is the beta function (replace $F(x)$ by 1 in the integral above). There are several important consequences of (8). One is an observation by Huang (1974), saying that the distribution of a single order statistic uniquely determines the population distribution $F(x)$. Another one is that it provides an easy method of computing given percentage points of $X_{r:n}$ in samples from a given population; namely, the incomplete beta function (the right hand side of (8) with a general variable $z$, say, for $F(x)$) has been tabulated extensively. One can also apply (8) for computing moments of $X_{r:n}$, which computations can be simplified by developing recursive formulas for the moments; we refer to David

(1981, pp. 46-48) for a good collection of such recursive formulas from the literature.

Assume now that $f(x) = F'(x)$ exists. Then, from (8),

$$f_{r:n}(x) = F'_{r:n}(x) = \{1/B(r,n-r+1)\}F^{r-1}(x)[1-F(x)]^{n-r}f(x).$$

Furthermore, the joint density $f_{r_1,r_2,\ldots,r_k:n}(x_1,x_2,\ldots,x_k)$ of $X_{r_j:n}$, $1 \le j \le k$, can easily be determined by a combinatorial argument. Namely, $f_{r_1,r_2,\ldots,r_k:n}(x_1,x_2,\ldots,x_k)$ represents the probability that, to every $t$, $1 \le t \le k$, there is a $j$ such that $x_t \le X_j < x_t + \Delta x_t$, and exactly $r_1-1$ of the $X_u$ fall below $x_1$, $r_2-r_1-1$ fall between $x_1$ and $x_2$, and in general, $r_t-r_{t-1}-1$ fall between $x_{t-1}$ and $x_t$ (besides the $X_j$ specified earlier). We thus have ($r_1 < r_2 < \ldots < r_k$)

$$f_{r_1,r_2,\ldots,r_k:n}(x_1,x_2,\ldots,x_k) =$$

$$n! \prod_{t=0}^{k} \frac{[F(x_{t+1})-F(x_t)]^{r_{t+1}-r_t-1}}{(r_{t+1}-r_t-1)!} \prod_{t=1}^{k} f(x_t),$$

where $-\infty = x_0 < x_1 < x_2 < \ldots < x_k < x_{k+1} = +\infty$, and $r_0 = 0$ and $r_{k+1} = n$. In particular, if $F(x) = 1-e^{-x}$, $x > 0$, and thus $f(x) = e^{-x}$ ($x > 0$), the substitution $r_t = t$ and $d_{t:n} = X_{t+1:n}-X_{t:n}$, $0 \le t \le k < n$, where $X_{0:n} = 0$, leads to the following result of Sukhatme (1937).

Theorem 1. If $F(x) = 1-e^{-x}$, $x > 0$, then the differences $d_{t:n}$ are independent random variables. The distribution of $(n-t)d_{t:n}$ is $F(x)$ itself. In particular, the distribution of $X_{t+1:n}$ coincides with that of

$$\frac{X_1}{n} + \frac{X_2}{n-1} + \ldots + \frac{X_t}{n-t}\ .$$

With the transformations at (1) - (4), several equivalent forms of Theorem 1 can be given. For example, in view of (4), the ratios $X_{t+1:n}/X_{t:n}$ are independent for uniform, or for Pareto families. This, of course, implies that the independence of the ratios of consecutive order statistics is insufficient to determine the population distribution. Interestingly, however, the independence of the differences $d_{t:n}$ uniquely determines the exponential family (the exponential distributions with location and scale parameters). In fact, much less is required to characterize exponentiality; see Chapter 3 in Galambos and Kotz (1978).

Characterization theorems are very important in connection with extreme value theory. Namely, (6) and (7) are very sensitive to small deviations in $F(x)$, and thus a non-exact model $(X,F(x))$ would give a useless result in applied fields, leading to one of the following two possibilities: either one would rely on (6) or (7), and dams would collapse, or (6) (or (7)) would be used as a first approximation, and one would build 10 times, or 50 times more reliable dams than the computations suggest. In other words, the distribution (6) and (7) of extremes can be utilized only if the model $(X,F(x))$ is exact; no approximation can be accepted (neither theoretical, nor empirical). We refer to Galambos (1978, p. 90) for numerical examples, or the reader can try to make computations in a Poisson model with a large parameter, when one is forced to use normal tables after a certain point (compare two tables, one which suggests the utilization of normal tables for values of the parameter exceeding 25, and another which still computes Posson values at this parameter value).

The whole essence of the asymptotic theory of extremes is to provide ways of avoiding the application of (6) and (7) by showing that, with suitable numbers $a_n$ and $b_n > 0$,

$$F^n(a_n + b_n z) \sim H(z),$$

where $H(z)$ is one of a limited number of possible distribution functions (in fact, one of three). Hence, operating with the limiting form $H(z)$, a problem of maxima reduces to a location and scale problem, for which very robust methods are available. The approach is, of course, the same when minima are involved. Details are given elsewhere in these same proceedings.

## 3. THE DISTRIBUTION OF ORDER STATISTICS OF DEPENDENT RANDOM VARIABLES.

The definitions and notations of the introduction remain unchanged when the random variables $X_1, X_2, \ldots, X_n$ are arbitrary, except that we have to expand the concept of the model we are dealing with. The model now has to include the distribution function $F_j(x)$ of $X_j$ for each j as well as the dependence structure of the $X_j$, $j \geq 1$. There are five major models for which the extremes $X_{1:n}$ and $X_{n:n}$ have been studied with some detail.

Model 1. The $X_j$ are independent, but the $F_j(x)$ are not identical.

It is easily established that

$$F_{n:n}(x) = F_1(x)F_2(x)\ldots F_n(x)$$

and

$$F_{1:n}(x) = 1-(1-F_1(x))(1-F_2(x))\ldots(1-F_n(x)).$$

Model 2. The $X_j$ are a finite segment of an infinite sequence of exchangeable random variables.

By the classical de Finetti theorem (de Finetti (1930)), there is a random variable $F(x, . )$ such that, almost surely, $F(x, . )$ is a distribution function in x, and

$$F_{n:n}(x) = E\{F^n(x, . )\}$$

and

$$F_{1:n}(x) = 1-E\{[1-F(x, . )]^n\}.$$

Model 3. The $X_j$ are exchangeable, but not known to have come from an infinite sequence.

There is a version of de Finetti's theorem applicable to this case, in which, instead of n-th powers, binomial coefficients take place in the expectations. When it was first established by Kendall (1967) (in another context), it looked promising for extreme value theory, but it was later found in Galambos (1973) that the same formula is applicable to arbitrary dependent sequences which certainly implies that, without further assumptions, the formula is too general for meaningful results. Criteria guaranteeing the extendibility of a sequence of n variables to a larger sequence of N exchangeable random variables proved very powerful for the asymptotic theory of extremes of dependent observations (see Chapter 3 in Galambos (1978)). However, the theory of extendibility of exchangeable variables into a larger set is at a very early stage (see Galambos (1982) and Spizzichino (1982)).

Model 4. The joint distribution of the $X_j$ is (multivariate) normal (so called Gaussian sequences).

Evidently

$$F_{n:n}(x) = N(x,x,\ldots,x),$$

where $N(x_1,x_2,\ldots,x_n)$ is the appropriate n-dimensional normal distribution function. In terms of the survival function of N, one can easily express $F_{1:n}(x)$ as well.

While these formulas seem simple, one actually has to know all correlation coefficients before they are of any use for computation. However, an asymptotic theory avoids this requirement.

Model 5. The $X_j$ form a stationary sequence.

Exact formulas for $F_{n:n}(x)$ and $F_{1:n}(x)$ cannot be provided in general, but the asymptotic theory of extremes for this model is well developed.

The asymptotic theory of extremes of each of these five models is discussed in Chapter 3 of Galambos (1978). The actual contributors to this field were S.M. Berman (Models 2,4 and 5), the present author (Models 2 and 3), M.J. Juncosa, D.G. Mejzler, J. Tiago de Oliveira, and I. Weisman (Model 1), Y. Mittal (Model 4), and M.R. Leadbetter, R.M. Loynes and G. O'Brien (Model 5). Exact references are given in the book by Galambos (1978), and actually several models are discussed in detail in these same Proceedings. See also Deheuvels (1981).

It was B. Epstein who advocated the development of asymptotic extreme value theory for dependent random variables. It was at a time when it was revolutionary to claim that distributions unrelated to the normal are sometimes better than the normal in engineering applications (in addition to Weibull, it was Epstein again whose work made the exponential distribution so prominent in applications). His work is summarized in the survey Epstein (1960). The inequalities used by him have been improved considerably; see Galambos (1984) for recent references.

Let us quote just one inequality of general appeal. Let $m_n = m_n(x)$ be the number of those $X_j$ for which $X_j \geq x$, $1 \leq j \leq n$. Let $s_{1,n}$ and $s_{2,n}$ be two statistics such that $E(s_{1,n}) = E(m_n)$ and $E(s_{2,n}) = E[m_n(m_n-1)]$. Then, with arbitrary integer $1 \leq k \leq n-1$,

$$1-H_n(x) \geq 2s_{1,n}/(k+1) - s_{2,n}/k(k+1)$$

is consistent in the sense, that the expected value of the right hand side is a lower bound for $1-H_n(x)$. Because k is arbitrary, one can preassign its value or it can also be computed from the data. The best lower bound is achieved above if k-1 is chosen as the integer part of $s_{2,n}/s_{1,n}$.

Other inequalities, providing both upper and lower bounds on $1-H_n(x)$, are available in terms of the binomial moments of $m_n$. They can also be restated in statistical forms, that is, instead of moments one can use statistics whose expected value coincides with the corresponding binomial moments of $m_n$. Such inequalities are always applicable when it is not sure which model is to be applied.

## 4. GROWTH RATES FOR EXTREMES: EXCEEDANCES

We return to the i.i.d. case, and we investigate the growth rate of $X_{r:n}$ from the following point of view. After the n independent observations $X_j$, $1 \leq j \leq n$, have been taken, and their order statistics $X_{r:n}$, $1 \leq r \leq n$, have been recorded, additional independent observations $X_{n+j}$, $1 \leq j \leq m$, are taken from the same population, whose distribution function F(x) is assumed to be continuous. We want to compare the sequences $\{X_{r:n}\}$ and $\{X_{r:n+m}\}$. For this purpose, we determine the probability that exactly k of the "future observations" $X_{n+j}$, $1 \leq j \leq m$, exceed $X_{r:n}$. We start with the fact that, for every j,

$$P(X_{t:n} < X_{n+j} < X_{t+1:n}) = 1/(n+1),\ 0 \leq t \leq n, \tag{9}$$

where $X_{0:n} = -\infty$ and $X_{n+1:n} = +\infty$. Hence, probabilistic statements on the positions of the future observations $X_{n+j}$ among the order statistics $X_{t:n}$ can be expressed in terms of placing m balls, each with equal probabilities, into n+1 urns. It should be noted, however, that even though the observations $X_{n+j}$ are assumed independent, in the urn model translation, the balls are not placed independently of each other into the urns. As a matter of fact, given that $X_{n+1}$ falls between $X_{t:n}$ and $X_{t+1:n}$, then $X_{t:n} = X_{t:n+1}$, $X_{n+1} = X_{t+1:n+1}$ and $X_{t+1:n} = X_{t+2:n+1}$, and thus, by (9),

$$P(X_{t:n} < X_{n+2} < X_{t+1:n} | X_{t:n} < X_{n+1} < X_{t+1:n}) = 2/(n+2),$$

while the unconditional value, again by (9), equals 1/(n+1). The symmetry of the problem in the variables $X_{n+j}$ can nevertheless be exploited by noticing that every outcome in the distribution of the m balls into the n+1 urns has the same probability. Hence, if we denote by $x_t$ the number of those $X_{n+j}$ which fall between $X_{t:n}$ and $X_{t+1:n}$, then the number of possible outcomes in the distribution of the balls into the urns is the same as the number of solutions of the equation

$$x_1+x_2+\ldots+x_{n+1} = m, \; x_t \geq 0 \text{ integer.} \tag{10}$$

Its solution is well known to be the binomial coefficient $\binom{m+n}{n}$. Now, for evaluating the probability that exactly k of the m future observations exceed $X_{r:n}$, we have to count the number of favorable distribution of the balls into the urns corresponding to this event. This means that we have to determine the number of solutions of

$$x_1+x_2+\ldots+x_r = m-k \text{ and } x_{r+1}+\ldots+x_{n+1} = k,$$

where, just as before, each $x_t \geq 0$ integer. From the solution of (10) we have the value $\binom{m+r-k-1}{r-1}\binom{k+n-r}{n-r}$. We thus have (Gumbel (1958) attributes the result to H.A. Thomas).

Theorem 2. The probability that exactly k of m future observations exceed the r-th order statistic $X_{r:n}$ of n past observations equals

$$\frac{\binom{m+r-k-1}{r-1}\binom{k+n-r}{n-r}}{\binom{m+n}{n}} . \tag{11}$$

In particular, k = 0, r = n = m yield

$$P(X_{n:n} = X_{2n:2n}) = \frac{1}{2} .$$

Other interesting and important consequences of Theorem 2 can be deduced. For easier reference, let us introduce N(n,m,r) as the number of m future observations which exceed $X_{r:n}$ of n past observations. Then (11) gives the distribution $P(N(n,m,r) = k)$, $0 \leq k \leq m$. Upon utilizing that the sum of the probabilities in (11) is one, we get

$$E(N(n,m,r)) = m(n-r+1)/(n+1) \tag{12}$$

and

$$V(N(n,m,r)) = m(n+m+1)r(n-r+1)/(n+1)^2(n+2). \tag{13}$$

(the actual computations are very similar to the more familiar hypergeometric case). Notice the dependence of the variance on r: it comes through r(n-r+1), which is the smallest for extremes and largest around 1/2 n (the median). In particular, with m = n+1 and r = n,

$$E(N(n,n+1,n)) = 1 \text{ and } V(N(n,n+1,n)) = 2n/(n+2),$$

which suggest the existence of a limiting distribution for N(n,n+1,n). It indeed follows from (11) that the distribution of N(n,m,n) is asymptotically geometric whenever m is proportional to n and n is large. Other asymptotic properties are summarized in Gumbel (1958, p. 73). Tables for the distribution in (11) are given in Epstein (1954). The urn model, which led to Theorem 2, is further exploited in Wenocur (1981), who utilizes this urn model and other variants in an analysis of waiting times and return periods related to order statistics.

## 5. GROWTH RATES FOR EXTREMES: LAWS OF LARGE NUMBERS

Another way of investigating the growth of extremes is to compare $X_{n:n}$ with some function g(n) of the sample size n. In particular, if $X_{n:n}-g(n)$ converges to zero in probability, we say that an additive law of large numbers applies to $X_{n:n}$, and if $(1/g(n))X_{n:n}$ converges to one in probability, we speak of a multiplicative law of large numbers. The present section is devoted to the investigation of these two laws of large numbers for i.i.d. samples and for the maxima. Other extremes can be investigated in the same manner without any change in the mathematical arguments. Some results directly extend to dependent models, in particular, to cases when a Borel-Cantelli lemma is available. But since the aim of the present paper is to give an introduction to some topics of the conference, the emphasis is on presenting the flavor of a topic rather than on a survey of results.

As before, we denote by F(x) the population distribution function. If $\omega(F) = \sup\{x:F(x) < 1\}$ is finite, then $X_{n:n}$ converges to $\omega(F)$ (in probability), and thus the laws of large numbers become trivial. We therefore assume that $\omega(F) = +\infty$, i.e., $F(X) < 1$ for all x, and $X_{n:n} \to +\infty$ with n.

Note that, for a population distribution F(x), an additive law of large numbers applies if, and only if, a multiplicative law of large numbers applies in the case when the population distribution is $F_1(x) = F(\log x)$, $x > 0$ (and zero for $x < 0$). Therefore, after finding a criterion for the validity of one law, it can easily be translated for the other.

The first results are due to Gnedenko (1943), which we formulate below.

Theorem 3. If $F(x) < 1$ for all $x$, then

(i) the multiplicative law of large numbers applies to $X_{n:n}$ if, and only if, for all $x > 1$, as $t \to +\infty$.

$$\frac{1-F(tx)}{1-F(t)} \to 0; \tag{14}$$

(ii) the additive law of large numbers applies to $X_{n:n}$ if, and only if, for all $x > 0$, as $t \to +\infty$ .

$$\frac{1-F(t+x)}{1-F(t)} \to 0. \tag{15}$$

In both laws of large numbers, one can take

$$g(n) = \inf\{x:1-F(x) \leq 1/n\} . \tag{16}$$

Proof. Because $X_{n:n} \to +\infty$ with probability converging to one, in the multiplicative law of large numbers, $g(n)$ must converge to $+\infty$ with $n$. Hence, for any fixed $x > 0$, $xg(n) \to +\infty$. Furthermore, by definition,

$$P(X_{n:n} < xg(n)) = F^n(xg(n)) \to \begin{cases} 1 & \text{if } x > 1, \\ 0 & \text{if } x < 1. \end{cases}$$

By assumption, for $n$ large $0 < F(xg(n)) < 1$, and thus we can write

$$F^n(xg(n)) = \exp\{n \log F(xg(n))\} = \exp\{n \log[1-(1-F(xg(n)))]\}$$

$$\sim \exp\{-n[1-F(xg(n))]\}.$$

We thus have, as $n \to +\infty$,

$$n[1-F(xg(n))] \to \begin{cases} 0 & \text{if } x > 1 \\ +\infty & \text{if } x < 1. \end{cases} \tag{17}$$

However, (14) and (17) are equivalent. Namely, if $t \to +\infty$ in an arbitrary manner, one can find an $n$ to every $t$ such that $g(n) \leq t < g(n+1)$, and thus, for $y,z > 0$,

$$\frac{1-F(zg(n+1))}{1-F(yg(n))} \leq \frac{1-F(tz)}{1-F(ty)} \leq \frac{1-F(zg(n))}{1-F(yg(n+1))} ,$$

from which, for $x > 1$ and $0 < y < 1$ such that $yx > 1$,

$$0 \leq \lim_{t=+\infty} \frac{1-F(tx)}{1-F(t)} = \lim_{t=+\infty} \frac{1-F(tyx)}{1-F(ty)} \leq \lim_{n=+\infty} \frac{1-F(yxg(n))}{1-F(yg(n+1))}$$

$$= \lim_{n=+\infty} \frac{n[1-F(yxg(n))]}{(n+1)[1-F(yg(n+1))]} = 0,$$

i.e. (17) implies (14). The converse is evident, namely, (14) with $t \sim g(n)$, where $g(n)$ is determined by (16), becomes (17).

Finally, part (ii) follows from part (i) in view of our earlier remark. The proof is completed.

Equivalent froms of (14) and (15) in terms of integrals of $1-F(x)$ are given in de Haan (1970). Geffroy (1958), recognizing that if $f(x) = F'(x)$ exists then

$$1-F(x) = \exp\{-\int_a^x \frac{f(y)}{1-F(y)}\,dy + \log\,[1-F(a)]\}$$

with any a for which the above expressions are defined, he conclude that, in view of

$$\frac{1-F(tx)}{1-F(t)} = \exp\{-\int_t^{tx} \frac{f(y)}{1-F(y)}\,dy\} = \exp\{-\int_1^x \frac{tf(ty)}{1-F(ty)}\,dy\},$$

(14) holds whenever

$$\lim_{z=+\infty} \frac{zf(z)}{1-F(z)} = +\infty. \tag{18}$$

The combination of (16) and (18) are the most convenient tools for a positive solution to the question of laws of large numbers.

One can immediately see that both laws of large numbers apply in the normal case, while only the multiplicative law applies in the exponential case. When the requirement of "convergence in probability" is replaced by "convergence with probability one", we speak of strong laws of large numbers. This case is well covere in Chapter 4 of Galambos (1978). In this same chapter, the role of extremes in the theory of sums also is discussed, for which see also the recent survey Mori (1981).

## References

David, H.A. (1981). Order statistics. 2nd ed., Wiley, New York.

Deheuvels, P. (1981). Univariate extreme values-theory and applications. 43rd Session, ISI, Buenos Aires.

Epstein, B. (1954). Tables for the distribution of the number of exceedances. Ann. Math. Statist. 25, 762-765.
Epstein, B. (1960). Elements of the theory of extreme values. Technometrics 2, 27-41.
de Finetti, B. (1930). Funzione carattenstica di un fenomeno aleatorio. Atti. Accad. Naz. Lincei Rend. Cl. Sci. Fiz. Mat. Nat. 4, 86-133.
Galambos, J. (1973). A general Poisson limit theorem of probabi-lity theory. Duke Math. J. 40, 581-586.
Galambos, J. (1978). The asymptotic theory of extreme order statistics. Wiley, New York.
Galambos, J. (1982). The role of exchangeability in the theory of order statistics. In: Exchangeability in probability and statistics (Eds.: G. Koch and F. Spizzichino), North Holland, Amsterdam, pp. 75-86.
Galambos, J. (1984). Order statistics. In: Handbook of Statistics, Vol. 4 (Eds.: P.R. Krishnaiah and P.K. Sen), North Holland, Amsterdam.
Galambos, J. and S. Kotz (1978). Characterizations of probability distributions. Springer Verlag, Lecture Notes in Mathematics, Vol. 675, Heidelberg.
Geffroy, J. (1958). Contributions a la theorie des valeurs extremes. Publ. Inst. Stat. Univ. Paris, 7/8, 37-185.
Gnedenko, B.V. (1943). Sur la distribution limite du terme maxi-mum d'une serie aleatoire. Ann. Math. 44, 423-453.
Gumbel, E.J. (1958). Statistics of extremes. Columbia Univ. Press.
de Haan, L. (1970). On regular variation and its application to the weak convergence of sample extremes. Math. Centrum Amsterdam.
Huang, J.S. (1974). Personal communication.
Kendall, D.G. (1967). On finite and infinite sequences of ex-changeable events. Studia Sci. Math. Hungar. 2, 319-327.
Mori, T. (1981). The relation of sums and extremes of random Variables. 43rd Session, ISI, Buenos Aires.
Spizzichino, F. (1982). Extendibility of symmetric probability distributions and related bounds. In: Exchangeability in probability and statistics (eds.: G. Koch and F. Spizzichi-no), North Holland, Amsterdam, pp. 313-320.
Sukhatme, P.V. (1937). Tests of significance for samples of the $\chi^2$-population with two degrees of freedom. Ann. Eugenics, London, 8, 52-56.
Wenocur, R.S. (1981). Waiting times and return periods related to order statistics: an application of urn models. In: Stat. Distr. in Sci. Work, Vol. 6, pp. 419-433.

# ASYMPTOTICS; STABLE LAWS FOR EXTREMES; TAIL PROPERTIES

Janos Galambos

Temple University, Philadelphia, Pa.

## ABSTRACT

When for a set of random variables the underlying univariate and multivariate distributions are only approximately known, they become useless even for estimating the values of the distribution function of the extremes. In some exact models, computational difficulties might arise. Both of these can be overcome in asymptotic models. The paper discusses asymptotic models with emphasis on the availability of several dependent extreme value models. For the classical model of independent and identically distributed random variables, a functional equation is deduced for the possible asymptotic (stable) distributions of normalized extremes. The solution of the functional equation is discussed, with emphasizing the necessity of assumptions on the domain of the equation in order to obtain the classical theory of extremes. Extensions to random sample sizes, and other related characterization theorems are also mentioned.

The tail properties of the stable distributions are discussed which lead to simple classification of population distributions in terms of their asymptotic extreme value distributions.

## 1. INTRODUCTION

When the distribution function $F(z_1,z_2,\ldots,z_n)$ of n random variables $X_1,X_2,\ldots,X_n$ is known, we say that the model $\{X_j, 1 \leq j \leq n, F(z_1,z_2,\ldots,z_n)\}$ has been specified. One particular



*J. Tiago de Oliveira (ed.), Statistical Extremes and Applications, 19–29.*

model is that of the classical model, when the $X_j$ are independent and identically distributed. In this case $F(z_1,z_2,\ldots,z_n) = F(z_1)F(z_2)\ldots F(z_n)$ with a univariate distribution function $F(z)$. The classical model will be abbreviated to $\{X,F(z),n\}$. Some other models can also be described in a simple way; for example, the independent model is determined by n univariate distribution functions, the Gaussian model, when $F(z_1,z_2,\ldots,z_n)$ is multivariate normal, by the expectation vector and the variance-covariance matrix, and the (infinite) exchangeable model, at least from the point of view of extremes, by two distribution functions which appear in the de Finetti representation. Other models are usually only partially known which make them impractical for finite n, but an asymptotic theory of extremes may provide quite an accurate result for the distribution of extremes when n is sufficiently large. We deal with the accuracy of such a theory in another paper in these same proceedings. The present paper is devoted to the asymptotic theory. The introductory paper by the present author in these same proceedings will frequently be referred to.

Some notations. The order statistics of the sequence $X_1,X_2,\ldots,X_n$ are denoted by $X_{1:n} \leq X_{2:n} \leq \cdots \leq X_{n:n}$. The focus of the present paper will be on $Z_n = X_{n:n}$, from which the results can easily be transformed to $W_n = X_{1:n}$ by the simple transformation of $X_j$ to $(-X_j)$. For some models, the behavior of $Z_n$ automatically determines the behavior of $X_{n-k:n}$ for all fixed k as $n \to +\infty$, but for most models, the asymptotic distribution of $Z_n$, with proper normalization, may exist without the existence of an asymptotic distribution for other order statistics (whatever be the normalization).

In the general model $\{X_j, 1 \leq j \leq n, F(z_1,z_2,\ldots,z_n)\}$,

$$H_n(z) = P(Z_n \leq z) = F(z,z,\ldots,z), \tag{1}$$

which may or may not be useful. For example, in the Gaussian model, even when all parameters are accurately known, nobody has ever tried to compute (1) with n = 250, say. In the classical model, (1) reduces to

$$H_n(z) = F^n(z), \tag{2}$$

which is completely satisfactory, if F(z) were completely known. However, if F(z) is only "almost known" (for example, its functional form is known from a characterization theorem but its parameters have to be estimated; or, when there are so many observations that the empirical distribution function gives a very accurate approximation to F), (2) becomes completely useless, due to the fact that $F^n$ is very sensitive to slight fluctuations in F (see the Introductory contribution of the author).

These computational difficulties can be avoided by developing an asymptotic theory by showing that, with some constants $a_n$ and $b_n > 0$,

$$P((Z_n-a_n)/b_n < x) = H_n(a_n+b_nx) \to H(x) \tag{3}$$

at every continuity point of H(x), where H(x) is a proper (and nondegenerate) distribution function. The added advantage of (3) to those already mentioned is that, when (3) can be solved, and the limit is sufficiently reliable as to its accuracy, then $F(z_1,z_2,...,z_n)$ (or F(z) in the classical model), become quite insignificant. Namely, the application of (3) to practical problems is that one assumes that the underlying distribution is H, and the maximal property of $Z_n$ is lost, because, in terms of H, $Z_n$ is not viewed any more as a function of the $X_j$. There is more to be gained from an asymptotic theory of extremes. Namely, there are practical problems such as strength which can be expressed as an extremal problem through the weakest link principle, in which we choose n arbitrarily, and thus we can let $n \to +\infty$ for getting the <u>exact</u> distribution of the strength of material. This approach is imbedded in the works of Epstein (see his survey Epstein (1960), and its references). The first mathematically precise model of this nature is developed in the present author's book, Galambos (1978, p. 189). That model gives a sound foundation for the distribution of the strength of a sheet of metal to be Weibull.

Let us go back to (3). A general theory for (1) and (3) cannot exist, because one could find a model to an arbitrary H(x) such that (1) and (3) would hold. There is, of course, one trivial example for this: one could choose all $X_j = X$, the same random variable whose distribution function is H(x). What is troublesome in this connection is that there are models, which are far from trivial and which can be faced both in theoretical and practical investigations, in which every H(x) can occur in (1) and (3). Such examples can be found among the Gaussian models as well as among the independent models (see Galambos (1978), Section 3.3). Therefore, if one wants to be general but wants to avoid trivial results at the same time, one has to make assumptions on the model in which a solution of (1) and (3) is sought. In the present paper, we actually adopt the following approach:

we start with the classical model from which we make two types of generalizations

(i) try to modify the model but only to the extent that the conclusions should remain the same as in the classical model;

and

(ii) try to find the possible limiting distributions $H(x)$ for specific models, with emphasis on their being different from the classical forms.

Extensions belonging to group (i), of course, do not add much to the applicability of extreme value theory, except that it justifies the approximation of a dependent model by a classical one. An important exception to this is when the asymptotic theory of extremes is utilized for determining the exact distribution of a random variable such as the quoted method of Galambos (1978) for the distribution of (random) strength. Group (ii), on the other hand, is very significant: it provides alternative when the conclusions of the classical model are clearly inappropriate (such as Bayesian statistical analysis, reliability of a piece of equipment, and others, about which more details are given in Section 3).

## 2. THE CLASSICAL MODEL

Given $\{X,F,n\}$, we seek conditions under which there exist constants $a_n$ and $b_n > 0$ such that (2) and (3) hold. Why do we need the normalization of $Z_n$? The major reason is the monotonicity of $Z_n : Z_n \leq Z_{n+1}$ and $Z_n \to \omega(F)$, where

$$\omega(F) = \sup\{z : F(z) < 1\}. \tag{4}$$

The choice of normalizing $Z_n$ linearly is arbitrary, and it expresses the fact that the origin and the unit of the coordinate system in which we plot $u = H_n(z)$ is changed in the hope that the escape of $Z_n$ to $\omega(F)$ can be followed in the new coordinate system.

Set $z_n = a_n + b_n x$. Then (2) and (3) yield

$$\left(1- \frac{n(1-F(z_n))}{n}\right)^n \to H(x), \tag{5}$$

which, when compared with the elementary relation

$$\left(1- \frac{u_n}{n}\right)^n \to e^{-u} \text{ whenever } u_n \to u, \tag{6}$$

gives the foundation of the asymptotic theory for $Z_n$ in the classical model. In fact, one gets from (5) and (6)

Theorem 1. In the classical model (3) holds if, and only if,

$$n[1-F(a_n+b_nx)] \to -\log H(x),$$

where $\log 0 = -\infty$.

The form of $H(x)$ can be obtained from a simple functional equation. Namely, writing $n = md$, with $d$ fixed, in (2) and (3), we get on the one hand,

$$F^{md}(a_{md}+b_{md}x) \to H(x), \tag{7}$$

but, on the other,

$$F^{md}(a_m+b_mx) \to H^d(x), \tag{8}$$

where, in both cases, $m \to +\infty$. Now, from a classical theorem of Hintchin (see Galambos (1978, p. 61) it follows that there are numbers $A_d$ and $B_d > 0$ such that

$$H^d(A_d+B_dx) = H(x). \tag{9}$$

Compare (9) with (2) and (3); we get that in the classical model $\{X,H(x),d\}$, for every $d > 1$, $Z_d$ can always be normalized (by $A_d$ and $B_d$) so that

$$H_d(A_d+B_d\ x) = H(x). \tag{10}$$

Because of this property, the limiting distributions $H(x)$ are called <u>stable</u> (<u>for the maxima in the classical models</u>). There are, therefore, as many stable distributions for the maxima in the classical models as many solutions the functional equation (9) has. It is not difficult to reduce (9) to the Cauchy functional equation and to prove the following result (see Galambos (1978, p. 71), Gnedenko (1943) and de Haan (1970)).

Theorem 2. There are three types of distribution functions $H(x)$ satisfying (9). These are (where $a > 0$ and $c$ are arbitrary constants):

(i) if $B_d > 1$ for one $d$, then $B_d > 1$ for all $d > 1$, and the solution of (9)

$$H_{1,\gamma}(ax+c) = \begin{cases} \exp(-x^{-\gamma}) & \text{if } x > 0 \\ 0 & \text{otherwise,} \end{cases}$$

where $\gamma > 0$;

(ii) if $B_d < 1$ for one $d > 1$, then $B_d < 1$ for all $d > 1$, and the solution of (9)

$$H_{2,\gamma}(ax+c) = \begin{cases} 1 & \text{if } x > 0 \\ \exp(-(-x)^{\gamma}) & \text{otherwise,} \end{cases}$$

where $\gamma > 0$;

and

(iii) in the remaining case ($B_d = 1$ for all $d \geq 1$), the solution of (9)

$$H_{3,0}(ax+c) = \exp(-e^{-x}).$$

We can go back to Theorem 1 and substitute $H(x)$ by the actual three limiting forms just obtained. We can, of course, choose $a_n$ and $b_n$ such that $a = 1$ and $c = 0$. With this additional assumption, Theorem 1 can be rewritten in three parts, separating the actual forms of $\log H(x)$. The cases $x = 0$ and $x = 1$ then reveal how one can choose $a_n$ and $b_n$. It is also immediate that $H_{1,\gamma}(x)$ comes up only if $\omega(F) = +\infty$, and $H_{2,\gamma}(x)$ only if $\omega(F) < +\infty$. In this latter case, one can transform the variables $X_j$ so that, for the transformed variables, $\omega(F) = +\infty$, and thus its extreme value problem can be settled by either case (i) or case (iii). We do not proceed on this line, however, because L. de Haan's contribution to these same proceedings will discuss variants of Theorems 1 and 2, which are stated in the form of necessary and sufficient conditions.

For purely mathematical reasons, one can ask whether at (9) we have to require that it should hold for all $d \geq 1$. The answer is no, but a single $d > 1$ does not suffice. It is, however, true that if (9) holds for $d = 2$ and 3, then it holds for all $d \geq 1$, and Theorem 2 thus follows. The complete solution to this problem is that if (9) holds for two values $d_1$ and $d_2$ then it holds for all $d \geq 1$ if, and only if, $(\log d_1)/(\log d_2)$ is irrational (see Sethuraman (1965), Mejzler (1965)). Although we posed this problem as a purely mathematical one, it has a very significant implication (both theoretical and practical). If we state (9) and (10) in terms of random variables, they become

$$(Z_d - A_d)/B_d = X_1 \quad \text{in distribution.} \tag{11}$$

Let now d be random. Which conditions on d would imply the validity of (11)? The mentioned solution to this problem by Sethuraman and Mejzler when d is preassigned tells us that there are several solutions of (11) with random d if log d is a lattice variable (i.e., if the set of values of d is a subset of $\{a^k, k \geq 1\}$, for some $a > 0$). Shimizu and Davies (1981) have shown that the solution of (11) is unique if log d is not lattice and if the normalizing constants $A_d$ and $B_d$ do depend on d. An important case, however, remains open: What are the solutions of (11) if both $A_d$ and $B_d$ are constants in d? Only a special case has been settled so far (Baringhaus (1980)). Its complete solution would be important in the theory of branching processes, where, as a limiting distribution, solutions of (11) come up (but nobody knows how they look) (see Cohn and Pakes (1978)).

Let us go back to Theorems 1 and 2. It is clear from Theorem 1 that not the distribution function F but only its tail 1-F(x) with x approaching $\omega(F)$ influences the behavior of $Z_n$. On the other hand, if we take both n and x "large" in Theorem 1, then from

$$-\log H(x) = -\log[1-(1-H(x))] \sim 1-H(x),$$

we get that the tail behaviours of F and H are, to some extent, comparable. Consequently, all distribution functions F for which Theorem 1 holds with one of the three possible H(x) have the same type of tail after the change of origin and scale of the coordinate system (but not changing the random variables $X_j$). In particular, if $\omega(F) = +\infty$ and if Theorem 1 holds, then the tail of F is either exponential (the case of $H_{3,0}(x)$, when $1-H_{3,0}(x) \sim -\log H_{3,0}(x) = e^{-x}$), or Pareto (the case of $H_{1,\gamma}(x)$, when $1-H_{1,\gamma}(x) \sim -\log H_{1,\gamma}(x) = x^{-\gamma}$, $x > 0$, $\gamma > 0$).

Let now F and G be two population distributions. First assume that there are sequences $a_n, b_n > 0$, $c_n$ and $d_n > 0$ such that both $F^n(a_n+b_nx)$ and $G^n(c_n+d_nx)$ converge to $H_{3,0}(x)$ for fixed x, as $n \to +\infty$. Then, by Theorem 1,

$$\lim_{n=+\infty} \frac{1-F(a_n+b_nx)}{1-G(c_n+d_nx)} = 1, \ x \text{ fixed.}$$

Conversely, let us assume that this last limit relation holds, and that $F^n(a_n+b_nx)$ converges to $H_{3,0}(x)$ for fixed x. Then, in view of

$$G^n(c_n+d_nx) = [1-\frac{1-G(c_n+d_nx)}{1-F(a_n+b_nx)}(1-F(a_n+b_nx))]^n,$$

Theorem 1 implies that $G^n(c_n+d_nx)$ converges to $H_{3,0}(x)$ as well. We can repeat the above argument with $H_{1,\gamma}(x)$ and $H_{2,\gamma}(x)$, in which cases x has to be restricted to a semi-line. We thus get that the tails of two population distributions, after a suitable change of the origin and the scale of the coordinate system, are asymptotically identical whenever the (linearly) normalized maxima from the two populations have the same limiting distribution functions.

Notice that the seemingly most important case of $\omega(F) < +\infty$, i.e. when the $X_j$ are bounded, is left out of the above classification. It is so because its asymptotic extreme value theory is developed in terms of the other two cases after a transformation. One would tend to consider the bounded case to be the most important one, because the view is wide spread among engineers that, since physical quantities are bounded, only bounded random variables are appropriate for their representation. This, however, is a misunderstanding of the fact that there is no contradiction between setting a high upper bound for a random variable X, and to assign a probability distribution F to it such that $\omega(F) = +\infty$, but $F(5) = 0.9999999$, say, where the unit is chosen so that 5 is far below the bound that one had in mind previously (but far above all values which are obviously reached by X). For example, if X is the highest daily water level at a given location of a river, then the above value for F(5) means that X will stay below 5 for a few hundred years (do not use laws of large numbers, but extreme value theory when counting those years).

The classical model seems to be so nice and complete, that its applicability seems to be almost unlimited. Why is it then that in so many practical problems the fit of all of the three asymptotic models is quite bad? It is certainly not the speed of convergence to the asymptotic distribution of the extremes (see the author's paper on this in these same proceedings). The reason lies in the fact that rarely are the basic random variables $X_j$, whose maximum is the governing law, independent and identically distributed. That is, the need exists to investigate dependent models in which the limiting behavior of the maxima is essentially different from that in the classical model. This is discussed in the next section.

## 3. DEPARTURES FROM THE CLASSICAL MODEL

The following reasons all speak against relying solely on the classical model in asymptotic extreme value theory:

(i) It is well known in reliability theory (see Barlow and Proschan (1975)) that the life distribution of a piece of equipment can always be represented as an extreme value distribution, which, when the number n of components is large, can be approximated by an asymptotic distribution of the extremes. However, if one insisted on applying here the classical models, the conclusion would be that every piece of equipment with a large number of components has as its life distribution one of the three possibilities for the maxima in the classical models. Data are evidently against it. It is clear that the assumptions of the classical models are violated in this case.

(ii) In Bayesian statistical thinking, even if observations are independent at a given parameter value, one has to sum (integrate) over the parameter space. Hence, exchangeable variables are as basic in Bayesian statistics as the i.i.d. variables are in classical statistics. Since the parameter (as a random variable) can serve as the mixing variable in the de Finetti representation, these exchangeable variables are always extendible to an infinite sequence without violating exchangeability.

(iii) Several characterization theorems conclude that specific random variables are normal. Therefore, when independence is violated, the need arises for the investigation of the extremes in general Gaussian sequences.

(iv) In determining the distribution of the strength of a sheet of metal, one divides the sheet hypothetically into smaller and smaller pieces, and, by the weakest link principle, the strength is exactly the minimum of the strengths of the pieces. Here, the pieces are definitely dependent, and the dependence structure changes with the number n of subdivisions. If all these dependent structures can be treated by a single limiting distribution of the minimum, then this limit is the exact distribution of the strength.

There is an appropriate model for each of these practical problems, in which the asymptotic theory of extremes has been investigated. In some cases this theory is well developed, although statisticians have not joined in yet to the extent as in the case of the classical models. We should add that the models available should also enlarge the horizon of those working in applied fields in which the classical models are quite routinely accepted. One frequently reads, and the founder of all such applications, Gumbel (1958) himself advocated, that the classical models are approximations at best to most problems. And yet, when a distribution does not fit some set of data, then, based on the fact that logarithmic transformation transforms one H to another in the classical models, the logarithm of the data is tested to fit one extreme value distribution, overlooking the fact that logarithmic transformations fit a curve not because it is the appropriate one but because log x is very slowly changing. A dependent model would add further possibilities, out of which one could select the best one for a particular problem.

The specific models, which can be utilized in cases (i) - (iv) have actually been mentioned in the Introductory part. Namely, in the reliability problem, the combination of the dependent model developed in Galambos (1972), and the independent models (without the restriction of identical distributions), investigated by Juncosa (1949), Mejzler (1956), Tiago de Oliveira (1976) and Mucci (1977), is very suitable due to the following conclusion for such models: if the random variables $X_1, X_2, \ldots, X_n$ form an $E_n$-sequence in the sense of Galambos (1978, p. 176), then the asymptotic distribution of the properly normalized minimum $W_n$ is a distribution function with monotonic hazard rate (combine Theorems 3.9.1 and 3.10.1 in Galambos (1978)). Surprisingly, this same technique leads to the Weibull character of strength.

The asymptotic theory of extremes of segments of infinite sequences of exchangeable variables is developed in Berman (1962) and Galambos (1975). An application of such a model, other than to Bayesian statistics, is presented in Dziubdziella and Kopocinski (1976). Out of the many papers dealing with Gaussian sequences, we mention Berman (1964), Pickands III (1967), Mittal and Ylvisaker (1975), and Ibragimov and Rozanov (1970, p. 250), whose combined works show both the flavor and the variety of techniques and results in this area.

Because, by a systematic development, Chapter 3 of Galambos (1978) combines and covers all available models (which have not been extended since the publication of that book), we do not reproduce actual details. However, for a newer survey, see Deheuvels (1981).

References

Baringhaus, L. (1980). Eine simultane Charakterisierung der geometrischen Verteilung und der logistischen Verteilung. Metrika 27, 237-242.

Barlow, R.E. and F. Proschan (1975). Statistical theory of reliability and life testing: Probability models. Holt, Rine-Hart and Winston, New York.

Berman, S.M. (1962). Limiting distribution of the maximum term in a sequence of dependent random variables. Ann. Math. Statist. 33, 894-908.

Berman, S.M. (1964). Limit theorems for the maximum term in

stationary sequences. Ann. Math. Statist. 35, 502-516.
Cohn, H. and A. Pakes (1978). A representation for the limiting random variable of a branching process with infinite mean and some related problems. J. Appl. Probability 15, 225-234.
Deheuvels, P. (1981). Univariate extreme values - theory and applications. 43rd Session, ISI, Buenos Aires.
Dziubdziela, W. and B. Kopocinski (1976). Limiting properties of the distance random variable. (in Polish). Przeglad. Stat. 23, 471-477.
Epstein, B. (1960). Elements of the theory of extreme values. Technometrics 2, 27-41.
Galambos J. (1972). On the distribution of the maximum of random variables. Ann. Math. Statist. 43, 516-521.
Galambos, J. (1975). Limit laws for mixtures with applications to asymptotic theory of extremes. Z. Wahrsch. verw. Gebiete 32, 197-207.
Galambos, J. (1978). The asymptotic theory of extreme order statistics. Wiley, New York.
Gnedenko, B.V. (1943). Sur la distribution limite du terme maximum d'une serie aleatoire. Ann. Math. 44, 423-453.
Gumbel, E.J. (1958). Statistics of extremes. Columbia University Press, New York.
Haan, L. de (1970). On regular variation and its application to the weak convergence of sample extremes. Math. Centre tracts, 32, Amsterdam.
Ibragimov, I.A. and Yu. A. Rozanov (1970). Gaussian stochastic processes. Nauka, Moscow.
Juncosa, M.L. (1949). On the distribution of the minimum in a sequence of mutually independent random variables. Duke Math. J. 16, 609-618.
Mejzler, D.G. (1956). On the problem of the limit distribution for the maximal term of a variational series (in Russian). L'vov Politechn. Inst. Naucn. Zp. (Fiz.-Mat.) 38, 90-109.
Mejzler, D. (1965). On a certain class of limit distributions and their domain of attraction. Trans. Amer. Math. Soc. 117, 205-236.
Mittal, Y. and D. Ylvisaker (1975). Limit distributions for the maxima of stationary Gaussian processes. Stoch. Proc. Appl. 3, 1-18.
Mucci, R. (1977). Limit theorems for extremes. Thesis for Ph.D., Temple University.
Pickands, J. III (1967). Maxima of stationary Gaussian processes. Z. Wahrsch. verw. Gebiete 7, 190-233.
Sethuraman, J. (1965). On a characterization of the three limiting types of the extreme. Sankhya A 27, 357-364.
Shimizu, R. and L. Davies (1981). General characterization theorems for the Weibull and the stable distributions. Sankhya A 43, 282-310.
Tiago de Oliveira, J. (1976). Asymptotic behavior of maxima with periodic disturbances. Ann. Inst. Statist. Math. 28, 19-23.

# SLOW VARIATION AND CHARACTERIZATION OF DOMAINS OF ATTRACTION

Laurens de Haan

Erasmus University Rotterdam

Suppose G is the distribution function of an extreme-value distribution. We shall develop necessary and sufficient conditions for a distribution function F such that for some choice of constants $a_n > 0$ and $b_n$ $(n = 1, 2, \ldots)$

(1) $\quad F^n(a_n x + b_n) \to G(x) \quad \text{weakly} \quad (n \to \infty).$

The left hand side is the distribution function of the normalized maximum of n i.i.d. random variables with distribution function F.

If (1) holds, we say that F is in the domain of attraction of G, notation: $F \in D(G)$. The constants $\{a_n\}$ and $\{b_n\}$ are called the attraction coëfficients or normalizing constants. We seek necessary and sufficient conditions on F such that (1) holds for a given extreme-value distribution G.



J. Tiago de Oliveira (ed.), Statistical Extremes and Applications, 31–48.

## 1. Conditions in terms of the inverse of the distribution function.

By taking logarithms both sides in (1) and using $t^{-1}\log(1-t) \to -1$ $(t\downarrow 0)$ we transform (1) into

$$(2) \quad n\{1 - F(a_n x+b_n)\} \to -\log G(x) \quad \text{weakly} \quad (n\to\infty).$$

Since G is continuous the convergence holds all x for which $0 < G(x) < 1$.

Next we define the function $U : \mathbb{R}_+ \to \mathbb{R}$ by

$$U(x) := \begin{cases} 0 & 0 \leq x < 1 \\ \left(\frac{1}{1-F}\right)^{\leftarrow}(x) & x \geq 1 \end{cases}$$

(The arrow denotes the inverse function defined e.g. by $U(x) := \inf\{y \mid 1/1-F(y) \geq x\}$). Relation (2) now reads

$$n^{-1}U^{\leftarrow}(a_n x+b_n) \to -\log G(x) \quad (n\to\infty).$$

The left hand side is a sequence of monotone function converging to a monotone function. This convergence is equivalent to the convergence of their inverse functions to the inverse function of the function at the right hand side. So (2) is equivalent to

$$(3) \quad \frac{U(nx)-b_n}{a_n} \to H_o(x) \quad \text{for } x > 0 \quad (n\to\infty)$$

with $H_o := (-\log G)^{\leftarrow}$. Taking into account the form of the extreme-value distribution we get

$$H_o(x) = A \cdot \frac{x^\zeta - 1}{\zeta} + B$$

for $A > 0$, $\zeta$ and B real (read Alogx+B for $\zeta=0$). In particular (3) holds for x=1. This gives (by subtraction)

$$(4) \quad \lim_{n\to\infty} \frac{U(nx) - U(n)}{a_n} = A \cdot \frac{x^\zeta - 1}{\zeta}.$$

We prefer to work with a continuous version of (4):

Lemma 1.

F is in a domain of attraction if and only if for some positive function a

$$(5) \quad \lim_{t\to\infty} \frac{U(tx) - U(t)}{a(t)} = \frac{x^\zeta - 1}{\zeta} \quad \text{for} \quad x > 0$$

where $\zeta$ is some real constant (read logx for $\zeta=0$).
If $\zeta > 0$, $F \in D(\Phi_{1/\zeta})$. If $\zeta = 0$, $F \in D(\Lambda)$.
If $\zeta < 0$, $F \in D(\Psi_{-1/\zeta})$.

Proof.

Use the inequalities (with $a(t) = A.a_{[t]}$)

$$\frac{U([t]x) - U([t])}{a_{[t]}} - \frac{U\big([t](1-\varepsilon)\big) - U([t])}{a_{[t]}}$$

$$\leq \frac{U(tx) - U(t)}{a_{[t]}} \leq \frac{U\big([t](x+\varepsilon)\big) - U([t])}{a_{[t]}}$$

which hold for sufficiently large t.

We are now going to study relation (5). An important result is the following.

Lemma 2.

Suppose (5) holds.

1. If $\zeta > 0$, then $\lim_{t\to\infty} U(t) = \infty$ and

$$(6) \quad \lim_{t\to\infty} \frac{U(t)}{a(t)} = \frac{1}{\zeta}.$$

2. If $\zeta < 0$, then $\lim_{t\to\infty} U(t) < \infty$ and, with $U(\infty) := \lim_{t\to\infty} U(t)$

$$(7) \quad \lim_{t\to\infty} \frac{U(\infty) - U(t)}{a(t)} = \frac{-1}{\zeta}.$$

3. If $\zeta = 0$, then

$$(8) \quad \lim_{t\to\infty} \frac{U(tx)}{U(t)} = 1 \qquad \text{for all } x > 0 \text{ if } U(\infty) = \infty$$

and

$$(9) \quad \lim_{t\to\infty} \frac{U(\infty) - U(tx)}{U(\infty) - U(t)} = 1 \text{ for all } x > 0 \text{ if } U(\infty) < \infty.$$

Moreover

$$(10) \quad \lim_{t\to\infty} \frac{a(tx)}{a(t)} = 1 \text{ for all } x > 0.$$

Corollary 3.

For $\zeta > 0$ (5) is equivalent to

$$(11) \quad \lim_{t\to\infty} \frac{U(tx)}{U(t)} = x^{\zeta} \text{ for all } x > 0$$

and for $\zeta < 0$ (5) is equivalent to: $U(\infty) < \infty$ and

$$(12) \quad \lim_{t\to\infty} \frac{U(\infty) - U(tx)}{U(\infty) - U(t)} = x^{\zeta} \text{ for all } x > 0.$$

Proof.

The corollary is obvious from (2) and Lemma 2. For the proof of Lemma 2 for $\zeta > 0$ first note that (5) implies for $z > 0$

$$\lim_{t\to\infty} \frac{a(tz)}{a(t)} = \lim_{t\to\infty} \left\{\frac{U(tzx) - U(t)}{a(t)} - \frac{U(tz) - U(t)}{a(t)}\right\}$$

$$/ \frac{U(tzx) - U(tz)}{a(tz)} = z^{\zeta}.$$

Hence for $z > 1$

$$\lim_{k\to\infty} \frac{U(z^{k+1}) - U(z^{k})}{U(z^{k}) - U(z^{k-1})} = \lim_{k\to\infty} \frac{U(z^{k+1}) - U(z^{k})}{a(z^{k})}$$

$$/ \frac{U(z^{k}) - U(z^{k-1})}{z^{\zeta}.a(z^{k-1})} = z^{\zeta}$$

i.e. for $k \geq n_0(\varepsilon)$

$$(13)\quad \{U(z^{k}) - U(z^{k-1})\}\, z^{\zeta}(1-\varepsilon) \leq U(z^{k+1}) - U(z^{k}) \leq$$

$$\leq \{U(z^{k}) - U(z^{k-1})\}\, z^{\zeta}(1+\varepsilon).$$

From this

$$\lim_{N\to\infty} U(z^{N+1}) - U(z^{n_0}) = \lim_{N\to\infty} \sum_{n=n_0}^{N} \left[\prod_{k=n_0}^{n} \frac{U(z^{k+1}) - U(z^{k})}{U(z^{k}) - U(z^{k-1})}\right]$$

$$\{U(z^{n_0}) - U(z^{n_0-1})\} \geq$$

$$\lim_{N\to\infty} \sum_{n=n_0}^{N} \left[\prod_{k=n_0}^{n} z^{\zeta}(1-\varepsilon)\right] \{U(z^{n_0}) - U(z^{n_0-1})\} = \infty$$

and hence $\lim_{t\to\infty} U(t) = \infty$.

In order to prove (6) add the inequalities (13) for $k = n_0, \ldots, n$; divide the result by $U(z^n)$ and take the limit $n \to \infty$. This gives

$$(14) \quad \lim_{n\to\infty} \frac{U(z^{n+1})}{U(z^n)} = z^\zeta .$$

For $x > 1$ define $n(x) \in \mathbb{N}$ such that $z^{n(x)} \leq x < z^{n(x)+1}$. We have for $t,x > 1$

$$\frac{U(z^{n(t)}\, z^{n(x)})}{U(z^{n(t)+1})} \leq \frac{U(tx)}{U(t)} \leq \frac{U(z^{n(t)+1}\, z^{n(x)+1})}{U(z^{n(t)})} .$$

Hence by (14) all limit points of $U(tx)/U(t)$ for $t \to \infty$ are in between $(x/z^2)^\zeta$ and $(z^2 x)^\zeta$. Letting $z \downarrow 1$ we get (11) and hence (6) (after using (5) again).

The other assertions are proved in a similar way.

Lemma 4.

Suppose (5) holds for $\zeta = 0$. For all $\epsilon > 0$ there exist $c > 0$, $t_0$ such that for $x \geq 1$, $t \geq t_0$

$$(15) \quad \frac{U(tx) - U(t)}{a(t)} \leq c.\, x^\epsilon .$$

Proof.

For all $z > 1$ there exists $t_0$ such that for $t \geq t_0$

$$\frac{U(te) - U(t)}{a(t)} \leq z \text{ and } \frac{a(te)}{a(t)} \leq z .$$

For $n = 1, 2, \ldots$ and $t \geq t_0$

$$\frac{U(te^n)-U(t)}{a(t)} = \sum_{k=1}^{n} \frac{U(te^k)-U(te^{k-1})}{a(te^{k-1})} \prod_{r=1}^{k-1} \frac{a(te^r)}{a(te^{r-1})} \leq$$

$$\leq \sum_{k=1}^{n} z \prod_{r=1}^{k-1} z \leq n . z^n .$$

Hence for $x > 1$ and $1 < z < e$

$$\frac{U(tx)-U(t)}{a(t)} \leq \frac{U(te^{[\log x]+1})-U(t)}{a(t)} \leq$$

$$\leq ([\log x]+1) z^{[\log x]+1} \leq z\, x^{2\log z}/\log z$$

Corollary 5.

If (5) holds for $\zeta = 0$, then $\int_1^\infty s^{-2} U(s)\, ds < \infty$ and

$$(16) \quad \lim_{t\to\infty} \frac{U_o(t)-U(t)}{a(t)} = 0$$

with

$$(17) \quad U_o(t) := e^{-1}.t. \int_{e^{-1}t}^{\infty} s^{-2}\, U(s)\, ds.$$

Proof.

Use (15) and dominated convergence.

Lemma 6.

Let $F_1$ and $F_2$ be two distribution functions with common upper endpoint $x_0$ (i.e. $F_i(x_0+\epsilon) = 1 > F_i(x_0-\epsilon)$ for $i=1,2$ and $\epsilon > 0$).

Equivalent are:

1) $F_1 \in D(\Lambda)$ and

$$(18) \quad \lim_{t\uparrow x_o} \frac{1-F_2(t)}{1-F_1(t)} = 1.$$

2) For some positive function a

$$(19) \quad \lim_{t\to\infty} \frac{U_1(tx) - U_1(t)}{a(t)} = \log x \quad \text{for all } x > 0$$

and

$$(20) \quad \lim_{t\to\infty} \frac{U_2(t) - U_1(t)}{a(t)} = 0.$$

Here $U_i := \left(\frac{1}{1-F_i}\right)^{\leftarrow}$.

Proof.

The inequalities (valid for sufficiently large t)

$$(1-\varepsilon)\{1-F_1(t)\} \leq 1-F_2(t) \leq (1+\varepsilon)\{1-F_1(t)\}$$

are equivalent to the following inequalities for their inverse functions (for sufficiently large s)

$$U_1((1-\varepsilon)s) \leq U_2(s) \leq U_1((1+\varepsilon)s)$$

which in turn are equivalent to

$$\frac{U_1((1-\varepsilon)s) - U_1(s)}{a(s)} \leq \frac{U_2(s) - U_1(s)}{a(s)} \leq \frac{U_1((1+\varepsilon)s) - U_1(s)}{a(s)}.$$

The extreme members tend to $\log(1-\varepsilon)$ and $\log(1+\varepsilon)$ respectively. By Lemma 1 the proof is complete.

Corollary 7.

Then $F_2 \in D(\Lambda)$ and $F_1$ and $F_2$ have the same normalizing constants.

## 2. Domains of attraction of $\Phi_\alpha$ and $\Psi_\alpha$, $(\alpha>0)$.

We are now able to give simple neccessary and suffient conditions for $D(\Phi_\alpha)$ and $D(\Psi_\alpha)$.

Theorem 8.

$F \in D(\Phi_\alpha)$ if and only if $F(x) < 1$ for all x and

$$(21)\quad \lim_{t\to\infty} \frac{1-F(tx)}{1-F(t)} = x^{-\alpha} \text{ for } x > 0.$$

Proof.

We know (Corollary 3) that $F \in D(\Phi_\alpha)$ if and only if

$$(22)\quad \lim_{t\to\infty} \frac{U(tx)}{U(t)} = x^{1/\alpha} \text{ for } x > 0.$$

From the definition of the inverse function, $U(x) := \inf\{y \mid 1/1-F(y) \geq x\}$ one sees that for any $\varepsilon > 0$

$$(23)\quad U\left(\frac{1-\varepsilon}{1-F(t)}\right) \leq t \leq U\left(\frac{1+\varepsilon}{1-F(t)}\right)$$

Suppose $F \in D(\Phi_\alpha)$. For $x > 0$ by (23)

$$\frac{U\left(\frac{x}{1-F(t)}\right)}{U\left(\frac{1+\varepsilon}{1-F(t)}\right)} \leq t^{-1}\, U\left(\frac{x}{1-F(t)}\right) \leq \frac{U\left(\frac{x}{1-F(t)}\right)}{U\left(\frac{1-\varepsilon}{1-F(t)}\right)}.$$

Hence by (22) (note $U(\infty) = \infty$ implies $F(x) < 1$ for all x)

$$\lim_{t\to\infty} t^{-1}\, U\left(\frac{x}{1-F(t)}\right) = x^{1/\alpha}.$$

The left hand side is a family of monotone functions converging to a monotone function. This is thus equivalent to the convergence of the family of inverse function to the inverse function of the right hand side. We find that $F \in D(\Phi_\alpha)$ implies (21).

Conversely suppose (21). As in the first part of the proof we can derive

$$\lim_{s\to\infty} s\{1-F(xU(s))\} = x^{-\alpha}.$$

The left hand side is a family of monotone functions converging to a monotone function. This is thus equivalent to the convergence of the family of inverse function to the inverse function of the right hand side. We find that (21) implies (22) which is equivalent to $F \in D(\Phi_\alpha)$.

Theorem 9.

$F \in D(\psi_\alpha)$ if and only if there exists $x_0$ such that $F(x_0+\epsilon) = 1 > F(x_0-\epsilon)$ for $\epsilon > 0$ and

$$\lim_{t\downarrow 0} \frac{1-F(x_0-tx)}{1-F(x_0-t)} = x^\alpha$$

Proof.

Similar to the previous proof.

Corollary 10a.

1) $F \in D(\Phi_\alpha)$ if and only if $\int_1^\infty x^{-1}\{1-F(x)\}dx < \infty$ and

$$\lim_{t\to\infty} \frac{\int_t^\infty x^{-1}(1-F(x))dx}{1-F(t)} = \frac{1}{\alpha}.$$

2) $F \in D(\psi_\alpha)$ if and only if $\int_{x_0-1}^{x_0} (x_0+x)^{-1}\{1-F(x)\}dx < \infty$ and

$$\lim_{t \uparrow x_0} \frac{\int_{x_0-t}^{x_0} (x_0+x)^{-1}\{1-F(x)\}dx}{1-F(x_0-t)} = \frac{1}{\alpha}.$$

Proof.

We only prove the first part, the second one is then obvious. Suppose $F \in D(\Phi_\alpha)$. Then (21) holds. The proof of the implication (5) $\Rightarrow$ (15) can easily be adapted to give: (21) implies that for all $\varepsilon > 0$ there exist $c > 0$, $t_0$ such that for $x \geq 1$, $t \geq t_0$

$$\frac{1-F(tx)}{1-F(t)} \leq c.x^{-\alpha+\varepsilon}.$$

The dominated convergence theorem then gives

$$\lim_{t\to\infty} \int_1^\infty \frac{1-F(tx)}{1-F(t)} \frac{dx}{x} = \frac{1}{\alpha}.$$

Next suppose

$$a(t) := \{1-F(t)\} \Big/ \int_t^\infty x^{-1}(1-F(x))dx \to \alpha \ (t\to\infty).$$

By partial integration one gets

$$1-F(t) = c.a(t) \exp(-\textstyle\int_0^t s^{-1}a(s)ds)$$

From this one easily checks (21).

Corollary 10b.

If $F \in D(\Phi_\alpha)$, then

$$\lim_{n\to\infty} F^n(a_n x) = \exp - x^{-\alpha} \text{ for } x > 0$$

with $a_n := U(n)$. If $F \in D(\Psi_\alpha)$, then

$$\lim_{n\to\infty} F^n(a_n x + x_0) = \exp - (-x)^\alpha \text{ for } x < 0$$

with $a_n = x_0 - U(n)$ and $x_0$ as above.

## 3. Domain of attraction of $\Lambda$.

**Theorem 11.**

$F \in D(\Lambda)$ if and only if there exists a positive function f such that for all x

$$(24) \quad \lim_{t\uparrow x_0} \frac{1-F(t+xf(t))}{1-F(t)} = e^{-x}.$$

Here $x_0 \leqq \infty$ such that $F(x_0+\varepsilon)=1 > F(x_0-\varepsilon)$ for $\varepsilon > 0$. If (24) holds for some f, then $\int_t^{x_0} (1-F(s)ds < \infty$ for $t < x_0$ and (24) holds with the choice

$$(25) \quad f(t) := \frac{\int_t^{x_0} (1-F(s))ds}{1-F(t)} .$$

**Proof.**

By Lemma 1 $F \in D(\Lambda)$ if and only if for some positive function a

$$(26) \quad \lim_{t\to\infty} \frac{U(tx) - U(t)}{a(t)} = \log x$$

where $U := \left(\frac{1}{1-F}\right)^{\leftarrow}$. We are going to translate this relation into a relation for $1/1-F$, the inverse function of U.

As in the the proof of theorem 8 we get

$$\frac{U(\frac{x}{1-F(s)}) - U(\frac{1}{1-F(s)})}{a(\frac{1}{1-F(s)})} - \frac{U(\frac{1+\varepsilon}{1-F(s)}) - U(\frac{1}{1-F(s)})}{a(\frac{1}{1-F(s)})} \leq$$

$$\leq \frac{U(\frac{x}{1-F(s)}) - s}{a(\frac{1}{1-F(s)})} \leq \frac{U(\frac{x}{1-F(s)}) - U(\frac{1-\varepsilon}{1-F(s)})}{a(\frac{1}{1-F(s)})}.$$

From these inequalities and (26) we get for $x > 0$

$$\lim_{s \uparrow x_0} \frac{U(\frac{x}{1-F(s)}) - s}{a(\frac{1}{1-F(s)})} = \log x.$$

The left hand side is a family of monotone functions converging to a monotone function. This is thus equivalent to the convergence of the family of inverse function to the inverse function of the right hand side. We find that $F \in D(\Lambda)$ implies (24) with $f(t) = a(\frac{1}{1-F(t)})$.
The converse implication is left to the reader.

We are now going to prove that if $F \in D(\Lambda)$, (24) holds with f defined by (25). By corollary 5 we have with

$$U_o(t) := e^{-1}t \int_{e^{-1}t}^{\infty} s^{-1}U(s)\, ds$$

$$(27) \quad \lim_{t\to\infty} \frac{U_0(t) - U(t)}{a(t)} = 0 \text{ and}$$

$$(28) \quad \lim_{t\to\infty} \frac{U_0(tx) - U_0(t)}{a(t)} = \log x \text{ for } x > 0.$$

Now there is a distribution function $F_o$ such that $U_0(t) = (\frac{1}{1-F_0})^{\leftarrow}(t)$ for sufficiently large t.
By Lemma 6 we get that $F_o \in D(\Lambda)$ and

(29) $1-F_0(t) \sim 1-F(t) \quad (t \uparrow x_0)$.

Also, inverting the limit relation (28), we have

(30) $$\lim_{t\uparrow x_0} \frac{1-F_0(t+xa(1/1-F_0(t)))}{1-F_0(t)} = e^{-x} \text{ for all } x.$$

Since (29) implies $\int_t^{x_0} (1-F(s))ds/1-F(t) \sim \int_t^{x_0} (1-F_0(s))ds/1-F_0(t)$ as $t\uparrow x_0$, it is now sufficient to prove theorem 11 for the continuous and strictly increasing distribution function $F_0$ (note that (30) holds locally uniformly).

Using (28), the inequality (15) and dominated convergence we get

$$\frac{t\int_t^\infty \frac{U_0(dv)}{v}}{a(t)} = \frac{t\int_t^\infty U_0(s)\frac{ds}{s^2} - U_0(t)}{a(t)} =$$

$$= \int_1^\infty \frac{U_0(tx)-U_0(t)}{a(t)} \cdot \frac{dx}{x^2} \to 1 \quad (t\to\infty).$$

With the substitution $s = U_0(t)$ i.e. $t = 1/(1-F_0(s))$ we then obtain

$$\lim_{s\uparrow x_0} \frac{\int_s^{x_0} (1-F_0(u))du}{a(1/1-F_0(s))(1-F_0(s))} = 1.$$

The theorem is proved.

In order to check the criterion of theorem 11 one has to verify (24) for all x. Such a parameter does not show up in the following criterion.

Theorem 12.

$F \in D(\Lambda)$ if and only if for $x < x_0$ (with $x_0$ such that for $\epsilon > 0$ $F(x_0+\epsilon) = 1 > F_0(x_0-\epsilon)$) $\int_x^{x_0}\int_t^{x_0} (1-F(s))ds\,dt < \infty$ and the function h defined by

$$(31)\quad h(x) := \frac{(1-F(x)) \int_x^{x_0}\int_t^{x_0} (1-F(s))\,ds\,dt}{\left\{ \int_x^{x_0} (1-F(s))\,ds \right\}^2}$$

satisfies

$$(32)\quad \lim_{x\uparrow x_0} h(x) = 1.$$

Proof.

Replacing t in (24) by $t' + yf(t') \uparrow x_0$ $(t'\uparrow x_0)$ one gets

$$\lim_{t'\uparrow x_0} \frac{1-F(t' + \{y + x\, f(t'+y\, f(t'))/f(t')\}\, f(t'))}{1-F(t')} =$$

$$= e^{-x} \cdot \lim_{t'\uparrow x_0} \frac{1-F(t'+y\, f(t'))}{1-F(t')} = e^{-x-y}.$$

Since by (24) also $\lim_{t'\uparrow x_0} [1-F(t'+(x+y)f'(t))]/[1-F(t')] = e^{-x-y}$ and the convergence is locally uniform, we must have

$$(33)\quad \lim_{t\uparrow x_0} \frac{f(t+xf(t))}{f(t)} = 1 \text{ for all } x.$$

Applying this to the function f given by (25) and combining

with (24) one gets that for all x

$$(34)\quad \lim_{t\uparrow x_o} \int_{t+xf(t)}^{x_0} (1-F(s))\,ds \Big/ \int_t^{x_0} (1-F(s))\,ds = e^{-x}$$

i.e. the distribution function

$F_1(x) := \max\left\{0, (1-\int_x^{x_0}(1-F(s))ds)\right\}$, is in $D(\Lambda)$.

Note that (34) holds with f defined by (25).

Since $F_1 \in D(\Lambda)$, we may apply theorem 11 to $F_1$ to find that (34) holds also for $F_1$ with the function f replaced by

$$(35)\quad f_1(t) = \frac{\int_t^{x_0} 1-F_1(s)\,ds}{1-F_1(t)} .$$

Since (34) holds with f defined by (25) and with $f_1$ defined by (35), we get by the local uniformity of relation (34) that $f_1(t) \sim f(t)$ $(t\uparrow x_0)$ i.e. (32).

Conversely suppose (32) holds. We first prove that the function $F_2$ defined for sufficiently large $t < x_0$ by

$$F_2(t) := 1-(\int_t^{x_0} (1-F(s))ds)^2 \Big/ \int_t^{x_0}\int_s^{x_0} (1-F(u)du\,ds$$

is a distribution function. Observe that by (32)

$$(36)\quad 1-F_2(t) \sim 1-F(t) \to 0\ (x\uparrow x_0).$$

Moreover one verifies easily that

$$(37)\quad \frac{d}{dt}\{-\log(1-F_2(t)\} = (2h(t)-1)\ /\ f_2(t)$$

with

$$f_2(t) := \int_t^{x_0}\int_s^{x_0} (1-F(u)\ du\ ds\ /\ \int_t^{x_0} (1-F(s))ds.$$

Now (37) shows (use (32)) that $F_2$ is increasing, hence $F_2$ is a distribution function.

By (36) and Lemma 6 it is sufficient to show $F_2 \in D(\Lambda)$.
Note that in the Radon-Nikodym sense

$$(38)\quad \frac{df_2(t)}{dt} = h(t) - 1.$$

Hence for $x \in \mathbb{R}$

$$\frac{f_2(t + xf_2(t)) - f_2(t)}{f_2(t)} = \int_0^x [h(t+uf_2(t))) - 1]du$$

It then follows from (32) that uniformly on bounded x-intervals

$$(39)\quad \lim_{t\to\infty} \frac{f_2(t + xf_2(t))}{f_2(t)} = 1.$$

For all real x by (37)

$$\frac{1-F_2(t+xf_2(t))}{1-F_2(t)} =$$

$$= \exp - \int_0^x (2h(t+uf_2(t))-1)\ \frac{f_2(t)}{f_2(t+uf_2(t))}du\ .$$

By (32) and (39) the right hand side converges to $e^{-x}(t\uparrow x_0)$

hence $F_2 \in D(\Lambda)$ by theorem 11.

Corollary 13.

If $F \in D(\Lambda)$, then

$$\lim_{n\to\infty} F^n(a_n x+b_n) = \exp - e^{-x} \qquad \text{for all } x$$

with

$$b_n := U(n)$$

$$a_n := f(U(n))$$

and f as in (25).

INTRODUCTION, GUMBEL MODEL

Leon Herbach

Polytechnic Institute of New York
333 Jay Street
Brooklyn, NY 11201, USA

ABSTRACT. Properties of the Gumbel distribution are given. Techniques are discussed for treating the special problems associated with this distribution when making the usual statistical decisions. These include point estimation, goodness-of-fit tests, hypothesis tests, interval estimation and prediction using the usual parameters plus others, such as return period, based on complete and censored samples.

## 1. INTRODUCTION

Let $X_{(1)}$, $X_{(n)}$ represent the smallest and largest value in a random sample of size n from a distribution F(x). We have seen that for most distributions there exist sequences of normalizing constants, $\delta_n > 0, \lambda_n$ such that for $V_n = (X_{(n)} - \lambda_n)/\delta_n$,

$$P(V_n \leqq x) = P(X_{(n)} \leqq \lambda_n + \delta_n x) = F^n(\lambda_n + \delta_n x),$$

approaches a nondegenerate and proper cdf $\underset{\sim}{L}(x)$ as $n \to \infty$ at all continuity points of $\underset{\sim}{L}(x)$. This limit $\underset{\sim}{L}(x)$ does not always exist. For example, it will not exist if F(x) is geometric or Poisson. When the limit distribution does exist, it can be only one of three distributions. A similar result holds for $X_{(1)}$. In fact, there exist similar but slightly more complicated results concerning the limit behavior for n large of any order statistic $X_{(i)}$, $i = 1,2,\ldots,n$.



*J. Tiago de Oliveira (ed.), Statistical Extremes and Applications, 49–80.*

This paper will treat only the case of $X_{(1)}$, $X_{(n)}$ for <u>one</u> of the three limiting distributions, the Gumbel distribution. This is the limit distribution when F(x) is, e.g., normal, log-normal, chi-square or exponential. Specifically, we shall treat the case when the limiting distributions of extreme order statistics, $X_{(1)}$ and $X_{(n)}$, in a large sample of size n, are the Gumbel distributions,

$$\underline{G}(x) = 1 - \exp(-\exp[(x-\lambda)/\delta]) \quad , \quad -\infty < x < \infty \quad , \tag{1}$$

and

$$\bar{G}(x) = \exp\{-\exp(-[(x-\lambda)/\delta])\} \quad , \quad -\infty < x < \infty \quad , \tag{2}$$

of the smallest and largest, respectively. These play a central role. Frequently they are still referred to as <u>the</u> extreme value distributions. Originally, this was probably due to the fact that the normal distribution belongs to the domain of attraction of G. However, with the increased interest in reliability during the past 20 years, the Weibull distribution (for smallest extremes) is used more frequently. Nevertheless, even in this case, the Gumbel distribution often plays a role. Just as if a random variable X has one of the two Gumbel distributions, -X has the other (so that results obtained for one of them can be applied easily to the other), other functions of X have the Fréchet or Weibull distribution. More specifically, if Y has a two-parameter (zero location parameter) Weibull distribution (for smallest extreme), then X = log Y has the distribution (1) [14,15] and $U=Y^{-1}$ has the two-parameter Fréchet distribution (for largest extreme). The relationships with X playing the central role are $Y=e^{X}$, $U=e^{-X}$.

Properties of the Gumbel distribution (2) will be given together with some techniques for making statistical decisions based on this distribution. No attempt will be made to be complete. However, it is hoped that procedures typically associated with extreme-value distributions and the Gumbel distribution, in particular, will be illustrated. Many of the methods (plus others) will be found in [10,24,27].

## 2. CHARACTERISTICS OF THE GUMBEL DISTRIBUTION

Since the distribution (2) is not symmetric, the natural location parameter will not be the mean. Also the convenient scale parameter will not be the standard deviation. Define the <u>standardized</u> or <u>reduced</u> variate by

$$Z = (X-\lambda)/\delta \quad \text{or} \quad X = \lambda + \delta X. \tag{3}$$

Then the relationship between the moment generating function of X, $m_X(t)$ and that of Z, $m_Z(t)$ is

$$m_X(t) = e^{\lambda t} m_Z(\delta t) , \tag{4}$$

where

$$m_Z(t) = \Gamma(1-t) .$$

When the mean $\mu$ exists, we have the relationship between moments about the mean,

$$\mu_i = \mu_{Z,i} = E(Z-\mu)^i \tag{5}$$

$$\mu_{X,i} = E(X-\mu)^i = \delta^i \mu_i \tag{6}$$

and the frequently used invariant measures of skewness and kurtosis,

$$\beta_1 = \mu_3/\sigma^3 \quad , \quad \beta_2 = \mu_4/\sigma^4 .$$

In addition, if we use $\dot{\mu}$ to represent one of the three most common averages of the reduced variate Z (the mode and median will be written as $\breve{\mu}$ and $\tilde{\mu}$, respectively). Clearly

$$\dot{\mu}_X = \lambda + \delta\dot{\mu} . \tag{7}$$

Note that (7) is also valid of $\dot{\mu}$ represents any specified percentile, $\xi_p$, satisfying

$$P(Z \leqq \xi_p) = p .$$

Then one can show [10, p. 174] that

$$\begin{aligned}
\mu &= \gamma && = .5772 = \text{Euler's constant} \\
\breve{\mu} &= 0 && \\
\tilde{\mu} &= -\log\log 2 && = 3.6651 \\
\xi_p &= -\log\log p^{-1} && \\
\sigma^2 = \mu_2 &= \pi^2/6 && = 1.6449 \\
\mu_3 &= && = 2.4041 \\
\mu_4 &= 3\pi^4/20 && = 14.6113 \\
\beta_1 &= 1.1396 && \\
\beta_2 &= 5.4 . &&
\end{aligned} \tag{8}$$

Thus using (3), (7), (8) and the fact that $\mu=0$, we see that $\lambda$ is the mode of X and $\delta^2$ is $6/\pi^2$ times the variance of X.

There are two other parameters which are very useful in studying extreme values. These are the return period and the characteristic value. For the sake of clarity, consider a specific example, the annual flood, and let

$$X_{(n)} = M_n = \text{maximum water level in } n=365 \text{ days}$$

$$\bar{G}_n(x) = P(M_n \leqq x) = F^n(x) , \tag{9}$$

where F(x) is the cdf of the daily maximum. For n large enough, we may assume that $\bar{G}_n(x)$ is given by (2)†. The designer of a dam is interested in some particular catastrophic value of x; perhaps the height the dam will withstand. In some sense, he would like to be reasonably sure that this dam will last m years or that in the next m years the annual flood (maximum water level per year) will not exceed x. The engineer is concerned with the event, $A = \{M_n > x\}$ and

$$p = P(M_n > x) = 1-\bar{G}(x) \tag{10}$$

$$M = \text{Number of years until A occurs} \tag{11}$$

for a fixed value of x, so that p is constant and M is a random variable. Clearly M has the geometric distribution,

$$f^*(m) = P(M=m) = p(1-p)^{m-1} , \quad m = 1,2,... \tag{12}$$

and

$$E(M) = p^{-1} = [1-\bar{G}(x)]^{-1} = \tau(x) = \tau, \text{ say.} \tag{13}$$

The quantity $\tau>1$ is the return period. It is a mean; on the average $\tau$ trials (years) will pass before $M_n$ exceeds x, once. Stated another way, it is the average life of the dam. The cdf of M is

$$F^*(m) = P(\text{event occurs before or during the } m^{th} \text{ year})$$

$$= 1 - (1-p)^m = 1 - (1-\tau^{-1})^m \tag{15}$$

---

† Note that the 365 observations used to obtain $X_{(n)}$ are not independent, so that the asymptotic distribution may not be given by (2). We shall ignore this now and assume that (2) is indeed valid. Other papers in the conference will show that this assumption is not unreasonable.

Thus, we have the probability that the event occurs before or at the return period $\tau$,

$$F^*(\tau) = 1 - (1-\tau^{-1})^{\tau} \sim 1 - e^{-1} = 0.63212 ,$$

since $\tau$ is large.

An estimate of $\tau$ using the geometric distribution is not very useful. The MLE of $\tau$ is $\hat{\tau} = N_k/k$, where $N_k$ is the time to the kth occurrence of A. this is useless since we would need to rebuild the dam k-1 times. Of course, we could use smaller values of x, say $x_1<x_2<x_3<\cdots x$; estimate $\tau(x_1),\tau(x_2),\ldots$; use these in (13) to get $\bar{G}(x_1),\bar{G}(x_2),\ldots$; which can then be used to estimate $\hat{\lambda},\hat{\delta}$ and thus get $\hat{G}(x)$ and finally $\hat{\tau}(x)$. As we shall see, the usual procedure is to estimate the parameters $\lambda$ and $\delta$ other ways. A useful conservative approximation for $\tau$ given by Fuller [34, p. 92] is

$$\tau(x) \cong [V(x)]^{-1}, \qquad V(x) = \exp[(x-\lambda)/\delta].$$

When $\tau(x_0)$ equals 100 in (13), $x_0$ is sometimes called "the largest flood in 100 years." The expression in quotes is a misnomer because the largest flood in 100 years is a random variable. In a sense, <u>on the average</u>, we expect to have to wait 100 years to observe a maximum yearly flood greater than or equal to $x_0$. The quantity $x_0$ is usually referred to as the <u>100 year flood</u>. Equation (13) may be written:

$$\bar{G}(x_0) = 1 - [\tau(x_0)]^{-1} , \tag{16}$$

so that for the 100 year flood, $\bar{G}(x_0) = .99$.

We now consider a closely related concept. Let F(x) be the cdf of the daily maximum level. Since 1-F(x) is the probability of observing a value not less than x, we expect n[1-F(x)] values in a random sample of size n to be equal to or greater than x. Define the <u>characteristic (largest) value</u> $\chi_n^*$ as the value of x with the property that, in n independent observations, the expected number of values equal to or greater than x is unity; i.e.

$$F(\chi_n^*) = 1-n^{-1} . \tag{17}$$

Using (9), we obtain

$$\bar{G}_n(\chi_n^*) = F^n(\chi_n^*) = (1-n^{-1})^n . \tag{18}$$

Note that this characteristic value depends on both the underlying distribution and the sample size. On the other hand, for large n we are dealing with the extreme value distribution and (18) becomes asymptotically

$$\bar{G}(\chi^*) = \lim_{n\to\infty}(1-n^{-1})^n = e^{-1} = .36788 \tag{19}$$

We will use (19) to define the <u>characteristic (largest) value</u>. It can be thought of as the 36.79 percentile of $\bar{G}$.

Other averages of M are of interest. The median of M, $\tilde{\tau}$ satisfies $F^*(\tilde{\tau})=.5$. Using (15), we have

$$(1-\tau^{-1})^{\tilde{m}} = .5 \quad \text{or} \quad \tilde{m}\log(1-\tau^{-1}) = -\log 2 \ ,$$

whence

$$\tilde{\tau} = \frac{\log 2}{-\log(1-\tau^{-1})} = \frac{.69315}{-\ln(1-\tau^{-1})} \sim .69315\tau + .34657$$

$$\sim .69315\tau \ .$$

Thus for independent trials, we have

P(at least one occurrence in m trials) ~ .632

P(at least one occurrence in $\tilde{m}$ trials) = .5

Gumbel called the return period of the mode, the <u>occurrence interval</u>. For (2), this becomes

$$\tau(\lambda) = [1-\bar{G}(\lambda)]^{-1} = 1-e^{-1} = 1.582 \ . \tag{20}$$

Similarly, the return period for the mean (annual flood) is

$$\tau(\mu) = \tau(.5772) = [1-\bar{G}(\mu)]^{-1} = 2.328 \tag{21}$$

and the return period of the median (annual flood) is 2.

The case of $X_{(1)} = L_n$, the minimum of n iid variates is treated similarly. In this case nF(x) is the expected number in a random sample of size n with values equal to or less than x. The characteristic (smallest) value of $\chi_1$ satisfies

$$nF(\chi_1) = 1 \tag{22}$$

the value of x with the property that, in n independent observations, the expected number of failures is 1. Using the fact that

$$1 - \underline{G}_n(x) = P(L_n > x) = [1-F(x)]^n$$

we obtain, analogously to (18) and (19),

$$\underline{G}_n(\chi_1) = 1 - [1-F(\chi_1)]^n = 1 - (1-n^{-1})^n \tag{23}$$

$$\underline{G}(\chi_1) = 1 - e^{-1} = .63212 \tag{24}$$

We use (24) to define the <u>characteristic (smallest) value</u>. The reason for the largest and smallest in parentheses is that they are frequently omitted. The characteristic value, which is being used depends on whether we are dealing with the largest or smallest value.

Occasionally another parametrization is used in (2), namely

$$\bar{G}^*(x) = \exp(-\exp[-\frac{x-\lambda}{\omega-\lambda}]) \quad , \quad -\infty < x < \infty \tag{25}$$

where $\lambda$ and $\delta = \omega-\lambda>0$ are still the location and scale parameters. Note that $\lambda$ is the characteristic value as defined in (21); as we have seen, it is also equal to $\breve{\mu}$, the mode of X. It is also clear that

$$\bar{G}(\lambda+\delta) = \bar{G}^*(\omega) = \exp(-e^{-1}) = .6922 . \tag{26}$$

Thus $\omega$, the 69.22 percentile seems to be a more "natural" parameter than $\delta$. Both $\lambda$ and $\omega$ may be though of as percentiles.

## 3. PROBABILITY PAPER

Probability papers in which the extremal distribution plots as a straight line have been used traditionally in analyzing data from all three of the extremal distributions. This is due to two reasons. Transformations which linearize the extremal distributions are easy to obtain; unlike the case of normal probability paper only logarithms are needed. In addition until the advent of computers, estimation of the parameters of the extreme distributions, without using graphical methods have been difficult.

Taking logarithms of (2) twice yields

$$y = -\ln(-\ln\bar{G}) = -\ln\ln(1/\bar{G}) = (x-\lambda)/\delta = z , \tag{27}$$

Although they are equivalent, we have distinguished between the two forms of the reduced variate. When the quantity is thought of as the inverse of $\bar{G}$, it is labeled y; when considered a function of $x,\lambda,\delta$ it is labeled z. Equation (27) may be written as

$$y = (x-\lambda)/\delta \quad \text{or} \quad x = \delta y + \lambda ,$$

which is linear. Of course the iterated logarithm is built into the paper so that one need only plot $x, \bar{G}$. Some prefer to plot x as the abscissa and some prefer to plot it as the ordinate, so that two forms of the plotting paper are used, as illustrated in Figures 1 and 2. Note that there are scales for $\bar{G}$ and y. In addition, the return period (usually called T on Gumbel probability paper) is given as another scale, representing (13). The return period of the mean and mode as given by (20) and (21) are labeled mean and mode on the paper.

We have ignored the problem of determining an estimate of $\bar{G}(x)$ to plot. One could use the plotting position to go with $x_{(i)}$, the ith order statistic in a sample of N from $\bar{G}(x)$ as i/N, the usual plotting position for a sample cdf, as used for example in applicaton of the Kolmogorov-Smirnov bounds. This is equivalent to estimating $\bar{G}(x_{(i)})$ as the proportion of x's less than or equal to $x_{(i)}$. Since 0 and 1 do not appear on the G-scale, we cannot plot $G(x_{(N)})$. If we estimate $\bar{G}(x_{(i)})$ by (i-1)/N or as the proportion of x's less than $x_{(i)}$ we cannot plot $G(x_{(1)})$. A "compromise" was suggested to "split the difference" and use $(i-\frac{1}{2})/N$ as the estimate of $\bar{G}(x_{(i)})$. Another procedure is to use the mean of $\bar{G}(x_{(i)})$ as a plotting position. This leads to plotting position i/(N+1). Blom [1] devoted almost an entire book to the problem of obtaining "optimum" plotting positions, based on an idea of Chernoff and Lieberman [2] that the plotting position should depend on the quantity being estimated. For example [1, p. 145] shows that for the special case of a <u>normal</u> <u>distribution</u> the plotting position $(i-\frac{3}{8})/(N+\frac{1}{4})$ leads to a <u>practically</u> unbiased estimate of $\sigma$ with a mean square error about the same as that of the unbiased best linear estimate, while the plotting position $(i-\frac{1}{2})/N$ leads to a biased estimate of $\sigma$ with nearly minimum mean square error.

Figure 3 taken from [9], where the plotting position was i/(N+1), illustrates the procedure for estimating $\delta$=778.5 and $\lambda$=2.839 for maximum atmospheric pressure at Bergen, Norway for the period between 1857 and 1926. (The quantity $\Phi(x)$ in the figure corresponds to our $\bar{G}(x)$). If the fitted straight line is extrapolated beyond the fitted values, one would "predict, for example, that a pressure of 793 mm corresponds to a probability of 0.994. That is, pressures of this magnitude have less than 1 chance in 100 of being exceeded in any particular year."

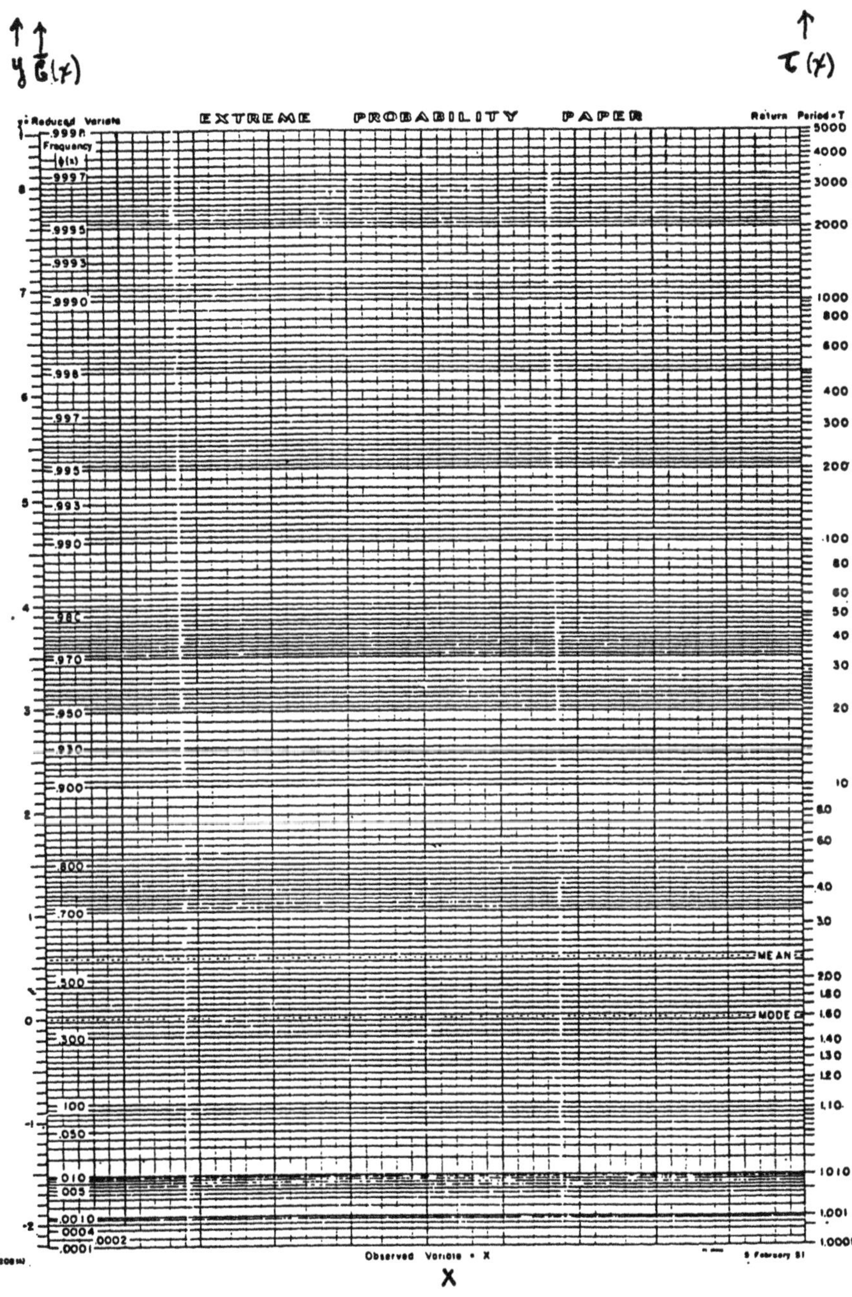


Fig. 1 - Specimen of Extreme Value Probability Paper Used by the Highway Materials Research Laboratory

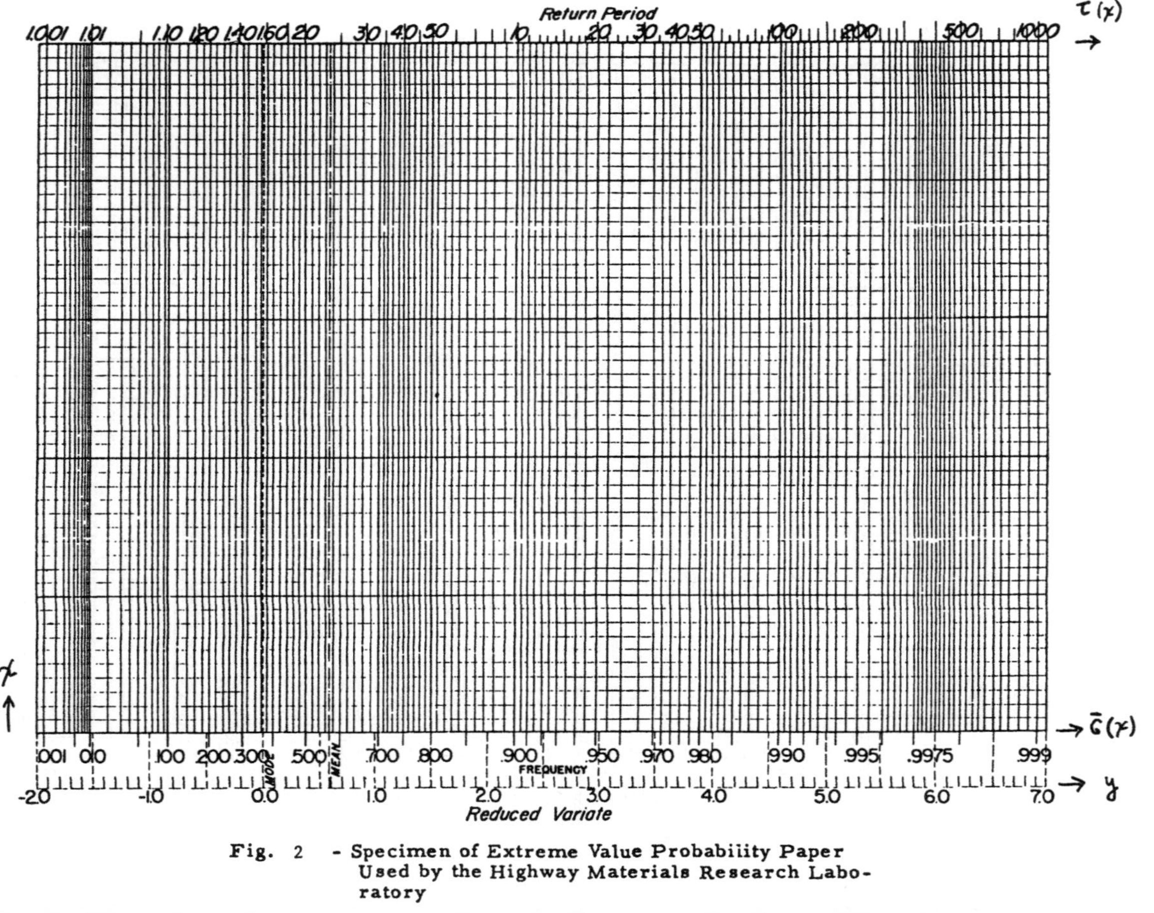


Fig. 2 - Specimen of Extreme Value Probabiiity Paper Used by the Highway Materials Research Laboratory

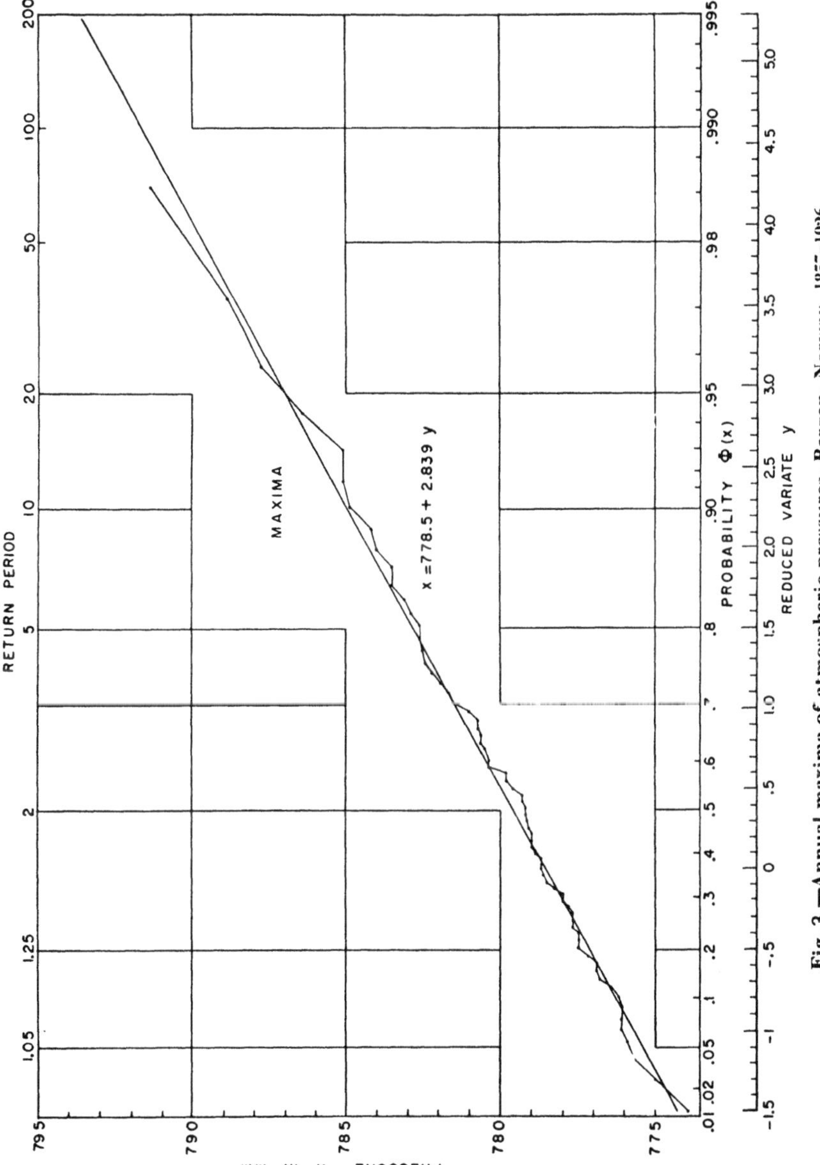


**Fig. 3 —Annual maxima of atmospheric pressures, Bergen, Norway, 1857–1926.**

## 4. (POINT) ESTIMATION

In statistics, the earliest methods of point estimation are the method of moments and the method of maximum likelihood. This is also true in the case of the Gumbel distribution. Thus these procedures will be given first, followed by more modern treatments involving best linear unbiased estimators and best linear invariant estimators.

### 4.1 Method of Maximum Likelihood (ML)

The logarithm of the likelihood of a random sample of N items from (2) is

$$L = N \log\alpha - \Sigma e^{-\alpha(x_i-\lambda)} - \alpha\Sigma(x_i-\lambda) \quad , \tag{28}$$

where $\alpha=\delta^{-1}$. Setting the partial derivates of L with respect to $\alpha$ and $\lambda$ equal to zero, yields the ML equations,

$$\hat{\alpha}^{-1} = \bar{x} - (\Sigma x_i e^{-\hat{\alpha}x_i}/\Sigma e^{-\hat{\alpha}x_i}) \tag{29}$$

$$N = e^{\hat{\alpha}\lambda}\, \Sigma e^{-\hat{\alpha}x_i} \quad , \tag{30}$$

where the summation goes over $i = 1,2,\ldots,N$. Equation (29) involves only $\alpha$, but is intractable without computers. One could use an iterative procedure, starting with the moment estimators for $\alpha,\lambda$ given in the next section. However, Kimball developed a simpler procedure to get approximate solutions. This is described by him in the sections of Gumbel's book which he wrote [10, pp. 229-235], summarizing some of his papers [16-18]. Panchang and Aggarwal in the "Poona Report" [31] were the first (and possibly the only) to use the iterative procedure indicated above, before the widespread use of computers. Maximum likelihood estimators are asymptotically efficient, unbiased with covariance matrix $\underset{\sim}{\Sigma}$ given by

$$\underset{\sim}{\Sigma}^{-1} = -NE[\frac{\partial^2}{\partial\lambda\partial\delta} \log \underset{\sim}{g}(x)] \ . \tag{31}$$

The estimates $\hat{\lambda}$, $\hat{\delta}$ are thus asymptotically normal with means $\lambda$, $\delta$ and

$$\mathrm{Var}(\hat{\lambda}) = [1 + 6\pi^{-2}(1-\gamma)^2]\delta^2/N$$

$$\mathrm{Var}(\hat{\delta}) = 6\pi^{-2}\delta^2/N \ .$$

$$\rho(\hat{\lambda},\hat{\delta}) = \{1 + \pi^2/[6(1-\gamma)^2]\}^{-1/2} \ .$$

Harter and Moore [11] also used the iterative procedure to solve the actual ML equations for the smallest value (1). Now computers entered the picture; they also included the case of doubly censored data (this could happen when the value of the annual flood is not recorded if it is either below or above some fixed value) and compared the asymptotic variances and covariances to those obtained if N=10 and 20 based on Monte Carlo sampling of 2000 samples for each case. For single censoring from above, they concluded that the mean square errors of the MLEs differed little from those of the best linear invariant estimators (see section 4.4) and when using strong asymmetric censoring, were much smaller than those of the best linear unbiased estimators.

Gomes [7, Chapter 3] investigated the maximum likelihood estimates of $\lambda,\delta$ using the ith largest observations in each of N samples from $\bar{G}(x)$ (Weismann [37] treated the case N=1). That is, for i=3, she considered $X_{(n-2)}, X_{(n-1)}, X_{(n)}$. She solved the more complicated version of (29),(30) and evaluated (31) in form of the digamma function and its derivative. The results agree with those of Kimball [17] and others for the use of i=1. She then investigated the properties of these estimates for i=1,2,3 and N = 5(5)30, by means of a Monte Carlo simulation based on between 4500 and 10,000 replicates for each case, and found that $\hat{\lambda}$ is positively biased for i=1 and negatively biased for i>1 and $\hat{\delta}$ is always negatively biased. In addition, she gave a similar treatment for estimating $\lambda,\delta$ using moment estimators, simple best linear unbiased estimators and simple best linear invariant estimators. For i=3, a simple linear estimator is one of the form

$$\sum_{j=1}^{N} \sum_{k=1}^{3} w_{kj} X_{kj} \, .$$

where $X_{kj}$ is the kth largest order statistic in the jth sample. The moment estimator of $\lambda$ is negatively biased for i=1 and positively biased for i>1 and that of $\delta$ is always positively biased. Thus the bias properties of maximum likelihood and of moment estimators are the reverse of one another. The method of moments gave a slightly better estimate of $\lambda$ for small values of N, but yields very poor estimates of $\delta$.

## 4.2 Method of Moments

The method of moments, as proposed by Gumbel [10, p. 227], referred to as the "simplified" method, starts with the fitted line (27),

$$y_i = -\ln(-\ln\bar{G}_i) \tag{32}$$

$$x_i = \lambda + \delta y_i \; . \tag{33}$$

Note that $y_i$ depends only on the plotting position $\bar{G}_i$. Tables [8, Table 2] giving $y_i$ as a function of $\bar{G}_i$ are available. The mean and variance of X are

$$EX = \lambda + \delta E(Y) \tag{34}$$

$$\text{Var } X = \delta^2 \text{ Var } Y \tag{35}$$

Since $Z \equiv Y$, using the population values given immediately after (7) and sample estimates for the moments of X, we obtain

$$\tilde{\lambda}_1 = \bar{x} - \tilde{\delta}_1\gamma \quad , \quad \tilde{\delta}_1 = s_x/(\pi^2/6)^{1/2} = s_x\sqrt{6}/\pi \tag{36}$$

where $\gamma$ is Euler's constant and

$$s_x^2 = \Sigma(x_i-\bar{x})^2/N \; .$$

For smaller values of N, Gumbel would replace $\gamma$, $\sqrt{6}/\pi$ in (36) by $\bar{y}$ and $s_y$.

Others, e.g. Posner [32] use a "modified" method of moments in which sample estimates for the moments of X are also used, yielding

$$\tilde{\lambda}_2 = \bar{x} - \tilde{\delta}_2\bar{y} \quad , \quad \tilde{\delta}_2 = s_x/s_y \; , \tag{37}$$

where

$$s_y = \Sigma(y_i-\bar{y})^2/N \; .$$

Godfrey [6] pointed out that these procedures are valid in the case of doubly censored data or that of suspended data if the missing observations are between specific known observations. The latter occurs when the <u>values</u>, but not the <u>number</u>, of floods between specific values are lost.

These estimates could be used to estimate any fractile (or percentile) by means of

$$\dot{x}_p = \dot{\lambda} + \dot{\delta}y_p \tag{38}$$

where $(\dot{\lambda},\dot{\delta})$ represents the particular estimator being used.

Lieblein [20] performed a Monte Carlo simulation, based on $N = 10,20,30$ and 12,000 values of $x_i$ to calculate the mean squared errors of both of these moment estimators for $\delta$, $x_{.95}$ and $x_{.99}$. He also compared these results with the exact variances of best linear unbiased estimates (BLUEs) which he also computed (see Section 4.3). The relationship among these results and others have been tabulated by Mann [24, p. 247]. She concluded that "the simplified moment estimators seem to given good results for very small N, but are probably never preferable to the BLUEs. The modified moment estimators appear to give poor results for small as well as large samples, except in estimating $\lambda$ ··· [the simplified moment estimator of $\lambda$] appears, however to have no particular advantage over MLEs or BLUEs, other than the fact that it can be calculated without the use of tables or the iterative procedures required for obtaining the maximum-likelihood estimates. The maximum-likelihood estimator seems to be preferable to both moment estimators for estimating $\delta$ and probably is preferable for estimating $x_R$ with $R \geqq .5$ for both large and small N."

## 4.3 Best Linear Unbiased Estimates (BLUEs)

Lieblein [20] was the first to recognize that since the Gumbel plots make use of order statistics, information is wasted if estimates do not use order statistics. He considered estimating the general parameter

$$\theta = \theta(a,b) = a\lambda + b\delta \ . \tag{39}$$

As special cases, we have

$$\lambda = \theta(1,0) \quad , \quad \delta = \theta(0,1) \tag{40}$$

and for the 100p, percentage point $x_p = \theta_p$

$$\theta_p = \theta(1,y(p)) = \lambda + \delta y(p) \tag{41}$$

where $y_p = y(p) = -\log(-\log p)$. He proposed to estimate $\theta$ by $\theta^*$, a linear combination of order statistics in which the weights are determined to yield minimum variance unbiased estimators. More specifically $\theta^*$ based on double censored data is given by

$$\theta^* = \sum_{i=r+1}^{N-s} w_i x_{(i)} \tag{42}$$

where, in view of (39),

$$w_i = aa_i + bb_i .$$

The values of $a_i$, for which $E\theta^* = \theta$ and $Var(\theta^*)$ is minimized are functions of N,r,s and may be determined by the generalized least squares procedure of Lloyd [22], which is based on the generalized Gauss-Markov theorem. This procedure uses the fact that the distribution of order statistics, $Y_{(i)}$, is parameter free and the mean, variance and covariances depend only on the standardized cdf and the sample size. For example

$$E(Y^k_{(i)}) = \frac{N!}{(i-1)!(N-i)!} \int_{-\infty}^{\infty} z^k \bar{G}^{(i-1)}(z)[1-\bar{G}(z)]^{N-i} \bar{g}(z)dz.$$

The method for obtaining these moments numerically is given by Lieblein [19].

We start with the fact, observed earlier, that

$$E(X_{(i)}) = \lambda + \delta\alpha_i \quad , \quad Cov(X_{(i)},X_{(j)}) = \delta^2 vij \ , \tag{43}$$

where $\alpha_i = E(Y_{(i)})$, $v_{ij}$ is the ijth element of V, the variance-covariance matrix of $\underset{\sim}{Y} = (Y_{(1)},Y_{(2)},\ldots,Y_{(N)})'$. The generalized least squares theorem then states that if $\underset{\sim}{X}$ is the vector of observations,

$$\underset{\sim}{X} = (X_{(1)}, X_{(2)},\ldots,X_{(N)})' \quad ,$$

with

$$E(\underset{\sim}{X}) = A\underset{\sim}{\theta} \quad , \quad Cov(\underset{\sim}{X}) = V\delta^2 \tag{44}$$

for A and V <u>known</u>, then the least-squares estimate of $\underset{\sim}{\theta}$ is the vector $\underset{\sim}{\theta}^*$ which minimizes

$$(\underset{\sim}{x} - A\underset{\sim}{\theta})'V^{-1}(\underset{\sim}{x} - A\underset{\sim}{\theta}) .$$

The minimum solution satisfies the "normal equations,"

$$A'V^{-1}A\underset{\sim}{\theta}^* = A'V^{-1}\underset{\sim}{x}$$

Thus

$$\underset{\sim}{\theta}^* = (A'V^{-1}A)^{-1}A'V^{-1}\underset{\sim}{x} \tag{45}$$

and the variance-covariance matrix of $\underset{\sim}{\theta}^*$ is

$$\mathrm{Var}(\underset{\sim}{\theta}^*) = (A'V^{-1}A)^{-1}\delta^2 \tag{46}$$

The estimates $\theta_i^*$ are unbiased and have minimum variance of all elements in the class of unbiased linear statistics. Note that (43) also gives the variances of $\theta_i^*$.

In our case $A = (\underset{\sim}{1},\underset{\sim}{\alpha})$, where $\underset{\sim}{1}$ is the N-dimensional column vector, all of whose elements are 1 and $\underset{\sim}{\alpha} = (\alpha_1,\alpha_2,\ldots\alpha_N)'$. Calculation of the means, variances and covariances of the order statistics $Y_{(i)}$ and the weights $a_i,b_i$ are tedious and require computers even for small values of N. They have been tabulated for N=1(1)6, r=s=0 in [20]; for N=2(1)6, r=0(1)N-2, s=0 in [21, pp. 300,301]; and have been extended by Mann [23, Appendix C] to N=2(1)25, r=0, s=0(1)N-2.

## 4.4 Best Linear Invariant Estimators (BLIEs)

Mann relaxed the unbiased condition of the previous section, defined loss as squared error divided by $\delta^2$ and considered weighted sums of $X_{(i)}$, $i=1,2,\ldots,r\leqq N$, as estimators of $\lambda,\delta,x_p$, with mean squared error, invariant under translations and multiplications. The estimators among these with smallest mean squared error she called the best linear invariant estimators. The BLUEs of Lieblein and and Zelen are also invariant. She showed [25] that these estimates $\lambda_1^*$, $\delta_1^*$ and $x_{p,1}^*$ are linear functions of the BLUEs of $\lambda$ and $\delta$. As expected, the mean squared errors of the BLIEs are uniformly less than those of the corresponding BLUEs [24, p. 237]. "In cases in which $r \ll N$, the ratios of expected losses of the BLUEs relative to those of the corresponding BLIEs are large.... For r=2, N=2(1)25 the expected losses of the BLUEs of $\lambda$ and $\delta$ increase with N from .659 to 8.24 and .712 to .980, respectively. The corresponding ranges of expected losses when the BLIEs are used are .657 to 4.49 and .416 to .495.... Furthermore, for m=2, the expected losses of the BLIEs of $x_p$ and $\delta$ are by definition less than or equal to those of the corresponding MLEs, which are also invariant. Tables of weights for obtaining the BLIEs and the expected losses of the estimators are available in [23, Appendix C]."

Asymptotically efficient linear estimators of location and scale parameters are derived and shown to be symptotically normal by Chernoff et al. [3]. "They are invariant and appear to be quite efficient with respect to the BLUEs, even for fairly small N [24]". Johns and Lieberman [15] used this procedure to determine the weights in the asymptotically efficient linear estimators for $\lambda$ and $\delta$. The order-statistic moment matrices are not required, as in the generation of BLIE weights." (For r,N large this is an advantage, since the moment calculations involve large rounding errors, requiring multiple-precision computing techniques.) "Thus the mean squared errors of these estimators have not been computed and it is difficult to compare these expected losses with those of BLUEs and BLIEs. A limited comparison based on hand computation is given in [24, p. 248]."

### 4.5 Some Other Linear Estimators

Other approximating methods exist for generating weights in linear estimators. As in the Johns and Lieberman procedure, they do not involve calculating order-statistic moments. The earliest procedure due to Blom [1, Section 9.2] and outlined by him in [33, Section 4B] is called the unbiased nearly best linear estimate (NBLUE) and was applied by him [33, p. 127] and Hassanein [12] to the Gumbel distribution (2). In this procedure one defines $\phi_i$ by

$$\bar{G}_Z(\phi_i) = i/(N+1) = p_i \; . \tag{47}$$

i.e., $\phi_i$ is the fractile of the reduced variable using the mean plotting position and $\psi_i = \bar{g}(\phi_i)$, where $\bar{g}_Z(x)$ is the density of Z. Then we have for N large

$$\mathrm{Cov}(\psi_i Z_{(i)}, \psi_j Z_{(j)}) \sim p_i(1-p_j)/(N+2) \; . \tag{48}$$

If we let

$$W_{(i)} = \psi_{i+1} Z_{(i+1)} - \psi_i Z_{(i)}; \; i=0,1,\ldots,n; \; \psi_0 = \psi_{n+1} = 0 \; ,$$

then

$$\mathrm{Var}(W_{(i)}) \sim N(N+1)^{-2}(N+2)^{-1} \tag{49}$$

$$\mathrm{Cov}(W_{(i)}, W_{(j)}) \sim -(N+1)^{-2}(N+2) \; . \tag{50}$$

The results in (49), (50) are independent of the distribution $\bar{G}(x)$. It need not even be an extreme-value distribution. Now the parameter $\theta$ given by (39) may be estimated, using an uncensored sample by a linear function,

$$\theta^{**} = \sum_{i=1}^{N} g_i x_{(i)} = \sum_{i=1}^{N} g_i(\lambda + \delta z_{(i)}) \; . \tag{51}$$

Here, of course, the weights may be different from those in

(42). It is convenient to replace $z_{(i)}$ by $w_{(i)}$ in (51) and to replace $g_i$ by $h_0, h_1, \ldots, h_n$ defined, apart from an additive constant, by

$$g_i = \psi_i(h_i - h_{i-1}) .$$

Then (51) becomes

$$\theta^{**} = \sum_{i=0}^{N} h_i[\lambda(\psi_i - \psi_{i+1}) - \delta w_{(i)}] .$$

For

$$i = 0,1,2,\ldots,N \quad ,$$

let

$$C_{1i} = \psi_i - \psi_{i+1}, \ C_{2i} = \psi_i E(Z_{(i)}) - \psi_{i+1}E(Z_{(i+1)}) = -E(W_{(i)}) \ ,$$

then

$$E(\theta^{**}) = \lambda \sum_{i=0}^{N} C_{1i}h_i + \delta \sum_{i=0}^{N} C_{2i}h_i \tag{52}$$

$$\mathrm{Var}(\theta^{**}) = \delta^2[\Sigma h_i^2 \mathrm{Var}(W_{(i)}) + \sum_{i \neq j}\sum h_i h_j \ \mathrm{Cov}(W_{(i)}, W_{(j)})] \tag{53}$$

Using the approximations (49) and (50) in (53), we obtain

$$\mathrm{Var}(\theta^{**}) \sim \frac{\delta^2}{(N+1)(N+2)} \sum_{i=0}^{N} (h_i - \bar{h})^2 \quad , \tag{54}$$

where $\bar{h} = (N+1)^{-1}(h_0 + h_1 + \ldots + h_N)$. Now using (39) and (52), it can be seen that $\theta^{**}$ will be unbiased if

$$\Sigma C_{1i}h_i = a \quad , \quad \Sigma C_{2i}h_i = b . \tag{55}$$

Now we seek weights $g_i$ (or $h_i$) which minimize (54), subject to (55), to obtain the NBLUE. It is called a "nearly" estimate because of the asymptotic approximations used in (49) and (50), but involves much less work than is expended in getting the BLUE. The result is

$$h_i = \bar{h} + a_1 C_{1i} + a_2 C_{2i} \quad ,$$

where

$$a_1 = d^{11}a + d^{12}b$$

$$a_2 = d^{21}a + d^{22}b$$

$$D^{-1} = (d^{\ell m}) \quad \text{is the inverse of} \quad D = (d_{\ell m})$$

$$d_{\ell m} = \sum_{i=0}^{N} C_{\ell i}C_{mi}; \quad \ell = 1,2; \quad m = 1,2$$

For the Gumbel distribution, the estimators $\lambda^{**} = \theta^{**}(1,0)$ and $\delta^{**} = \theta^{**}(0,1)$ are asymptotically efficient and asymptotically normal. Hassanein [33] used this procedure to estimate $\lambda, \delta, x_p$ for N=2(1)10(5)25 and all possible single and double censoring.

Blom also considered several modifications of NBLUEs. For example [1, Chapter 6 and Sections 9.3,10.7] he considered one called the nearly unbiased nearly best estimate based on transformed beta variables. It is well known that if $U_i = \bar{G}(X_{(i)})$, $i=1,2,\ldots,N$ is the ith order statistic in a sample of N independent observations from the uniform distribution on (0,1), for any cdf $\bar{G}$ (not necessarily the Gumbel distribution), then $U_i$ has the beta distribution. The quantity $X_{(i)} = \bar{G}^{-1}(U_i)$ is called a transformed beta variable. Blom replaced the relationship $E[\bar{G}(X_{(i)})] = i/(N+1)$ by

$$E(X_{(i)}) = \bar{G}^{-1}[i/(N+1)] + R_i \tag{56}$$

and showed that $R_i = 0(N^{-1})$, hence tends to zero as $N\to\infty$. He then approximated (56) by

$$E(X_{(i)}) \sim \bar{G}^{-1}(\pi_i) , \qquad \pi_i = (i-\alpha_i)/(N-\alpha_i-\beta_i+1) \tag{57}$$

and showed how to choose $\alpha_i, \beta_i$ appropriately. (If $\alpha,\beta$ are independent of i and $\bar{G}$ is the normal cdf, one gets the result given in Section 3.) Then, the nearly unbiased nearly best estimator is obtained when one uses

$$\phi_i' = \bar{G}(\pi_i) ,$$

where $\pi_i$ is given by (57) in place of $\phi_i$ in (47) and uses (57) in place of the known $E(X_{(i)})$, so that we are approximating both the mean and the variance of the linear estimator of $\theta$.

Lieblein [20] also suggested using quantile estimators, based on a method proposed by Mosteller [29] for normal samples, when N is large and one wishes to estimate the parameters using a few selected observations. The procedure uses observations $X_{(i)}$, $X_{(j)}$, $X_{(k)}$ where $i=\ell N$, $j=mN$ and $k=nN$†, $0<\ell<m<n<1$, with $\ell,m,n,a,b$ suitably chosen so that

$$\hat{\theta}_p^* = ax_{(j)} + b(x_{(k)}-x_{(i)}) \tag{58}$$

is an (asymptotically) unbiased estimator of the parameter $x_p = \lambda+\delta y_p$ given by (41). The mean and variance of $\hat{\theta}_p^*$ may be computed from the moments of the order statistics in (58). For N large and $\ell,m,n$ not too close to 0 or 1, the quantities $X_{(i)},X_{(j)},X_{(k)}$ are jointly normal with

$$E(X_{(i)}) = t_\ell \tag{59}$$

$$\mathrm{Var}(X_{(i)}) = \ell(1-\ell)/\{N[\bar{g}(t_\ell)]^2\} \quad , \tag{60}$$

where $t_\ell$ is defined by $\ell = \int_{-\infty}^{t_\ell} \bar{g}(x)dx$. Similar results hold for the means and variances of $X_{(j)}$ and $X_{(k)}$. The covariance of $X_{(i)}$ and $X_{(j)}$, $i<j$ is given by

$$\mathrm{Cov}(X_{(i)},X_{(j)}) = \ell(1-m)/[N\bar{g}(t_\ell)\bar{g}(t_m)] \quad . \tag{61}$$

Similar results hold for $\mathrm{Cov}(X_{(i)},X_{(k)})$ and $\mathrm{Cov}(X_{(j)},X_{(k)})$. For $\hat{\theta}_p^*$ in (58) to be unbiased, we have

$$E(\hat{\theta}_p^*) = x_p = \lambda + \delta y(p) \tag{62}$$

where $x_p$ is the 100 pth percentile, i.e.

$$x_\ell = t_\ell = \lambda + \delta y_\ell \quad . \tag{63}$$

Using (59) and (62) we obtain

$$at_m + b(t_n-t_\ell) = \lambda + \delta y(p) \tag{64}$$

Finally, using (63) in (64) we obtain

$$a(\lambda+\delta y_m) + b(y_n-y_\ell)\delta \equiv \lambda + \delta y_p \quad . \tag{65}$$

---

† Since the quantities $\ell N,mN,mN$ will generally not be integers, i,j,k are defined by Lieblein as the integers closest to these quantities, and by Dubey [4] and Hassanein [13] as the closest integer, not less than the quantities.

Equating coefficients of $\lambda$ and $\delta$ yields

$$a = 1 \quad , \quad b = (y_p - y_m)/(y_n - y_\ell) \tag{66}$$

so that (58) becomes

$$\hat{\theta}^*_p = x_{(j)} + \frac{y_p - y_m}{y_n - y_\ell}(x_{(k)} - x_{(i)}) \ . \tag{67}$$

In principle, the quantities $\ell$,m,n might be determined to minimize the variance of (67). This was not done because it would involve extensive computation and N is large and so efficiency is not too important. Instead, he used estimators

$$\hat{\theta}^{**}_p = \hat{\lambda}^{**} + \delta^{**} y_p \ , \tag{68}$$

where $\lambda^{**}$ and $\delta^{**}$ are reasonably efficient estimators of $\lambda$ and $\delta$, involving the smallest number of order statistics. Gumbel showed [20, p. 58] that the value of m for which $Var(X_{(mN)})$ is least is m=0.20319, so that Lieblein used

$$\hat{\lambda}^{**} = x_{(0.2N)} \ . \tag{69}$$

The scale parameter $\delta$ may be estimated by $C[x_{(nN)} - x_{(\ell N)}]$, the difference of two order statistics, multiplied by an unbiasing factor. Trial and error indicated that $\ell \doteq 0.03$, n=0.85 gave an estimate of $\delta$ with efficiency close to the maximum: Using these values, yielded

$$\hat{\theta}^{**}_p = x_{(.2N)} + 0.3256(y_p + 0.4759)(x_{(.85N)} - x_{(.03N)}) \ . \tag{70}$$

The variance of this estimator is

$$Var(\hat{\theta}^{**}_p) = 8.6916d^2 - 0.0681d + 1.5442 \quad , \tag{71}$$

where

$$d = 0.3256y_p + 0.1549 \quad .$$

Since $\hat{\theta}^{**}$ is unbiased, its efficiency may be measured by the ratio of the Cramer-Rao lower bound variance and (71). For p=.95,.99,1.00 (the limiting value), the efficiency measure is .645, .649, .660 respectively. Mann [24, p. 243] indicated that for p=.90, the efficiency measure is .645.

Hassanein [13] increased the number of quantiles used in the estimation of $\lambda$ and $\delta$. He used $X_{(1)}, X_{(2)}, \ldots, X_{(k)}$, where

$k=2(1)10$, using results of Ogawa [30] to determine the quantiles to be used which maximize the joint efficiency of the estimators.

Englehardt and Bain [5] obtained an unbiased estimator of $\delta$ for the Gumbel smallest distribution (1) based on the first r of N order statistics,

$$\delta^{***} = \sum_{i=1}^{r} |X_{(s)} - X_{(i)}|/(Nk_{r,N}) , \tag{72}$$

$$k_{r,N} = N^{-1}\Sigma_1^r E|Z_{(s)}-Z_{(i)}| ,$$

where s is a function of r and N and $Z_{(i)}$ is the reduced ith order statistic. The corresponding estimators of $\lambda, x_p$ are

$$\lambda^{***} = X_{(s)} - \delta^{***}EZ_{(s)} \tag{73}$$

$$x_p^{***} = \lambda^{***} + \delta^{***}y_p \tag{74}$$

$$y_p = \ln[-\ln(1-p)]$$

## 5. (CONFIDENCE) INTERVAL ESTIMATION

The estimates given in Section 4 can be used to obtain interval estimates or to test hypotheses about the parameters. Most of them are asymptotically normal with known asymptotic covariances so that one can compute asymptotic confidence intervals, using the t-distribution (joint confidence regions using the F-distribution). To use the others one resorts to Monte Carlo simulations. For example, using (33) and the Gumbel method of moments (36), the moment estimate of $x_p$ is

$$\tilde{x}_{p,1} = \bar{x} + (y_p-\gamma)\frac{\sqrt{6}}{\pi}s_x , \tag{75}$$

whence

$$E(\tilde{X}_{p,1}) = \lambda + \gamma\delta + (y_p-\gamma)\sqrt{6}\pi^{-1}\delta E(S_Y)$$

$$= x_p + [\sqrt{6}\pi^{-1}E(S_Y)-1](y_p-\gamma)\delta .$$

which (incidentally) gives the bias of $\tilde{x}_{p,1}$. The variance of (75) is [20, p. 42].

$$\mathrm{Var}(\tilde{X}_{p,1}) = \delta^2[\frac{\pi^2}{6N} + \frac{6}{\pi^2}(y_p-\gamma)^2\sigma^2(S_Y) + 2(y_p-\gamma)\frac{\sqrt{6}}{\pi}\mathrm{Cov}(\bar{Y},S_Y)], \quad (76)$$

$$= S_p^2\delta^2 , \quad \text{say,}$$

where $\sigma^2(S_Y)$ is the variance of $S_Y$, the sample standard deviation of a sample from the reduced $\bar{G}$ distribution $\Phi(y)=\exp(-e^{-y})$. The simulation using 1200 observations from the reduced distribution was used to "compute" $\sigma^2(S_Y)$ and $\mathrm{Cov}(\bar{Y},S_Y)$ and thus the factor $S_p^2$ in (76) for N=10,20,30 for p=.95,.99,.999. Lieblein obtained confidence limits on $x_p$ by assuming that $\tilde{X}_{p,1}$ may be approximated by a Gumbel distribution with scale parameter $\delta_1 = B_p\delta$, i.e. $\delta_1$ is some multiple of the scale parameter of the underlying distribution. If so then

$$\mathrm{Var}(\tilde{X}_{p,1}) = \pi^2\delta_1^2/6 . \quad (77)$$

and (76),(77) imply that

$$\delta_1 = \sqrt{6}S_p\delta/\pi = B_p\delta ,$$

where

$$B_p = \frac{\sqrt{6}}{\pi}S_p = \frac{\sqrt{6}}{\pi}[\frac{\mathrm{Var}(X_{p,1})}{\delta^2}]^{1/2} .$$

The 100 C percent confidence limits on $x_p$ are given by

$$\tilde{x}_{p,1} \pm \tilde{\Delta}_{C,p}\delta_1 \quad \text{or} \quad \tilde{x}_{p,1} \pm \tilde{\Delta}_{C,p}B_p\delta \quad (78)$$

where

$$\tilde{\Delta}_{C,p} = \Phi(y_p) - \Phi(-y_p) .$$

For C=.68,.95, we have $\Delta_{C,p}$=1.141,3.067. Lieblein [20, p. 82] gave the "half-width intervals" $\tilde{\Delta}_{C,p}$ for 68 and 95 percent confidence bounds for $x_p$ for the three values of N and p indicated and also for the BLUEs given in Section 4.3. The limits in (78) get wider as p increases. One can plot the locus of limits (78) on Figure 3 to give 68% and 95% confidence bands on the plotted line as the set of points $(\tilde{x}_{p,1},p)$ in order to be able to make confidence statements about predictions outside the region covered by the data [20, p. 86]. Confidence bounds on $x_p$

are also referred to as tolerance bounds on the distribution $\bar{G}(x)$.

Johns and Lieberman [14,15] used their linear estimators to get a lower confidence limit on the reliability function

$$R(x_0) = P(X>x_0) ,$$

when X has the smallest Gumbel distribution (1), so that

$$R(x_0) = \exp[-\exp(-\mu/\delta)] , \quad \mu = \lambda - x_0 .$$

Since $R(x_0)$ is an increasing function of $\mu/\delta$, a lower confidence limit on $\mu/\delta$ can be used to determine a lower confidence limit on $R(x_0)$. That is,

$$P(L < \mu/\delta) = \beta \Rightarrow$$

$$P(L^* = \exp\{-\exp(-L)\} < \exp\{-\exp(-\mu/\delta)\} = R(x_0)) = \beta . \quad (79)$$

They proposed to use their asymptotically efficient linear estimators of $\mu$ and $\delta$ based on the first r order statistics in a sample of N mentioned in Section 4.4. Let $Y_{(i)} = X_{(i)}/x_0$. Then use as estimates of $\mu,\delta$,

$$Z_a = \sum_{i=1}^{r} a_i Y_{(i)} , \quad Z_b = \sum_{i=1}^{r} b_i Y_{(i)} .$$

Consider

$$V_a = \delta^{-1}(Z_a-\mu) , \quad V_b = \delta^{-1} Z_b .$$

The joint distribution of $V_a,V_b$ is parameter free. They next define a function $L(t)$ with the property that for fixed $\beta$, $0<\beta<1$,

$$P(L(t) < tV_b-V_a) = \beta \quad \text{for all } t \quad (80)$$

and assume $P(V_b>0) = 1$. Finally

$$P\{L(Z_a/Z_b) < \mu/\delta\} = P\{(Z_a/Z_b) < L^{-1}(\mu/\delta)\}$$

$$= P[\frac{\delta V_a+\mu}{\delta V_b} < L^{-1}(\tfrac{\mu}{\delta})] = P\{V_a + \tfrac{\mu}{\delta} < L^{-1}(\tfrac{\mu}{\delta})V_b\}$$

$$= P[\tfrac{\mu}{\delta} < L^{-1}(\tfrac{\mu}{\delta})V_b-V_a] = P\{L[L^{-1}(\tfrac{\mu}{\delta})] < L^{-1}(\tfrac{\mu}{\delta})V_b-V_a\}$$

$$= \beta$$

because of (80). Thus $L(Z_a/Z_b)$ is a lower confidence limit for $\mu/\delta$ with confidence coefficient $\beta$ and the corresponding $L^*(Z_a/Z_b)$ given by (79) is a lower confidence limit for $R(x_0)$. They show that asymptotically L and L* are efficient bounds for $\mu/\delta$ and $R(x_0)$. For N=10,15,20,30,50,100 and four values of r for each they computed Monte Carlo simulations of the distribution of $tV_b-V_a$ for <u>fixed</u> values of $t=Z_a/Z_b$. Thus they generated tables from which one can obtain the bounds for $R(x_o)$ for specific $x_0$ for $\beta$=.5,.75,.90,.95 and .99. To get a lower bound on $x_R$, "one finds in the table a value of $L^*(Z_a/Z_b)$ corresponding to a specified R. Then the bound on $x_R$ is $Z_a-tZ_b$, where t is the tabled value of $Z_a/Z_b$ associated with $L^*(Z_a/Z_b)$ for the appropriate confidence level" [24, p. 253].

The Englehardt-Bain estimates (72)-(74) may be used to get an approximate interval for $\delta$. The mean and variance of

$$Nk_{r,N}\delta^{***}/\delta = \sum_{i=1}^{r} (X_{(r)}-X_i)/\delta$$

are each approximately equal to $Nk_{r,N}$ for N large and $r/N \leqq .5$. Thus they suggest approximating the distribution of $2Nk_{r,N}\delta^{***}/\delta$ by a chi-square with $2Nk_{r,N}$ degrees of freedom. A better approximation is due to Montfort [28] who considered

$$V_i = \frac{X_{(i+1)}-X_{(i)}}{EZ_{(i+1)}-EZ_{(i)}} \tag{81}$$

and noted that, when $N \geqq 20$, $H_i = \delta^{-1}V_i$, $i=1,2,\dots,N-1$ is approximately an exponential variate with mean unity and $Cov(H_iH_j)$ is nearly zero for $i \neq j$. Thus $2H_i$ may be approximated by independent chi-squares with two degrees of freedom. For $r \leqq .9N$

$$2Nk_{r,N}\delta^{***}/\delta = W_{r,N} = 2\sum_{i=1}^{r}(X_{(r)}-X_{(i)})/\delta$$

$$= \sum_{1}^{r-1} i[EZ_{(i+1)}-EZ_{(i)}]2H_i ,$$

a sum of weighted chi-square variates. Methods of Patnaik-Satterthwaite-Welch-Cochran can be used to approximate sums of weighted chi-squares by a chi-square [27, p. 177]. Thus 2mW/v is approximately chi-square with $2m^2/v$ degrees of freedom, where m and v are the mean and variance, respectively of $W_{r,N}$. For r/N small 2mW/v W and the degrees of freedom are approximately $2Nk_{r,N}$. If we designate the variance of $\delta^{***}/\delta$ by $\ell_{r,N}$, we have

$$\ell_{r,N} = v/m^2 .$$

Since $2m^2/v$ will probably not be integral, Mann suggests using the Wilson-Hilferty transformation of the chi-square variate to a normal one [27, p. 177]. Thus we have all the information needed to get confidence limits on $\delta$.

"For $N \geqq 10$, $\beta = 1-p \geqq .9$, it can be shown [27, p. 249] that

$$Q = \frac{\lambda^{***} - x_p - \delta^{***} B_{r,N}/\ell_{r,N}}{(-y_p - B_{r,N}/\ell_{r,N})\delta}$$

has an approximate $\chi_d^2/d$ distribution, with

$$y_p = \log[-\log(1-p)] \quad ;$$

Q and $\delta^{***}/\delta$ are independent. Thus

$$F_x = \frac{\lambda^{***} - x_p - \delta^{***} B_{r,N}/\ell_{r,N}}{\delta^{***}(-y_p - B_{r,N}/\ell_{r,N})}$$

has approximately an F distribution with degrees of freedom

$$[2(y_p + B_{r,N}/\ell_{r,N})^2/(A_{r,N} - B^2_{r,N}/\ell_{r,N}),\ 2/\ell_{r,N}] ,$$

where

$$A_{r,N} = \mathrm{Var}(\lambda^{***}/\delta) \quad \text{and} \quad B_{r,N} = \mathrm{Cov}(\lambda^{***}/\delta,\ \delta^{**}/\delta)$$

Thus one can get confidence bounds on $\lambda$.... To obtain a lower confidence bound on $R(x_0)$ for specified $x_0$ one can use $x_p = x_0$ and $y_p = \log(-\log R(x_0))$."

## 6. TESTING OF HYPOTHESES

Hypothesis tests on parameters can be handled, in the usual way, by relating the acceptance region to the confidence region. One testing procedure that is different and hence interesting is the test of whether the data actually come from a Gumbel distribution. Montfort [28] used a function of (81), $V_i$ the difference of order statistics to give a goodness-of-fit test for the smallest Gumbel distribution, i.e. a test of $\underline{G}(x)$ against the smallest Weibull extreme distribution. This is based on the fact, noted above, that when the underlying distribuiton is $\underline{G}(x)$, the $H_i$ are distributed approximately exponentially with parameter unity and zero covariances. Mann et al. [26] used another function of the $V_i$ to test this hypothesis. These will be omitted here because it is believed the procedure will be discussed in the next paper in this series.

Another interesting test of the hypothesis that the largest extreme is Gumbel against the alternative that it is either Weibull or Fréchet is due to Tiago de Oliveira [35,36]. This asymptotic procedure is based on the Mises-Jenkinson form [10, pp. 163,262] of the cdf of the largest reduced extreme,

$$\bar{\underset{\sim}{L}}(z|k) = \exp[-(1+kz)^{-1/k}] \quad , \quad 1+kz > 0 \; . \tag{82}$$

We have

$$\bar{\underset{\sim}{L}}(z|0^+) = \bar{\underset{\sim}{L}}(z|0^-) = \exp(-e^{-z}) = \bar{G}(z)$$

$$\bar{\underset{\sim}{L}}(z|k) = 1 \quad , \quad k < 0 \quad , \quad z > -1/k > 0$$

$$\bar{\underset{\sim}{L}}(z|k) = 0 \quad , \quad k > 0 \quad , \quad z < -1/k < 0$$

and $\underset{\sim}{L}$ has a Weibull distribution if $k < 0$ and a Fréchet distribution if $k > 0$ with shape parameters $-k^{-1}$ and $k^{-1}$ respectively. He developed a locally most powerful test of $k=0$ vs. $k>0$ and $k<0$ and thus a test for the trilemma decision problem, Gumbel-Weibull-Fréchet, based on a random reduced sample $(z_1, z_2, \ldots, z_N)$ and the statistic $\bar{v} = \Sigma v(z_i)/N$, where

$$v(z) = \frac{\partial \log \bar{\underset{\sim}{L}}'(z|k)}{\partial k}\Big|_{k=0} = \frac{z^2}{2} - z - \frac{z^2}{2}e^{-z}$$

in which

$$\bar{v} \leqq b_N \quad \Longrightarrow \quad \text{Weibull}$$

$$b_N < \bar{v} \leqq a_N \quad \Longrightarrow \quad \text{Gumbel}$$

$$a_N < \bar{v} \quad \Longrightarrow \quad \text{Fréchet}$$

and gave properties of $a_N, b_N$ to minimize the three probabilities of making the wrong decision, for large N. For the case of the non-reduced sample $(x_1, x_2, \ldots, x_N)$ he made use of MLEs of the $\lambda, \delta$.

REFERENCES

1. Blom, G. 1958, Statistical Estimates and Transformed Beta-Variables, John Wiley, New York.

2. Chernoff, H. and Lieberman, G.J. 1954, "Use of normal probability paper." J. Amer. Statist. Assoc. 49, pp. 778-785.

3. Chernoff, H., Gastwirth, L. and Johns, M.V., Jr. 1967, "Asymptotic distribution of linear combinations of functions of order statistics with applications to estimation," Ann. Math. Statist. 38, pp. 52-71.

4. Dubey, S.D. 1967, "Some percentile estimators for Weibull parameters," Technometrics 9, pp. 119-129.

5. Englehardt, M. and Bain, L.J. 1973, "Some complete and censored sampling results for the Weibull or extreme-value distribution," Technometrics 15, pp. 541-549.

6. Godfrey, M.L. 1958, "Theory of extreme values applied in tests," Industrial Laboratories, August.

7. Gomes, M.I. 1978, "Some probabilistic and statistical problems in extreme value theory." Unpublished doctoral dissertation, Univ. of Sheffield.

8. Gumbel, E.J. 1953, Probability Tables for the Analysis of Extreme-Value Data, (U.S.) National Bureau of Standards, Applied Mathematics Series, No. 22, pp. 19-25.

9. Gumbel, E.J. and Lieblein, J. 1954, "Some applications of extreme-value theory." American Statistician 8, Dec., p. 16.

10. Gumbel, E.J. 1958, Statistics of Extremes, Columbia Univ. Press, New York.

11. Harter, H.L. and Moore, A.H. 1968, "Maximum-likelihood estimation, from doubly censored samples of the parameters of the first asymptotic distribution of extreme values," J. Amer. Statist. Assoc. 63, pp. 889-901.

12. Hassanein, K.M. 1964, "Estimation of the parameters of the extreme-value distribution by order statistics," AD 622257, Univ. of North Carolina, Chapel Hill.

13. Hassanein, K.M. 1968, "Analysis of extreme-value data by sample quantiles for very large samples," J. Amer. Statist. Assoc. 63, pp. 877-888.

14. Herbach, L. 1970, "An exact asymptotically efficient confidence bound for reliability in the case of the Gumbel distribution." Technometrics 12, pp. 700-701.

15. Johns, M.V., Jr. and Lieberman, G.J. 1966, "An exact asymptotically efficient confidence bound for reliability in the case of the Weibull distribution." Technometrics 8, pp. 135-175.

16. Kimball, B.F. 1946, "Sufficient statistical estimation functions for the parameters of the distribution of maximum values." Ann. Math. Statist. 17, pp. 299-305.

17. Kimball, B.F. 1949, "An approximation to the sampling variance of an estimated maximum value of given frequency based on fit of doubly exponential distribution of maximum values." Ann. Math. Statist. 20, pp. 110-113.

18. Kimball, B.F. 1956, "The bias in certain estimates of the parameters of the extreme-value distribution." Ann. Math. Statis. 27, pp. 758-767.

19. Lieblein, J. 1953, "On the exact evaluation of the variances and covariances of order statistics in samples from the extreme-value distribution," Ann. Math. Statist. 24, pp. 282-287.

20. Lieblein, J. 1954, A New Method of Analyzing Extreme Value Data, (U.S.) National Advisory Committee for Aviation, Tech. Note 3053.

21. Lieblein, J. and Zelen, M. 1956, "Statistical analysis of the fatigue life of deep groove ball bearings," J. Res. Nat. Bur. Stand., 57, pp. 273-316.

22. Lloyd 1952, "Least-squares estimation of location and scale parameters using order statistics," Biometrika, 39, pp. 88-95.

23. Mann, N.R. 1967, "Results on location and scale parameter estimation with application to the extreme-value distribu-

tion." Aerospace Research Laboratories Report ARL 67-0023, Wright-Patterson Air Force Base, Ohio.

24. Mann, N.R. 1968, "Point and interval estimation procedures for the two-parameter Weibull and extreme-value distributions." Technometrics 10, pp. 231-256.

25. Mann, N.R. 1969, "Optimum estimators for linear functions of location and scale parameters," Ann. Math. Statist. 40, pp. 2149-2155.

26. Mann. N.R., Scheuer, E.M. and Fertig, K.W. 1973, "A new goodness-of-fit test for the two-parameter Weibull or extreme-value distribution with unknown parameters," Communications in Statistics, 2, pp. 383-400.

27. Mann, N.R., Schafer, R.E. and Singpurwalla, N.D. 1974, Methods for Statistical Analysis of Reliability and Life Data, John Wiley, New York.

28. Montfort, M.A.J. van 1970, "On testing that the distribution of extremes is of Type I when Type II is the alternative," Jour. of Hydrology, 11, pp. 421-427.

29. Mosteller, F. 1946, "On some useful 'inefficient' statistics," Ann. Math. Statis. 17, pp. 377-408.

30. Ogawa, J. 1951, "Contributions to the theory of systematic statistics I," Osaka Math J. 3, pp. 175-213.

31. Panchang, G.M. and Aggarwal, V.P. 1962, "Peak flow estimations by method of maximum likelihood," Technical Memorandum HLO 2, Central Water and Power Research Station, Poona, India.

32. Posner, E.C. 1965, "The application of extreme-value theory to error-free communication." Technometrics 7, pp. 517-529.

33. Sarhan, A.E. and Greenberg, B.G. (eds.) 1962, Contributions to Order Statistics, John Wiley, New York.

34. Tiago de Oliveira, J. 1972, "Statistics for Gumbel and Fréchet distributions" in A. Freudenthal (ed.), Structural Safety and Reliability, Pergamon, New York.

35. Tiago de Oliveira, J. 1981, "Statistical choice of univariate extreme models" in C. Taillie et al. (eds.), Statistical Distributions in Scientific Work, 6, pp. 367-387, D. Reidel, Boston.

36. Tiago de Oliveira, J. 1982, "Decision and modeling for extremes" in J. Tiago de Oliveira and B. Eptstin (eds.), Some Recent Advances in Statistics, pp. 101-110. Academic Press, London.

37. Weissman, I. 1978, "Estimation of parameters and large quantiles based on the k largest observations," J. Amer. Statist. Assoc. 73, pp. 812-815.

# STATISTICAL ESTIMATION OF PARAMETERS OF THE WEIBULL AND FRECHET DISTRIBUTIONS

Nancy R. Mann

Department of Biomathematics, UCLA
Los Angeles, CA 90024

ABSTRACT: Estimation procedures for parameters of the Frechet and the three-parameter Weibull distribution are reviewed, after the relationships between these two distributions and the Gumbel distribution are discussed. Included in the review are maximum-likelihood and moment estimators as well as linearly based estimators and some that are extremely simple, involving only a few order statistics. Large- and small-sample properties are discussed.

## INTRODUCTION

The Weibull distribution is an asymptotic distribution of smallest extremes, with the initial underlying distribution being bounded below. Thus, any Weibull variate, X, is larger than some lower threshold value $\lambda$, which may be equal to zero. Early investigators such as Fisher and Tippett (1928), who originally derived the distribution called the Weibull a "type-III asymptotic distribution" of smallest extremes. Gnedenko (1943) and certain others designated it "type-II".

Waloddi Weibull (1939) gave an empirical derivation of this distribution in an analysis of dynamic breaking strengths by requiring only that it meet certain practical criteria. That article and other related articles of Weibull dealing with the modeling of failure (or survival) data seem to have found a large audience among those concerned with reliability analysis, particularly after the advent of the space age. Thus, the name Weibull has become firmly attached to the distribution, and its use as a model for "time-to-failure" has been widespread. It arises naturally also



*J. Tiago de Oliveira (ed.), Statistical Extremes and Applications, 81–89.*

in the analysis of droughts. See, for example, Gumbel (1954).

The mirror image of the Weibull distribution, the Fisher-Tippett type-III distribution of largest extremes, has been used to model maximum temperatures, maximum wind speeds and maximum earthquake magnitudes. See Jenkinson (1955) and Yegulalp and Kuo (1974).

The distribution function of a Weibull variate X with threshold (or location) parameter $\lambda$ is given by

$$F_X(x) = \begin{cases} 1 - \exp[-(\frac{x-\lambda}{\delta})^{\beta}] , & x \geq \lambda \\ 0 , & \text{otherwise.} \end{cases} \quad (1)$$

The parameters $\beta > 0$ and $\delta > 0$ are associated with shape and scale, respectively. If $\lambda$ is equal to zero, then $Z = \ell nX$ has a Gumbel, or Fisher-Tippett Type-I, distribution with location parameter $\eta = \ell n\ \delta$ and scale parameter $\xi = 1/\beta$. Many estimation procedures for the two-parameter Weibull have been developed as techniques for estimating location and scale parameters of the Gumbel distribution. See Mann, Schafer, Singpurwalla (1974) and Mann and Singpurwalla (1982).

The Frechet distribution is an asymptotic distribution of largest extremes, derived by Frechet under the condition that the initial variates be non-negative. The distribution function for a Frechet variate Y has the form

$$F_Y(y) = \begin{cases} \exp[-(\frac{\gamma}{y})^{\alpha}] , & y \geq 0 \\ 0, & \text{otherwise,} \end{cases} \quad (2)$$

with $\alpha > 0$ a shape parameter and $\gamma > 0$ a scale parameter.

The Frechet distribution is a special case of a Fisher-Tippett type-II distribution of largest extremes, i.e., one with threshold parameter $\mu$ equal to zero. In the case of $\mu \neq 0$, y is replaced by $y - \mu$ in the right side of Equation (2). The Frechet distribution has been widely used as a model for floods and maximum rainfalls. See Gumbel (1954) and Jenkinson (1955).

It is fairly evident that if Y has a Frechet distribution with shape and scale parameters $\alpha$ and $\gamma$, respectively, then $Y^{-1}$ has a two-parameter Weibull distribution with shape parameter $\alpha$ and scale parameter $\gamma^{-1}$. Thus, when data have not been censored, estimation procedures that have been derived for the two-parameter Weibull distribution can be used to estimate parameters of the Frechet distribution by simply taking reciprocals of the observed values. Methods applicable for Weibull censoring on the right will be appropriate for censoring Frechet data sets on the left, i.e., the smallest data values.

Very little, if any, research has addressed the problem of estimation for Fisher-Tippett type-II distributions with non-zero location (or threshold) parameters. This is perhaps a reflection of the perceived lack of real-world application for distributions of largest extremes with non-zero lower thresholds, or for distributions of smallest extremes with non-zero upper thresholds.

## ESTIMATION PROCEDURES

Since the reciprocal of a random variate having a Frechet distribution is a two-parameter Weibull variate - and since the natural logarithm of such a variate has a Gumbel distribution, point and interval estimation procedures developed for the Gumbel, the Frechet or the two-parameter Weibull distribution may be applied to either of the other two. Such procedures can also be used for the mirror images of each of these three distributions.

For the mirror image of any classical asymptotic extreme-value distribution, the random variate is the negative of the original variate, the scale parameter is unchanged, as well as any shape parameter, and the location parameter is the negative of the original location parameter. We see then that all five of the other classical extreme-value distributions can be transformed to a Gumbel distribution either by a simple change of variable or by a change of variable with a setting of a location parameter equal to zero.

From previous discussion, one can infer that problems in obtaining point and interval estimates can occur when the censoring model used in developing estimation methods for one of these distributions is uncharacteristic of the way data arise for another. Maximum-likelihood (ML) procedures usually will present no problems in this regard since ML computer codes written for iterative estimation of the parameters ordinarily allow for fairly general censoring patterns. Linear estimators, which can be obtained without iteration, specify specific censoring patterns.

Suppose one would like to obtain linear point estimates of Frechet parameters when the largest $n - r \geq 0$ of n sample observations are censored. The ordered negative logarithms of the observed values are realizations of order statistics from a Gumbel distribution with location parameter $- \ell n\ \gamma$ and scale parameter $\alpha^{-1}$. The ordering of the observations is clearly the reverse of the original order, and the censoring now involves the $n - r$ smallest observations.

For such censorings, weights, for obtaining best invariant estimates of $-\ell n\ \gamma$ and $\alpha^{-1}$ among those linear in the Gumbel order statistics (logarithms of the Frechet order statistics in reverse order) have been calculated and tabulated by Laue (1974) for $n = 1(1)25$, $r = 2(1)n$. For zero censoring they are equal to the weights calculated and tabulated by Mann (1967) for best linear invariant (BLI) estimation from the smallest $r \leq n$ Gumbel order statistics.

From BLI estimates of $\alpha^{-1}$ and $-\ \ell n\ \gamma$ and tabulated values that can be converted to values proportional to the variances and covariances of the best linear unbiased (BLU) estimators, one can obtain BLU estimates of the location and scale parameters and BLU and BLI estimates of any quantile of the distribution. BLI estimators and ML equations yield estimates of $\alpha^{-1}$ and $-\ \ell n\ \gamma$ that tend to be very nearly equal in value, and the constants in the variance-covariance matrix of the BLU estimators can be used effectively to remove the bias from ML estimates if unbiased estimates are desired. All three types of estimators are asymptotically unbiased, asymptotically efficient and asymptotically normal.

The Three Parameter Weibull Distribution

For the Weibull distribution (or for its mirror image, the Fisher-Tippett type-III distribution of largest values), point and interval parameter estimation is considerably more difficult when a location parameter $\lambda$ (or $-\lambda$) is unknown. The degree of difficulty is a function of the value of the Weibull shape parameter $\beta$.

When $\beta$ is known to be less than 2, regularity conditions (Cramér, 1946) fail for maximum-likelihood estimation of $\lambda$ and $\delta$. For $1 < \beta < 2$, the ML estimate of $\lambda$ is not asymptotically normal, and whether or not it is efficient is an open question. See Woodroofe (1974). For $\beta$ known to be $\leq 1$ or for $\hat{\beta}$, the ML estimate of $\beta$, less than or equal to unity, the sample minimum is a hyperefficient estimator of $\lambda$, as shown by Dubey (1966); and any positive value taken for the estimate of the scale parameter maximizes the likelihood function.

When $\beta$ equals 1, the two-parameter exponential emerges as a special case having estimators

$$\hat{\lambda} = (nX_{(1)} - \bar{X})/(n - 1)$$

and

$$\hat{\delta} = n(\bar{X} - X_{(1)})/(n - 1)$$

that are minimum-variance unbiased. See Epstein and Sobel (1954). Here $X_{(1)}$ is the smallest observable variate.

For $\beta \leq 1$, the distribution is reverse J-shaped, and for $\beta > 1$ there is a single mode at $\lambda + \delta(1 - 1/\beta)^{1/\beta}$. Computational difficulties are likely to be encountered if $\beta$ is close to 1, even though it actually exceeds 1. Cohen and Whitten (1982) advise that except for the special case of known $\beta = 1$, one should not attempt to obtain ML estimates for the three-parameter Weibull distribution unless there is reason to expect $\beta > 2.2$ .

On the other hand, Harter and Moore (1965), who have also investigated maximum-likelihood estimation for this distribution, suggest for $\hat{\beta} \leq 1$ (or for $\beta$ known to be $< 1$) censoring the smallest observation and any others equal to it. They use the rule of false position (iterative linear interpolation) for determining the value, if any, of a parameter being estimated at any given step that satisfies the appropriate likelihood equation into which the latest estimates (or known values) of the other two parameters have been substituted. Iterative estimation of the three parameters, one at a time, is in the cyclic order $\delta$, $\beta$ and $\lambda$ after initial estimates are selected.

Lemon (1975) has modified the likelihood equations so that one need iteratively solve only two equations for estimates of $\lambda$ and $\beta$, which together then specify an estimate of $\delta$. Cohen and Whitten (1982) suggest other modifications to the ML methods which involve simplification of one of the equations.

For estimation of the three Weibull parameters, the method of moments is one that works well, in general, in that estimates can always be obtained. Tables of Dubey (1967) applying to the two-parameter distribution and Monte Carlo results of Cohen and Whitten (1982) applying to the three-parameter distribution indicate that the moment estimators are very efficient for $\beta = 4.7$ and become progressively less efficient as $\beta$ tends towards zero or towards $\infty$. Cohen and Whitten (1983) provide a table which facilitates estimation of $\beta$ from the distribution third standardized sample moment. These authors also give several modifications to the moment equations which simplify estimation of all three parameters.

Graphical plotting techniques for use with Weibull probability paper provide a rather speedy method for estimation of the Weibull parameters. For unknown $\lambda$, however, the plotting must be iterative. An initial guess $\tilde{\lambda} \leq x_{(1)}$ is made for $\lambda$ and this value is subtracted from each $x_{(i)}$, $i = 1,\ldots,n$. Trial and error or some systematic technique employed to find a value of $\lambda$ such that the transformed data points form a linear plot. One decreases the estimate $\tilde{\lambda}$ if the plotted curve is concave downward and increases it if the curve is concave upward. See Kao (1959) or Mann, Schafer, Singpurwalla (1974) for discussion of plotting conventions and techniques for estimation of the other two parameters.

A graphical plot provides a representation of the data that is easily understood by the layman. It is extremely useful in detecting outliers, in estimating a nonzero Weibull location parameter $\lambda$ and in helping with a decision as to whether observed failure times are from a Weibull distribution. One would not expect, however, that parameter estimators defined by graphical procedures that are subjective, would be as efficient for estimation as the linear estimators they approximate. One method of removing the subjectivity from this method of estimation has been to use the estimate of $\lambda$ that gives the smallest sum of squared residuals about a fitted line. Properties of estimates of $\lambda$ obtained this way are not known.

Another iterative technique for estimating $\lambda$ is one of Mann and Fertig (1973). If $\beta$ is not too large, as a function of sample size, this method can also provide a lower confidence bound for $\lambda$. In any case, one can use the proposed statistic to test $H_o$: $\lambda = 0$ against $H_1$: $\lambda > 0$ .

The basic statistic for the test of $H_o$ is a ratio of selected sums of differences of successive ordered logarithms of sample observations, each difference being divided by its expected value. For obtaining a point estimate of $\lambda$ (which is median unbiased) or a lower confidence bound, one subtracts from each observation an initial guess for $\lambda$ and iterates until the modified statistic is numerically equal to an appropriate percent point of the statistic's distribution. This method has two advantages, the first being that the test statistic, under $H_o$, has a Beta distribution, and the second, that it is appropriate for use with censored data. The latter advantage can also be claimed for a similar statistic of Weissman (1978) and for certain linear estimators of $\xi = \beta^{-1}$ and $\eta = \ell n\ \delta$ that can be applied to the ordered logarithms of the observed data after an estimate of $\lambda$ is subtracted from each of the original values.

Wyckoff, Bain and Engelhardt (1980) suggest the use of an approximation of the BLU estimator, a simplified linear estimator of $\xi$ (See Bain, 1972), after having substituted an estimate for $\lambda$ based on a simple estimator. This seems to yield appropriate and quite efficient estimates for $\xi$, and ultimately for $\eta$, if $\beta$ is

3.0 or less when there is no censoring (or if $\beta$ is somewhat less than 3.0 when there is censoring). For larger values of $\beta$, estimates of $\xi$ obtained by this method have a substantial negative bias unless sample size is extremely large.

Besides the linear estimators of $\xi$ and $\eta$, there are other simple estimators of the Weibull parameters that are appropriate for use under certain conditions. For $\beta \leq 1$, a special case suggested by Zanakis (1979) of an estimator suggested earlier by Dubey (1966), namely, $\lambda' = (X_1 X_n - X_2^2)/(X_1 + X_n - 2X_2)$, has relatively small bias and mean squared error. For $\beta > 1$, a statistic investigated by Kappenmann (1981), has smaller bias and mean squared error than $\lambda'$ for estimating $\lambda$, but is also considerably more difficult to compute, involving a weighted average of all the order statistics.

Zanakis and Mann (1982) compared two simple estimators of $\beta$ based on a few order statistics:

$$\tilde{\xi}^* = 3.643/\ell n[(X_{(k)} - X_{(j)})/(X_{(j)} - X_{(i)})]$$

and

$$\tilde{\xi}' = 2.989/\ell n[(X_{(m)} - \lambda')/(X_{(\ell)} - \lambda')],$$

with i, j, k, $\ell$ and m defined appropriately.
They found both estimators to be good for small values of $\beta$, but having more negative bias than estimators based on the procedures prescribed by Wyckoff, Bain and Engelhardt. The simple estimators are useful for first guesses in maximum-linelihood estimation and in other situations in which quick results are necessary.

For r = n and known $\beta \leq 2$, Tiago de Oliveira (1983) shows that $\alpha^* = \sum_{i=1}^{n} (X_{(i)} - X_{(1)})^{\beta}/n$, with $\alpha = \delta^{\beta}$, has all the asymptotic properties of the ML estimate of $\alpha$. If $\beta$ is known but $\lambda$ and $\delta$ are unknown, the tables of Harter and Dubey (1967) can be used to test hypotheses concerning the distribution mean and variance.

## REFERENCES

Bain, L.J., 1972, Technometrics 14, pp. 831-840.

Cohen, A.C., Jr. and Witten, B.J., 1982. Commun. in Statist. 11, pp. 2631-2656.

Cramér, H., 1946, *Mathematical Methods of Statistics*. Princeton University Press, Princeton, N.J.

Dubey, S.D., 1966, Naval Res. Logist. Quart. 13, pp. 253-263.

Dubey, S.D., 1966, Naval Res. Logist. Quart. 14, pp. 261-267.

Epstein, B. and Sobel, M., 1954, Ann. Math. Statist. 25, pp. 373-381.

Fisher, R.A. and Tippett, L.H.C., 1928, Proc. Camb. Philos. Soc. 24, pp. 180-190.

Gnedenko, B.V., 1943, Ann. Math. 44, pp. 423-453.

Gumbel, E.J., 1954, *Statistical Theory of Extreme Values and Some Practical Applications. A Series of Lectures.* National Bureau of Standards Applied Mathematics Series, 33, U.S. Government Printing Office, Washington, D.C.

Harter, H.L. and Dubey, S.D., 1967, *Aerospace Research Laboratories Report ARL 67-0059*, Office of Aerospace Research, Wright-Patterson A.F.Base, Ohio.

Harter, H.L. and Moore, A.H., 1965, Technometrics, 7, pp. 639-643.

Jenkinson, A.F., 1955, Quart. J. Royal Meteor. Soc. 33, pp. 349-363.

Kao, J.H.K., 1959, Technometrics 1, pp. 389-407.

Kappenman, R.F., 1981, J. Statist. Comput. Simul. 13, pp. 245-254.

Laue, R.V., 1974, Unpublished paper, Bell Laboratories, Holmdel, N.J.

Lemon, G.H., 1975, Technometrics 17, pp. 247-254.

Mann, N.R., 1967, Technometrics 9, pp. 629-645.

Mann, N.R., Schafer, R.E. and Singpurwalla, N.D., 1974, *Methods for Statistical Analysis of Reliability and Life Data*, John Wiley, New York.

Mann, N.R. and Singpurwalla, N.D., 1982, in *Encyclopedia of Statistical Sciences*, Eds. S. Kotz, N.L. Johnson and C.B. Read, Vol 2, pp. 606-613.

Tiago de Oliveira, J., 1983, Unpublished paper.

Weibull, W., 1951, J. Appl. Mechanics 18, pp. 293-297.

Weissman, I., 1978, J. Amer. Statist. Assoc. 73, pp. 812-815.

Wyckoff, J., Bain, L.J. and Engelhardt, M.D., 1980, J. Statist. Comput. Simul. 11, pp. 139-151.

Woodroofe, M., 1974, Ann. Statist. 2, pp. 474-488.

Yegulalp, T.M. and Kuo, J.T., 1974, Bull. Seism. Soc. Amer. 64, pp. 393-414.

Zanakis, S.H., 1979, J. Statist. Comput. Simul. 9, pp. 101-116.

Zanakis, S.H. and Mann, N.R., 1982, Naval Res. Logist. Quart. 29, pp. 419-428.

ACKNOWLEDGMENT

The writing of this article and the necessary research were supported by the Office of Naval Research under Contract No. N00014-82-K-0023.

# UNIVARIATE EXTREMES; STATISTICAL CHOICE

J. Tiago de Oliveira

Academy of Sciences of Lisbon
Faculty of Sciences of Lisbon
Center for Statistics and Applications
(I.N.I.C.)

ABSTRACT. Different procedures for choosing between Fréchet, Gumbel and Weibull models to fit data are described in the paper. The first one is the one presented by the author which using von Mises-Jenkinson unifying form, has a locally optimal behaviour at the neighbourhood of Gumbel distribution. After, approaches developed by van Montfort, van Montfort and Otten, Galambos and Gomes are described, as well as a modified use of probability paper.

## INTRODUCTION

It is well known that the three limiting distributions of univariate extremes, under reduced form $\Psi_\theta(z)$ (Weibull distribution), $\Lambda(z)$ (Gumbel distribution) and $\Phi_\theta(z)$ (Fréchet distribution) can be imbedded in the general von Mises-Jenkinson formula

$$G(z|k) = \exp(-(1+kz)^{-1/k}), \quad -\infty<k<+\infty,$$

if $1+kz>0$, with $G(z|0^+)=G(z|0^-)=\Lambda(z)$, completing the definition by $G(z|k)=1$ if $z\geq -1/k$ when $k<0$ and by $G(z|k)=0$ if $z\leq -1/k$ when $k>0$. It is immediate that for $k<0$ we have $\Psi_{-1/k}(z) = G(-\frac{1+z}{z}|k)$, Weibull distribution with $\theta = -1/k$ and for $k>0$

we have $\Phi_{1/k}(z) = G(\frac{z-1}{k}|k)$, Fréchet distribution with $\theta = 1/k$; evidently, for $k=0$, we obtain Gumbel distribution $\Lambda(z)$ as $G(z|0^+) = G(z|0^-) = \Lambda(z)$.



*J. Tiago de Oliveira (ed.), Statistical Extremes and Applications, 91–107.*

The purpose of this paper is to describe a statistic to choose between the three models ($k < 0$, $k = 0$ or $k > 0$), first for reduced values ($z$) and, after for general random variables ($x = \lambda + \delta z$), with unknown location and dispersion parameters. We will begin by giving the finite choice approach, as in Tiago de Oliveira (1981) and (1982), enlarged to a more general set-up in Tiago de Oliveira (1983) and, after, a new approach, using a minimum variance statistic, the two ways being coincident in the statistical technique used, which illuminates diversely the problem.

After, we will sketch other analysis of the problem by van Montfort (1973), van Montfort and Otten (1978), Galambos (1980), M. Ivette Gomes (1982) and Pickands (1975) as well as the use of probability paper.

For other details see Tiago de Oliveira (1981) and (1982).

## THE LOCALLY OPTIMAL APPROACH

Consider a sample of (independent) reduced values $z = (z_1,\ldots,z_n)$ with distribution function $G(z|k)$ from which we want to decide for the Weibull model ($k < 0$), the Gumbel model ($k = 0$) or Fréchet model ($k > 0$).

Let $g(z|k) = G'(z|k)$ denote the density,

$$\mathcal{L}_n(z|k) = \mathcal{L}(z_1,\ldots,z_n|k) = \prod_1^n g(z_i|k)$$

the likelihood of the sample and $v(z) = \left.\dfrac{\partial \log g(z|k)}{\partial k}\right|_{k=0} = \dfrac{z^2}{2} - z - \dfrac{z^2}{2} e^{-z}$.

As it is well known, the locally optimal test of Gumbel versus Fréchet models ($k = 0$ vs $k > 0$) is given by the rejection region $V_n(z) = \sum_1^n v(z_i) \geq a'_n$, the locally optimal test of Gumbel versus Weibull models ($k = 0$ vs $k < 0$) is given by the rejection region $V_n(z) = \sum_1^n v(z_i) \leq b'_n$ and the asymptotical unbiased locally optimal test of Gumbel versus Fréchet or Weibull models ($k = 0$ vs $k \neq 0$) is given by the rejection region $V_n(z) \leq b_n$

or $V_n(z) \geq a_n$ .

It is, thus, natural to search quantities $\bar{b}_n \leq \bar{a}_n$ such that for the significance level $\alpha$, we have

$$\text{Prob } \{V_n(z) \leq \bar{b}_n\} = \alpha/2$$

$$\text{Prob } \{V_n(z) \geq \bar{a}_n\} = \alpha/2$$

and $\text{Prob } \{\bar{b}_n < V_n(z) < \bar{a}_n\}\ 1-\alpha$

As for $k = 0$, $v(z)$ has mean value zero and variance $\sigma^2 = \Gamma^{(4)}(1)/4 + \Gamma^{(3)}(1) + \Gamma''(1) = 2.42361$ we know that $V_n(z)/\sqrt{n}\,\sigma$ is asymptotically standard normal and, thus, for large $n$ we can take $-\bar{b}_n = \bar{a}_n = \sqrt{n}\,\sigma\,\lambda\,\alpha/2$ where $\lambda\,\alpha/2$ is the solution of the equation $N(x) = 1-\alpha/2$, $N(.)$ denoting the standard normal distribution function.

Let us denote by $\mu(k)$ and $\sigma^2(k)$ the mean value and the variance of $v(z)$ with respect to the distribution function $G(z|k)$; $\mu(0) = 0$ and $\sigma^2(0) = \sigma^2$. It can be shown that $\mu(k)$ exists only when $-1 < k < 1/2$ and the variance exists only when $-1 < k < 1/4$. As $\mu'(0) = \sigma^2(>0)$ we see that $\mu(k)$ increases in the neighbourhood of $k = 0$. We have $\mu(k) = 2.42361\,k + 4.15362\,k^2 + O(k^3)$ and $\sigma^2(k) =$ $= 2.42361 + 25.26470\,k + 275.03597\,k^2 + O(k^3)$ .

The probabilities of correct decision are:

- for the Weibull model $(k < 0)$,

$$P_n(k) = \text{Prob } \{V_n(z) \leq -\sqrt{n}\,\sigma\,\lambda\,\alpha/2 | k\} =$$
$$= \text{Prob } \{\frac{V_n(z) - n\mu(k)}{\sqrt{n}\,\sigma(k)} \leq \frac{-\sqrt{n}\,\sigma\,\lambda_{\alpha/2} - n\,\mu(k)}{\sqrt{n}\,\sigma(k)} | k\} \text{ if } -1<k<0$$

which is approached asymptotically by

$$\tilde{P}_n(k) = N(\frac{-\sqrt{n}\,\sigma\,\lambda_{\alpha/2} - n\,\mu(k)}{\sqrt{n}\,\sigma(k)}) ;$$

as for small values of $k < 0$ we have $\mu(k) < 0$ we see that $\tilde{P}_n(k) \to 1$;

- for the Gumbel model ($k = 0$),

$$P_n(0) = \text{Prob}\,\{-\sqrt{n}\,\sigma\,\lambda_{\alpha/2} \leq V_n(z) \leq \sqrt{n}\,\sigma\,\lambda\,\alpha/2 \;|k = 0\}$$

which is approached asymptotically by

$$\tilde{P}_n(0) = N(\lambda_{\alpha/2}) - N(-\lambda_{\alpha/2}) = 1-\alpha;$$

- for the Fréchet model ($k > 0$),

$$P_n(k) = \text{Prob}\,\{V_n(z) \geq \sqrt{n}\,\sigma\,\lambda_{\alpha/2}\;|k\} =$$

$$= \text{Prob}\{\frac{V_n(z) - n\,\mu(k)}{\sqrt{n}\,\sigma(k)} \geq \frac{\sqrt{n}\,\sigma\,\lambda_{\alpha/2} - n\,\mu(k)}{\sqrt{n}\,\sigma(k)}\;|k\} \text{ if } 0<k<1/4$$

which is approached asymptotically by

$$\tilde{P}_n(k) = 1-N(\frac{\sqrt{n}\,\sigma\,\lambda_{\alpha/2} - n\,\mu(k)}{\sqrt{n}\,\sigma(k)})\;;$$

as for small values of $k > 0$ we have $\mu(k) > 0$ we see that
$\tilde{P}_n(k) \to 1$ .

The decision procedure is thus 'consistent' in the same way as tests are said to be consistent, Gumbel model ($k = 0$) having here a central position. Recall that the procedure is unbiased and locally optimal at $k = 0$.

But the decision procedure can be changed to be 'strongly consistent', i.e., to be such that the probability of correct decision converges to 1, whatever may be $k$, in the neighbourhood of $k = 0$.

Consider now a sequence $\{d_n\}$ ($d_n > 0$) and take the decision procedure as follows:

decide for Weibull model ($k < 0$) if $V_n(z) \leq -\sqrt{n}\,\sigma\,d_n$, decide for Gumbel model ($k = 0$) if $|V_n(z)| < \sqrt{n}\,\sigma\, d_n$

and decide for Fréchet model ($k > 0$) if $V_n(z) \geq \sqrt{n}\,\sigma\, d_n$ .

If we take $d_n \to \infty$, as for $n > n(M)$ we have $d_n > M(>0)$, we know that for $n > n(M)$ we have

$$\text{Prob}\ \{|V_n(z)| < \sqrt{n}\,\sigma\, d_n | k=0\} > \text{Prob}\ \{|V_n(z)| < \sqrt{n}\,\sigma\, M | k = 0\}$$

and as the last expression converges to $N(M) - N(-M)$ we see that

$$\text{Prob}\ \{|V_n(z)| > \sqrt{n}\,\sigma\, d_n | k = 0\} \to 1 \ .$$

We have, now, to condition $\{d_n\}$ in such way that

$$\text{Prob}\ \{V_n(z) \leq -\sqrt{n}\,\sigma\, d_n | k < 0\} \to 1$$

and

$$\text{Prob}\ \{V_n(z) \geq \sqrt{n}\,\sigma\, d_n | k > 0\} \to 1 \ ,$$

in a neighbourhood of $k=0$.

But as $\text{Prob}\ \{V_n(z) \leq \sqrt{n}\,\sigma\, d_n | k<0\} =$

$$= \text{Prob}\ \left\{ \frac{V_n(z) - n\,\mu(k)}{\sqrt{n}\,\sigma(k)} \leq \frac{-\sqrt{n}\,\sigma\, d_n - n\mu(k)}{\sqrt{n}\,\sigma(k)} \middle| k<0 \right\} \quad (\text{if } -1<k<0)$$

this last probability converges to 1 if $\dfrac{-\sqrt{n}\,\sigma\, d_n - n\,\mu(k)}{\sqrt{n}\,\sigma(k)} \to +\infty$ .

The last(sufficient) condition can be written as $\sqrt{n}\left(-\dfrac{\sigma\, d_n}{\sqrt{n}} - \mu(k)\right) \to$ $\to +\infty$ . So $d_n/\sqrt{n} \to 0$, as $\mu(k)$ is continuous, $\mu(0) = 0$ and $\mu(k)<0$ for $k<0$, is thus a sufficient condition. The same condition $d_n/\sqrt{n} \to 0$ can be shown to imply $\text{Prob}\ \{V_n(z) \geq \sqrt{n}\,\sigma\, d_n | k>0\} \to 1$. The condition is, also, necessary.

Evidently the condition $d_n/\sqrt{n} \to 0$ does not imply "over rejection" of Gumbel model, as shows the strong consistency proven. The technique is asymptotically unbiased.

Optimization of $\{d_n\}$ has not yet been solved. A solution for the obtention of $\{d_n\}$ is, after choosing a (moving) level of significance $\alpha_n$ $(0<\alpha_n<1,\ \alpha_n \to 0)$, to define $d_n$ by Prob $\{|V_n(z)| < \sqrt{n}\sigma\, d_n \mid k = 0\} \simeq N(d_n)-N(-d_n) = 1-2\,\alpha_n$ or, asymptotically, as it is known

$$d_n \sim \sqrt{-2 \log \alpha_n} - \frac{\log(-\log \alpha_n)+\log(4\pi)}{2\sqrt{-2 \log \alpha_n}}$$

As $\alpha_n \to 0$ we see that $d_n \to +\infty$; the condition $d_n/\sqrt{n} \to 0$ leads to $\dfrac{\log \alpha_n}{n} \to 0$. A simple solution is to take $\alpha_n = 1/n$, leading thus to a probability of incorrect rejection of k=0 of order of 2/n

Let us, now, generalize the procedure, introducing location $\lambda$ and dispersion $\delta(>0)$ parameters.

When the underlying distribution is the general Gumbel distribution $\Lambda((x-\lambda)/\delta)$, the maximum likelihood estimators $\hat{\lambda}$ and $\hat{\delta}$, for the sample $(x) = (x_1,\ldots,x_n)$ are given by the equations

$$\hat{\delta} = \bar{x} - \frac{\sum_1^n x_i\, e^{-x_i/\hat{\delta}}}{\sum_1^n e^{-x_i/\hat{\delta}}}$$

and

$$\hat{\lambda} = -\hat{\delta} \log\left(\frac{\sum_1^n e^{-x_i/\hat{\delta}}}{n}\right).$$

Then we will use 'estimated' reduced values $\hat{z}_i = (x_i-\hat{\lambda})/\hat{\delta}$ and compute the 'estimated' $V_n(z)$ as

$$\hat{V}_n(x) = \sum_1^n v((x_i-\hat{\lambda})/\hat{\delta} = \frac{1}{2\hat{\delta}^2}\{\sum_1^n x_i^2 - 2\,\hat{\delta}\sum_1^n x_i - \frac{n\sum_1^n x_i^2\, e^{-x_i/\hat{\delta}}}{\sum_1^n e^{-x_i/\hat{\delta}}}\}$$

$$= \frac{1}{2}\{\sum_1^n ((x_i-\bar{x})/\hat{\delta})^2 - n\frac{\sum_1^n ((x_i-\bar{x})/\hat{\delta})^2 \exp(-(x_i-\bar{x})/\hat{\delta})}{\sum_1^n \exp(-(x_i-\bar{x})/\hat{\delta})}\} \qquad (1)$$

Using the $\delta$-method, described in Tiago de Oliveira (1982'), with a lengthy algebra, or following Tiago de Oliveira (1981), it can be shown that, for Gumbel model, $V_n(x)/n$ is asymptotically normal with mean value zero and variance $\hat{\sigma}^2/n$ with $\hat{\sigma}^2 = 2.09757$. Remark that $\hat{\sigma}^2 < \sigma^2$ with a reduction of 13.4%. More generally $\hat{V}_n(x)/n$ is asymptotically normal with mean value $\hat{\mu}(k)$ and variance $\hat{\sigma}^2(k)/n$, if the underlying distribution is $G(z|k)$; $\hat{\mu}(0) = 0$, $\hat{\sigma}^2(0) = \hat{\sigma}^2$ and $\hat{\mu}(k)$ is increasing in the neighbourhood if $k = 0$. The expansions of $\hat{\mu}(k)$ and $\hat{\sigma}(k)$ are $\hat{\mu}(k) = 2.09797\,k + 4.58653\,k^2 + O(k^3)$ and $\sigma(k) = 2.09797 + 19.91015\,k + 248.60225\,k^2 + O(k^3)$.

Thus we can apply previous reasoning used with the reduced sample and formulate the decision procedure as follows: choose $\{d_n\}$ such that $d_n \to \infty$ and $d_n/\sqrt{n} \to 0$ and decide for Weibull model ($k<0$) if $\hat{V}_n(x) \le \sqrt{n}\,\hat{\sigma}\,d_n$, decide for Gumbel model ($k=0$) if $|\hat{V}_n(x)| < \sqrt{n}\,\hat{\sigma}\,d_n$ and decide for Fréchet model ($k>0$) if $\hat{V}_n(x) \ge \sqrt{n}\,\hat{\sigma}\,d_n$.

The decision procedure is, evidently, strongly consistent in the neighbourhood of $k=0$.

Fòr more details and complements see Tiago de Oliveira (1981).

Let us make few remarks.

The method of analysis described above can be related to the likelihood ratio test of $k=0$ vs $k\neq 0$.

It was shown, in Tiago de Oliveira (1982'), using the $\delta$-method, that - $\ell_n$ denoting the likelihood ratio - $(\sqrt{n}\,\tau(\theta_0)(\theta-\theta_0))^2$ and $-2\log \ell_n$ are asymptotically equivalent and so $-2\log \ell_n$ has, asymptotically, a $\chi^2$ distribution with 1 degree of freedom. But, Cramer (1946), it is well known that $\sqrt{n}\tau(\theta_0)\ (\hat{\theta}-\theta_0)$ and $\frac{1}{\sqrt{n}\,\tau(\theta_0)}\sum_1^n \frac{\partial \log f(x_i|\theta)}{\partial\theta}\Big|_{\theta=0}$ are asymptotically equivalent and standard normal, when $\tau(\theta_0^2)$ is the variance of $\frac{\partial \log f(x_i|\theta)}{\partial\theta}\Big|_{\theta=\theta_0}$

As in our case $\frac{\partial \log g(z|k)}{\partial k}\Big|_{k=0} = v(z)$ and $\tau^2(\theta_0)=\sigma^2$, the variance of $v(z)$ for $k=0$, we get that $-2\log \ell_n$ is asymptotically equivalent to $(V_n(z)/\sqrt{n}\,\sigma)^2$ and the result coincides, asymptotically, with the two-sided test given. This result is, already, weaker than the one given, using the technique above, because we can, thus, solve a statistical trilemma.

We can extend this result to the (real) case of existence of location and dispersion parameters, but the computation is cumbersome.

Note that as we are taking $k=0$ vs $k\neq 0$, so a test in the neighbourhood of $k=0$, the likelihood ratio technique can be applied because for $k<0$ (Weibull model) the maximum likelihood estimator exists for $-1/k>2$ or $k>-1/2$.

As $\hat{V}_n(x)$ can be written in the form

$$\frac{1}{2}\left\{\sum_1^n \left(\frac{x_i-\bar{x}}{\hat{\delta}}\right)^2 - \frac{n}{\sum_1^n e^{-(x_i-\bar{x})/\hat{\delta}}}\sum_1^n \left(\frac{x_i-\bar{x}}{\hat{\delta}}\right)^2 e^{-(x_i-\bar{x})/\hat{\delta}}\right\}$$

and

$$\frac{1}{n}\sum_1^n e^{-(x_i-\bar{x})/\hat{\delta}}$$

can be expected to be close to

$$\int_{-\infty}^{+\infty} e^{-(x-(\lambda+\gamma\delta))/\delta}\, d\,\Lambda\left(\frac{x-\lambda}{\delta}\right) = e^{\gamma}$$

we could try to substitute $\frac{1}{n}\sum_{1}^{n} e^{-(x_i-\bar{x})/\hat{\delta}}$ by $e^{\gamma}$ and so use the new statistic

$$\tilde{V}_n(x) = \frac{1}{2}\left\{ \sum_{1}^{n} \left(\frac{x_i-\bar{x}}{\hat{\delta}}\right)^2 - e^{-\gamma} \sum_{1}^{n} \left(\frac{x_i-\bar{x}}{\hat{\delta}}\right) e^{-(x_i-\bar{x})/\hat{\delta}} \right\}.$$

This statistic is not asymptotically equivalent to $\hat{V}_n(x)$ but it can be shown using the $\delta$-method, that $\tilde{V}_n(x)/\sqrt{n}$ is asymptotically normal with mean value zero and finite variance, if the underlying distribution is the Gumbel one.

We could, also, search a quasi-linear method of decision, in the lines of Tiago de Oliveira (1966), in an way analogous to the previous one. The optimal decision statistic, independent of the location and dispersion parameters would, then, be

$$\frac{\int_{-\infty}^{+\infty} d\lambda \int_{-\infty}^{+\infty} d\delta\, \delta^{n-2}\, V_n(\lambda+\delta\, x_i)\, \mathcal{L}_n(\lambda\ \delta\ x_i \mid k=0)}{\int_{-\infty}^{+\infty} d\lambda \int_{-\infty}^{+\infty} d\delta\, \delta^{n-2} \mathcal{L}_n(\lambda\, \delta\ x_i \mid k=0)}$$

which, even after integration in $\lambda$, leads to an unmanageable expression.

## THE MINIMUM VARIANCE APPROACH

Let us begin, once more, to consider the reduced case. From a sample $(z) = (z_1,\ldots,z_n)$ we want to find a statistic $t_n(z) = t(z_1,\ldots,z_n)$ which approach in mean zero (0) if $k=0$ and deviates locally from zero at the quickest speed (positive) if $k\neq 0$ ($k \ast 0$) The basic idea is to use $t_n(z)$ to decide for Weibull model of $t_n(z) \leq B_n$, for Gumbel model if $B_n < t_n(z) < A_n$ and for Fréchet model if $A_n \leq t_n(z)$; we can, evidently, expect $B_n < 0 < A_n$.

With $\mathcal{L}_n(z|k) = \mathcal{L}(z_1,\ldots,z_n|k) = \prod_1^n g(z_i|k)$ we will denote the mean value and variance of $t_n(z)$ for the value $k$ of the shape parameter by

$$\nu_n(k) = \int_{-n}^{+\infty} t_n(z)\mathcal{L}_n(z|k)$$

and

$$\tau_n^2(k) = \int_{-\infty}^{+\infty} t_n^2\,(z)\mathcal{L}_n(z|k) - \nu_n^2(k)\,,$$

when they exist.

The conditions described above, with a scaling condition on the variance at k=0, for commodity, are

$$\nu_n(0) = \int_{-\infty}^{+\infty} t_n(z)\,\mathcal{L}_n(z|0) = 0$$

$$\tau_n^2(0) = \int_{-\infty}^{+\infty} t_n^2(z)\,\mathcal{L}_n(z|0) = C \quad \text{(a positive constant)}$$

and $\nu_n'(0) = \left.\dfrac{d\,\nu_n(k)}{dk}\right|_{k=0} = \int_{-\infty}^{+\infty} t_n(z)\,\sum_1^n v(z)\,\mathcal{L}_n(z|0) = \max(>0)$

where $v(z) = \left.\dfrac{\delta \log g(z|k)}{\delta k}\right|_{k=0} = \dfrac{z^2}{2} - z - \dfrac{z^2}{2}\,e^{-z}$, as before.

The Lagrangean of the Calculus of Variations, with Lagrange multipliers A and B for the side conditions $\nu_n(0)=0$ and $\tau_n^2(0) = C$ is

$$L(t_n) = \{t_n(z) + A\,t_n(z) + B\,t_n^2(z)\}\mathcal{L}_n(z|0)$$

and the Euler-Lagrange equation for $t_n$ is

$$\frac{\partial\, L(t_n)}{\partial\, t_n} = 0$$

giving thus

$$\sum_1^n v(z_i) + A + 2B\, t_n(z) = 0, \quad \text{as } \mathcal{L}_n(z|0) > 0$$

for $-\infty < z < +\infty$ .

Multiplying by $\mathcal{L}_n(z|0)$ and integrating the equation for $t_n(z)$ we get $A=0$ and, thus, $t_n(z)$ is proportional to

$$V_n(z) = \sum_1^n v(z_i).$$

Denoting, as before, by $\mu(k)$ and $\sigma^2(k)$ the mean value and the variance of $v(z)$ with respect to $g(z|k)$ we see that taking $t_n(z) = V_n(z)$ we have

$$\nu_n(k) = n\,\mu(k) \quad (\nu_n(0) = 0)$$

and

$$\tau_n^2(k) = n\,\sigma^2(k) \quad (\tau_n^2(0) = n\,\sigma^2) .$$

Then as $\nu_n'(0) = \tau_n^2(0) = n\,\sigma^2 > 0$, the variance increases with $k$.

This result is thus coincident with the previously obtained and the analysis can proceed in the same lines as before.

We could continue the variational approach, searching now a statistic $t_n(x) = t(x_1,\ldots,x_n)$, independent of the location and dispersion parameters (i.e., $t(\alpha+\beta\, x_1,\ldots,\alpha+\beta\, x_n) = t(x_1,\ldots,x_n)$) but the result does not seem workable.

## OTHER ANALYTIC APPROACHES

Let us describe now, sketchily, the approaches tried by van Montfort (1973), van Montfort and Otten (1978), Galambos (1980) and Gomes (1982), and also Pickands (1975).

van Montfort, after describing a statistic $W_1$, analogous to the t-test (invariant for location and dispersion transformations) develops the following test statistic $W$ for Gumbel against

Fréchet models. Define $W = \frac{1}{2}\log\frac{1+r}{1-r}$, where $r$ is the correlation coefficient between $\Psi_n(i+1/2) = -\log(-\log((i+1/2)/n))$ and $\ell_i = (X'_{i+1} - X'_i)/(\eta_{i+1} - \eta_i)$ $(i=1,2,\ldots,n-1)$ where the $X'_i$ are the order statistics of a sample $(X_1,\ldots,X_n)$ and $\eta_i$ are the mean values of $X'_i$ for Gumbel distribution under reduced form $(\lambda=0, \delta=1)$. Simulation has shown that $W$ is approximately normal with the mean value and variance to be fitted to the simulated critical points. The rejection region is $W>a$.

The test proposed by van Montfort and Otten, for the same models is given by the statistic $A = \sqrt{n}\sum_2^n \frac{(\Delta_i - \bar{\Delta})}{\sigma_\Delta} W_i$ where $W_i = \frac{L_i}{L_2 + \ldots + L_n}$, $L_i = (X'_i - X'_{i-1})/(\eta_i - \eta_{i-1})$, $X'_i$ and $\eta_i$ with the same meaning as before, $\Delta_i = -\gamma - (S^*_i - S^*_{i-1})/(S_i - S_{i-1})$ $(i=2,\ldots,n)$ $\gamma = 0.57722$ (the Euler constant),

$$S_i = i\binom{n}{i}\sum_{j=o}^{n-i}(-1)^j\binom{n-i}{j}\frac{\log(i+j)}{i+j}, \quad S^*_i = i\binom{n}{i}\sum_{j=o}^{n-i}(-i)^j\binom{n-i}{j}\frac{\log^2(i+j)}{2(i+j)}, \quad \bar{\Delta} = \frac{1}{n-1}\sum_2^n \Delta_i \text{ and } \sigma^2_\Delta = \frac{1}{n-1}\sum_2^n(\Delta_i - \bar{\Delta})^2.$$

The rule is to decide for $k>0$ (Fréchet) if $A$ is large, for $k<0$ (Weibull) if $A$ is small and for $k=0$ (Gumbel) if $A$ is intermediate. To this statistic $A$, a normal distribution fitted gives conservative values, as shown by simulation.

In Galambos another approach was presented, for the same type of decision for Gumbel and Fréchet models. It can be shown that the maximum of an IID sequence of random variables $\{X_j\}$, unbounded to the right and with mean value, is attracted to $\Lambda(.)$ if $\text{Prob}\{X > t+\mu(t)x\}/P(X>t) \to e^{-x}$ as $t \to +\infty$, where $\mu(t) = \int_t^{+\infty}(1-F(y))dy/(1-F(t)$ is the conditional mean of $X-t$ if $X>t$.

Thus for large $t$, the $Y_j = (X_j - t)/\mu(t)$ are asymptotically standard exponential, and $\mu(t)$ can be estimated by $\mu^*(t)$, the average of the values $X_j - t > 0$. Then Galambos suggests to use for $t$ the m-th maximum $(t = X'_{n+1-m})$ and to proceed to the common tests for exponentiality; in the case of rejection of $\Lambda(.)$ we will accept $\Phi_\theta(.)$.

The approach by M.I. Gomes uses a statistic proposed by Gumbel for the estimation of the shape parameter in Fréchet distribution. She studies, by simulation, the behaviour of the statistic $\frac{X'_n - X'_{[n/2]+1}}{X'_{[n/2]+1} - X'_1}$ where $X'_n$ and $X'_1$ are obviously the maximum and the minimum of the sample and $X'_{[n/2]+1}$ the median of the sample.

Jenkinson (1955) suggested the following decision statistic: consider a sample of $2n$, split it in $n$ pairs whose maxima are taken and use the ratio of the variances of the $2n$ observations and the (produced) sample of $n$ maxima of pairs. This statistic is, evidently, independent of the location and dispersion parameters and converges to a function of the shape parameter. Neither the distributions of the two last statistics nor the asymptotic distributions have been obtained; but the asymptotic distribution of the 1st and of a modification of the last are under study.

Pickands (1975) gives a procedure to estimate the upper tail of a distribution with an approach close to Galambos (1980) testing procedure.

For large $n$ he shows that the conditional distribution $\frac{1-F(x+a)}{1-F(a)}$, if $F$ is continuous, is close to $1-G(x|a,c) = \exp -\int_0^{x/a} [\max(0,1+ct)]^{-1} dt$ if $F$ is attracted to a limiting extreme value distribution, $c<0$, $c=0$, $c>0$ corresponding to asymptotics to $k>0$ (Fréchet), $k=0$ (Gumbel) and $k<0$ (Weibull) distributions. Then, letting $X''_1 \geq X''_2 \geq \ldots \geq X''_k \geq \ldots \geq X''_n$ be the (descending) order statistics of a sample of $n$ observations IID with distribution function $F(x)$ and taking the upper $4M$ observations $(X''_1 \geq X''_2 \geq \ldots \geq X''_{4M})$

$$\hat{c} = \log\left((X''_M - X''_{2M})/(X''_{2M} - X''_{4M})\right)/\log 2$$

and $\hat{a}=(X''_{2M}-X''_{4M})/\int_0^{\log 2} e^{\hat{c}n}\,du$ are estimators of c and a if $M \overset{P}{\to} \infty$ but $M/n \overset{P}{\to} 0$. A choice of M is to take $d_M = \min_{1\le \ell \le n/4} d_\ell$ where $d_\ell = \sup_{0<x<+\infty} |\hat{F}_\ell(x)-G_\ell(x|\hat{a},\hat{c})|$ and $\hat{F}_\ell(x)$ is the empirical distribution function of $X''_i - X''_{4\ell}$ $(i=1,2,\ldots,4\ell-1)$ and $G_\ell(x|\hat{a},\hat{c})$ is computed with $\hat{a}$, $\hat{c}$ corresponding to $M=\ell$.

$1-F(x)$ is then approximated by $1-G_M(x|\hat{a},\hat{c})$.

## THE USE OF PROBABILITY PAPER

Let us consider a sample of reduced values of extremes $(z_1,\ldots,z_n)$ and suppose that we want to decide graphically, and quickly, what is the form of extremes model that fits better the data. We will describe a graphical technique, based in Tiago de Oliveira (1983').

As it is well known the probability paper for Gumbel distribution is made as follows: as $p=\Lambda(z)$ can be written as $\Lambda^{-1}(p)=z$ if we note by $y=\Lambda^{-1}(p)$ and graduate the y-axis in the functional scale p, the relation $p=\Lambda(z)$ can be written as the 1st diagonal $y=z$. On the other side, if the true distribution is $G(z|k)$, the relation $p=\Lambda(y)=G(z|k)$ may be written as $y=\Lambda^{-1}G(z|k)$ and the relation between $(z,y)$ (or $(z,p)$) is the curve in the plane $y=\log(1+kz)/z$.

Let us recall now the plotting positions. If $p_{i,n}$ is the plotting position for the order statistic $z'_i$ ($=z'_{i,n}$ with greater rigour) the distance between the plotted point $(z'_i, p'_{i,n})$ and the 1st diagonal is $|\Lambda^{-1}(p_{i,n})-z'_i|$ horizontally or vertically or $1/\sqrt{2}$ of it orthogonally. The distance $|\Lambda^{-1}(p_{i,n})-z'_i|$ is minimized in mean if $\Lambda^{-1}(p_{i,n})$ is the median of $z'_i$ (with Gumbel distribution) or $p_{i,n}$ is the median of a Beta $(i, n-1-i)$ distribution; a good approximation is to take $p_{i,n} \approx \frac{i-0.3}{n+0.4}$; remark that instead of using the least squares approach we are minimizing $\max_i |\Lambda^{-1}(p_{i,n})-z'_i|$.

Once chosen the plotting positions, let us consider the general case, i.e. observations with location and dispersion parameters $\lambda$ and $\delta$, so that $z_i = (x_i - \lambda)/\delta$ are the reduced values. Then if $\chi_k(p)$ is the p-quantile of $G(z|k)$, i.e. $G(\chi_k(p)|k) = p$, the p-quantile of $G((x-\lambda)/\delta|k)$ is, evidently, $\lambda + \delta\, \chi_k(p)$. Consider then the random variable

$$v = \frac{x-(\lambda+\delta\, \chi_k(r))}{(\lambda+\chi_k(s|\delta)-(\lambda+\chi_k(r|\delta)} = \frac{z-\chi_k(r)}{\chi_k(s)-\chi_k(r)}$$

with $0 < r < s < 1$ ; $v$ is independent of the location and dispersion parameters.

Evidently we are going to plot estimated $v_i'$ where the $r$ and $s$ quantiles are estimated from the sample by $X'_{[nr]+1}$ and $X'_{[ns]+1}$.

The relation $p = \Lambda(y) = G(z|k)$ can then be written as

$$y = \bar{\Lambda}^1\, G(\chi_k(r) + (\chi_k(s)-\chi_k(r))v|k) = y_k(v),$$

for $r$ and $s$ fixed. Taking $v=0$ and $v=1$ we see that the curve $y_k(v)$ passes through the points $(0, \chi_o(r))$ and $(1, \chi_u(s))$, or $(0,r)$ and $(0,s)$ if we use the functional scale for $y$. Thus between $v=0$ and $v=1$ the curves $y_k(v)$ will with difficulty be separated from the 1st diagonal, chiefly taking in account that the plotted points, even using real and not estimated reduced observations, do not fall exactly in the 1st diagonal. In summary, the points between $v=0$ and $v=1$ are lost for separation of models; it is, thus, natural to use about 1/3 of the observations for the zone $v \in [0,1]$ and 2/3 outside; a smaller percentage in this zone would introduce instability in the implicit estimation of $\lambda$ and $\delta$ by the quantiles, as the denominator would be small.

With this rule of thumb, we have taken $\chi_o(r) = 0$ and $\chi_o(s) = 1$ or $r = \bar{\Lambda}^1(0) = e^{-1} = 0.3678794$ and $s = \bar{\Lambda}^1(1) = \exp(-e^{-1}) = 0.6922006$ (note that, as said before, $r \approx 1/3$ and $s \approx 2/3$; pratically we can take $r = 0.368$ or even $r = 0.37$ and $s = 0.692$ or even $s = 0.69$.

Table I

| | | |
|---|---|---|
| 1941 | 129 | Q(.367)= 96.00 |
| 1942 | 117 | |
| 1943 | 100 | |
| 1944 | 100 | |
| 1945 | 132 | Q(.692)= 108.00 |
| 1946 | 94 | |
| 1947 | 108 | |
| 1948 | 113 | |
| 1949 | 96 | $\lambda^*=$ 96.00 |
| 1950 | 113 | |
| 1951 | 96 | |
| 1952 | 72 | |
| 1953 | 98 | $\delta^*=$ 12.00 |
| 1954 | 85 | |
| 1955 | 124 | |
| 1956 | 108 | |
| 1957 | 102 | $\hat{\lambda}=$ 94.71 |
| 1958 | 102 | |
| 1959 | 112 | |
| 1960 | 107 | |
| 1961 | 86 | $\hat{\delta}=$ 12.49 |
| 1962 | 91 | |
| 1963 | 96 | |
| 1964 | 89 | |
| 1965 | 90 | |
| 1966 | 89 | Vn = -9.63 |
| 1967 | 89 | |
| 1968 | 84 | |
| 1969 | 107 | |
| 1970 | 111 | Vn/(2.09797xn)=-1.21 |

Let us now compute the curves $y_k(v)$. As $\chi_k(p) = \frac{(-\log p)^{-k}-1}{k}$ and so $\chi_k(e^{-1}) = 0$ and $\chi_k(\exp(-e^{-1})) = (e^k-1)/k$ we get

$$y_k(v) = \frac{\log(1+(e^k-1)v)}{k}$$

which is evidently defined when $1+(e^k-1)v>0$; note that $y_k(v)$ is convex for $k<0$ (Weibull model) and concave for $k<0$ (Fréchet model) $y_0(v)$ (Gumbel model) being a straight line.

REFERENCES

Galambos, J. (1980), "A Statistical Test for Extreme Value Distributions", Coll. Math. Soc. Janos Bolyai 32, Budapest, pp. 221--230.

Gomes, M. Ivette (1982), "A Note on Statistical Choice of Univariate Extreme Models", Act. IX, Jorn. Hispano-Lusas Mat., to be publ.

Jenkinson, A.F. (1955), "The Frequency Distribution of the Annual Maximum (or Minimum) Values of Meteorological Elements, Quat. J. Roy. Meteorol. Soc. 81, pp. 158-171.

Pickands, James, III (1975), "Statistical Inference Using Extreme Order Statistics", Ann. Stat., vol. 3, pp. 119-131.

Tiago de Oliveira, J. (1981), "Statistical Choice of Univariate Extreme Models", Stat. Distr. in Scient. Work, C. Taillie et al. (eds.), vol. 6, pp. 367-307, D. Reidel Publ. Co.

(1982), "Decision and Modelling for Extremes", Some Recent Advances in Statistics, J. Tiago de Oliveira and B. Epstein (eds.), pp. 101-110, Academic Press.

(1982'), "The δ-method for Obtention of Asymptotic Distributions; Applications", Publ. Inst. Stat. Univ. Paris, XXVII, pp. 49-70.

(1983), "Statistical Choice of Non-Separated Models", to be publ.

(1983'), "Probability Paper and Plotting Positions - a New Approach", to be publ.

von Montfort, M.A.J. (1973), "An Asymmetric Test of the Type of Distribution of Extremes", Meded. Landbouwhogeschool Wageningen (73-18), pp. 1-15.

von Montfort, M.A.J. and Otten, Albert (1978), "On Testing a Shape Parameter in Presence of a Location and a Scale Parameter", Math. Operations forsch , Stat., Ser. Statistics, vol. 9, nº1, pp. 91-104.

# STATISTICAL ESTIMATION IN EXTREME VALUE THEORY

Ishay Weissman
Technion - Israel Institute of Technology

ABSTRACT

If the maximum (or minimum) of a sample, properly standardized, has a nondegenerate limit distribution (as the sample size tends to ∞) then for every fixed $k$, the joint distribution of the $k$ upper (lower) sample extremes converges to a nondegenerate limit. Based on this limiting distribution, if only sample extremes are available (e.g. in life testing situations) tail parameters can be consistently estimated.

## 1. INTRODUCTION

Let $X_{(1)} \geq X_{(2)} \geq \dots \geq X_{(n)}$ be the order statistics in a sample of size $n$ from a distribution function $F$ and suppose that $F \in D(G)$. That is

$$F^n(a_n x + b_n) \to G(x) \quad (n \to \infty) \tag{1}$$

for some constants $a_n > o$, $b_n$. We are concerned here with drawing inference about quantities associated with the right tail of $F$ when $n$ is large. Quantities of interest are e.g.

$F(x)$ for large $x$

$\eta(p) = F^{-1}(p)$, $p$ of the form $1 - c/n$ ($c$ constant)

$\mu = \sup\{x : F(x) < 1\}$

$a_n$, $b_n$



J. Tiago de Oliveira (ed.), Statistical Extremes and Applications, 109–115.

Similarly, when

$$[1 - F(a_n x + b_n)]^n \to 1 - G(x) \quad (n \to \infty) \tag{1*}$$

we are concerned with analogous quantities associated with the left tail of F.

A commonly used estimation method, developed at length by Gumbel (1958), is to divide the original (unordered) data into subsamples. The maxima (or minima) of the subsamples are then used to estimate the parameters of G, one of the three extreme value limit distributions. This method comes natural when some seasonality exists in the data (e.g. environmental series) and a natural subsample is one season's data. However, in general there is no natural seasonality in the data and the subsample method appears artificial and wasteful.

Example. Suppose the life distribution of a certain product is exponential, $F(x) = 1 - e^{-x/\theta}$ $(x>o)$ and we want to estimate the mean life $\theta$. We put $n$ items on test, but we cannot wait until all items fail. Two possible schemes are (1) Gumbel: divide the sample into $k$ subsamples (of size $m = n/k$ each), stop the experiment as soon as there is at least one failure in each subsample and then use the first failure-times (from each subsample) for estimation. (2) Type II censoring: stop the experiment as soon as there are $k$ failures (in the entire sample) and use the $k$ failure-times for estimation. The two schemes have the following properties:

| | (1) | (2) |
|---|---|---|
| Data | $Y_i$ = min of subsample i | $0 = \tilde{X}_o \leq \tilde{X}_1 \leq \dots \leq \tilde{X}_k$ |
| $\hat{\theta}$ (MLE or UMVUE) | $\frac{1}{k} \sum_1^k (mY_i)$ | $\frac{1}{k} \sum_1^k (n+1+i)(\tilde{X}_i - \tilde{X}_{i-1})$ |
| $E(\hat{\theta})$ | $\theta$ | $\theta$ |
| $Var(\hat{\theta})$ | $\theta^2/k$ | $\theta^2/k$ |
| Expected duration of experiment | $EY_{max} = \frac{\theta k}{n} [1 + \frac{1}{2} + \dots + \frac{1}{k}]$ | $E\tilde{X}_k = \theta[\frac{1}{n} + \frac{1}{n-1} + \dots + \frac{1}{n-k+1}]$ |
| For $n \to \infty$, $k \to \infty, k/n \to 0$ | $\sim \frac{\theta}{n} \cdot k \ln k$ | $\sim \frac{\theta}{n} \cdot k$ |

If $k$ is determined so as to satisfy a certain precision requirement, then the relative efficiency of the subsamples-method (in terms of experiment duration) tends to 0. In other words, if we waited so long (until each subsample has at least one failure) - we might as well use all failure-times available from the entire sample and thereby obtain a much smaller variance of estimate. In what follows we shall deal with estimation based on the $k$ largest (or smallest) order statistics when the sample size $n$ is large. Other relevant references are Weiss (1971), Hill (1975) and Pickands (1975).

## 2. ASYMPTOTIC THEORY

Define $X_{ni} = (X_{(i)} - b_n)/a_n$ , then under (1) we have

$$\{X_{ni}: i = 1,\dots,k\} \overset{D}{\to} \{M_i = \lambda^{-1}(Z_1+\dots+Z_i): i=1,\dots,k\} , \quad (2)$$

where $\lambda(x) = -\log G(x)$ and the $\{Z_i\}$ is a sequence of iid exponential variates with mean 1 (see David (1981), p. 266). Hence for large $n$, the $k$ largest order statistics, up to a linear transformation, are distributed approximately as $(M_1,\dots,M_k)$. There are only 3 possible models to consider

$$\text{Gumbel:} \quad \lambda(x) = e^{-x} \quad (3.1)$$

$$\text{Frechet:} \quad \lambda(x) = x^{-\alpha} \quad (x > o), \ (\alpha > o) \quad (3.2)$$

$$\text{Weibull:} \quad \lambda(x) = (-x)^{\alpha} \quad (x < o), \ (\alpha > o) . \quad (3.3)$$

In case (3.1), $\lambda^{-1}(y) = -\ln y$ thus we have

$$D_i = M_i - M_{i+1} \overset{D}{=} -\log \frac{Z_1 + \dots + Z_i}{Z_1 + \dots + Z_i + Z_{i+1}}$$

$$\overset{D}{=} -\log \beta\,(i,1) \overset{D}{=} \frac{Z_i}{i} .$$

Namely, the spacings $D_i$ are independent exponential with means $1/i$, independent of $M_k$ $(i = 1,\dots,k-1)$. In cases (3.2) and (3.3) the ratios $M_i/M_{i+1}$ are independent. Independent and exponential spacings among order statistics is a famous characterization of the exponential distribution. The latter is an important member of the domain of attraction of the Gumbel distribution (case 3.1).

One can plot $X_{(i)}$ vs. $E(M_i)$ $(i = 1,\dots,k)$ for each one of the three models and decide which one (if any) fits the data, i.e. forms a straight line. If $k$ is at our disposal - it can be determined as the largest value for which $X_{(1)}, \dots, X_{(k)}$ fit the model. Once the model is determined the parameters can be estimated. Maximum likelihood estimators (MLE) and minimum variance estimators (MVE) are discussed in Weissman (1978). In case (3.1), the MLE are

$$\hat{a}_n = \frac{1}{k} \Sigma_{i=1}^{k-1} (X_{(i)} - X_{(k)}) = \frac{1}{k} \Sigma_{i=1}^{k-1} i\ (X_{(i)} - X_{(i+1)})$$

$$\hat{b}_n = \hat{a}_n \ln k + X_{(k)}$$

and the MVE are

$$a^*_n = \frac{1}{k-1} \Sigma_{i=1}^{k-1} (X_{(i)} - X_{(k)}) = \frac{k}{k-1} \hat{a}_n$$

$$b^*_n = a^*_n (S_k - \gamma) + X_{(k)}$$

where $S_1 = 0$, $S_k = 1 + \frac{1}{2} + \ldots + \frac{1}{k-1}$ $(k > 1)$ and $\gamma = .5772$ is Euler's constant. Note that $2(k-1)\ a^*_n/a_n$ is (asymptotically) a $\chi^2(2k-2)$ variate and thus confidence intervals for $a_n$ can be constructed. The distribution of $U_k = (X_{(k)} - b_n)/a^*_n$ is parameter free and thus confidence intervals for $b_n$ can be constructed provided the percentage points are available. The latter are tabulated in Weissman (1978).

In case (3.1) we have

$$(\eta(1 - \frac{c}{n}) - b_n)/a_n \to - \ln c \quad (n \to \infty)$$

Thus

$$\hat{\eta}(1 - \frac{c}{n}) = \hat{a}_n(-\ln c) + \hat{b}_n \tag{4}$$

is the MLE for $\eta(1-c/n)$.

For large values of $y$, under (1)

$$F(y) \sim G^{\frac{1}{n}} ((y - b_n)/a_n) = e^{-\frac{1}{n}\lambda((y-b_n)/a_n)} \tag{5}$$

thus estimation of $F(y)$ is obtained by substitution of $\hat{a}_n$, $\hat{b}_n$ in (5).

Similar estimates are obtained in Weissman (1980) for type I censoring. Namely, only values larger than some $x_o$ (fixed) are

observable—k is then random. It turns out that $x_o$ plays the role of $X_{(k)}$ in the formulas for $\hat{a}_n$, $\hat{b}_n$ and $\hat{\eta}$.
In principle, these asymptotic results hold for every distribution function $F \in D(\exp(-e^{-x}))$ (exponential, gamma, normal, lognormal, logistic, Weibull - to name a few). But for finite n, better approximations are obtained for distributions with exponential tail. Recently Boos (1983) used (4) to estimate large quantiles $\eta(p)$ (with $p \geq .95$) for a number of distributions. He found e.g. that for $n = 100$, $k = 10$, $p = .95, .99, .995$ our method is more efficient than sample quantiles in terms of mean square error) for the normal, Weibull (shape = 4), $\chi_4^2$, $t_8{}^{(1)}$ and lognormal ($\sigma=\frac{1}{2}$) distributions. It is less efficient for distributions with heavier than exponential tails as $t_3$ and lognormal ($\sigma=1$).

In case (3.2), the transformation $Y_{(i)} = \ln X_{(i)}$ transforms the model to the first one. In case (3.3) the transformation $Y_{(i)} = -\ln(\mu - X_{(i)})$ does the job. Here $\mu$ is finite and if unknown - it must be estimated. In this case the estimation of the three parameters ($a_n$, $b_n$, $\mu$) is most complicated.

## 3. THE THREE-PARAMETER CASE

The three-parameter case is encountered mostly in life testing or strength distributions, where the available data are

$$T_i = X_{(n+1-i)} \qquad i = 1,\dots, k$$

with $k << n$. Here we assume (1*) instead of (1) with $1 - G(x) = \exp\{-x^\alpha\}$ $(x>o)$. For example if $F(t) = 1-\exp\{((t-\mu)/\beta)^\alpha\}$ $(t > \mu)$, the Weibull distribution then (1*) holds with $a_n = \beta n^{-1/\alpha}$ and $b_n = \mu$, and 3 parameters must be estimated $(\alpha,\beta,\mu)$. The maximum likelihood equations are

$$\beta = (n/k)^{1/\alpha}\,(T_k - \mu) \tag{6.1}$$

$$\alpha^{-1} = k^{-1}\,\Sigma_{i=1}^{k} \ln\{(T_k - \mu)/(T_i - \mu)\} \tag{6.2}$$

$$\alpha^{-1} = 1-k\,(\Sigma_{i=1}^{k}\,(T_k - \mu)/(T_i - \mu))^{-1}\,. \tag{6.3}$$

Any solution of (6) necessarily yields $\hat{\alpha} > 1$, so the method does not work for $\alpha < 1$. On the other hand, $T_1 \xrightarrow{a.s.} \mu$ and hence $T_1$ is a strongly consistent estimator of $\mu$. However, if $T_1$ is used to estimate $\mu$ then (6.1) and (6.2) give $\hat{\beta} = \infty$, $\hat{\alpha} = 0$. Maximum likelihood estimation is discussed by Hall (1982) and Smith and Weissman (1983).

Confidence intervals for $\mu$ are suggested by Weissman (1981,1982). Using (2) one gets

$$R_{kn} = \frac{T_1 - \mu}{T_k - T_1} \xrightarrow{D} R_k ,$$

where $R_k$ is a random variable whose df is

$$H_{k,\alpha}(x) = 1 - \left(1 - \left(\frac{x}{1+x}\right)^{\alpha}\right)^{k-1} \qquad (x > 0) .$$

If $\alpha$ is known then the quantiles $\gamma_{k,\alpha}(p)$ of $H_{k,\alpha}$ can be determined in a closed form. For large values of $n$, $R_{k,n}$ is used as a pivotal function to construct confidence intervals for $\mu$. If $\alpha$ is unknown we use the pivotal functions

$$W_k^{(n)} = \log \frac{T_k - \mu}{T_1 - \mu} \Big/ \sum_{i=1}^{k-1} \log \frac{T_k - \mu}{T_i - \mu} \qquad (k \geq 3)$$

or

$$Q_{k,m}^{(n)} = \log \frac{T_m - \mu}{T_1 - \mu} \Big/ \log \frac{T_k - \mu}{T_m - \mu} \qquad (1 < m < k \leq n)$$

to construct confidence intervals for $\mu$. Under (1*) both have limiting distributions which do not depend on any parameter. The quantiles of these distributions are tabulated in Weissman (1981, 1982). When more information is known about $F$ - corrections for finite $n$ are possible. $W_k^{(n)}$ performs better than $Q_{k,m}^{(n)}$ but in cases of multiple censoring, where only 3 order statistics are available, the latter can be used. Solving $W_k^{(n)} = w_k(.5)$, where $w_k(p)$ is the p-quantile of the limiting distribution of $W_k^{(n)}$, we get an estimate for $\mu$ which is median unbiased. Simulation studies showed good performance when $\alpha < 1$, and this is helpful since in this case $(\alpha < 1)$ the maximum likelihood method fails.

Notes

(1) $\chi^2$ - distribution and t-distribution with 4 and 8 degrees of freedom, respectively.

## References

Boos, D.D., 1983, Using extreme value theory to estimate large percentiles, Technical report, Department of Statistics, North Carolina State University.

Gumbel, E.J., 1958, Statistics of Extremes, Columbia University Press, New York.

Hall, P., 1982, On estimating the endpoint of a distribution, Ann. of Statist. 10, pp. 556-568.

Hill, B.M., 1975, A simple general approach to inference about the tail of a distribution, Ann. of Statist. 3, pp. 1163-1174.

Pickands, J. III, 1975, Statistical inference using extreme Order Statistics, Ann. of Statist. 3, pp. 119-131.

Smith, R.L. and Weissman, I., 1983, Maximum likelihood estimation of the lower tail of a probability distribution. (Unpublished).

Weiss, L., 1971, Asymptotic inference about a density function at the end of its range, Nav. Res. Log. Quart. 18, pp. 111-114.

Weissman, I., 1978, Estimation of parameters and large quantiles based on the k largest observations, JASA 73, pp. 812-815.

Weissman, I., 1980, Estimation of tail parameters under type I censoring, Commun. Statist. A9(11), pp. 1165-1175.

Weissman, I., 1981, Confidence intervals for the threshold parameter, Commun. Statist. 10(6), pp. 549-557.

Weissman, I., 1982, Confidence intervals for the threshold parameter II: Unknown shape parameter, Commun. Statist. A11(21), pp. 2451-2474.

# PROBABILISTIC ASPECTS OF MULTIVARIATE EXTREMES

Paul Deheuvels

Université Paris VI

ABSTRACT. The probabilistic theory of multivariate extreme values is reviewed, with emphasis on the limiting asymptotic distributions obtained for the extremes of i.i.d. sequences of random vectors.

OUTLINE OF CONTENTS.



1. MULTIVARIATE EXTREME VALUES. Let $X_n=(X_n(1),\ldots,X_n(d))$, $n=1, 2,\ldots$ be an i.i.d. sequence of random vectors in $R^d$. The upper (resp. lower) multivariate extremes of $X_1,\ldots,X_n$ are defined respectively by

$$Z_n = (Z_n(1),\ldots,Z_n(d)) = (\max_{1\le i\le n} X_i(1),\ldots, \max_{1\le i\le n} X_i(d)),$$

and

$$Y_n = (Y_n(1),\ldots,Y_n(d)) = (\min_{1\le i\le n} X_i(1),\ldots, \min_{1\le i\le n} X_i(d)).$$

In the sequel, we shall consider the following problems:



*J. Tiago de Oliveira (ed.), Statistical Extremes and Applications, 117–130.*

- What are the possible limiting types of $Z_n$ (resp. $Y_n$) as $n\to\infty$ ?

- Under which conditions does $Z_n$ (resp. $Y_n$) have a specific limiting type ?

We shall call any limiting type of $Z_n$ (resp. $Y_n$) a <u>multiva--riate extreme value distribution for maxima</u> (resp. <u>minima</u>).

In general, a multivariate random vector $T_n=(T_n(1),\dots,T_n(d))$ is said to <u>converge in type</u> toward a limiting distribution L iff there exists two non-random sequences $a_n=(a_{n1},\dots,a_{nd})$ and $b_n=(b_{n1},\dots,b_{nd})$, where $a_{n1} > 0,\dots,a_{nd} > 0$, such that

$$a_n^{-1}(T_n-b_n)=\left(\frac{T_n(1)-b_{n1}}{a_{n1}},\dots,\frac{T_n(d)-b_{nd}}{a_{nd}}\right) \overset{W}{\to} L \text{ as } n\to\infty.$$

This implies (as a necessary but not sufficient) condition) that each coordinate converges in type to the corresponding marginal distribution of L. If we limit ourselves (without loss of generality since $\min X_i = -\max(-X_i)$) to the case of $Z_n$, then (see Galambos, 1978), for any $j=1,\dots,d$, the limiting distribution of $a_{nj}^{-1}(Z_n(j)-b_{nj})$ as $n\to\infty$ must be (if it exists and if non degenerate, with proper normalizing constants) one of the classical univariate extreme value distributions, given by their distribution functions (for $a > 0$):

$$\Lambda(x) = e^{-e^{-x}} \quad , -\infty < x < +\infty \text{ (the Gumbel law)} ,$$

$$\Phi_a(x)=\begin{cases}\exp(-x^{-a}) & , x > 0\\ 0 & , x \leq 0\end{cases} \quad \text{(the Fréchet law)},$$

$$\Psi_a(x) =\begin{cases}1 & , x \geq 0\\ \exp(-(-x)^{a}) & , x < 0\end{cases} \quad \text{(the Weibull law)}.$$

The problem of finding the proper limiting type and the corresponding domain of attraction for each coordinate has recieved an extensive treatment and will not be considered here. It is also well known that the joint distribution of d r.vs. is not characterized by the mere knowledge of their marginal distributions. We shall therefore assume that, for any $j=1,\ldots,d$,

$$a_{nj}^{-1}(Z_n(j)-b_{nj}) \stackrel{W}{\to} G_j \in \{\Lambda,\Phi_a \text{ or } \Psi_a\},$$

and specialize in the problem of finding the possible joint limit distributions (i.e. the multivariate extreme value distributions) $G$ of $a_n^{-1}(Z_n-b_n)$ with marginals $G_1,\ldots,G_d$.

2. COPULAE AND DEPENDENCE FUNCTIONS. The discussion is greatly simplified by the introduction of dependence functions (or copulae, not to be confused with the dependence functions in Tiago de Oliveira's sense discussed in §3), a concept due to Fréchet, 1951.

Let $T_n=(T_n(1),\ldots,T_n(d))$, $n=1,2,\ldots$ be a sequence of $R^d$-valued r.vs., whose distribution function and marginal d.fs. will be denoted by

$$F_n(x_1,\ldots,x_d)=P(T_n(1)\leq x_1,\ldots,T_n(d)\leq x_d),$$
$$F_{nj}(x_j)=P(T_n(j)\leq x_j),\ j=1,\ldots,d.$$

We have then the following essential lemmas, whose proofs are detailed in Deheuvels, 1978.

Lemma 1. There exists a probability measure on $(0,1)^d$ with uniform margins and distribution function $D_n(u_1,\ldots,u_d)$ such that, for any $x_1,\ldots,x_d$, continuity points of $F_{n1},\ldots,F_{nd}$, we have

$$F_n(x_1,\ldots,x_d)=D_n(F_{n1}(x_1),\ldots,F_{nd}(x_d)).$$

Conversely, if $D_n^{\star}(u_1,\ldots,u_d)$ is an arbitrary distribution function of a probability measure on $(0,1)^d$ with uniform margins, then

$$F_n(x_1,\ldots,x_d)=D_n^{\star}(F_{n1}(x_1),\ldots,F_{nd}(x_d))$$

defines the distribution function of a probability measure with marginal distribution functions $F_{n1},\ldots,F_{nd}$.

$D_n$ is called a <u>dependence function</u> (or <u>copula</u>) of $F_n$. It is unique iff $F_{n1},\ldots,F_{nd}$ are continuous.

<u>Lemma</u> 2. Let $F_n$ be defined for n=1,2,... as in Lemma 1, and let $D_n$, n=1,2,... be associated dependence functions. Let also $F_\infty(x_1,\ldots,x_d)$ be a distribution function with continuous marginal d.fs. $F_{\infty 1},\ldots,F_{\infty d}$ and (unique) dependence function $D_\infty$.

Then, $F_n \overset{W}{\to} F$ as $n\to\infty$ if and only if:

- For any $j=1,\ldots,d$, $F_{nj} \overset{W}{\to} F_{\infty j}$ ;
- $D_n \overset{W}{\to} D_\infty$.

<u>Lemma</u> 3. Let $F_n(x_1,\ldots,x_d)$ be defined as above. Let for j=1, ...,d, $x_j \to \phi_j(x_j)$ be a nondecreasing mapping of R on itself.

Then, if $D_n$ is a dependence function of $F_n$, it is also a dependence function of $F_n(\phi_1(x_1),\ldots,\phi_d(x_d))$.

In the specific case of the upper extreme $Z_n$, if we denote by $F(x_1,\ldots,x_d)=P(X_n(1)\le x_1,\ldots,X_n(d)\le x_d)$ the d.f. of $X_n$, then the distribution function of $Z_n$ is

$$P(Z_n(1)\le x_1,\ldots,Z_n(d)\le x_d)=F^n(x_1,\ldots,x_d).$$

It follows that if D is a dependence function of F, then

$$D^n(u_1^{1/n},\ldots,u_d^{1/n}),\ 0\le u_j\le 1,\ 1\le j\le d,$$

is a dependence function of $F^n$ (or $Z_n$). A direct consequence of Lemma 1, Lemma 2 and Lemma 3 is then that:

<u>Theorem</u> 1. Let $Z_n=(\max_{1\le i\le n} X_i(1),\ldots,\max_{1\le i\le n} X_i(d))$, where $X_n=(X_n(1),\ldots,X_n(d))$ is an i.i.d. sequence of random vectors with distribution function F, marginal distribution functions $F_{(1)}$, ...,$F_{(d)}$, and dependence function D.

Then, in order that $a_n^{-1}(Z_n-b_n)=(a_{n1}^{-1}(Z_n(1)-b_{n1}),\ldots,a_{nd}^{-1}(Z_n(d)-b_{nd}))$ converges as $n\to\infty$ toward a limiting distribution with non degenerate margins (with $a_{nj}>0$, $1\le j\le d$, $n\ge 1$), it is necessary and sufficient that:

(i) For any $j=1,\dots,d$, $F^n_{(j)}(a_{nj}x_j+b_{nj})$ converges as $n\to\infty$ toward a limiting distribution (of type $\Lambda, \Phi_a$ or $\Psi_a$) $G_j$.

(ii) There exists a limiting dependence function $D_\infty$ such that

$$D_\infty(u_1,\dots,u_d)=\lim_{n\to\infty} D^n(u_1^{1/n},\dots,u_d^{1/n}),\ 0\le u_j\le 1,\ 1\le j\le d.$$

Under these conditions, the limiting distribution of $a_n^{-1}(Z_n-b_n)$ as $n\to\infty$ is $G_\infty(x_1,\dots,x_d)=D_\infty(G_1(x_1),\dots,G_d(x_d))$.

The main conclusion of all this is that the problem of the characterization of the multivariate extreme value distributions can be separated in two completely distinct subproblems:

- Finding the limiting types of the coordinates;
- Finding the limit (if it exists) of $D^n(u_1^{1/n},\dots,u_d^{1/n})$.

It is also clear that one may modify arbitrarily each marginal distribution without changing D. Hence the choice of specific marginals for $X_n(1),\dots,X_n(d)$ can be made following a matter of pure convenience.

Historically, the limiting structure of bivariate extremes was described by Tiago de Oliveira, 1958, 1962/63, 1975, Geffroy, 1958, and Sibuya, 1960.

The limiting structure of multivariate extremes for $d\ge 3$ has been characterized by De Haan and Resnick, 1977, Deheuvels, 1978, 1981, Pickands, 1980, 1981 and 1978 (in Galambos, 1978).

Tiago de Oliveira uses Gumbel marginals, while De Haan and Resnick use Frechet margins, and Deheuvels, uniform margins. Pickands, working on the minimum $Y_n$ rather than on the maximum $Z_n$, uses exponential margins.

We shall, in the following, make a brief survey of some essential characterizations, starting with the bivariate case, which gives rise to specific developments.

2. BIVARIATE EXTREME VALUE DISTRIBUTIONS. We shall assume here that d=2. The characterization of bivariate extreme value distributions due to the simultaneous work of Geffroy, Sibuya and Tiago de Oliveira in 1958, completed by the latter in a series of papers (1975, 1980), can be summarized in the following:

Theorem 2. Let $\Lambda(x,y)$ be the distribution of a bivariate extreme value distribution with Gumbel margins:

$$\Lambda(t,+\infty)=\Lambda(+\infty,t)=\exp(-e^{-t}),\ -\infty < t < +\infty.$$

Then, there exists a function $\theta(t)$, $-\infty < t < +\infty$, such that

$$\Lambda(x,y) = \exp(-(e^{-x}+e^{-y})\theta(x-y)),\ -\infty < x,y < +\infty.$$

Conversely, in order that $\theta(.)$ be such that $\Lambda(x,y)$ defined as above be an extreme value distribution, it is necessary and sufficient that $\psi(t)=(1+t)\theta(\mathrm{Log}\ t)$ be convex on $(0,+\infty)$ and such that

$$\max(t,1) < \psi(t) < 1+t.$$

Proof. See Tiago de Oliveira, 1975. See also Galambos, 1978, Pickands, 1981, and Deheuvels, 1983a.

The function $\theta(.)$ has been named by Tiago de Oliveira the dependence function of $\Lambda(.,.)$. Since no confusion can be made with the dependence function-copula of §1, we shall use this name in the sequel. (Note that $\theta(-u)$ corresponds to $\Lambda(y,x)$).

Even though the construction of $\psi$ and $\theta$ can be made quite easily by the biais of Theorem 2, a more tractable representation, due to Pickands, 1981 yields somewhat simpler results.

Let $\Lambda(x,y)$ stand for the distribution function of $(X,Y)$, and put $V=e^{-X}$, $W=e^{-Y}$. We have then, for $u,v,w>0$,

$$P(V>v,\ W>w) = G(v,w) = \Lambda(-\mathrm{Log}v,-\mathrm{Log}\ w),$$
$$P(V>u) = P(W>u) = e^{-u}.$$

The distribution of $(V,W)$ is then a bivariate extreme value distribution for minima with exponential margins. In this case, the following result is true:

Theorem 3. $G(v,w)=P(V>v,W>w)$ is the survivor function of a bivariate extreme value distribution with exponential margins if and only if there exists a bounded positive Radon measure $dH$ on $[0,1]$ such that

(i) $\int_0^1 u\,dH(u) = \int_0^1 (1-u)\,dH(u) = 1.$

Furthermore, if we put successively

(ii) $H(v) = \int_0^v dH(u),\ 0 \le v \le 1,$

(iii) $A(v)=\int_0^1 \max(u(1-v),v(1-u))\,dH(u)=1+\int_0^v (H(u)-1)\,du,$

(iv) $B(s,t) = (s+t)A(\frac{t}{s+t}),$

Then, for any $s,t>0$, we have

(v) $G(s,t) = \exp(-B(s,t)).$

Conversely, if $B(s,t)=-\mathrm{Log}\ G(s,t)$ is given, then

(vi) $A(v) = B(1-v,v),\ 0 \le v \le 1,$

which enables to evaluate $H$ from (iii).

Under the notations above, the dependence function $\theta(.)$ of $\Lambda(x,y)=G(e^{-x},e^{-y})=\exp(-(e^{-x}+e^{-y})\theta(x-y))$ is given by

(vii) $\theta(t) = A(\frac{1}{1+e^{-t}}),\ -\infty < t < +\infty.$

Proof. See Pickands, 1981, and Deheuvels, 1983b.

Even though the formulae (i-vii) of Theorem 3 may appear involved at first sight, they give a powerful tool for the description of these distributions.

We give now some specific examples.

Examples. 1°) If $V$ and $W$ are independent, then $B(s,t)=s+t$, $A(v)=1$, $dH=\delta_0+\delta_1$, where $\delta_u$ is the Dirac measure at point $u$.

2°) If $V=W$, then $B(s,t)=\max(s,t)$, $A(v)=\max(v,1-v)$, $dH=2\delta_{1/2}$.

3°) The "mixed model": $B(s,t)=s+t+\alpha st/(s+t)$, $0 \le \alpha \le 1$.

4°) The "biextremal model": $B(s,t)=\max(s+(1-\alpha)t,\ t)$, $0 \le \alpha \le 1$,

$A(v)=\max(1-\alpha(1-v),1-v)$.

5°) The "Gumbel model": $B(s,t)= \max(s,t)+(1-\alpha)(s+t)$.

6°) The "natural model", introduced by Tiago de Oliveira,1980, consists in defining $(V,W)$ by $V=\min(a\xi_1,b\xi_2)$, $W=\min(c\xi_2,d\xi_3)$, where $\xi_i$,$i=1,\ldots,4$ are i.i.d. exponentially distributed random variables. We get here $dH=d^{-1}\delta_0+(b^{-1}+c^{-1})\delta_{c/(b+c)}+a^{-1}\delta_1$.

7°) Some other models (see Tiago de Oliveira, 1980, Deheuvels, 1983b) can be generated by noting that if for $i=1,\ldots,n$, $P(V_i>v,W_i>w)=\exp(-B_i(v,w))$ is the survivor function of an extreme value distribution with exponential margins, then it is also the case for $\exp(-\sum_{i=1}^n\lambda_iB_i(v,w))$, where $\sum_{i=1}^n\lambda_i=1,\lambda_i\geq 0,1\leq i\leq n$.

Such a distribution may be called a <u>mixture</u>.

Pickands, 1981, gives to $dH$ the name of <u>characteristic distribution</u> of $G(.,.)$. $H(.)$ is called characteristic distribution function, and $A(.)$ dependence function. This last name will not be used here, in order to avoid possible confusion with Tiago de Oliveira's $\theta(t)=A(t/(1+e^{-t}))$.

We shall not discuss here inference problems for such distributions, refering to Tiago de Oliveira, 1975, 1980, Pickands, 1981, and Deheuvels, 1980b, for various approaches to the problem.

4. MULTIVARIATE EXTREME VALUE DISTRIBUTIONS. As we have seen in §2, all we need is to find all possible limits as $n\to\infty$ of $D^n(u_1^{1/n},\ldots,u_d^{1/n})$, where $D$ is a dependence function. For this, if we put $u_i=\exp(-z_i)$, $i=1,\ldots,d$, it can be seen that, as $n\to\infty$,

$$D^n(u_1^{1/n},\ldots,u_d^{1/n}) \sim \exp(n\mathrm{Log}\ D(1-\frac{z_1}{n},\ldots,1-\frac{z_d}{n})) \sim$$

$$\sim \exp(n(1-D(1-\frac{z_1}{n},\ldots,1-\frac{z_d}{n})),$$

and hence, that

$$D_\infty(u_1,\ldots,u_d)=\lim_{n\to\infty} D^n(u_1^{1/n},\ldots,u_d^{1/n})=$$

$$=\exp(-\lim_{n\to\infty} n(1-D(1+\frac{\mathrm{Log}\ u_1}{n},\ldots,1+\frac{\mathrm{Log}\ u_d}{n})).$$

By using the fact that D is the d.f. of a probability measure, the following representation (Deheuvels, 1978) is obtained:

Theorem 4. $D_\infty$ is the dependence function of a multivariate extreme value distribution with nondegenerate margins iff there exists for $k=2,\dots,d$, $1\leq i_1<\dots<i_k\leq d$, a positive measure $\mu_{k;i_1,\dots,i_d}$ on the simplex $S_{k;i_1,\dots,i_d}=\{(v_1,\dots,v_d);\sum v_{i_j}=1,$ $v_{i_j}\geq 0,\ v_\ell=0, \ell\in\{i_1,\dots,i_k\}\}$, such that

$$D_\infty(u_1,\dots,u_d)=u_1\dots u_d\exp\Big(-\sum_{k=2}^{d}(-1)^k\sum_{1\leq i_1<\dots<i_k\leq d}$$

$$\int_{S_{k;i_1,\dots,i_d}}\max(v_{i_j}\operatorname{Log} u_{i_j})d\mu_{k;i_1,\dots,i_k}(v)\Big).$$

The important element of this representation lies in the signs of the component measures $\mu_{k;i_1,\dots,i_k}$. One may see (as in De Haan and Resnick, 1977) that we have

$$D_\infty(u_1,\dots,u_d)=\exp\Big(-\int_0^{-\operatorname{Log} u_1}\dots\int_0^{-\operatorname{Log} u_d}dM\Big),$$

where M is an exponent measure, or (as in Galambos, 1978), that

$$D_\infty(u_1,\dots,u_d)=u_1\dots u_d\exp\Big(-\sum_{k=2}^{d}(-1)^k D_k(u_1,\dots,u_d)\Big),$$

where $D_k\geq 0$ for $k=2,\dots,d$.

It is clear from Theorem 4 that the explicit structure of the multivariate extreme value distributions is somewhat involved as d gets large if expressed in terms of dependence functions.

A simpler representation, quoted in Galambos, 1978, and due to Pickands, 1980, 1981 (see Deheuvels, 1983) uses exponential margins and is given as follows.

Theorem 5. $G(v_1,\dots,v_d)=P(V_1>v_1,\dots,V_d>v_d)$ is the survivor function of a multivariate extreme value distribution for minima with exponential margins iff there exists a positive measure $\mu$ on the simplex $S_d=\{(v_1,\dots,v_d),\sum v_i=1, v_i\geq 0\}$, such that

$$G(v_1,\ldots,v_d)=\exp(-\int_{S_d} \max_{1\leq j\leq d} (u_j v_j)d\mu(v)).$$

The main interest of this representation lies in the fact that the choice of $\mu\geq 0$ is arbitrary. Here $\mu$ coincides in the bivariate case (d=2) with the characteristic measure introduced in §2. However it is to be noted that the assumption that the margins $V_1,\ldots,V_d$ follow standard exponential distributions imposes further restrictions on $\mu$.

The preceding results yield a complete descriptive and constructive approach of the probabilistic structure on multivariate extreme value distributions. We shall now review various extensions and applications of this theory (see also Gumbel, 1960, Galambos, 1975, Tiago de Oliveira, 1980).

5. MODELS FOR POINT PROCESSES WITH EXPONENTIAL MARGINS. Let $\{\xi_n, -\infty < n < +\infty\}$ be an i.i.d. sequence of exponentially distributed random variables. Let $\{\alpha_n, -\infty < n < +\infty\}$ be an infinite sequence of numbers such that, for any n, $0<\alpha_n\leq\infty$, and $\sum 1/\alpha_n=1$, with $1/\infty=0$. Put

$$T_n= \min\{\alpha_i \xi_{n-i}, -\infty < i < +\infty\}.$$

The following results have been proved in Deheuvels,1983a,b.

Theorem 6. Under the hypotheses above, $\{T_n, -\infty < n < +\infty\}$ is a stationary process with marginal exponential distribution, and such that all its finite dimensional distributions are multivariate extreme value distributions for minima.

$\{T_n\}$ has been called a Moving Minimum (MM) process.

Theorem 7. Let $G(v_1,\ldots,v_d)$ be the survivor function of an arbitrary multivariate extreme value distribution for minima with exponential margins. Then there exists a sequence of MM processes $\{T_n^{(m)}\}$ such that if $G_m(v_1,\ldots,v_d)$ is the survivor function of $(T_1^{(m)},\ldots,T_d^{(m)})$, then $\lim_{n\to\infty} G_m = G$.

In other words, G can be approximated as closely as wished by moving minimum MM models.

These processes possess finite dimensional distributionss whose characteristic measures (in Pickand's sense) are finite linear combinations of Dirac measures. Similar distributions and processes have been introduced with Gumbel margins by Tiago de Oliveira, 1980. They appear to be a very interesting general model to modelize and to analyze empirical data. In particular, the $\{T_n\}$ can be used for the interarrival times of point processes, giving natural extensions of the Poisson model.

6. MAX-INFINITE DIVISIBLE DISTRIBUTIONS. The concept of max-infinite divisibility is due to Balkema and Resnick, 1977. It was extended in Deheuvels, 1980 in the case of dependence functions. We shall say in general (see Deheuvels, 1980) that the order of a distribution function $F(x_1,\ldots,x_d)$ is the maximal value of $r\geq 1$ such that $F^{1/r}$ is still the distribution function of a probability measure. If $r=\infty$, then F is said to be max-infinitely divisible.

Every multivariate extreme value distribution (for maxima) is max-infinitely divisible. A parallel can be made between these notions and the case of stable and infinitely divisible distributions in the classical sense, where the discussion is made in terms of characteristic functions instead of distribution functions. In general we have (Balkema and Resnick, 1977, Deheuvels, 1980):

Theorem 8. If $F(x_1,\ldots,x_d)$ is max-infinitely divisible with marginals df's $F_{(1)},\ldots,F_{(d)}$, then

$$\max\{0,1-d+\sum_{j=1}^{d}F_{(j)}(x_j)\}\leq F(x_1,\ldots,x_d)\leq \min_{1\leq j\leq d} F_{(j)}(x_j).$$

Let D be the dependence function of F. Then there exists for $k=2,\ldots,d$ and $1\leq i_1<\ldots<i_k\leq d$, positive measures $\nu_{k;i_1,\ldots,i_k}$ on $R^k$ such that

$$D(u_1,\ldots,u_d)=u_1\ldots u_d\exp\Big(-\sum_{k=2}^{d}(-1)^k \sum_{1\le i_1<\ldots<i_k\le d} \int_0^{-\mathrm{Log}\,u_{i_1}} \ldots \int_0^{-\mathrm{Log}\,u_{i_k}} d\nu_{k;i_1,\ldots,i_k}\Big).$$

This representation is similar to the one given in Theorem 4, where, in addition, the measures $\nu_{k;.}$ have a specific structure due to the fact that

$$D_\infty^r(u_1^{1/r},\ldots,u_d^{1/r}) = D_\infty(u_1,\ldots,u_d), \text{ for any } r>0.$$

7. APPLICATIONS. The interest of multivariate extreme value distributions does not need to be emphasized. They provide a very adequate model to analyze meteorological data (for instance the extreme value of meteorological measures collected at d different neighboring sites of observation), or failure and life-time expreiments. One must insist on the fact that, even though the structure of bivariate extremes in known since 1960, the corresponding results for higher dimensions were only obtained in the 1980's, and hence, that the statistical theory corresponding to these results is very new.

Multivariate extreme value distributions appear also naturally as the finite dimensional distributions of extremal processes, and also as the limiting joint distributions of the maximum and the minimum of stationary sequences (Davis, 1982). Their use in point processes has been allready mentioned in §5.

Domains of attraction have been discussed by Deheuvels, 1978, Marshall and Olkin, 1982, and Berman, 1961, 1962.

Other definitions for multivariate extremes are discussed in Barnett, 1976. We shall not develop here the corresponding theories, refering also to David, 1981, for further details.

REFERENCES.

The following list is not intended to cover the subject, and corresponds to the papers that have been cited in our discussion.

Balkema, A. and Resnick, S.I. (1977) Max-infinite divisibility, J. App. Probab. 14 309-319

De Haan, L. and Resnick, S.I. (1977) Limit theory for multivariate sample extremes, Z. Wahrscheinlichkeit. verw. Gebiete 40 317-337

Berman, S.M. (1961/62) Convergence to bivariate limiting extreme value distributions, Ann. Inst. Statist. Math. 13 217-223

Fréchet, M. (1951) Sur les tableaux de corrélation dont les marges sont données, Annales de l'Université de Lyon A3, 14 53-77

Geffroy, J. (1958/59) Contributions à la théorie des valeurs extrêmes, Publ. Instit. Statist. Univ. Paris, 7-8 37-184

Gumbel, E.J. (1960) Distributions des valeurs extrêmes en plusieurs dimensions, Publ. Instit. Statist. Univ. Paris 9 171-3

Sibuya, M. (1960) Bivariate extreme statistics I, Ann. Inst. Statist. Math. 11 195-210

Pickands, J. III (1980) Multivariate negative exponential and extreme value distributions, Unpublished manuscript, University of Pennsylvania

Pickands, J. III (1981) Multivariate extreme value distributions Proceedings 43d Session of the I.S.I. (Buenos Aires) 859-878

Marshall A.W. and Olkin I. (1982) Domains of attraction of multivariate extreme value distributions, Annals of Probability 10 168-177

Deheuvels, P. (1978) Caractérisation complète des lois extrêmes multivariées et de la convergence des types extrêmes, Publ. Instit. Statist. Univ. Paris 23 1-36

Deheuvels, P. (1980) The decomposition of infinite order and extreme multivariate distributions, in: Asymptotic Theory of Statistical Tests and Estimation, I.M. Chakravarti edit., 259-286, Academic Press, New York

Deheuvels, P. (1983) Point processes and multivariate extreme values, J. Multivariate Analysis 13 257-271

Deheuvels, P. (1983) Point processes and multivariate extreme values (II), to be published in Multivariate Analysis VI , P.R. Krishnaiah edit., North Holland, Amsterdam

Deheuvels, P. (1980) Some applications of the dependence functions to statistical inference, Coll. Math. J. Bolyai 32, 183-201, North Holland, Amsterdam

Galambos, J. (1978) The Asymptotic Theory of Extreme Order Statistics, Wiley, New York

Tiago de Oliveira J. (1958) Extremal distributions, Revista da Fac. Ciencias Univ. Lisboa A7 215-227

Tiago de Oliveira J. (1975) Bivariate and multivariate extreme value distributions, in: Statistical Distributions in Scientific work, Patil, G.P. and al. edit., Reidel, Dordrecht 355-61

Tiago de Oliveira (1980) Bivariate extremes: foundations and Statistics, in: Multivariate Analysis V , P.R. Krishnaiah ed. North Holland, Amsterdam, 349-368

Galambos, J. (1975) Order statistics of samples from multivariate distributions, J.A.S.A. 70 674-680

Davis, R.A. (1982) Limit laws for the maximum and minimum of stationary sequences, Z. Wahrscheinlichkeit. verw. Gebiete 61 31-42

Barnett, V. (1976) The ordering of multivariate data, J. Royal Stat. Soc. A, 139, 318-354

David, H. Order Statistics, Wiley, New York

# BIVARIATE MODELS FOR EXTREMES; STATISTICAL DECISION

J. Tiago de Oliveira

Academy of Sciences of Lisbon
Faculty of Sciences of Lisbon
Center for Statistics and Applications
(I.N.I.C)

ABSTRACT. After describing the asymptotic distribution of bivariate samples of maxima, with Gumbel margins, some important correlation and regression results are given. The five models known until now (logistic, mixed, Gumbel, biextremal and natural) are described and statistical decision related to them is presented. Finally, a triextremal model is described.

## INTRODUCTION

Consider a sequence of IID random pairs $\{(X_i, Y_i)\}$, $i=1,2,\ldots,n$ with distribution function $F(x,y) = \text{Prob}\{X_i \leq x, Y_i \leq y\}$. The distribution function of the (virtual) point $(\max_1^n X_i, \max_1^n Y_i)$ is evidently $\text{Prob}\{\max_1^n X_i \leq x, \max_1^n Y_i \leq y,\} = F^n(x,y)$. This point is observed with $\text{Prob}\{(\max_1^n X_i, \max_1^n Y_i) \in \{(X_1,Y_1),\ldots,(X_n,Y_n)\}\} =$

$= n \iint_{-\infty}^{+\infty} F^{n-1}(x,y)\, dF(x,y)$ which can oscilate between 0 for $\underline{F}(x,y) = \max(0, F(x,+\infty) + F(+\infty,y) - 1)$ (complete negative association) and 1 for $\bar{F}(x,y) = \min(F(x,+\infty), F(+\infty,y))$ (complete positive



*J. Tiago de Oliveira (ed.), Statistical Extremes and Applications, 131–153.*

association), with the value $1/n$ for the independence case $F_0(x,y) = F(x,+\infty)\ F(+\infty,y)$. Dually if $S(x,y) = \text{Prob}\{X_i>x, Y_i>y\} =$ $=1+F(x,y)-F(x,+\infty)-F(+\infty,y)$ (or, equivallently, $F(x,y)=1+S(x,y)-$ $-S(x,-\infty)-S(-\infty,y)$) we see that $\text{Prob}\{\min_1^n X_i>x, \min_1^n Y_i>y\}=S^n(x,y)$ or, yet, $\text{Prob}\{\min_1^n X_i\leq x, \min_1^n Y_i \leq y\} = 1^1+S^n(x,y)-S^n(x,-\infty)-S^n(-\infty,y)$.

From the duality of the formulae or from the relation $\min_1^n X_i =$ $=-\max_1^n (-X_i)$ we can convert maxima results into minima results; as a consequence, for sake of simplicity and commodity, we will deal, only, with maxima results, from now on.

As the manipulation of the exact (finitary) distribution of bivariate maxima is, in general, difficult, for practical applications we will use asymptotic distributions of bivariate maxima, even if we know $F(x,y)$, unless it is very manageable. For basic results in the area see Geffroy (1958/59), Sibuya (1960), Tiago de Oliveira (1958) and (1962/63).

## THE ASYMPTOTIC RESULTS

As it is done in general, for univariate extremes we will search if there exist linear transforms of $\max_1^n X_i$ and $\max_1^n Y_i$, with coefficients dependent on $n$, such that the transformed pair has a non-degenerate (i.e., different from an almost sure random pair) and proper (i.e., taking finite values with probability 1) distribution. In other words, we will search attraction coefficients $\lambda_n$ and $\delta_n(>0)$ and $\lambda'_n$ and $\delta'_n(>0)$, not univocally defined, such that

$$\text{Prob}\{(\max_1^n X_i-\lambda_n)/\delta_n\leq x,\ (\max_1^n Y_1-\delta_n)/\delta'_n\leq y\} =$$

$$= F^n(\lambda_n+\delta_n x,\ \lambda'_n+\delta'_n y) \overset{w}{\to} L(x,y), \text{ where } L(.,.)$$

is non-degenerate and proper.

As both margins must be of one of the forms $\Psi_\theta(.)$, $\Lambda(.)$ or

$\Phi_\theta(.)$ given by

$$\Psi_\theta(z) = \exp(-(-z)^\theta) \quad \text{for} \quad -\infty < z < 0, \text{ with } \theta>0,$$

$$= 1 \quad \text{for} \quad z \geq 0 \quad \text{(Weibull distribution);}$$

$$\Lambda(z) = \exp(-e^{-z}) \quad \text{for} \quad -\infty < z < +\infty \quad \text{(Gumbel distribution)}$$

and $$\Phi_\theta(z) = 0 \quad \text{for} \quad -\infty < z < 0,$$

$$= \exp(-z^{-\theta}) \quad \text{for} \quad 0 \leq z < +\infty, \text{ with } \theta>0$$
(Fréchet distribution)

which are easily transformable between them, we will suppose, from now on, that the margins have Gumbel distributions, i.e. that $L(x, +\infty) = \Lambda(x)$ and $L(+\infty, y) = \Lambda(y)$; we will then denote $L(x,y)$ by $\Lambda(x,y)$. Remark that $\Lambda(x,y)$ it continuous, as the margins are continuous.

As $F^n(\lambda_{nm} + \delta_{nm}\, x, \lambda'_{nm} + \delta'_{nm} y) \overset{w}{\to} \Lambda^{1/m}(x,y)$ and $F^n(\lambda_n + \delta_n x, \lambda'_n + \delta'_n y) \overset{w}{\to} \Lambda(x,y)$ by Khintchine convergence of types theorem, we see that exist $\alpha_m$, $\beta_m > 0$, $\alpha'_n$ and $\beta'_m > 0$ such that the stability relation

$$\Lambda(\alpha_m + \beta_m\, x, \alpha'_m + \beta'_m\, y - \Lambda^{1/m}(x,y)$$

is true and as we have Gumbel margins we have

$$\Lambda(\alpha_m + \beta_m x) = \Lambda^{1/m}(x)$$

and $$\Lambda(\alpha'_m + \beta'_m\, y) = \Lambda^{1/m}(y)$$

so that $\alpha_m = \alpha'_m = \log m$ and $\beta_m = \beta'_m = 1$. Thus the stability relation reduces to

$$\Lambda(x,y) = \Lambda^m(x + \log m, y + \log m) \; .$$

Passing now from integers to rationals and after to reals and using the continuity of $\Lambda(.,.)$, we get the stability relation in final form

$$\Lambda(x,y) = \Lambda^t(x + \log t, y + \log t), \quad x,y, \; \forall t>0$$

Putting now $x + \log t = 0$ we get

$$\Lambda(x,y) = \Lambda^{e^{-x}}(0, y-x)$$

which can be written in the form

$$\Lambda(x,y) = \exp\{-(e^{-x}+e^{-y})\, k(y-x)\}$$

where $k(w) = -\log \Lambda(0,w)/(1 + e^{-w})$

is called the <u>dependence function</u> .

Recall that a standard Gumbel random variable has all moments, the mean value being $\gamma = 0.57722$ (Euler constant) and the variance $\pi^2/6$.

A basic inequality can immediately be written. From Boole-Fréchet inequality $\max(0, \Lambda(x)+\Lambda(y)-1) \leq \Lambda(x,y) \leq \min(\Lambda(x),\Lambda(y))$, written at the point $(x+\log t, y+\log t)$, raised to power $t(>0)$ and using the stability relation we get

$$\max{}^t(0,\Lambda(x+\log t)+\Lambda(y+\log t)-1) \leq \Lambda(x,y) \leq \min{}^t\{\Lambda(x+\log t), \Lambda(y+\log t)\} = \min(\Lambda(x), \Lambda(y))$$

Letting $t \to \infty$ we get finally

$$\Lambda(x)\, \Lambda(y) \leq \Lambda(x,y) \leq \min(\Lambda(x), \Lambda(y))$$

or $$\frac{\max(1, e^w)}{1+e^w} \leq k(w) \leq 1 .$$

The relation $\Lambda(x)\,\Lambda(y) \leq \Lambda(x,y)$ means that we have <u>positive association</u>, i.e., that, for instance, if $X$ has a large (small) observed value the corresponding observed value of $Y$ is large (small) with higher probability for $\Lambda(x,y)$ than in the case of independence $\Lambda(x)\,\Lambda(y)$. This fact will be confirmed later as it will be shown that all correlation coefficients are non-negative. Sibuya (1960) proved, the first time, the positive association.

The dependence function $k_I(w) = 1$ corresponds obviously to independence. The case $k_D(w) = \dfrac{\max(1,e^w)}{1+e^w}$ , corresponding to the distribution function $\min(\Lambda(x), \Lambda(y))$ expresses that $\text{Prob}\{Y=X\}=1$; it will be called diagonal case. Recall that we are, always, dealing with reduced values: in practice we will have always to introduce the location and dispersion parameter for both margins. In the diagonal case $X$ and $Y$ are related linearly with probability 1, the straight line having positive slope.

It can be shown that the dependence function corresponds to a distribution function if and only if $k(.)$ satisfies the following conditions.

$k(-\infty)=k(+\infty)=1$ (a consequence of $k_D(w) \leq k(w) \leq k_I(w)$ ;

$(1+e^{w})\ k(w)$ a non-decreasing function;

$(1+e^{-w})\ k(w)$ a non-increasing function

and $\Delta^2_{x,y}\ \{e^{-x}+e^{-y})\ k(y-x)\} \leq 0$ .

In the case of existence of a planar density (i.e., if $k''(.)$ exists) the dependence function must satisfy the following conditions

$$k(-\infty)=k(+\infty)=1$$

$$((1+e^{w})\ k(w))' \geq 0$$

$$((1+e^{-w})\ k(w))' \leq 0$$

and

$$(1+e^{-w})\ k''(w)+(1-e^{-w})\ k''(w) \geq 0 .$$

It is immediate, from the conditions given above, that if $k_1(w)$ and $k_2(w)$ are dependence functions also $\theta\ k_1(w)+(1-\theta)k_2(w)$ $(0 \leq \theta \leq 1)$ are dependence functions: they correspond to the distribution functions $\Lambda_1^{\theta}\ (x,y)\ \Lambda_2^{1-\theta}\ (x,y)$. Also, if $k(w)$ is a dependence function, $k(-w)$ is a dependence function; the case when $(X,Y)$, is exchangeable i.e. $\Lambda(x,y) = \Lambda(y,x)$, we have $k(w)=k(-w)$.

Another function which is important is $D(w) = \text{Prob}\{X-Y \leq w\}$, the distribution function of the (reduced) difference $W=Y-X$.

It is very easy so show that

$$D(w) = \frac{e^{w}}{1+e^{w}} + \frac{k'(w)}{k(w)} .$$

For the independence case $k_I(w)=1$ we have the standard logistic: in the diagonal case $k_D(w)$ we have $D(w)=H(w)$, the Heavside jump function $(H(w)=0$ if $w<0$, $H(w)=1$ if $w\geq 0)$.

Note that $D(w)$ has mean value (zero) and variance, as $X$ and $Y$ have, separately, the same distribution, with variance. $k(w)$ can be expressed in $D(w)$ as

$$k(w) = \frac{\exp \int_{-\infty}^{w} D(t)\,dt}{1+e^{w}}$$

The distribution function of $W=Y-X$ is characterized, in the case of existence of a planar density, by the two conditions

$$\int_{-\infty}^{+\infty} w \, d\, D(w) = 0$$

and $D'(w)\geq D(w)\,(1-D(w))$.

In the same conditions it can be shown that the joint distribution of $V=e^{-X}+e^{-Y}$ and $W=Y-X$ can be written as

$$\text{Prob}\{V\leq v,\ W\leq w\} = (1-e^{-v\,k(w)})\,D(w)-$$

$$-v\int_{-\infty}^{w} dt\,\frac{k(t)}{1+e^{t}}\,e^{-vk(t)}D(t).$$

A question that will be relevant later, for modelizing, is the generation of bivariate extreme models.

Evidently, the convexity of the set of dependence functions, leading from the distribution functions $\Lambda_1(x,y)$ and $\Lambda_2(x,y)$ to $\Lambda_1^{\theta}(x,y)\,\Lambda_2^{1-\theta}(x,y)$ $(0\leq\theta\leq 1)$ is one of the generation techniques.

But, aside from the use of convexity, there is, also, a max--technique that will be helpful. Let us consider the (random) extremes pair $(X,Y)$ with distribution function $\Lambda(x,y)=\exp\{-(e^{-x}+e^{-y})\,k(y-x)\}$ and consider the new pair

$$\bar{X} = \max(X-a,\ Y-b)$$

$$\bar{Y} = \max(X-c,\ Y-d)$$

with $a$, $b$, $c$, $d$ to be fixed.

As $\text{Prob}\{\bar{X}\leq x\}=\text{Prob}\{X\leq a+x,\ Y\leq b+x\}=\exp\{-(e^{-a}+e^{-b})\,k(b-a)e^{-x}\}$ must be a standard Gumbel margin we must have $(e^{-a}+e^{-b})\,k(b-a)=1$. As this relation can be written as $(1+e^{-(b-a)})\,k(b-a)=e^{a}$ and $(1+e^{-w})\,k(w)$ is non-increasing from $+\infty$ to 1 we must have $e^{a}\geq 1$ or $a\geq 0$. Analogous by we have $b\geq 0$ and $(e^{-c}+e^{-d})\,k(d-c)=1$ with $c\geq 0$ and $d\geq 0$. The new dependence function is given by

$$\bar{k}(w|a,b,c,d) = \frac{\{\max(e^{-a},\ e^{-c-w})+\max(e^{-b},\ e^{-d-w})\}}{1+e^{w}}\ \times$$

$$\times\ k(\min(b,d+w)-\min(a,c+w))$$

with $a,b,c,d\geq 0$ and $(e^{-a}+e^{-b})\,k(b-a)=1$, and $(e^{-c}+e^{-d})\,k(d-c)=1$. We will apply this, later, with $k_I(w)=1$ (independence). Let us make, now, some complementary remarks.

A characterization of $\Lambda(x,y)$ can be as follows. As used in the max-technique of generation of bivariate extremes model, if $(X,Y)$ has distribution function $\Lambda(x,y)$ then the random variable $Z=\max(X-a,Y-b)$ has the distribution function $\text{Prob}\{Z\leq z\}=\text{Prob}\{X-a\leq z,Y-b\leq z\}=\Lambda(z+a,\ z+b)=\Lambda(z-\lambda(a,b))$ with $\lambda(a,b)=\log(e^{-a}+e^{-b})+\log k(b-a)$. Let us show, reversely, that if $(X,Y)$ is a random pair with distribution function $F(x,y)$ and $\max(X-a,\ Y-b)$ has Gumbel distribution with a location parameter $\lambda(a,b)$, whatever $a$ and $b$ are, then we must have $F(x,y)=\Lambda(x,y)$.

As $\text{Prob}\{\max(X-a,\ Y-b)\leq z\}=F(z+a,\ z+b)=\Lambda(z-\lambda(a,b))$, if we put $z+a=\xi$, $z+b=\eta$ we have $F(\xi,\eta)=\Lambda(z-\lambda(\xi-z,\ \eta-z))$ independent of $z$. Putting now $z=\xi$ and $\lambda(0,\ \eta-\xi)=\log(1+e^{-(\eta-\xi)})$ $\times\log k(\eta-\xi)$ we get the desired result. Note that $\lambda(a,b)=\lambda(0,b-a)-a$.

An important point is the search of the conditions under which an (initial) distribution function $F(x,y)$ is attracted, for maxima, to independence and diagonal cases. Let $P(u,v)$ be a function defined by $\text{Prob}\{X>x,Y>y\}=P(F(x,+\infty),\ F(+\infty,y))$. Sibuya (1960) has shown that the necessary and sufficient condition for limiting independence is that $P(1-s,\ 1-s)/s\to 0$ as $s\to 0$ and the necessary and sufficient condition for having the diagonal case as limit is $P(1-s,\ 1-s)/s\to 1$ as $s\to 0$. With the first result he has shown that maxima pairs of the binormal distribution are asymptotically independent if $\rho \neq \pm 1$.

Also Geffroy (1958/59) did show that a sufficient condition for limiting independence is that

$$\frac{1+F(x,y)-F(x,w_y)-F(w_x,y)}{1-F(x,y)} \to 0$$

as $x \to w_x$ and $y \to w_y$, $w_x$ and $w_y$ being the right-end points of the supports of X and Y respectively.

Sibuya conditions (and Geffroy condition) are easy to interpret : we have limiting independence if Prob{X>x, Y>y} is a vanishing summand of Prob(X>x or Y>y) and the diagonal case as limit if Prob{X>x,Y>y} is the leading summand of Prob{X>x or Y>y}.

REGRESSION AND CORRELATION RESULTS

Using the fact that the covariance can have the general expression

$$\mathrm{cov}(X,Y) = \iint_{-\infty}^{+\infty} \left[F(x,y)-F(x,+\infty)\ F(+\infty,y)\right] dx\ dy$$

it is immediate that

$$\mathrm{cov}(X,Y|k) = \iint_{-\infty}^{+\infty} \left[\Lambda(x,y)-\Lambda(x)\ \Lambda(y)\ \right]dx\ dy = -\int_{-\infty}^{+\infty} \log\ k(w)dw$$

so that the correlation coefficient is

$$\rho(k) = -\frac{6}{\pi^2} \int_{-\infty}^{+\infty} \log\ k(w)dw.$$

$\rho(k)$ is a functional of the dependence function; when $k_\theta(.)$ is a parameterized family, by $\theta$ say, we will denote the functional $\rho(k_\theta)$ which is, a function of $\theta$ by the simpler notation $\rho(\theta)$.

$$\text{As}\quad k_D(w) = \frac{\max(1,\ e^w)}{1+e^w} \leq k(w) \leq k_I(w) = 1 \quad \text{we see that}$$

$0 = \rho(k_I) \leq \rho(k) \leq \rho(k_D) = 1$. For the diagonal case we have $\rho(k_D)=1$ as could be expected; the fact that $\rho(k) \geq 0$ is consistent with the positive association, mentioned before. As $k(w) \leq 1$, $\rho(k) = 0$

implies $k(w) = 1$ and, thus, null correlation is equivalent to independence. Also as $\rho(k_D)=1= -\frac{6}{\pi^2}\int_{-\infty}^{+\infty} \log k_D(w)\, dw$ we can write $\rho(k)=1 - \frac{6}{\pi^2}\int_{-\infty}^{+\infty} \log \frac{k(w)}{k_D(w)}\, dw$ and thus $\rho(k)=1$ implies $\int_{-\infty}^{+\infty} \log \frac{k(w)}{k_D(w)}\, dw = 0$ and as $k_D(w) \leq k(w)$ we must have $k(w)=k_D(w)$: unit correlation is, then, equivalent to the diagonal case.

The last expression for $\rho(k)$ may, in some cases, be more convenient for computation, as it happens in the natural model where the tails of $k(.)$ are identical with the tails of $k_D(.)$.

Another way of obtaining $\rho(k)= -\frac{6}{\pi^2}\int_{-\infty}^{+\infty} \log k(w)\, dw$ is the following: as $W=Y-X$ has mean value zero we have

$$\operatorname{var}(Y-X) = \frac{\pi^2}{3}(1-\rho(k)) = \int_{-\infty}^{+\infty} w^2\, dD(w),$$

which by integration by parts gives the expression of $\rho(k)$.

In an analogous way we can show that the difference - sign correlation (Kendall's $\tau$) is

$$\tau(k) = 4 \iint_{-\infty}^{+\infty} \Lambda(x,y)\, d\,\Lambda(x,y) - 1$$

$$= \int_{-\infty}^{+\infty} \left(\frac{k'(w)}{k(w)}\right)^2 dw - 2\int_{-\infty}^{+\infty} \frac{e^w}{(1+e^w)^2} \log k(w)\, dw =$$

$$1 - \int_{-\infty}^{+\infty} D(w) \quad (1-D(w))\, dw.$$

The last expression, as $D'(w) \geq D(w)\ (1-D(w))$, shows $0 \leq \tau(k) \leq 1$ in connexion with the positive association; it is immediate, for the independence case, $\tau(k_I) = 0$ as $D(w)=e_w/(1+e^w)$ and for the diagonal case, that $\tau(k_D)=1$ as $D(w)=H(w)$.

For the grade correlation $\chi(k)$ we have

$$\chi(k) = 12 \iint_{-\infty}^{+\infty} \Lambda(x,y)\, d\,\Lambda(x)\, d\,\Lambda(y) - 3 = 12 \int_{-\infty}^{+\infty} \frac{e^w}{(1+e^w)^2}$$

$\frac{1}{(1+k(w))^2}\,dw-3$; as from the inequalities $k_D(w)\leq k(w)\leq k_I(w)$ we see that $0=\chi(k_I)\leq\chi(k)\leq\chi(k_D)=1$, also in accordance with the positive association.

Let us discuss now some results on regression. For commodity we will only consider the regression of $Y$ on $X$, the regression of $X$ on $Y$ being dealt with in the same way with the substitution of $k(w)$ by $k(-w)$. The linear regression line is evidently $Ly(x)=\gamma+\rho(k)\,(x-\gamma)$.

As $\text{Prob}\{Y\leq y|X=x\}=\Lambda(y|x)$ is given, in the case of density, by

$$\Lambda(y|x) = \frac{1}{\Lambda'(x)}\;\frac{\partial\,\Lambda(x,y)}{\partial\,x} = \exp\{e^{-x}-(e^{-x}+e^{-y})k(y-x)\}\times\{k(y-x)+ +(1+e^{-(y-x)})k'(y-x)\}$$

the regression line is

$$\bar{y}(x|k)=\int_{-\infty}^{+\infty} y d\,\Lambda(y|x)$$

and as $\gamma=\int_{-\infty}^{+\infty} y\,d\,\Lambda(y)$ we get, by integration by parts,

$$\bar{y}(x|k) = \gamma+\int_{-\infty}^{+\infty}(\Lambda(y)-\Lambda(y|x))\,dy$$

$$= \gamma+\int_{-\infty}^{+\infty}\{\exp(-e^{-x}e^{-w})-\left[k(w)+(1+e^{-w})k'(w)\right]\times$$

$$\times\exp\{e^{-x}\left[(1+e^{-w})\;k(w)-1\right]\}\}dw$$

$$=\gamma+\int_{-\infty}^{+\infty}\{(1+e^{-w})\exp(-e^{-x}.e^{-w})-(1+e^{-w})k(w)\exp(-e^{-x}((1+e^{-w})$$

$$k(w)-1)))\}dw$$

The correlation ratio is given by

$$R^2(y|x;k)=\frac{6}{\pi^2}\int_{-\infty}^{+\infty}(\bar{y}(x|k)-\gamma)^2\,dy =$$

$$=\frac{6}{\pi^2}\iint_{-\infty}^{+\infty} dv\;dw\Big[\frac{e^{-v}e^{-w}}{1+e^{-v}+e^{-w}} - 2\,\frac{e^{-v}((1+e^{-w})k(w)-1)}{e^{-v}+(1+e^{-w})k(w)-1} +$$

$$+\frac{((1+e^{-v})k(v)-1)\;((1+e^{-w})k(w)-1)}{(1+e^{-v})k(v)+(1+e^{-w})k(w)-1}\Big].$$

Median regression can be defined as the solution of the equation $\Lambda(\tilde{y}|x) = 1/2$; it takes the form $\tilde{y}(x|k)=x+\Phi(x)$ where $\Phi(x)$ is given by $e^{-x}(1+e^{-\Phi})k(\Phi)-\log|k(\Phi)+(1+e^{-\Phi})k'(\Phi)|=\log 2+e^{-x}$. In two of the models (logistic and mixed) described below the curves are approximately linear-see Gumbel and Mustafi (1968).

## THE DIFFERENTIABLE MODELS

The first models to be introduced were the differentiable ones, i.e. with a planar density. With reduced Gumbel margins, they are

the logistic model $\Lambda(x,y|\theta)=\exp\{-(e^{-x/(1-\theta)}+e^{-y/(1-\theta)})^{1-\theta}\}$

$0\leq\theta\leq 1$ (for $\theta=0$ we have independence and for $\theta=1^-$ we get the diagonal case, which is the only situation without density) and the

mixed model $\Lambda(x,y|\theta)=\exp(-e^{-x}-e^{y}+\theta/(e^{x}+e^{y}))$, $0\leq\theta\leq 1$.

Let us consider the logistic model. The dependence function is $k(w|\theta) = (1+e^{-w/(1-\theta)})^{1-\theta}/(1+e^{-w})=k(-w|\theta)$ - the model is exchangeable - and we have $D(w|\theta)=(1+e^{-w/(1-\theta)})^{-1}=1-D(-w|\theta)$. The correlation coefficient is $\rho(\theta)=\theta(2-\theta)$ and the difference-sign correlation coefficient is $\tau(\theta)=\theta$. Linear regression is, evidently, $Ly(x|\theta)= \gamma+\theta(2-\theta)(x-\gamma)$.

Owing to the importance of independence, underlined before, we can formulate a L.M.P. test of $\theta=0$ vs $\theta>0$. For the sample $((x_1,y_1),\ldots,(x_n,y_n))$ the likelihood is

$$L(x_i,y_i|\theta)= \prod_1^n \frac{\partial^2 \Lambda(x_i,y_i|\theta)}{\partial x_i\, \partial y_i}$$

and the rejection region has the form,

$$\frac{\partial}{\partial\theta}\left(\sum_1^n \log \frac{\partial^2 \Lambda(x_i,y_i|\theta)}{\partial x_i\, \partial y_i}\right)\Bigg|_{\theta=0} > c_n$$

or $$\sum_1^n v(x_i,y_i)\geq c_n$$

where

$$v(x,y)=x+y+x\ e^{-x}+y\ e^{-y}+(e^{-x}+e^{-y}-2)\ \log(e^{-x}+e^{-y})+\frac{1}{e^{-x}+e^{-y}}\ .$$

It can be shown that $v(X,Y)$, in case of independence, has mean value zero but infinite variance; the central limit theorem, in the usual terms, can not be applied and seems difficult to apply it, under a more general setting, owing to the form of $v$. Consequently, the only hope seems to be the recourse to simulation of the behaviour of $\frac{1}{n}\sum_{1}^{n} v(x_i,y_i)$ (which converges a.s. to zero, in case of independence, by Khintchine's theorem). Estimation by the maximum likelihood is possible but it has, at least at $\theta=0$, the difficulties arising from the infinite variance of $v$.

As $\theta$ is a dependence parameter, it is natural to use the correlation coefficients whose estimators are, as it is well known, independent of the location and dispersion parameters of the margins, to test $\theta=0$ is $\theta>0$. It was shown, in Tiago de Oliveira (1964) and (1975), that the most efficient is the correlation coefficient. Recall that in the case of independence $\sqrt{n}\ r$ is asymptotically standard normal. As $\rho(\theta)=\theta(2-\theta)$ a point estimator of $\theta$ is given by $r=\theta^*(2-\theta^*)$ or $\theta^*=1-\sqrt{1-r}$ .

A quick approach may be the quadrants method (see annex): As the medians of the margins, in reduced form, are $\tilde{\mu}=-\log\log 2=$ $=0.3665129$ we have $p(\theta)=\text{Prob}\{X\leq\tilde{\mu},Y\leq\tilde{\mu}\}=\exp(-2^{-\theta}\log 2)$ so the probability that both components are at the same side of the medians ($(X\leq\tilde{\mu},Y\leq\tilde{\mu})$ or $(X>\tilde{\mu},Y>\tilde{\mu})$) is $2p(\theta)$. If $N$ denotes the number of sample pairs with components both smaller or both larger than the margin medians $\tilde{\mu}$ we know that $\sqrt{n}\ \frac{N/n-2p(\theta)}{\sqrt{2p(\theta)\ (1-2\ p(\theta))}}$ is asymptotically standard normal. The point estimator of $\theta$ is given by

$$2^{-\theta^{**}}=\frac{\log(2n)-\log N}{\log 4}\ .$$

As $p(\theta)$ increases from $p(0)=1/4$ to $p(1)=1/2$ the solution is well defined if $N/n\geq 1/2$. The use of the $\delta$-method shows that

$$\sqrt{\frac{n\ N}{n-N}}\ \log 2.\log\frac{2n}{N}\ (\theta^{**}-\theta)$$

is asymptotically standard normal.

The mixed model has the dependence function $k(w|\theta) =$ $= 1-\theta \dfrac{e^w}{(1+e^w)^2} = k(-w|\theta)$, and so the model is exchangeable, and we have

$$D(w|\theta) = \frac{e^w}{1+e^w} \cdot \frac{(1+e^w)^2-\theta}{(1+e^w)^2-\theta\, e^w} = 1-D(-w|\theta).$$

Remark that $k(w|\theta)$ is the $(1-\theta,\theta)$ mixture of dependence functions $1$ and $1-\dfrac{e^w}{(1+e^w)^2}$.

The correlation coefficient has the expression $\rho(\theta) = \dfrac{6}{\pi^2}$ $(\text{arc cos } (1-\theta/2))^2$ which increases from $\rho(0)=0$ to $\rho(1)=2/3$. The L.M.P. test of $\theta=0$ vs $\theta>0$ is given by the rejection region

$$\sum_1^n v(x_i,y_i) \geq c_n$$

with

$$v(x,y) = 2\,\frac{e^{2x}.e^{2y}}{(e^x+e^y)^3} - \frac{e^{2x}+e^{2y}}{(e^x+e^y)^2} + \frac{1}{e^x+e^y}.$$

As before the situation is marred by the fact that for independence although $v(X,Y)$ has mean value zero the variance is infinite. The usual central limit theorem can not be applied and a more general result seems difficult to prove. $c_n$ seems only obtainable by simulation of $\frac{1}{n}\sum_1^n v(x_i y_i)$, as this average converges to zero a.s. in case of independence.

Correlation coefficients, independent of the margin parameters, are then to be used to test $\theta=0$ vs $\theta>0$. Once more the usual correlation coefficient is the best. The point estimator of $\theta$, if $0\leq r \leq 2/3$, is given by $\theta^* = 2(1-\cos \sqrt{\frac{\pi^2}{6}\, r})$.

A quick approach may also be the quadrants method. As the medians of the margins are, also, $\tilde{\mu}$ we have $p(\theta) = \text{Prob}\{X\leq\tilde{\mu}, Y\leq\tilde{\mu}\} =$ $=\exp(-2 \log 2(1-\theta/4))$ which varies from $p(\theta)= 1/4$ to $p(1)=\sqrt{2}/4$. If $N/n$ denotes the fraction of sample pairs both smaller or both larger than the margin medians we know that $\sqrt{n}\,\dfrac{N/n-2p(\theta)}{\sqrt{2p(\theta)/(1-2p(\theta))}}$ is asymptotically standard normal. If $1/2 \le N/n \le \sqrt{2}/2$ the estimator of $\theta$ is given by $\theta^{**}=2(1+\dfrac{\log N/n}{\log 2})$ and, once more by the $\delta$-method, we see that $\dfrac{\log 2}{2}\sqrt{\dfrac{n\,N}{n-N}}(\theta^{**}-\theta)$ is asymptotically standard normal.

Let us,finally,discuss the statistical choice of the two non--separated models (logistic and mixed), non-separated because they converge in the independence case ($\theta=0$ in both cases). As developed in Tiago de Oliveira (1983) the decision rule (with $v_L$ and $v_M$ denoting the $v$ functions of the logistic and mixed models, given before) is:

decide for independence if $\sum_1^n v_L(x_i,y_i) \le a_n$, $\sum_1^n v_M(x_i,y_i) \le b_n$;

decide for the logistic model if $\sum_1^n v_L(x_i,y_i) > a_n$, $\sum_1^n \dfrac{v_L(x_i,y_i)}{a_n} >$

$$> \sum_1^n \frac{v_M(x_i,y_i)}{b_n}$$

decide for the mixed model if $\sum_1^n v_M(x_i,y_i) > b_n$, $\sum_1^n \dfrac{v_M(x_i,y_i)}{b_n} >$

$$> \sum_1^n \frac{v_L(x_i,y_i)}{a_n}$$

The asymptotic distribution of the pair $(\sum_1^n v_L(x_i,y_i), \sum_1^n V_M(x_i,y_i))$ is not known, even in the independence case and, thus, $a_n$ and $b_n$ must be computed by simulation, in such way that the probability of deciding wrongly for logistic or mixed models, in independence case, should be half of the significance level.

## THE NON-DIFFERENTIABLE MODELS

Until now we know three forms of the non-differentiable models - i.e., with a singular component, not having thus a planar density- the Gumbel model, the biextremal model and the natural model. They are characterized, respectively, by the distribution functions

$\Lambda(x,y|\theta)=\exp\{-(e^{-x}+e^{-y}-\theta\min(e^{-x},e^{-y})\}, 0\leq\theta\leq 1$, $\Lambda(x,y|\theta)=$

$=\exp\{-\max(e^{-x}+(1-\theta)e^{-y}, e^{-y})\}$, $0\leq\theta\leq 1$, and $\Lambda(x,y|\alpha,\beta)=$

$=\exp\{-|(e^{\beta}-1)\max(e^{-x},e^{\alpha-y})+(1-e^{\alpha})\max(e^{-x},e^{\beta-y})|/(e^{\beta}-e^{\alpha})\}$,

$-\infty<\alpha\leq 0\leq\beta<+\infty$.

Let us consider the Gumbel model. Its dependence function is $k(w|\theta)=1-\theta\,\dfrac{\min(1,e^{-w})}{1+e^{-w}}$ which is the $(1-\theta,\theta)$ mixture of $k_I(w)=1$ (independence) and $k_D(w)=\dfrac{\max(1,e^{-w})}{1+e^{-w}}$ (diagonal case). The model is exchangeable as $k(w|\theta)=k(-w|\theta)$ and we have $D(w|\theta)=\dfrac{1-\theta}{1-\theta+e^{-w}}$ if $w<0$, $=\dfrac{e^{w}}{1-\theta+e^{w}}$ if $w\geq 0$, with a jump of $\dfrac{\theta}{2-0}$ at $w=0$, which is the probability $P(Y=X)$; if $w\neq 0$ we have $D(w|\theta)+D(-w|\theta)=1$. This model is a conversion to Gumbel margins of a bivariate model with exponential margins, due to Marshall and Olkin (1967).

The correlation coefficient is $\rho(\theta)=\dfrac{12}{\pi^2}\int_0^{\theta}\dfrac{\log(2-t)}{1-t}\,dt$, increasing from $\rho(0)=0$ to $\rho(1)=1$ and the difference sign correlation is $\tau(\theta)=\theta(2-\theta)$.

The use of given formulae shows that

$$\bar{y}(x|\theta)=\gamma+\log(1-\theta)+\int_{(1-\theta)e^{-x}}^{+\infty}e^{-t}/t.dt-(1-\theta)\exp(\theta\,e^{-x})$$

$$\int_{e^{-x}}^{+\infty}e^{-t}/t.dt=x+\int_0^{(1-\theta)e^{-x}}\frac{1-e^{-t}}{t}dt-(1-\theta)\exp(\theta\,e^{-x})\int_{e^{-x}}^{+\infty}\frac{e^{-t}}{t}dt$$

Although the form of $\bar{y}(x|\theta)$ is very different from the linear regression $\gamma+\rho(\theta)(x-\gamma)$, the non-linearity index $\frac{R^2-\rho^2}{1-\rho^2}$ is very small (between 0 and 0.006) which shows that the reduction of the variance of the prediction (or estimation) of the other component, when we pass from linear regression to general regression, is very small: in fact, the large difference between the two regressions only happens for very small and very large values of X, whose probability is very small: see Tiago de Oliveira (1974).

To avoid the estimation of the margin parameters we can use the correlation coefficients. The test for independence can be done, as before, based on the fact that if $\theta=0$ (independence) we know that $\sqrt{n}\,r$ is asymptotically standard normal.

The method of quadrants can be used, once more, with $p(\theta)=\text{Prob}\{X\leq\tilde{\mu},Y\leq\tilde{\mu}\}=2^{\theta}/4$, increasing from $p(0)=1/4$ to $p(1)=1/2$.

Then $\sqrt{n}\,\frac{N/n-2p(\theta)}{\sqrt{2p(\theta)(1-2p(\theta))}}$ is asymptotically standard normal and

$\theta^{**}=\frac{\log(N/n)}{\log 2}+1$ (supposing that $N/n\geq 1/2$) is an estimator of $\theta$.

Also it is immediate to show $\sqrt{n}\,\sqrt{\log 2}\,\sqrt{\frac{e^{-1+\theta}}{1-2^{-1+\theta}}}(\theta^{**}-\theta)$ and

$\sqrt{n}\,\sqrt{\log 2}\,\sqrt{\frac{N}{n-N}}(\theta^{**}-\theta)$ are asymptotically standard normal.

The biextremal model, which appears naturally in the study of extremal processes, see Tiago de Oliveira (1968) and (1969), can be obtained directly as follows: consider $(X,Z)$ independent reduced Gumbel variables and form the new pair $(X,Y)$ with $Y=\max(X-a,Y-b)$, with a and b such that $\text{Prob}\{Y\leq y\}=\text{Prob}\{\max(X-a,Z-b)\leq y\}=\Lambda(y)$ which implies $e^{-c}+e^{-d}=1$. Putting $e^{-c}=\theta$, $e^{-d}=1-\theta$ $(0\leq\theta\leq 1)$ we have $Y=\max(X+\log\theta,\ Z+\log(1-\theta))$. This was the initial use of the max-technique to generate new models and is a particular case of a natural model with $a=0$, $b=+\infty$, $c=-\log\theta$ and $d=-\log(1-\theta)$. The dependence function is $k(w|\theta)=1-\frac{\min(\theta,e^{w})}{1+e^{w}}$

and we have $D(w|\theta)=0$ if $w<\log\theta$ and $=\frac{1}{1+(1-\theta)e^{-w}}$ if $w=\log\theta$

with a jump of $\theta$ at $\log\theta$, which is $\text{Prob}\{Y=X+\log\theta\}$. Remark that $k(w|\theta)\neq k(-w|\theta)$ and so the model is not exchangeable.

The correlation coefficient is $\rho(\theta) = -\frac{6}{\pi^2}\int_0^\theta \frac{\log t}{1-t}\,dt$, increasing from $\rho(0)=0$ to $\rho(1)=1$ and the difference-sign correlation coefficient is $\tau(\theta)=\theta$. The linear regression is, as always, $Ly(x) = \gamma+\rho(\theta)(x-\gamma)$ .

As before, the general regressions

$$\bar{y}(x|\theta) = x+\log\theta + \int_0^{\frac{1-\theta}{\theta}e^{-x}} \frac{1-e^{-t}}{t}\,dt =$$

$$= \gamma+\log(1-\theta) + \int_{\frac{1-\theta}{\theta}e^{-x}}^{+\infty} \frac{e^{-t}}{t}\,dt$$

and

$$\bar{x}(y|\theta) = y-\log\theta-(1-\theta)\exp(\theta\, e^{-x}) \int_{\theta e^{-x}}^{+\infty} \frac{e^{-t}}{t}\,dt$$

do introduce a small reduction of variance in comparison with the linear regressions $Ly(x|\theta)=\gamma+\rho(\theta)(x-\gamma)$ and $Lx(y|\theta)=\gamma+\rho(\theta)(y-\gamma)$, the non-linearity index varying from 0 to 0.068 in the first case and from 0 to 0.007 in the second regression; see Tiago de Oliveira (1974).

The test of independence ($\theta=0$ is $\theta>0$) can be made by using the correlation coefficient as usual.

The quadrants method can be applied, using the fact that $p(\theta) = \text{Prob}\{X\leq\tilde{\mu}, Y\leq\tilde{\mu}|\theta\}=2^{\theta}/4$, which is equal to the results for the Gumbel model.

Finally, let us consider the natural model. Using the max--technique, from a pair $(X,Y)$ of independent reduced Gumbel random variables we get a new (dependent) random pair $\bar{X} = \max(X-a,Y-b)$, $\bar{Y} = \max(X-c,Y-d)$ with Gumbel margins if and only if $e^{-a}+e^{-b}=1$ and $e^{-c}+e^{-d}=1$. The dependence function is $\bar{k}(w|a,b,c,d) = (e^{-\min(a,c+w)} + e^{-\min(b,d+w)})/(1+e^{-w})$, whose tails coincide with the tails of the diagonal case $k_D(w) = \dfrac{\max(1,e^w)}{1+e^w}$ . It is easy to see that $\bar{W} = \bar{Y}-\bar{X} = \min(a+w,b)- \min(c+w,d)$ is such that $a-c\leq\bar{W} \leq b-d$ if $a+d \quad b+c$ and $b-d\leq \bar{w} \leq a-c$ if $b+c< a+d$. The case $a+d=b+c$ is irrelevant because $\bar{W}$ is then an almost sure random variable with $\bar{W} = a-c = b-d$ and as $\bar{W}$ must be zero (its mean value is zero) we have $a=c$, $b=d$, $\bar{X} = \bar{Y}$ .

From now on we will suppose $a+d<b+c$, the case $b+c<a+d$ being dealt with by exchange of $\bar{X}$ and $\bar{Y}$. Let us put $\alpha=a-c$, $b-d=\beta$; by the fact that $\bar{W}$ has a mean value zero we have $\alpha<0<\beta$. Using the relations $e^{-a}+e^{-b}=1$ and $e^{-c}+e^{-d}=1$, with $a-c=\alpha$ and $b-d=\beta$ we get the final expression

$$k(w|\alpha,\beta) = \frac{(e^{\beta}-1)\max(1,e^{\alpha-w})+(1-e^{\alpha})\max(1,e^{\beta-w})}{(e^{\beta}-e^{\alpha})(1+e^{-w})}$$

or, in more detail

$$
\begin{aligned}
k(w|\alpha,\beta) &= \frac{1}{1+e^{w}} && \text{if } w<\alpha \\
&= \frac{e^{\beta}-1+(1-e^{\alpha})\,e^{\beta-w}}{(e^{\beta}-e^{\alpha})(1+e^{-w})} && \text{if } \alpha\leq w\leq\beta \\
&= \frac{1}{1+e^{-w}} && \text{if } \beta<w .
\end{aligned}
$$

Note that $k(w|-\beta, -\alpha) = k(-w|\alpha,\beta)$; we get an exchangeable model if $\alpha=-\beta$. Note also that for $\alpha=\log\theta$, $\beta= +\infty$ we get the biextremal model and for $\alpha= -\infty$, $\beta=-\log\theta$ its dual.

We have also $D(w|\alpha,\beta)$ given by

$$
\begin{aligned}
D(w|\alpha,\beta) &= 0 && \text{if } w < \alpha \\
&= \frac{e^{\beta}-1}{e^{\beta}-1+(1-e^{\alpha})\,e^{\beta-w}} && \text{if } \alpha \leq w < \beta \\
&= 1 && \text{if } \beta \leq w ;
\end{aligned}
$$

thus we have jumps of $\dfrac{(e^{\beta}-1)\,e^{\alpha}}{e^{\beta}-e^{\alpha}}$ at $w=\alpha$ and of $\dfrac{(1-e^{\alpha})}{e^{\beta}-e^{\alpha}}$ at

$w=\beta$.

The fact that $\alpha \leq w \leq \beta$ expresses if $\alpha>-\infty$, $\beta<+\infty$ a strong stochastic connexion between $\bar{X}$ and $\bar{Y}$.

The correlation coefficient can be written as

$$\rho(\alpha,\beta) = 1- \frac{6}{\pi^2} \int_{-\infty}^{+\infty} \log \frac{k(w|\alpha,\beta)}{k_D(w)} \, dw =$$

$$= 1 - \frac{6}{\pi^2} \int_{\alpha}^{\beta} \log \frac{e^{\beta}-1+(1-e^{\alpha})\, e^{\beta-w}}{(e^{\beta}-e^{\alpha})\, \max(1,e^{-\alpha})} \, dw$$

and the difference-sign correlation coefficient is

$$\tau(\alpha,\beta)= 1-\int_{\alpha}^{\beta} D(w|\alpha,\beta)(1-D\,(w(\alpha,\beta))\, dw =$$

$$= 1- \frac{(e^{\beta}-1)(1-e^{\alpha})}{e^{\beta}-e^{\alpha}} = \frac{e^{\beta+\alpha}-2\, e^{\alpha}+1}{e^{\beta}-e^{\alpha}} \; .$$

When dealing with reduced random variables $\bar{w}_i = \bar{y}_i-\bar{x}_i$ as $\alpha\leq\bar{w}_i\leq\beta$ we can take as estimators of $\alpha$ and $\beta$ the statistics $\alpha^* = \min(0, \bar{y}_i-\bar{x}_i)$ and $\beta^* = \max(0, \bar{y}_i-\bar{x}_i)$ .

The general regression is given by

$$\bar{y}(x|\alpha,\beta)=x+\beta- \frac{e^{\beta}-1}{e^{\beta}-e^{\alpha}} \exp\left(\frac{1-e^{\alpha}}{e^{\beta}-e^{\alpha}}\, e^{-x}\right) \times$$

$$\times \int_{\frac{1-e^{\alpha}}{e^{\beta}-e^{\alpha}} e^{-x}}^{\frac{e^{\beta}}{e^{\alpha}} \frac{1-e^{\alpha}}{e^{\beta}-e^{\alpha}} e^{-x}} \frac{e^{-t}}{t} \, dt$$

In case of simmetry $(\alpha=-\beta)$ we have

$$k(w|\alpha,\beta) = \frac{1}{1+e^{w}} \qquad \text{if} \quad w<\alpha$$

$$= \frac{e^{\beta}}{1+e^{\beta}} \qquad \text{if} \qquad \beta \leq w \leq \beta$$

$$= \frac{1}{1+e^{-w}} \qquad \text{if} \qquad \beta \leq w$$

and

$$D(w|\beta) = 0 \qquad \text{if} \qquad w<\beta$$

$$= \frac{1}{1+e^{-w}} \qquad \text{if} \qquad -\beta<w<\beta$$

$$= 1 \qquad \text{if} \qquad \beta<\alpha$$

with jumps of $(1+e^{\beta})^{-1}$ at $-\beta$ and $\beta$.

Also

$$\rho(\beta) = 1- \frac{12}{\pi^2} \int_0^{\beta} \frac{w}{1+e^{w}}\, dw \,,$$

$$\tau(\beta) = \frac{2}{1+e^{\beta}} \,,$$

$$\bar{y}(x|\beta) = x+\beta - \frac{e^{\beta}}{1+e^{\beta}} \exp\left(\frac{e^{-x}}{1+e^{\beta}}\right) \times$$

$$\times \int_{\frac{e^{-x}}{1+e^{\beta}}}^{\frac{e^{2\beta}}{1+e^{\beta}}} \frac{e^{-t}}{t}\, dt$$

and $\beta^{*} = \max_i |w_i|$

## THE TRIEXTREMAL MODEL

Consider 3 independent random variables $X,Y,Z$ with Gumbel distribution and consider the triple $(\bar{X},\bar{Y},\bar{Z})$ defined as $\bar{X}=X, \bar{Y}= \max(X-a, Y-b)$, $\bar{Z} = \max(\bar{Y}-c, Z-d)$. To impose $(\bar{X},\bar{Y})$ to have a biextremal distribution $\Lambda(x,y|\alpha)$ we must have $e^{-a}+e^{-b}=1$ and $e^{-a}=\alpha(0\leq\alpha\leq 1)$; to impose to $(\bar{Y},\bar{Z})$ to have, then, a biextremal distribution $\Lambda(x,y|\beta)$ we must have $e^{-c}+e^{-d}=1$ and $e^{-c}=\beta(0\leq\beta\leq 1)$.

Then the joint distribution of $(\bar{X},\bar{Y},\bar{Z})$ is

$$\Lambda(x,y,z|\alpha,\beta)=\text{Prob}\{\bar{X}\leq x,\bar{Y}\leq y,\bar{Z}\leq z\}=$$
$$=\text{Prob}\{X\leq x, X\leq a+y, Y\leq b+y, X\leq a+c+z, Y\leq b+c+z, Z\leq d+z\}=$$
$$=\exp\{-\max(e^{-x},e^{-a}\,e^{-y},e^{-a}\,e^{-c}\,e^{-z})-\max(e^{-b}\,e^{-y},e^{-b}\,e^{-c}\,e^{-z})$$
$$-e^{-d}\,e^{-z}\}$$
$$=\exp\{-\max(e^{-x},\alpha\max(e^{-y},\beta\,e^{-z}))-(1-\alpha)\max(e^{-y},\beta\,e^{-z})-(1-\beta)e^{-z}\}$$

and we get obviously $\Lambda(+\infty,y,z|\alpha,\beta)=\Lambda(y,z|\beta)$ and $\Lambda(x,y,+\infty)=\Lambda(x,y|\alpha)$ and $\Lambda(x,+\infty,z)=\Lambda(x,z|\alpha\beta)$. $(0\leq\alpha\beta\leq 1)$.

Other models and general results are given in Tiago de Oliveira (1962/63).

## ANNEX: THE QUADRANTS METHOD

Consider a sample of $n$ IID random pairs $\{(x_1,y_1),\ldots,(x_n,y_n)\}$ whose distribution function is $F(x,y|\theta)$, where $\theta$ is an (unique) dependence parameter and the margins $A(x)=F(x,+\infty|\theta)$ and $B(y)=F(+\infty,y|\theta)$ are parameter-free.

Let us denote by $\xi$ and $\eta$ the margin medians ($A(\xi)=1/2$, $B(\eta)=1/2$) by $N_1,N_2,N_3,N_4$ ($N_1+N_2+N_3+N_4=n$) the sample cardinals $N_1=\#(x_i>\xi,y_i>\eta)$, $N_2=\#(x_i\leq\xi,y_i>\eta)$, $N_3=\#(x_i\leq\xi,y_i\leq\eta)$ and $N_4=\#(x_i>\xi,y_i\leq\eta)$ and by $p_1(\theta)$, $p_2(\theta)$, $p_3(\theta)$, $p_4(\theta)$ the probabilities $p_1(\theta)=P(X>\xi,Y>\eta)=p(\theta)$, $p(\theta)=1/2-p(\theta)$, $p_3(\theta)=p(\theta)=F(\xi,\eta|\theta)$ and $p_4(\theta)=1/2-p(\theta)$.

The maximum likelihood estimator of $\theta$ is given by the equation $p(\theta^{**})=\dfrac{N_1+N_3}{2n}$; we will denote by $N=N_1+N_3$.

As $\sqrt{n}\,\dfrac{N/n-2p(\theta)}{\sqrt{2p(\theta)\,(1-2p(\theta)}}$ is asymptotically standard normal we know, by the δ-method - see Tiago de Oliveira (1982)-, that $\sqrt{2n}\,\dfrac{p'(\theta)}{\sqrt{p(\theta)\,(1-2p(\theta))}}\,(\theta^{**}-\theta)$ is also asymptotically standard normal.

When we consider $F(x,y|\theta)=\Lambda(x,y|\theta)$ we have $\xi=\eta=-\log\log 2$ and we get $p(\theta)=\Lambda(\xi,\eta|\theta)=\exp\{(-e^{-\xi}+e^{-\eta})k(\eta-\xi|\theta)\}=$ $=\exp(-2\log 2\times k(0|\theta))=4^{-k(0|\theta)}$. Thus the estimator is given by the equation $p(\theta^{**})=N/2n$ or $k(0|\theta^{**})=\dfrac{\log(2n)-\log N}{\log 4}$.

Note that as $1/2\leq k(0)\leq 1$, from the basic inequality, we see that we must have $n/2\leq N\leq n$, the last inequality being trivial. A test of $\theta=0$ (independence) vs $\theta>0$ is thus given by the rejection region $\sqrt{n}(2N/n-1)\geq\lambda_\alpha$ where $\lambda_\alpha=N^{-1}(1-\alpha)$, $N(.)$ denoting the standard normal distribution function.

We can, evidently, extend the method to use other quantiles in the margins or with estimated parameters in the margins.

This method is one to Gumbel and Mustafi (1967), but with modifications.

## REFERENCES

Geffroy, J. (1958/59), "Contributions à l'Étude des Valeurs Extrêmes". Publ. Inst. Stat. Univ. Paris, vols. 7 and 8, vol. 62.

Gumbel, E.J. and Mustafi, C.K. (1967), "Some Analytical Properties of Bivariate Extremal Distributions". J. Amer. Stat. Assoc., (1968), "Conditional Medians for Bivariate Extremal Distributions". Columbia Univ., Dept. Industrial Engineering, Techn. Report.

Marshall, Albert W. and Olkin, Ingram. (1967), "A Multivariate Extremal Distribution". J. Amer. Stat. Assoc., vol. 62.

Sibuya, M. (1960), "Bivariate Extremal Distributions,I". Ann. Inst. Stat. Math., vol. XI.

Tiago de Oliveira, J. (1958), "Extremal Distributions". Rev. Fac. Ciências Lisboa, 2 ser., A, Mat, vol. VII.

(1962/63), "Structure Theory of Bivariate Extremes; Extensions. Estudos Mat., Estat. e Econometua, vol. VII.
(1964), "Statistical Decision for Bivariate Extremes". Port. Math., vol. 24.
(1968), "Extremal Processes; Definition and Properties". Publ. Inst. Stat. Univ. Paris, vol. XVII.
(1969), "Biextremal Distributions: Statistical Decision". Trab. Estad. y Inv. Oper., vol. XXI, Madrid.
(1974), "Regression in the Non-differentiable Bivariate Extreme Models". J. Amer. Stat. Assoc., vol. 69.
(1975), "Statistical Decision for Extremes". Trab. Estad. y Inv. Oper., vol. XXVI, Madrid.
(1982), "The δ-method for the Obtention of Asymptotic Distributions ".Publ. Inst. Stat. Univ. Paris, vol. XXVI.
(1983), "Statistical Choice for Non-separated Models", to be publ.

# EXTREMES IN DEPENDENT RANDOM SEQUENCES

M.R. Leadbetter

University of North Carolina

This (largely expository) paper concerns the extension of some of the main distributional results of classical extreme value theory to apply to dependent - and in particular stationary - sequences. Central ideas of the classical theory are first discussed along with the broad organization of and motivation for their derivations. It is then shown how weak restrictions on dependence structure allow the theory to be generalized to include stationary (and more general) sequences. Implications for the asymptotic distributions of extreme order statistics and related Poisson convergence theory of high level exceedances are also described.

## 1. INTRODUCTION

Classical extreme value theory is concerned with asymptotic statistical properties of quantities such as the maximum $M_n = \max(\xi_1, \xi_2, \ldots, \xi_n)$ of independent identically distributed (i.i.d.) random variables (r.v.) as n becomes large. The theory has a long history of useful application in a wide variety of situations where the i.i.d. assumptions provide a reasonable basic model.

This paper will be primarily concerned with the relaxation of the independence assumption to see how the classical theory may still be applied under weak restrictions indeed. This obviously provides for a vast increase in potential applications for the theory and, of course, a justification for previous applications of the classical theory in cases where the independence assumptions were only approximately valid. Brief comments only will be made (in Section 6) concerning the effects of changes in distribution - especially due to the addition of seasonal components and trends.



*J. Tiago de Oliveira (ed.), Statistical Extremes and Applications, 155–165.*

The discussion will focus primarily on the central result of the classical theory here called the "Extremal Types Theorem," which limits the possible asymptotic distributions for the maximum $M_n$ to just three essentially different "types," and to results surrounding this theorem. Section 2 contains a discussion of the specific classical results and the central ideas of their proof. In Section 3 it is shown how the Extremal Types Theorem may be extended to apply to dependent situations, with a discussion of domains of attraction, related results and examples in Section 4. Section 5 is concerned with point processes formed e.g. by the instants at which high levels are exceeded by a stationary sequence --giving Poisson and related results in such cases. Finally in Section 6 some very brief comments are made about non-stationary cases and extremes in continuous time--the latter being a topic dealt with in detail in a companion paper by Rootzén [13]. Further details concerning all these topics may be found in the recent volume [9].

## 2. CLASSICAL RESULTS

As noted above, classical extreme value theory is especially concerned with asymptotic distributional and related properties of the maximum $M_n = \max(\xi_1,\ldots,\xi_n)$ as $n \to \infty$ where $\xi_1,\xi_2,\ldots$ are assumed to be i.i.d. If F is the common distribution function (d.f.) of the $\xi_i$, then clearly $M_n$ has the d.f. $F^n(x)$, so that if F is known, so is the d.f. of $M_n$ for any n. It is however obviously useful to have an asymptotic result which is less dependent on the precise form of the d.f. F. Specifically the following theorem was proved in varying degrees of completeness by the early workers (e.g. [3],[4],[5]) showing that $M_n$ may have only three essentially different types of limiting behavior under linear normalizations.

Theorem 2.1 (Extremal Types Theorem)

If $$P\{a_n(M_n - b_n) \le x\} \to G(x) \tag{2.1}$$

for some constants $a_n > 0$, $b_n$ and a non-degenerate d.f. G, then G must be one of the three types:

Type I $\quad G(x) = \exp(-e^{-x}) \qquad -\infty < x < \infty$

Type II $\quad G(x) = \exp(-x^{-\alpha}) \qquad x > 0, \alpha > 0$

Type III $\quad G(x) = \exp(-(-x)^{\alpha}) \qquad x < 0, \alpha > 0$ □

In this, replacement of x by ax + b for any a > 0, b is permitted. More specifically if two d.f.'s $G_1, G_2$ are said to be *of the same type* if $G_1(x) = G_2(ax + b)$ for some a > 0, b, then the theorem says that only three essentially different types may occur as non-degenerate limiting distributions for maxima. (The word "essentially" is used since Types II and III are really families of types with parameter $\alpha$).

As a familiar example, a standard normal sequence has a Type I limit with $a_n = (2\log n)^{\frac{1}{2}}$, $b_n = (2\log n)^{\frac{1}{2}} - \frac{1}{2}(2\log n)^{-\frac{1}{2}}(\log\log n + \log 4\pi)$. Common examples for Types II and III are provided by Cauchy and uniform sequences respectively.

A detailed proof of Theorem 2.1 may be found e.g. in [9], and here we just indicate how it may be broadly organized to bring out the features which allow its extension to dependent cases. First we note a well known basic result of Khintchine regarding convergence in distribution (cf. [2],[9]).

Lemma 2.2 (Khintchine)

If $\{F_n\}$ is a sequence of d.f.'s such that $F_n(a_n^{-1}x + b_n) \xrightarrow{d} G(x)$ for some $a_n > 0$, $b_n$ and non-degenerate d.f. G, whereas $F_n(\alpha_n^{-1}x + \beta_n) \xrightarrow{d} G'(x)$ for some $\alpha_n > 0$, $\beta_n$, non-degenerate G', then G and G' are of the same type; i.e. $G'(x) = G(ax + b)$ for some $a > 0, b$.

We note also a further basic concept of classical extremal theory--namely that of "max-stability." Specifically a non-degenerate d.f. G is called *max-stable* if for each $n = 2,3,\ldots$, $G^n$ is of the same type as G, i.e. if $G^n(a_n x + b_n) = G(x)$ for some $a_n > 0, b_n$.

It is easy to see that if G is a non degenerate limit law for maxima of i.i.d. random variables (i.e. (2.1) holds) then it is max-stable. For (2.1) may be rewritten as $F^n(x/a_n + b_n) \to G(x)$ from which it follows, by replacing n by nk for fixed k and taking kth roots, that $F^n(x/a_{nk} + b_{nk}) \to G^{1/k}(x)$. Identifying $F^n$ with $F_n$ of Lemma 2.1 it then follows that $G^{1/k}$ is of the same type as G so that G and $G^k$ are of the same type for each k, showing max-stability of G. The converse is just as easily proved leading to the following result.

Lemma 2.3

The class of d.f.'s G which are limit laws for maxima of i.i.d. sequences, coincides precisely with the class of max-stable d.f.'s.

This result is one "compartment" of the proof of the classical Extremal Types Theorem. The other part is to identify the max-stable distributions with the class of Type I,II, and III extreme value d.f.'s. It is a very easy exercise to show that the latter are max-stable (e.g. if $G(x) = \exp(-e^{-x})$ then $G^n(x) = G(x - \log n)$). The converse requires proof but is most readily accomplished by using techniques of de Haan involving applications of inverse functions.

This identification of max-stable and extreme value d.f.'s is an analytic exercise and has nothing to do with independence or dependence of the sequence $\{\xi_n\}$ whose maxima are considered. Thus to extend the Extremal Types Theorem to a class of r.v.'s which are not i.i.d. it is only necessary to extend Lemma 2.3 to apply to that class.

The appropriate extension along these lines will be carried out in Section 3. However here we note some other related matters in the i.i.d. case. The first of these concerns domains of attraction. That is if each $\xi_i$ has a d.f. F, which (if any) of the three types applies to $M_n = \max(\xi_1, \xi_2, \ldots, \xi_n)$? Necessary and sufficient conditions are known for each of the three types. These concern the tail behavior of F, i.e. the asymptotic form of $1 - F(x)$ as x increases. For example, if $F(x) < 1$ for all x and if $(1 - F(t))/(1 - F(tx)) \to x^{\alpha}$, as $t \to \infty$, each $x > 0$, then a Type II limit applies. (More details and corresponding criteria for the other types may be found in [6],[9]).

A result which is useful in discussing domains of attraction and for other purposes is the following.

Lemma 2.4

If $\{\xi_n\}$ is an i.i.d. sequence and $0 \le \tau \le \infty$, then with the usual notation $P\{M_n \le u_n\} \to e^{-\tau}$ if and only if $n[1 - F(u_n)] \to \tau$. □

This result is trivially proved (e.g. in one direction by writing $P\{M_n \le u_n\} = F^n(u_n) = [1 - (1 - F(u_n))]^n = (1 - \tau/n + o(1/n)) \to e^{-\tau}$ if $n[1 - F(u_n)] \to \tau$). For sequences which are not i.i.d. it is also important, and true under weak conditions, but the proof, while not difficult, is no longer trivial.

Finally in the i.i.d. context suppose that $\{u_n\}$ is a sequence of constants and let $S_n$ denote the number of $\xi_i$, $1 \le i \le n$, which exceed $u_n$ (i.e. $S_n$ may be called the number of exceedances of the "level" $u_n$). Then $S_n$ is a binomial r.v. with parameters $n, p_n = 1 - F(u_n)$. If $n[1 - F(u_n)] \to \tau$ for some finite $\tau > 0$, it follows that $S_n$ has a Poisson limit, $P\{S_n = r\} \to e^{-\tau}\tau^r/r!$ This result has interesting consequences. For example, if $M_n^{(2)}$ denotes the *second* largest of $\xi_1, \ldots, \xi_n$ then the events $\{M_n^{(2)} \le u_n\}$ and $\{S_n \le 1\}$ are clearly identical. If the maximum $M_n$ has the limiting distribution G, then writing $u_n = x/a_n + b_n$ in (2.1) it follows that $P\{M_n \le u_n\} \to G(x)$. Writing also $e^{-\tau} = G(x)$ it further follows from Lemma 2.4 that $P\{M_n^{(2)} \le u_n\} = P\{S_n \le 1\} \to e^{-\tau} + \tau e^{-\tau}$, which may at once be rewritten as $P\{a_n(M_n^{(2)} - b_n) \le x\} \to G(x)[1 - \log G(x)]$, giving the asymptotic distribution of the second largest value. Asymptotic distributions for the third, fourth, etc. largest values may be similarly obtained.

## 3. EXTREMES OF DEPENDENT SEQUENCES

We show now how the Extremal Types Theorem may be stated in a rather general setting. To keep the discussion notationally simple we use a form which is best suited for (though not restricted to) stationary sequences where the statistical properties of $\{\xi_n\}$ do not change with the "time" parameter n. However stationarity will not be assumed in the general result, which will include "modest" forms of non-stationary behavior. An even more general result may be given to cover greater departures from stationarity though we do not do so here.

If $p, i_1, i_2, \ldots, i_p$ are integers, write $F_{i_1 \ldots i_p}$ for the joint d.f. of the r.v.'s $\xi_{i_1} \ldots \xi_{i_p}$ from the sequence $\{\xi_n\}$ i.e.

$$F_{i_1 \ldots i_p}(x_1, \ldots, x_p) = P(\xi_{i_1} \le x_1, \ldots, \xi_{i_p} \le x_p)$$

and when $x_1 = x_2 \ldots = x_p = u$, $F_{i_1 \ldots i_p}(u) = F_{i_1 \ldots i_p}(u, u \ldots u)$ as a notational convenience.

If the maximum $M_n = \max(\xi_1, \ldots, \xi_n)$ has a non-degenerate asymptotic distribution G i.e. if (2.1) holds, then $P(M_n \le x/a_n + b_n) \to G(x)$, which may be written as $F_{1 \ldots n}(u_n) \to G(x)$ in which $u_n = x/a_n + b_n$. Suppose now that for each integer k

$$F_{1 \ldots n}(u_n) - F^k_{1 \ldots [n/k]}(u_n) \to 0 \text{ as } n \to \infty \tag{3.1}$$

where [n/k] denotes the integer part of n/k. It then follows simply by (3.1) that $F_{1 \ldots [n/k]}(a_n^{-1}x + b_n) = F_{1 \ldots [n/k]}(u_n) \to G^{1/k}(x)$ and hence, replacing n by nk, that

$$F_{1 \ldots n}(a^{-1}_{nk}x + b_{nk}) \to G^{1/k}(x) \text{ as } n \to \infty$$

for each k. Identifying $F_{1 \ldots n}$ with the $F_n$ of Khintchine's result (Lemma 2.2) we see that $G^{1/k}$ is of the same type as G for each k, leading as in the i.i.d. case, to the fact that G is max-stable. Thus we have the following rather general result.

Theorem 3.1

Let $\{\xi_n\}$ be a sequence and $M_n = \max(\xi_1, \ldots, \xi_n)$ satisfying (2.1), i.e. $P\{a_n(M_n - b_n) \le x\} \xrightarrow{d} G(x)$ for some non-degenerate G, $a_n > 0, b_n$. If the d.f.'s $F_{1 \ldots n}$ satisfy (3.1) for each $k = 2, 3, \ldots$ then G is max-stable, and hence one of the extreme value types. □

The condition (3.1) involves the behavior of all of the r.v.'s $\xi_1, \ldots, \xi_n$. In fact, though, it can be shown that (3.1) holds if the dependence of the terms of the sequence decays sufficiently rapidly, i.e. in that terms *far apart* tend to be independent in a precise sense. Although the more general setting may be kept as it is convenient to consider a *stationary* sequence $\{\xi_n\}$. That is it will be assumed that the probabilistic properties of the $\xi_i$ do not change with time in the sense that the joint d.f. $F_{i_1+m,i_2+m\ldots i_p+m}$ of $\xi_{i_1+m}\ldots\xi_{i_p+m}$ is independent of m (and hence the same as $F_{i_1\ldots i_p}$) for any choice of $p, i_1 \ldots i_p$.

The assumption (3.1) may be interpreted as meaning that if the r.v.'s $\xi_1 \ldots, \xi_n$ are divided into k successive groups of approximately n/k each, then the k sub-maxima from the groups have a form of asymptotic independence. It turns out that a rather weak condition will guarantee this property.

To see this it is convenient to use an approach of dependent central limit theory, employed in the extremal context by Loynes [10]. The integers 1,2,..., n are divided into 2k consecutive groups ("intervals") $I_1, I_1^*, I_2, I_2^* \ldots I_k, I_k^*$ all becoming larger as n increases but the $I_j$ being much larger than the $I_j^*$. Specifically except possibly for $j = k$, $I_j, I_j^*$ have size $[n/k] - \ell_n, \ell_n$ respectively, where $\ell_n \to \infty$ but $\ell_n = o(n)$.

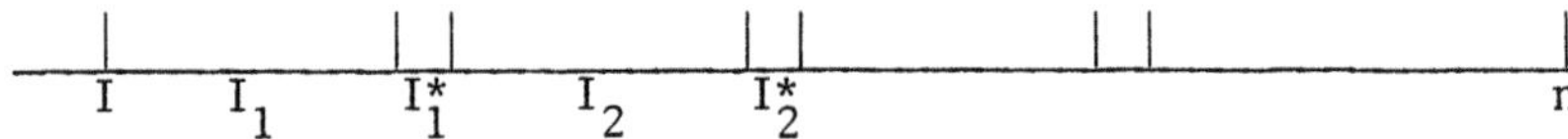


Since the maximum of $\xi_1, \ldots, \xi_n$ is likely to occur on an $I_j$ rather than the smaller $I_j^*$, one may hope to approximate $M_n$ by the maximum on the $I_j$ intervals. However, since the intervals $I_j^*$ are also becoming larger, the $I_j$ are separated by increasingly large distances and the maxima on these intervals may tend to become independent if the dependence in the sequence falls off appropriately. This then leads readily to the approximate independence required for (3.1)

Various dependence restrictions will suffice for this purpose, including "independence," "strong mixing" or for normal sequences the decay rate condition $r_n \log n \to 0$ of Berman [1] for the covariance sequence $\{r_n\}$. A very weak general condition involves the approximate factorization of the joint distribution function of widely separated groups of the $\xi_j$ as follows.

The condition $D(u_n)$ will be said to hold (for a given real sequence $\{u_n\}$) if for any integers $1 \le i_1 < i_2, \ldots, < i_p < j_1, \ldots < j_{p'} \le n$ with $j_1 - i_p \ge \ell$, we have

$$|F_{i_1\ldots i_p j_1\ldots j_{p'}}(u_n) - F_{i_1\ldots i_p}(u_n) \cdot F_{j_1\ldots j_{p'}}(u_n)| < \alpha_{n,\ell}$$

where $\alpha_{n,\ell_n} \to 0$ as $n \to \infty$ for some sequence $\ell_n = o(n)$.

If $D(u_n)$ holds it may be shown that (3.1) holds (and indeed even in a more general form - cf. [9]). Thus the following form of the Extremal Types Theorem holds for the stationary sequences.

Theorem 3.2

Let $\{\xi_n\}$ be a stationary sequence such that $M_n = \max(\xi_1, \ldots, \xi_n)$ has a non-degenerate limiting distribution G as in (2.1) for some constants $a_n > 0, b_n$. Suppose that $D(u_n)$ holds for all sequences given by $u_n = x/a_n + b_n$, $-\infty < x < \infty$. Then G is one of the classical extreme value types. □

It should be noted that if $\{\xi_n\}$ is a stationary *normal* sequence, then $D(u_n)$ holds for *any* sequence $\{u_n\}$ if Berman's covariance condition $r_n \log n \to 0$ holds. Thus in that case $D(u_n)$ is at least as weak as the already weak covariance condition.

## 4. DOMAINS OF ATTRACTION

In this section we shall see that under $D(u_n)$ conditions the introduction of dependence does not alter the limiting distributional type of the maximum in many cases of interest. Indeed in many cases - perhaps most - even the normalizing constants are the same as the classical ones.

We focus again on stationary sequences (though extensions are possible) and write F for $F_1$, the common d.f. of all $\xi_i$. Corresponding to $\tau > 0$ a sequence $u_n(\tau)$ is defined satisfying $n[1 - F(u_n(\tau))] \to \tau$.

Such a $u_n(\tau)$ obviously may be defined (by equality) when F is continuous and in fact may also be defined in all discontinuous cases where F leads to a non-degenerate limiting extreme value d.f. in the classical i.i.d. setting. The following result is basic to the domain of attraction discussion (for proof, see [8],[9]).

Lemma 4.1

Let $\{\xi_n\}$ be a stationary sequence and $u_n(\tau)$ as above and such that $D(u_n(\tau_0))$ holds for some $\tau_0 > 0$ and such that $P\{M_n \le u_n(\tau_0)\}$ converges. Then there exists $\theta$, $0 \le \theta \le 1$ such that $P\{M_n \le u_n(\tau)\} \to e^{-\theta\tau}$ for all $\tau$, $0 \le \tau \le \tau_0$. □

If $P\{M_n \le u_n(\tau)\} \to e^{-\theta\tau}$ for all $\tau > 0$ we say that $\{\xi_n\}$ has *extremal index* $\theta$. The extremal index is useful in comparing maxima of dependent and i.i.d. sequences. Specifically if $\{\xi_n\}$ is a stationary sequence let $\{\hat{\xi}_n\}$ denote the associated *independent* sequence i.e. having the same d.f. F as each $\xi_i$ but being independent. Write $\hat{M}_n = \max(\hat{\xi}_1, \ldots, \hat{\xi}_n)$. Then the following result holds (cf. [8],[9] for proof).

Lemma 4.2

Suppose that the stationary sequence $\{\xi_n\}$ has extremal index $\theta$. Let $\{v_n\}$ be any sequence of constants. Then if $\theta > 0$

$$P\{\hat{M}_n \le v_n\} \to \rho \text{ if and only if } P\{M_n \le v_n\} \to \rho^\theta . \qquad \square$$

This result enables us to give conditions in terms of the extremal index under which $M_n$ has a limiting distribution if and only if $\hat{M}_n$ does and to see how the classical attraction criteria may be used in dependent cases. For example if $P\{a_n(\hat{M}_n - b_n) \le x\} \to G(x)$, non-degenerate and $\theta > 0$, Lemma 4.2 shows (with $v_n = x/a_n + b_n$) that $P\{a_n(M_n - b_n) \le x\} \to G^\theta(x)$ which is easily checked to be of the same type as G for each of the three possible forms of G. This and an equally obvious converse statement lead at once to the following result.

Lemma 4.3

Let the stationary sequence $\{\xi_n\}$ have extremal index $\theta > 0$. Then $M_n$ has a non-degenerate limiting distribution if and only if $\hat{M}_n$ does, and these are of the same type with the same normalizing constants. In the case $\theta = 1$, the limiting distribution of $M_n$ and $\hat{M}_n$ are identical. $\square$

It follows from this result that if $\theta > 0$, the limiting distributions of $M_n$ and $\hat{M}_n$ are of the same type, and indeed if the limiting distribution for $\hat{M}_n$ is G, then that for $M_n$ is $G^\theta$, with the same normalizing constants. Thus the classical criteria for domains of attraction may be used for a dependent sequence of this type. In particular for stationary normal sequences with $r_n \log n \to 0$ the Type I limit still holds with the classical constants.

The above discussion relies on the existence of a (non-zero) extremal index. A simple general criterion for this is given by the conditions of Lemma 4.1 (where $\theta > 0$ provided $\lim P(M_n \le u_n(\tau_0)) < 1$). However sufficient criteria of more practical value may be given in terms of the quantities

$$E_{n,k}^{(r)} = k \sum_{1 \le i_1 < \dots < i_r \le [n/k]} P\{\xi_{i_1} > u_n, \xi_{i_2} > u_n \dots \xi_{i_r} > u_n\}$$

as the following result shows.

Theorem 4.4

Let the stationary sequence $\{\xi_n\}$ satisfy $D(u_n(\tau))$ for each $\tau > 0$.

(i) If $\limsup_{n\to\infty} E_{n,k}^{(2)} \to 0$ as $k\to\infty$, then $\{\xi_n\}$ has extremal index $\theta = 1$.

(ii) More generally if for some $\tau_0 > 0$, $\limsup_{n\to\infty} |E_{n,k}^{(2)} - \tau_0(1-\theta)| \to 0$ as $k\to\infty$ and $\limsup_{n\to\infty} E_{n,k}^{(3)} \to 0$ as $k\to\infty$ then $\{\xi_n\}$ has extremal index $\theta$. □

More general results of this type also hold but are of course more complicated for application.

The case $\theta = 1$ is perhaps the most common - applying obviously to i.i.d. sequences but also to normal sequences satisfying the covariance condition $r_n \log n \to 0$ (which implies (i) of Theorem 4.4). However all values of $\theta$ in $[0,1]$ are possible. For example a simple case where $\theta = \frac{1}{2}$ is given by taking an i.i.d. sequence $\eta_1, \eta_2, \eta_3, \ldots$ and writing $\xi_i = \max(\eta_i, \eta_{i+1})$.

## 5. ASSOCIATED POINT PROCESSES

As noted at the end of Section 2, for an i.i.d. sequence the number $S_n$ of exceedances by $\xi_1 \ldots \xi_n$ of a level $u_n$ satisfying $n[1 - F(u_n)] \to \tau$, has an asymptotic Poisson distribution with mean $\tau$. Indeed it may further be shown that the suitably normalized "time instants" at which exceedances occur, is asymptotically Poisson. More specifically, if $N_n$ denotes the point process on $[0,1]$ consisting of points $j/n$ for which $\xi_j > u_n$, then $N_n \xrightarrow{d} N$ as $n\to\infty$ where $N$ is a Poisson Process on $[0,1]$ with intensity $\tau$.

This result leads in particular to the asymptotic distribution of the second, third (etc.) maxima in i.i.d. sequences. However one may obtain even more general results by simultaneous consideration of exceedances of more than one level - showing that the point processes of exceedances of higher levels are "thinned" versions of those of the lower levels. This yields, in particular, *joint* asymptotic distributions of the various rth largest maxima (cf. [9]).

An alternative and aesthetically appealing approach in the i.i.d. case is to plot, in the plane, the points $(j/n, a_n(\xi_j - b_n))$ where $a_n$ and $b_n$ are the normalizing constants in (2.2) for the maximum. It may then be shown (cf. [12],[14]) that these points - regarded as a point process in (a subset of) the plane converge to a two dimensional Poisson Process. Further the same joint distributional results for rth largest maxima follow as before.

For dependent cases with extremal index $\theta = 1$, generalization of these results are possible, leading to the same asymptotic (joint) distributional forms as for i.i.d. sequences. This, as noted previously, is the most commonly encountered situation. However cases when $\theta < 1$ do have some interest. As may be noticed from the previous discussion the case $\theta < 1$ tends to arise when there is strong local dependence (e.g. between $\xi_j$ and $\xi_{j+1}$) in the sequence. In such cases exceedances of a high level $u_n$ tend to occur in clusters rather than singly. This does not significantly affect the asymptotic distribution of the *maximum*, as we have seen, but does substantially alter the distributions for the rth largest values when $r \geq 2$. The reason for this is simply that when $\theta = 1$ e.g. the second largest occurs at a point "significantly distinct" from the maximum, whereas when $\theta < 1$ they may occur in the same cluster.

If, instead of looking at all exceedances, one considers only the positions of clusters, then it is possible to obtain a Poisson limit again (consistent with the asymptotic distribution of the maximum (cf. [8]). It is also possible as shown recently by T. Hsing and by J. Hüsler to demonstrate that a compound Poisson limit holds (under stronger assumptions) for the exceedances. However, the multilevel results and the two dimensional point process result are much more complicated. (The class of possible point process limits in the latter case has been discussed in general terms by Mori [11] and further representations have been recently developed by T. Hsing).

## 6. FURTHER COMMENTS

As noted while the focus has been on stationary cases, some of the results given can be extended to include important forms of non-stationarity. One such important case is that in which a seasonal component or trend is superimposed on a stationary sequence. This must obviously alter the discussion to some non-trivial extent in that e.g. the maximum is more likely to occur where the added component is high. This is discussed in detail for the case of normal sequences in [9] where it is shown how the normalizing constants must be altered to take account of added components which may be rather general. Related results in a more general (non-normal) setting are given in [7].

## 7. REFERENCES

[1] Berman, S.M. "Limiting distribution of the maximum term in sequences of dependent random variables," 1962, Ann. Math. Statist. 33, pp. 894-908.

[2] Feller, W. *An Introduction to Probability Theory and Its Applications*, Vol. 2, 2nd ed. New York: Wiley, 1971.

[3] Fisher, R.A. and Tippett, L.H.C. "Limiting forms of the frequency distribution of the largest or smallest member of a sample," 1928, Proc. Cambridge Phil. Soc. 24, pp. 180-190.

[4] Fréchet, M. "Sur la loi de probabilité de l'écart maximum," 1927, Ann. Soc. Math. Polon. 6, pp. 93-116.

[5] Gnedenko, B.V. "Sur la distribution limite du terme maximum d'une série aléatoire," 1943, Ann. Math. 44, pp. 423-453.

[6] Haan, L. de. "On regular variation and its application to the weak convergence of sample extremes," 1970, Amsterdam Math. Centre Tracts 32, pp. 1-124.

[7] Hüsler, J. "Asymptotic approximation of crossing probabilities of stationary random sequences," to appear in Z. Wahrsch. verw. Geb.

[8] Leadbetter, M.R. "Extremes and local dependence in stationary sequences," to appear in Z. Wahrsch. verw. Geb.

[9] Leadbetter, M.R., Lindgren, G., and Rootzén, H. *Extremes and Related Properties of Random Sequences and Processes*, New York: Springer-Verlag, 1983.

[10] Loynes, R.M. "Extreme values in uniformly mixing stationary stochastic processes," 1965, Ann. Math. Statist. 36, pp. 993-999.

[11] Mori, T. "Limit distribution of two-dimensional point processes generated by strong-mixing sequences," 1977, Yokohama Math. J. 25, pp. 155-168.

[12] Pickands, J. III. "The two-dimensional Poisson process and extremal processes," 1971, J. Appl. Probab. 8, pp. 745-756.

[13] Rootzén, H. "Extremes in continuous parameter stochastic processes," invited paper, NATO Advanced Study Institute on Statistical Extremes and Applications, Vimeiro, Portugal, 1983. (Companion paper in this volume).

[14] Weissman, I. "Multivariate extremal processes generated by independent non-identically distributed random variables," 1975, J. Appl. Probab. 12, pp. 477-487.

# EXTREMES IN CONTINUOUS STOCHASTIC PROCESSES

Holger Rootzén

University of Copenhagen and UNC, Chapel Hill

The by now well developed extremal theory for stationary sequences, including the Extremal Types Theorem and criteria for convergence and domains of attraction, has interesting extensions to continuous parameter cases. One way of using discrete parameter results to obtain these extensions--this is the approach to be reviewed here--proceeds by the simple device of expressing the maximum over an expanding interval of length $T = n$, say, as the maximum of n "submaxima" over fixed intervals, viz $M(T) = \max\{\zeta_1, \ldots, \zeta_n\}$ where $\zeta_i = \sup\{\xi(t);\ i-1 \le t \le i\}$. The proofs involve three main ingredients, (i) results on the tail of the distribution of one $\zeta$, i.e. of the maximum over a fixed interval, (ii) mixing conditions; which limits the dependence between extremes of widely separated $\zeta$'s and (iii) clustering conditions; which specify the dependence between neighboring $\zeta$'s. Each of (i)-(iii) also involves rather elaborate "discretization" procedures to enable probabilities to be calculated from finite-dimensional distributions. Methods and results for the Gaussian case will be stressed, following the extensive treatment in Part III of Leadbetter, Lindgren and Rootzén: Extremes and related properties of stationary sequences and processes, Springer (1983). Finally, alternative approaches will be briefly commented on.

## 0. INTRODUCTION

In many practical situations, interest is focused on extreme values of processes which have an inherently continuous parameter. To mention just one example, extreme wind- and wave-loads on constructions are of obvious importance and clearly concern processes depending on a continuous time-parameter. The continuous nature



*J. Tiago de Oliveira (ed.), Statistical Extremes and Applications, 167–180.*

of the parameter may increase the size of extremes significantly. In Fig. 1.1 this is illustrated by a series of measurements of extreme temperatures in Uppsala, Sweden. Before 1839 measurements were made only three or four times a day, and sometimes at irregular intervals, so that only a "sampled version" of the temperature process was observed. After 1839 observations of true (continuous) maxima and minima were made possible by installation of a max-min thermometer. Clearly recorded maxima tend to be higher and recorded minima tend to be lower after 1839. The effect is particularly clear for the minimum because it is usually attained very early in the morning, while recordings normally were made later in the day.

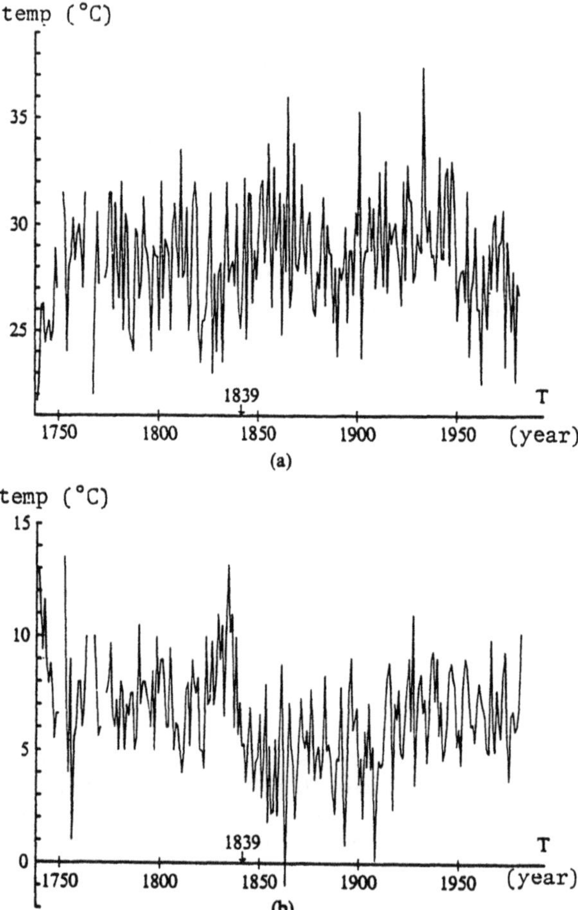


Fig. 1.1. (a) Maximum and (b) minimum recorded temperature in Uppsala, Sweden, during July; note the introduction of a max-min thermometer in 1839. (Data provided by S. Hellstrom.)

This paper contains a survey of some recent results from the extreme value theory for continuous parameter processes, mainly following the exposition in [11] and [12]. In particular we attempt to highlight some of the issues and techniques for continuous parameter processes, as opposed to the discrete parameter case. The reader is assumed to have some acquaintance with the basic results from extreme value theory for discrete parameter processes, as set forth e.g. in the companion paper [9], and we will freely use results and concepts from [9].

Thus, we are concerned with the maximum $M(T) = \sup\{\xi(t);\ 0 \le t \le T\}$ and related quantities, in a continuous parameter stochastic process $\{\xi(t);\ t \ge 0\}$, which throughout will be assumed to have continuous sample paths and to satisfy $P(\xi(t) = x) = 0$, for all x and t. One of the most impressive results is for the case when $\{\xi(t)\}$ is normal, and is due to Pickands (1969a,b), with corrections, alternate proofs and complements being added in [3],[10],[15],[16] and [20]. To state the theorem, for $H_\alpha$ a constant to be discussed further in Section 1, let

$$(0.1)\qquad a_T = (2 \log T)^{\frac{1}{2}},$$

$$b_T = a_T + \{\frac{2-\alpha}{2\alpha} \log \log T + \log(C^{1/\alpha} H_\alpha (2\pi)^{-\frac{1}{2}} 2^{(2-\alpha)/2\alpha})\}/a_T.$$

<u>Theorem 0.1</u>. Let $\{\xi(t);\ t \ge 0\}$ be a stationary normal process, with mean zero and covariance function $r(t) = E(X_s X_{s+t})$ which satisfies

$$(0.2)\qquad r(t) \log t \to 0, \text{ as } t \to \infty,$$

and

$$(0.3)\qquad r(t) = 1 - C|t|^\alpha + o(|t|^\alpha), \text{ as } t \to 0$$

for some $C > 0$ and $\alpha \in (0,2]$.

Then $P(a_T(M(T) - b_T) \le x) \to \exp\{-e^{-x}\}$, as $T \to \infty$, for $a_T, b_T$ given by (0.1). □

Theorem 0.1 closely resembles the limit theorem for maxima of stationary normal sequences (the main difference being that $b_T$ is larger than its discrete parameter counter part) and reinforces the idea that the extreme value theory for dependent sequences should be possible to extend to continuous parameter processes.

One way of doing this proceeds by the simple device of writing $M(T)$ for $T = n$ as

$$(0.4)\qquad M(n) = \max\{\zeta_1, \ldots, \zeta_n\},$$

for

$$(0.5) \qquad \zeta_i = \sup\{\xi(t);\ i-1 \le t \le i\},$$

and leads to a rather satisfactory general theory, including an extremal types theorem and criteria for convergence and domains of attraction. (For technical reasons it is sometimes, also in the normal case, convenient to instead write $M(nh) = \max\{\zeta_1', \ldots, \zeta_n'\}$ for $\zeta_i' = \sup\{\xi(t);\ (i-1)h \le t \le ih\}$, for some fixed $h > 0$. However, to keep the exposition simple, we will not consider this further.) The proofs use three main steps, (i) results on the tail of the distribution of the maximum over fixed intervals, (ii) mixing conditions, which limit the dependence between widely separated $\zeta_i$'s and (iii) clustering conditions, which specify the dependence between neighbouring $\zeta_i$'s. Each of these further use rather elaborate discretization procedures in order to calculate probabilities from finite-dimensional distributions. Here we only discuss a few of the major issues in (i)-(iii), using the normal processes from Theorem 0.1 as prominent examples, and refer the reader to [11] for detailed proofs.

The steps (i)-(iii) are discussed in Sections 1-3, respectively. Section 4 contains comments on more detailed convergence results, concerning point-processes of upcrossings and of local maxima, kth largest local maxima, etc. Finally, Section 5 contains some brief comments on other approaches to extremal results for continuous parameter processes.

## 1. MAXIMA OVER FIXED INTERVALS, DISCRETIZATION

With a slight change of emphasis, the conclusion of Theorem 0.1 can be written as $P(M(T) \le u_T) \to \exp\{-e^{-x}\}$, for $u_T = x/a_T + b_T$. One may then consider $\{u_T\}$ as a family of increasing "levels," and is interested in the asymptotic probability that $\xi(t)$ does not exceed $u_T$ in the (expanding) interval $[0,T]$. This will then also include cases in which $u_T$ is not a function of some parameter $x$, or when $u_T$ depends on $x$ in some more complicated (non-linear) way.

Most of the results here will be couched in terms of a function $\psi(u)$ which in an appropriate way will describe the probability that $\{\xi(t)\}$ exceeds a high level $u$ in a *fixed* interval $[0,h]$, and which then will be used to find the probability that $\{\xi(t)\}$ exceeds $u_T$ in the *expanding* interval $[0,T]$. Thus, often we will assume that

$$(1.1) \qquad P(M(h) > u) \sim h\psi(u) \quad , \text{ as } u \to \infty,$$

for $0 < h \le 2$, but sometimes weaker conditions will suffice. It will also be shown that sometimes $\psi(u)$ can be identified with the mean number of upcrossings of the level u in unit time, $\mu(u)$, in important cases when this is finite. In any case, it is possible to define $\psi(u)$ to be $P(M(h_0) > u)/h_0$ for some fixed $h_0 > 0$, and then attempt to verify (1.1) for other values of h.

The condition (1.1) is satisfied in many cases of interest, but it may require a major effort to prove this. E.g. in Theorem 0.1, the condition (0.3) is sufficient to imply that (1.1) holds, with

$$\psi(u) = C^{1/\alpha} H_\alpha u^{(2-\alpha)/\alpha} (2\pi)^{-\frac{1}{2}} e^{-u^2/2}, \tag{1.2}$$

but the proof of this involves quite intricate computations and constitutes the main part of the proof of the theorem. Perhaps one indication of the complexity of the proof is that it establishes the existence of the constant $H_\alpha$, but does not show how to compute it, and indeed, the value of $H_\alpha$ is only known for the most important cases, $\alpha = 1,2$, with $H_1 = 1$, $H_2 = \pi^{-\frac{1}{2}}$. Further (1.1) has the perhaps less obvious consequence that

$$P(\zeta_1 > u, \zeta_2 > u) = o(\psi(u)), \text{ as } u \to \infty, \tag{1.3}$$

which later will be useful in connection with step (iii) in the proofs. In fact, let $I_1 = [0,1]$, $I_2 = [1,2]$ so that, writing $M(I) = \sup\{\xi(t); t \in I\}$ for arbitrary sets I, we have $\zeta_1 = M(I_1), \zeta_2 = M(I_2)$. By (1.1) and stationarity

$$P(\{M(I_1) > u\} \cup \{M(I_2) > u\}) = P(M(I_1 \cup I_2) > u) \sim 2\psi(u), \text{ as } u \to \infty,$$

$$P(M(I_1) > u) = P(M(I_2) > u) \sim \psi(u), \text{ as } u \to \infty, \text{ and hence}$$

$$P(M(I_1) > u, M(I_2) > u) = P(M(I_1) > u) + P(M(I_2) > u)$$

$$- P(\{M(I_1) > u\} \cup \{M(I_2) > u\}) = o(\psi(u)), \text{ as } u \to \infty.$$

To be able to compute probabilities in terms of finite-dimensional distributions, a condition relating "continuous and discrete maxima" is needed. For this, let $\{q = q(u)\}$ be a family of constants which tend to zero as u tends to infinity. We will then assume that for any fixed $h > 0$, and for $q = q(u) \to 0$ to be specified later, (for normal processes q should be chosen so that $qu^{2/\alpha} \to 0$, sufficiently slowly),

$$\frac{P(M(h) > u, \xi(jq) \le u;\ 0 \le jq \le h)}{\psi(u)} \to 0, \text{ as } u \to \infty, \tag{1.4}$$

and use this to approximate M(h) by $M_q(h) = \max\{\xi(jq);\ 0 \le jq \le h\}$ in an appropriate way.

The function $\psi(u)$ plays an important role in the conditions above, and is used to describe the tail of the distribution of the maximum over a fixed interval. As already noted, $\psi(u)$ can sometimes also be identified with $\mu(u)$, the mean number of upcrossings of u in unit time, so that (1.1) holds for $\psi(u) = \mu(u)$. For normal processes satisfying (0.3) this is so if $\alpha = 2$, while for $0 < \alpha < 2$ the sample paths of $\{\xi(t)\}$ are much less smooth than for $\alpha = 2$, which in particular leads to $\mu(u) = \infty$ for any u. For $\alpha = 2$, $\mu(u)$ is given by Rice's formula,

$$\mu(u) = (C/2)^{\frac{1}{2}}\pi^{-1}e^{-u^2/2} , \tag{1.5}$$

and for nonnormal processes $\mu(u)$ may under weak conditions be computed as

$$\mu(u) = \int_0^\infty zp(u,z)dz, \tag{1.6}$$

where $p(u,z)$ is the joint density function of $\xi(t)$ and $\xi'(t)$. (Of course, (1.6) reduces to (1.5) for normal processes.) The relations (1.5),(1.6) can be proved rather simply by expressing $\mu(u)$ as a limit of $J_q(u) = q^{-1}P(\xi(0) \le u < \xi(q))$, $(q > 0)$, i.e. $\mu(u) = \lim_{q\to 0} J_q(u)$. In the normal case, this holds not only for fixed u, but also when $u\to\infty, q\to 0$ in a suitably coordinated way. Here we will use (1.7) as an assumption:

$$J_q(u)/\mu(u) \to 1, \tag{1.7}$$

and also that

$$P(M(q) > u) = o(\mu(u)), \text{ as } u\to\infty. \tag{1.8}$$

It can be readily proved that (1.4), for $\psi(u) = \mu(u)$, follows from (1.7),(1.8), which in concrete situations (e.g. for normal processes with $\alpha = 2$) may be much easier to check than (1.4).

Just to indicate the flavor of the proof, let $N_u(q)$ be the number of upcrossings of u by $\{\xi(t);\ 0 \le t \le q\}$, and note that then

$$\begin{aligned}
&(q\mu(u))^{-1}P(\xi(0) \le u, \xi(q) \le u, M(q) > u) \\
&= (q\mu(u))^{-1}\{P(\xi(0) \le u, M(q) > u) - P(\xi(0) \le u < \xi(q))\} \\
&\le (q\mu(u))^{-1}\{P(N_u(q) \ge 1) - qJ_q(u)\} \le (q\mu(u))^{-1}\{q\mu(u) - qJ_q(u)\} \\
&= 1 - J_q(u)/\mu(u) \quad \to 0.
\end{aligned}$$

which then together with (1.8) easily leads to (1.4). Further, in "regular" cases (1.1) is often satisfied with $\psi(u) = \mu(u)$, since

$$P(M(h) > u) \ge P(N_u(h) \ge 1),$$

$$P(M(h) > u) \le P(N_u(h) \ge 1) + P(\xi(0) > u)$$
$$\le h\mu(u) + o(\mu(u)),$$

by (1.8), and since if "there asymptotically is only zero or one upcrossings of u in [0,h]" then furthermore $P(N_u(h) \ge 1 \sim h\mu(u)$.

## 2. THE EXTREMAL TYPES THEOREM

To avoid repetition, we will say that the extremal types theorem holds (for the process $\{\xi(t)\}$ and constants $\{a_T > 0\}$ and $\{b_T\}$) if

$$(2.1) \qquad P(a_T(M(T) - b_T) \le x) \to G(x), \text{ as } T \to \infty,$$

with G(x) nondegenerate, implies that G(x) is one of the three types

$$\text{Type I:} \quad G(x) = \exp(-e^{-x}) \quad , -\infty < x < \infty,$$

$$\text{Type II:} \quad G(x) = \exp(-x^{-\alpha}) \quad , x > 0,\ \alpha > 0,$$

$$\text{Type III:} \quad G(x) = \exp(-(-x)^{\alpha}) \quad , x \le 0,\ \alpha > 0.$$

In this, replacement of x by $ax + b$ in the right hand side is permitted for any $a > 0, b$.

A preliminary form of the extremal types theorem is obtained by simply assuming that the sequence $\{\zeta_n\}$ satisfies the mixing condition $D(u_n)$ as stated in [9], Section 3.

Theorem 2.1. Suppose that the sequence $\{\zeta_n\}$ defined by (0.5) satisfies $D(u_n)$ whenever $u_n = x/a_n + b_n$, with $a_n, b_n$ as in (2.1). Then the extremal types theorem holds. □

The proof is trivial; choosing $T = n$ in (2.1) it follows that

$$P(a_n(\max\{\zeta_1, \dots, \zeta_n\} - b_n) \le x) = P(a_n(M(n) - b_n) \le x) \to G(x),$$

and the result is then an immediate consequence of the extremal types theorem for dependent sequences ([9], Theorem 3.3). In particular, Theorem 2.1 shows that the three extreme value distributions are the only possible distributional limits for $a_T(M(T) - b_T)$ if $\{\xi(t)\}$ is strongly mixing (since then $\{\zeta_n\}$ is strongly mixing and hence satisfies $D(u_n)$).

However, it is also desirable to have conditions for the extremal types theorem directly in terms of the finite-dimensional distributions $\{F_{t_1,\dots,t_n}\}$ of $\{\xi(t)\}$. To this end we introduce a continuous parameter mixing condition, to be called $C(u_T)$, analogous to the condition $D(u_n)$ for the discrete parameter case. For brevity, we will write $F_{t_1,\ldots t_n}(u)$ for $F_{t_1,\ldots t_n}(u,\ldots,u)$ and we will let the $t_i$'s be members of a discrete set $\{jq; j=1,2,\ldots\}$. *The condition* $C(u_T)$ *will be said to hold for the process* $\{\xi(t)\}$ *and the family of constants* $\{u_T; T>0\}$ *with respect to the family* $\{q=q(u)=q(u_T)\to 0\}$ *if for any points* $s_1<\ldots<s_p<t_1<\ldots<t_{p'}$ *belonging to* $\{kq; 0\le kq\le T\}$ *and satisfying* $t_1-s_p\ge\tau$ *we have that* $|F_{s_1,\ldots,s_p,t_1,\ldots t_{p'}}(u)-F_{s_1,\ldots,s_p}(u)F_{t_1,\ldots,t_{p'}}(u)|\le\alpha_{T,\tau}$, *where* $\alpha_{T,\lambda T}\to 0$ *as* $t\to\infty$ *for any* $\lambda\in(0,1)$.

It can in particular be shown, using the normal comparison lemma ("Slepian's lemma"), that the normal processes from Theorem 0.1 satisfy $C(u_T)$ if $u_T$ satisfies $T\psi(u_T)\to\tau\ge 0$, with $\psi$ given by (1.2) for $\{q(u)\}$ such that $qu^{2/\alpha}=q(u)u^{2/\alpha}\to 0$ sufficiently slowly. Furthermore these q's can be chosen so that (1.1) and (1.4) hold, and for $\alpha=2$ one may instead of (1.4) verify (1.7),(1.8).

The condition $D(u_n)$ for the sequence $\{\zeta_n\}$ can now be obtained from $C(u_T)$, and leads to the following general continuous version of the extremal types theorem. It should perhaps be pointed out that (1.1) is not assumed to hold in this result.

Theorem 2.2. Suppose there is a function $\psi(u)$ with $P(\xi(0)>u)=o(\psi(u))$ such that $T\psi(u_T)$ is bounded for $u_T=x/a_T+b_T$, for each real x, with $a_T,b_T$ as in (2.1) and that $C(u_T)$ holds, for some family of constants $\{q=q(u)\}$ which satisfies (1.4). Then the extremal types theorem holds for $\{\xi(t)\}$. □

## 3. CRITERIA FOR CONVERGENCE AND DOMAINS OF ATTRACTION

So far we have been concerned with the possible limits of $P(a_T(M(T)-b_T)\le x)$, which was rewritten as $P(M(T)\le u_T)$. We turn now to the question of convergence of $P(M(T)\le u_T)$ for families $u_T$ which are not necessarily linear functions of a parameter x. These results are of interest in their own right, but also since they make it possible to simply modify the classical criteria for domains of attraction to the three limiting distributions, to apply in this continuous parameter context.

The main purpose is to obtain the equivalence of $P(M(1)>u_T)\sim\tau/T$ and $P(M(T)\le u_T)\to e^{-\tau}$ under appropriate conditions. In Section 1 was shown that (1.1) bounds the probability that $\zeta_i$ and $\zeta_{i+1}$ exceed u simultaneously, and $C(u_T)$ insures that $\zeta$'s which are far apart are almost independent. However, we also need a "clustering condition" which limits the probability that $\zeta$'s which are moderately far apart are simultaneously large. The following simple sufficient condition called $C'(u_T)$ is analogous to $D'(u_n)$ for sequences, see [11], p. 58 (and which in turn corresponds to the condition in [9], Theorem 4.2 (i)).

*The condition* $C'(u_T)$ *will be said to hold for the process* $\{\xi(t)\}$ *and the family of constants* $\{u_T;\ T>0\}$, *with respect to the constants* $\{q=q(u_T)\to 0\}$, *if* $\limsup_{T\to\infty}(T/q)\sum_{1<jq<\varepsilon T} P(\xi(0)>u_T,\ \xi(jq)>u_T)\to 0$ *as* $\varepsilon\to 0$, *for some* $h>0$.

For normal processes, $C'(u_T)$ is easily shown to hold under the conditions of Theorem 0.1, e.g. using the normal comparison lemma in a similar, but simpler way as for $C(u_T)$).

The relations between $C'(u_T)$ for $\{\xi(t)\}$ and $D'(u_n)$ for $\{\zeta_n\}$ are not quite as straightforward as between $C(u_T)$ and $D(u_T)$. Nevertheless, using similar calculations as in the discrete case (rather than proving D,D' for $\{\zeta_n\}$ and then using the discrete result) it is possible to prove the following theorem.

Theorem 3.1. Suppose that (1.1) holds for some function $\psi$, and let $\{u_T\}$ be a family of constants such that $C(u_T)$, $C'(u_T)$ hold with respect to the family $\{q(u)\}$ of constants satisfying (1.4). Then

$$T\psi(u_T)\to\tau>0 \tag{3.1}$$

if and only if

$$P(M(T)\le u_T)\to e^{-\tau}. \tag{3.2}$$

□

As indicated above, all the conditions of Theorem 3.1 are implied by the conditions of Theorem 0.1, if $u_T$ is chosen as in (3.1), with $\psi$ given by (1.2). Hence (3.2) holds for normal processes which satisfy (0.2) and (0.3). It is easily checked that if $u_T=x/a_T+b_T$, for $a_T,b_T$ given by (0.1), then (3.1) holds with $\tau=e^{-x}$, and hence Theorem 0.1 follows from Theorem 3.1.

By virtue of (1.1) one may replace (3.1) in the theorem by

$$TP(M(1)>u_T)\to\tau>0, \tag{3.3}$$

in complete analogy with the sequence result for $\{\zeta_n\}$ since $M(1)=\zeta_1$ (cf. [9], Theorem 4.2 (i)). In particular, one should note that it is the distribution of $M(1)=\zeta_1$ which is important, rather than the marginal d.f. of $\{\xi(t)\}$ as perhaps might be naively expected. However, as noted in Section 1, it may sometimes also be convenient to have a different interpretation of the function $\psi(u)$ than through (3.3), e.g. as $\psi(u)=\mu(u)$, the mean number of upcrossings in unit time.

Next, we will say that any independent identically distributed sequence $\hat\zeta_1,\hat\zeta_2,\ldots$ whose marginal d.f. F satisfies $1-F(u)\sim P(M(1)>u)$, as $u\to\infty$, is an *independent sequence associa-*

*ted with* $\{\xi(t)\}$. If (1.1) holds, this is clearly equivalent to

(3.4) $\qquad 1-F(u) \sim \psi(u)$, as $u \to \infty$.

Theorem 3.1 can now be applied to show how the function $\psi$ may be used in the classical criteria for domains of attraction, to determine the asymptotic distribution (if any) of $M(T)$. We write $\mathcal{D}(G)$ for the domain of attraction to the (extreme value) d.f. G, i.e. for the set of all d.f.'s F such that $F^n(x/a_n' + b_n') \to G(x)$, for some sequences $\{a_n' > 0\}$, $\{b_n'\}$.

<u>Theorem 3.2.</u> Suppose that the conditions of Theorem 3.1 hold for all families $\{u_T\}$ of the form $u_T = x/a_T + b_T$, where $a_T > 0$ and $b_T$ are given constants, and that

(3.5) $\qquad P(a_T(M(T) - b_T) \le x) \to G(x)$.

Then (3.4) holds for some $F \in \mathcal{D}(G)$. Conversely, suppose (1.1) holds and (3.4) is satisfied for some $F \in \mathcal{D}(G)$, let $a_n' > 0, b_n'$ be constants such that $F^n(x/a_n' + b_n') \to G(x)$, and define $a_T = a_{[T]}', b_T = b_{[T]}'$. Then (3.5) holds, provided the conditions of Theorem 3.1 are satisfied for each family $\{u_T = x/a_T + b_T\}$. □

## 4. POINT PROCESSES OF UPCROSSINGS AND LOCAL MAXIMA

Corresponding to the Poisson limit theorem for exceedances by a stationary sequence, one expects that under suitable conditions the number of upcrossings of high levels will be Poisson in character. Since upcrossings of high levels tend to become rare, a normalization is needed in this. In the case of a finite expected number $\mu(u)$ of upcrossings in unit time, it is natural to choose the normalization so that the expected number of upcrossings in $[0,T]$ is asymptotically constant, i.e. to let $u = u_T$ and T tend to infinity in such a way that $T\mu(u_T) \to \tau > 0$, as $T \to \infty$. A normalized point process, $N_t^*$, of upcrossings of $u_T$ can then be defined as $N_T^*(I) = \#\{$upcrossings of $u_T$ by $\xi(t)$ for $t/T \in I\}$. It is then relatively easy to show that if the conditions of Theorem 3.1 are satisfied, with $\psi(u) = \mu(u)$, then $N_T^*$ converges in distribution to a Poisson process with intensity $\tau$. Thus, this in particular holds for the normal processes from Theorem 0.1, for $\alpha = 2$.

Next, suppose that $\xi(t)$ has a continuous derivative a.s., and define $N_u(t)$ to be the number of upcrossings of u by $\{\xi(t);\ 0\le t\le T\}$ for which the process value exceeds u. Clearly $N_u'(T) \ge N_u(T) - 1$, since at least one local maximum above u occurs between two upcrossings of u. It is also reasonable to expect that if the sample functions are not too irregular, there will tend to be just one local naximum above u between successive upcrossings of u, when u is large, so that $N_u'(T)$ and $N_u(T)$ will tend to be approximately equal. This is in fact so under appropriate further conditions (involving the fourth spectral moment in the normal case). In particular this leads to the asymptotic distribution of $M_n^{(k)}(T)$, the kth largest local maximum in $[0,T]$.

Clearly this can be generalized to establish convergence of the point process of local maxima of heights above u, and this can be further generalized to consider several levels simultaneously, and indeed to "complete Poisson convergence" of the point process in the plane consisting of heights and locations of local maxima, ([11], Section 9.5).

Furthermore, it is also possible to obtain Poisson limits in cases with irregular sample paths for which $\mu(u) = \infty$, (as for normal processes with $0<\alpha<2$) by the simple device of using so-called $\varepsilon$-upcrossings and $\varepsilon$-maxima instead of ordinary upcrossings and local maxima. (In fact, Pickands' original proof of Theorem 0.1 used the notion of $\varepsilon$-upcrossings.) Here, given an $\varepsilon > 0$, the function $f(t)$ is said to have an $\varepsilon$-upcrossing of the level u at $t_0$ if $f(t) \leq u$ for $t \in (t_0 - \varepsilon, t_0)$ and if for each $\eta > 0$, $f(t) > u$ for some $t \in (t_0, t_0 + \eta)$, so that clearly the number of $\varepsilon$-upcrossings in $[0,T]$ is bounded by $T/\varepsilon$. $\varepsilon$-maxima are defined similarly, by excluding local maxima which are less than a distance $\varepsilon$ apart.

## 5. OTHER APPROACHES TO EXTREME VALUE THEORY FOR CONTINUOUS PARAMETER PROCESSES

In this section we review, without attempt at completeness, some of the literature on extreme values for continuous parameter stationary processes. A general approach which is rather different than the one discussed above is taken by Berman in a series of papers [4],[5],[6]. In this interest is centered on the "sojourn of $\xi(t)$ above u in the interval $[0,T]$" defined as the integral $L_T(u) = \int_0^T I\{\xi(t) > u\}dt$, where $I\{\cdot\}$ denotes the indicator function, which is one if the event within curly brackets occurs, and zero otherwise. Using suitable conditions, Berman finds the asymptotic behavior of the distribution of $L_T(u)$ as $u \to \infty$, for fixed T, and a compound Poisson limit theorem for $L_T(u)$ when u and T both tend to infinity, in a coordinated way. Under additional conditions, which include that the marginal d.f. of $\{\xi(t)\}$ belongs to the Type I domain of attraction, he moreover uses the equivalence of the events $\{M(T) > u\}$ and $\{L_T(u) > 0\}$ to find the tail of the distribution of $M(T)$, for fixed T, and a double exponential limit for $M(T)$ as $T \to \infty$. In the proofs an important role is played by the identity $\int_x^\infty P(L_T(u) > y)dy = \int_0^T P(L_T(u) > x, \xi(0) > u)dt$, which can be shown to hold for stationary processes, by a relatively straightforward use of Fubini's theorem. Berman also applies his general results to special classes of processes, in particular obtaining the asymptotic behavior of the distribution of the maximum and of the time spent over u in a fixed interval for wide classes of Gaussian processes, of random Fourier sums, of Markov processes and of so-called $\chi^2$-processes (to be discussed below). (In a sense these results are complementary to the ones in this paper, since they can be taken as specifying the function $\psi(u)$ in (1.1)). Finally, in [1],[7],[8], Berman also studies the extremal behavior of diffusion processes, partly using special methods involving various rescalings.

Let $\{\xi_1(t)\},\ldots,\{\xi_k(t)\}$ be independent continuous stationary normal processes. In analogy with the ordinary $\chi^2$-distribution, the process $\xi(t) = \sum_{i=1}^k \xi_i(t)^2$ can be called a $\chi^2$-process with k degrees of freedom. Using methods which are closely related to the approach in this paper, Sharpe (1978) and Lindgren (1980a,b) among other results obtain a limiting Type I distribution for such processes. In this the use of the mean number of upcrossings and Rice's formula is particularly convenient since the normal variables $\xi_i(t)$ are independent of the derivatives $\xi_i'(t)$, and hence the joint distribution of $\xi(t)$ and $\xi'(t) = \sum_{i=1}^k 2\xi_i(t)\xi_i'(t)$ is amenable to analysis. Lindgren further obtains limit results for $\xi(t) = g(\xi_1(t),\ldots,\xi_k(t))$ for more general functions g than $g(x_1,\ldots, x_k) = \sum_1^k x_i^2$. In this the results for the $\chi^2$-process plays a central role.

Mittal and Ylvisaker (1975) consider stationary normal processes with strong dependence, which do not satisfy (0.3), but where instead $r(t) \log t \to \gamma$ with $0<\gamma\leq\infty$. For a class of such processes they show that the limiting distribution of the maximum is a convolution of a Type I distribution and a normal distribution. The main feature of their proofs is a comparison with stationary normal processes with *constant* correlation, via the normal comparison lemma. Of course, these strongly dependent normal processes do not satisfy $C(u_T)$.

Finally the extremal behavior, including the asymptotic form of sample paths near extremes, of moving averages $\xi(t) = \int a(\lambda - t)dZ(\lambda)$ of independent increments stable processes $\{Z(\lambda)\}$ are studied in Rootzén (1978). In this situation, extreme values of $\xi(t)$ are caused by one very large jump in $\{Z(\lambda)\}$, and as a consequence the limiting distribution of maxima are of Type II, rather than Type I, and $C(u_T)$ holds, but not $C'(u_T)$.

References

[1] Berman, S.M. (1964). Limiting distribution of the maximum of a diffusion process. Ann. Math. Statist. 35, 319-329.

[2] Berman, S.M. (1971a). Excursions above high levels for stationary Gaussian processes. Pacific J. Math. 36, 63-79.

[3] Berman, S.M. (1971b). Maxima and high level excursions of stationary Gaussian processes. Trans. Amer. Math. Soc. 160, 65-85.

[4] Berman, S.M. (1974). Sojourns and extremes of Gaussian processes. Ann. Probab. 2, 999-1026 (Correction (1980), 8, 999).

[5] Berman, S.M. (1980). A compound Poisson limit for stationary sums, and sojourns of Gaussian processes. Ann. Probab. 8, 511-538.

[6] Berman, S.M. (1982a). Sojourns and extremes of stationary processes. Ann. Probab. 10, 1-46.

[7] Berman, S.M. (1982). Sojourns and extremes of a diffusion process on a long interval. Adv. Appl. Probab. 14, 811-832.

[8] Berman, S.M. (1983a). High level sojourns of a diffusion process on a fixed interval. Z. Wahrsch. verw. Geb. 62, 185-199.

[9] Leadbetter, M.R. (1984). Extremes in dependent random sequences. Invited paper, NATO Advanced Study Institute on Statistical Extremes and Applications, Vimeíro, Portugal.

[10] Leadbetter, M.R., Lindgren, G. and Rootzén, H. (1978) Conditions for the convergence in distribution of maxima of stationary normal processes. Stochastic Process. Appl. 8, 131-139.

[11] Leadbetter, M.R., Lindgren, G. and Rootzén, H. (1983). *Extremes and related properties of random sequences and processes*. New York: Springer.

[12] Leadbetter, M.R. and Rootzén, H. (1982). Extreme value theory for continuous parameter stationary processes. Z. Wahrsch. verw. Geb. 60, 1-20.

[13] Lindgren, G. (1980a). Point processes of exits by bivariate Gaussian processes and extremal theory for the $\chi^2$-process and its concomitants. J. Multivariate Anal. 10, 181-206.

[14] Lindgren, G. (1980b). Extreme values and crossings for the $\chi^2$-process and other functions of multidimensional Gaussian processes, with reliability applications. Adv. Appl. Probab. 12, 746-774.

[15] Lindgren, G., de Maré, J. and Rootzén, H. (1975). Weak

convergence of high level crossings and maxima for one or more Gaussian processes. Ann. Probab. 3, 961-978.

[16] Mittal, Y. (1979). A new mixing condition for stationary Gaussian processes. Ann. Probab. 7, 724-730.

[17] Mittal, Y. and Ylvisaker, D. (1975). Limit distributions for the maxima of stationary Gaussian processes. Stochastic Process. Appl. 3, 1-18.

[18] Pickands, J. III (1969a). Upcrossing probabilities for stationary Gaussian processes. Trans. Amer. Math. Soc. 145, 51-73.

[19] Pickands, J. III (1969b). Asymptotic properties of the maximum in a stationary Gaussian process. Trans. Amer. Math. Soc. 145, 75-86.

[20] Qualls, C. and Watanabe, H. (1972). Asymptotic properties of Gaussian processes. Ann. Math. Statist. 43, 580-596.

[21] Rootzén, H. (1978). Extremes of moving averages of stable processes. Ann. Probab. 6, 847-869.

[22] Sharpe, K. (1978). Some properties of the crossings process generated by a stationary $\chi^2$-process. Adv. Appl. Probab. 10, 373-391.

COMPARISON TECHNIQUE FOR HIGHLY DEPENDENT STATIONARY GAUSSIAN PROCESSES

Yashaswini Mittal

Department of Statistics
Virginia Polytechnic Institute and State University
Blacksburg, VA 24061

Abstract: Extreme value theory for stationary Gaussian processes crucially depends on comparison techniques. Slepian's inequality and Berman's Lemma are two well known comparison techniques that have been used in the independent and midly dependent Gaussian processes. These techniques need extensions to deal with strongly dependent cases. A method (called third comparison technique (TCT) here) was introduced by Mittal and Ylvisaker in 1975. It is explained and illustrated here with examples of various results in the literature dealing with strongly dependent stationary Gaussian processes.

## 1. Introduction

The beginning of classical extreme value theory is generally taken to be Gnedenko's paper [3] eventhough there had been earlier work on the problem including that of Fisher and Tippett [2]. The particular case of Stationary Gaussian processes began its momentum with the pioneering work of Berman. If $\{X_n, n \geq 0\}$ is a stationary Gaussian sequence with $EX_n = 0$, $EX_n^2 = 1$ and $EX_nX_0 = r_n$, $n \geq 0$, then Berman [1] showed that $r_n \ell n\, n \to 0$ is a sufficeint condition for

$$\ell t_{n\to\infty}\ p\{c_n(M_n - b_n) \leq x\} = \exp(-\exp(-x)) . \quad (1.1)$$

for all x where $M_n = \max_{1\leq i\leq n} X_i$ $c_n = (2\ell nn)^{1/2}$ and $b_n = c_n - \ell n(4\pi\ell nn)/(2c_n)$. In the following we review the paper of Mittal and Ylvisaker [12] which gave results of the type (1.1) for highly



J. Tiago de Oliveira (ed.), Statistical Extremes and Applications, 181–195.

dependent stationary Gaussian sequences. The exact results are stated in Section 3.

The most powerful tool in the stationary Gaussian Extrem value Theory (SGEVT) has been comparison techniques. Berman [1] invented a comparison technique (called Berman's Lemma) for his results (1.1). Berman's Limma is an easy extention of Slepian's comparison technique [15] (called Slepian's inequality). In Section 2 we will review these techniques and point out the difficulties of using them for highly dependent sequences. (All processes and sequences in this paper are taken to be Stationary Gaussian real valued with mean zero.) This section also explains the extensions (henceforth called the third comparison technique (TCT)) developed in [1]) to overcome these difficulities.

In Section 3 we give the statement and the proof of the main result in [12]. Section 4 gives examples of the use of TCT.

## Section 2: Comparison Techniques

Since the finite dimensional distributions of a zero mean stationary Gaussian process (SGP) is determined by its covariance function, it seems intuitively reasonable to think that smaller covariances produce more erratic behavior of the process and hence fatter tails for the distribution of $M_T = \max_{0\leq t\leq T} X_t$ and higher covariances "bind" the process together and thus produce smaller values of $M_T$. Slepian [15] first gave the result

$$P\{X(t) \leq 0;\ 0 \leq t \leq T\} \geq P\{Y(t) \leq 0;\ 0 \leq t \leq t\} \tag{2.1}$$

where both $\{X(t)\}$ and $\{Y(t)\}$ are SGP with mean zero variance one and covariance functions $\{r(t)\}$ and $\{\gamma(t)\}$ respectively with $r(t) \geq \gamma(t)$ for $0 \leq |t| \leq T$. Slepian's results are based on rather easy computation of partial derivatives of multivariate normal distribution functions. Eventhough Slepian derived his results independently, the mathematical results have been given by Plackett [14] before him. Both Slepian and Plackett's papers depend on the work of Schlafli which dates back to 1858. The profound effect created by Slepian's paper (as evidenced by Berman's paper, Lemma and the subsequent rapid development of SGEVT) is due largely to the point of view of [15] rather than the actual mathematical results which are simple enough. While as Plackett was concerned with the reduction formulae for the multivariate normal integrals, Slepian was obviously interested in the use of the technique for computation of bounds of the type indicated in (2.1). There exists some controversy regarding "who invented "Slepian's inequality"?" (Obviously due to the earlier date of Placketts's paper). The above statements are no attempt to either stir or settle it but a modest explanation for referring to (2.1) and its derivates as "Slepian's inequality". Fortunately no such controversy surrounds Berman's Lemma eventhough partial derivatives very similar to the ones considered by Berman are used in Plackett's paper (and probably many other places as well).

Let $\Phi_n(u,P)$ be $P(X_1 \leq u,\ldots,X_n \leq u) = P(M_n(P) \leq u)$ where $X_1,\ldots,X_n$ are multivariate normal with mean zero and covariance $p_{ij} = EX_iX_j$, $P = ((p_{ij}))$. It is easy to see by using the partial derivatives of $\Phi_n(u, )$ w.r.t. $\lambda_{ij}$ that $|\Phi_n(u,P) - \Phi_n(u,Q)|$ is a function of $|p_{ij}-q_{ij}|$ $(Q = ((q_{ij})))$ and some multivariate normal probabilities. Finding an upper bound for these probabilities Berman gets

<u>Lemma 2.1</u> (Berman)

$$\begin{aligned} |P(M_n(R)\leq u)-P(M_n(S)\leq u)| &= |P(M_n(R)>u)-P(M_n(S)>u| \\ &\leq \sum_{j=1}^{n-1} (n-j)|r_i-s_j|\phi_2(u,u,w_j) \end{aligned} \tag{2.2}$$

where we take $R = \{r_i\}$, $S = \{s_j\}$ to be stationary covariance sequences and

$$\phi_2(u,u,r) = (2\pi)^{-1}(1-r^2)^{-\frac{1}{2}}\exp\{-u^2/(1+r)\} \tag{2.3}$$

and $w_j = \max(|r_j|,|s_j|)$. Taking $s_j = 0 \ \forall j \neq 0$, we have the R.H.S. of (2.2) as

$$\sum_{j=1}^{n-1} (n-j)|r_j|\phi_2(u,u,|r_j|) \tag{2.4}$$

Result (1.1) is true for $M_n = M_n^*$ = maximum of i.i.d. N(0,1) variables and now we have a comparison technique which gives

$$\begin{aligned} |P(M_n\leq u_n(x)) - P(M_n^*\leq u_n(x))| &\leq n^{1+\alpha}\delta(1)\ \phi_2(u_n(x),u_n(x),\delta(1)) \\ &+ n^2\delta(n^\alpha)\phi_2(u_n(x),u_n(x),\delta(n^\alpha)) \end{aligned} \tag{2.5}$$

for $0<\alpha<1$; $\delta(i) = \max_{j\geq i}|r_j|$ and $u_n(x) = b_n+xc_n^{-1}$ with $b_n$ and $c_n$ as in (1.1). By substituting the values of $u_n(x)$ and choosing $\alpha < (1-\delta(1))/(1+\delta(1))$ we see that R.H.S. of (2.5) tends to zero as $n \to \infty$ if $r_n \ell nn \to 0$ (and hence $\delta(n^\alpha)\ell nn \to 0$).

Two major difficulties arise in trying to find the behavior of $M_n$ for a strongly dependent process. (1) What is the "right" normalizing expression $u_n(x)$ and the "right" S to which we want to compare $P(M_n \le u_n(x))$? and (2) How can we make the R.H.S. of (2.2) "less" dependent on $w_i = \max(r_j, s_j)$ which makes comparisons of "big" R and S very difficult if not impossible.

Solution to the first difficulty arises by coming to a general understanding of "how a SGP becomes dependent?" One way to creat a group of dependent Gaussian variables from i.i.d. ones is by "binding" them by an additional independent N(0,1) variable. That is

$$(1-\rho)^{\frac{1}{2}}Z_i + \rho^{\frac{1}{2}}U \quad i = 1,2,\dots,n \tag{2.6}$$

has a multivariate normal distribution with constant correlation $\rho$ when $Z_i$ are i.i.d. N(0,1). A rather startling discovery is to find (vaguely) that "almost all" dependent SGP are created in the above fashion viz for every SGP $\{X_n\}$ there exists some $\rho_n$ such that the behavior of $X_1,\dots,X_n$ closely resembles that of (2.4) with $\rho = \rho_n$ for all n. In case $r_n (= EX_nX_0)$ is monotone non-increasing, we can take $\rho_n = r_n$. Eventhough the above statement is false in the complete generality in which it is started, it is true enough that many aspects of the behavior of $M_n$ were discovered that were unknown before adoption of this point of view. Let us now take $\{r_n\}$ montone and believe that $\{X_1,\dots,X_n\}$

behaves like $(1-r_n)^{\frac{1}{2}}Z_i + r_n^{\frac{1}{2}}U$ $1 \le i \le n$ for all n, then $M_n$ can be approximated by

$$(1-r_n)^{\frac{1}{2}}M_n^* + r_n^{\frac{1}{2}}U \tag{2.7}$$

where $M_n^* = \max_{1\le i\le n} Z_i$. Normalizing $M_n^*$ by constants $\alpha_n^{-1}$ and $\beta_n$ we see that

$$(1-r_n)^{\frac{1}{2}}\alpha_n(M_n^* - \beta_n) + r_n^{\frac{1}{2}}\alpha_n U \sim \alpha_n(M_n - (1-r_n)^{\frac{1}{2}}\beta_n) \tag{2.8}$$

Thus we see that if $b_n (=\beta_n)$ is the right location constant for $M_n^*$ then $(1-r_n)^{\frac{1}{2}}b_n$ is the right one for dependent sequences ($b_n$ and $(1-r_n)^{\frac{1}{2}}b_n$ are of course equivalent if $r_n \ell nn \to 0$). The rôle of the scaling constants is also apparent. We know that for $M_n^*$ we have to take $\alpha_n = c_n = (2\ell nn)^{\frac{1}{2}}$. If $r_n^{\frac{1}{2}}c_n \to 0$, the limiting distribution of $c_n(M_n - (1-r_n)^{\frac{1}{2}}b_n)$ is the classical extreme value distribution (EVD) of $c_n(M_n^* - b_n)$, if $r_n^{\frac{1}{2}}c_n \to \gamma > 0$, it should be the convolution of EVD and a normal distribution and a change in the scaling constant is warranted if $r_n^{\frac{1}{2}}c_n \to \infty$. In such a case, the obvious choice for $\alpha_n$ is $r_n^{-1/2}$. Thus in case $r_n^{1/2}c_n \to \infty$,

$$r_n^{-1/2}(M_n - (1-r_n)^{1/2}b_n) \sim (1-r_n)^{1/2}(r_nc_n)^{-1/2}(c_n(M_n^* - b_n)) + U \tag{2.9}$$

and shows that the limiting distribution should be normal. We will look at more examples of this method in Section 4.

The second difficulty in comparing dependent sequences is much more technical and the solution is more or less algebraic in nature eventhough the actual computations are rather complex. Now we know that the right normalizing sequence is $u_n'(x) = (1-r_n)^{1/2}b_n + r_n^{1/2}x$ and the right S for every stage n is $s_j = r_n$, $1 \le j \le n$. If we tried to use (2.2) without modification, we have the R.H.S. equal to (for monotone $\{r_i\}$)

$$\sum_{j=1}^{n-1} (n-j)\phi_2(u_n'(x),u_n'(x),r_j) \tag{2.10}$$

Notice that

$$-(u_n'(X))^2/(1+r_j) \sim -(1-r_j)2\ell nn + (1-r_j)\ell n\ell nn + (1-r_j)r_nb_n^2 + o(1)$$

and the R.H.S. of (2.10) is at least

$$(\text{constant})\ r_nn^2\{\exp -2\ell nn + \ell n\ell nn + 2r_nb_n^2(1 + o(1))\} \to \infty \tag{2.11}$$

as $n \to \infty$ if $r_n\ell nn \to \infty$. Thus (2.2) by itself is inadequate for comparison.

Going back to the representation (2.9), we first "subtract off" $r_n$ from $\{r_i, 1 \le i \le n\}$ to get the first order terms of the convolution, and express the error in terms of probabilities of $M_n(R')$ at suitable levels $u_n''$, where $R' = ((r_i-r_n)/(1-r_n))$. Suitable bounds for the error then are found by one more "subtraction" and judicious use of (2.2).

The "subtraction" poses no problems if $\{r_n, n \ge 0\}$ is taken to be convex since $\{(r_i-r_n)/(1-r_n), 1 \le i \le n\}$ is then positive definite and can serve as a covariance function. In case $\{r_n\}$ is taken to be only monotone, one first has to create a suitable upper bound for $\{r_i, 1 \le i \le n\}$ from which successive "subtractions" are possible. Actual such procedure is long and tedious and hence was not included in [12]. Details however, can be found in McCormick and Mittal [8]. For even weaker mixing conditions see Mittal [11] and Leadbetter, Lindgren and Rootzen [4].

## Section 3: Details of TCT

As a first example of the use of TCT, we state the Theorem of Mittal and Ylvisaker [12] and prove part B of it.

**Theorem 3.1:** Let $\{X_n\}$ be a SGP with mean zero variance one and $r_n = EX_0X_n$. Let $M_n = \max_{1\le i\le n} X_i$ with $b_n$ and $c_n$ as in (1.1).

(A) If $r_n \ell nn \to \gamma > 0$ as $n \to \infty$

$$\ell t_{n\to\infty} \; P[M_n \le b_n + xc_n^{-1}] = \int_{-\infty}^{\infty} \exp[-\exp[-x+\gamma-(2\gamma)^{1/2}y)]]\phi(y)dy \quad (3.1)$$

where $\phi(y) = (2\pi)^{-1}\exp(-y^2/2)$.

(B) If (i) $r_n$ is convex for $n \ge 0$
(ii) $r^n \to 0$ as $n \to \infty$
and (iii) $r_n^n \ell nn \uparrow$ as $n \to \infty$

then

$$\ell t_{n\to\infty} \; P(r_n^{1/2}(M_n - (1-r_n)^{1/2}b_n) \le x) = \Phi(x) \quad (3.2)$$

where $\Phi(x) = \int_{-\infty}^{x} \phi(y)dy$ .

**Proof of part B: Part I: The first subtraction:**

Because $\{r_n, n \ge 0\}$ is convex, there exist Gaussian r.v.s. $\{X_1', 1 \le i \le n\}^n$ with mean zero and covariance $EX_i'X_j' = (r(i-j)-r_n)/(1-r_n)$ for each n. Write for fixed n

$$M_n = (1-r_n) \max_{1\le i\le n} X_i' + r_n^{1/2}U \quad (3.3)$$

where U is N(0,1) and independent of $X_i'$. Thus

$$\begin{aligned} P(M_n \le (1-r_n)^{1/2}b_n + r_n^{1/2}) &= P((1-r_n)^{1/2}M_n' + r_n^{1/2}U \le (1-r_n)^{1/2}b_n + r_n^{1/2}x \\ &= \int_{-\infty}^{\infty} P(M_n' \le b_n + (r_n/(1-r_n))^{1/2}(x-u))\phi(u)du \, . \end{aligned} \quad (3.4)$$

where $M_n' = \max_{1\le i\le n} X_i'$. We notice that the expression

$$P(M_n' \le b_n + (r_n/(1-r_n))^{1/2}(x-u)) \quad (3.5)$$

in monotone decreasing (in u) from one to zero. Thus the R.H.S. of (3.4) is at least

$$\int_{-\infty}^{x-\varepsilon} P(M_n' \le b_n + r_n^{\frac{1}{2}}\varepsilon)\phi(u)du = \Phi(x-\varepsilon)P(M_n' \le b_n + \varepsilon r_n^{\frac{1}{2}}) \quad (3.6)$$

and at most

$$\int_{-\infty}^{x+\varepsilon} 1.\phi(u)du + P(M_n' \le b_n - r_n^{\frac{1}{2}}\varepsilon) \cdot 1 \cdot \quad (3.7)$$

$$= \Phi(x+\varepsilon) + P(M_n' \le b_n - r_n^{\frac{1}{2}}\varepsilon) \quad .$$

for every $\varepsilon > 0$. Using Slepian's inequality, we know that

$$P(M_n' \le b_n + \varepsilon\, r_n^{\frac{1}{2}}) \ge P(M_n^* \le b_n + c_n^{-\frac{1}{2}}(\varepsilon c_n r_n^{\frac{1}{2}}) \quad (3.8)$$

$$\sim \exp(-\exp(-\varepsilon\; c_n r_n^{\frac{1}{2}})) \to 1 \text{ as } n \to \infty \; .$$

Thus (3.2) follows if we show that

$$P(M_n' \le b_n - c\, r_n^{\frac{1}{2}}) \to 0 \quad \text{as } n \to \infty \quad . \quad (3.9)$$

Part II: Second Subtraction:

First we will illustrate huristically that a direct use of (2.2) would still not lead to the desired result. Let us denote $\rho_i^{(n)} = \rho_i = (r_i - r_n)/(1-r_n)$ $1 \le i \le n$, $t(n) < n$ as some function of n and notice that as a consequence of (B) (iii) we have (ignoring o(1) terms)

$$\rho_{t(n)} < r_n\left(\frac{\ell nn - \ell n\; t(n)}{\ell nn}\right) \; . \quad (3.10)$$

The tail of the sum in the R.H.S. of (2.2) is

$$n \sum_{j=t(n)}^{n} \rho_j \exp\left(- \frac{(b_n - \varepsilon r_n^{\frac{1}{2}})^2}{(1-r_n)(1+\rho_j)}\right) \quad .$$

$$\le n(n-t(n))\rho_{t(n)} \exp(-b_n^2 + \rho_{t(n)} b_n^2 (1 + 0(1)) \; .$$

If $t(n)$ is so large that $\rho_{t(n)} b_n^2 \to 0$ then $n/t(n) \to 1$ and we would need to split the sum in $O(n)$ parts which is not much better than the starting point. And if we choose $t(n)$ such that $t(n)/n \to 0$ then $\rho_{t(n)} b_n^2 \to \infty$. A second subtraction could simplify these matters considerably. (Note: The first subtraction is essential for the proof and yields the first order terms of the limit. The second subtractin is merely an algebraic convenience. It is possible that some judicious choice of $t(n)$ or more stringent conditions on $r_n$ may eliminate the need for the second subtraction. The proof can also be completed by successive subtractions instead of appealing to (2.2) after the second one. This is the reason why the proof can be trivially (philosophically trivial, algebraically messy) extended to the monotone decreasing (instead of convex) $r_n$). Let us choose $t(n) = n \exp(-\ell nn)^{\frac{1}{2}})$.

Using Slepian's inequality we know that

$$P(M_n' \leq b_n - \varepsilon r_n^{\frac{1}{2}}) \leq P(M_n^U \leq b_n - \varepsilon r_n^{\frac{1}{2}}) \qquad (3.11)$$

where the covariance sequences from which $M_n'$ and $M_n^U$ arise are given pictorially in Figure 1a and 1b respectively

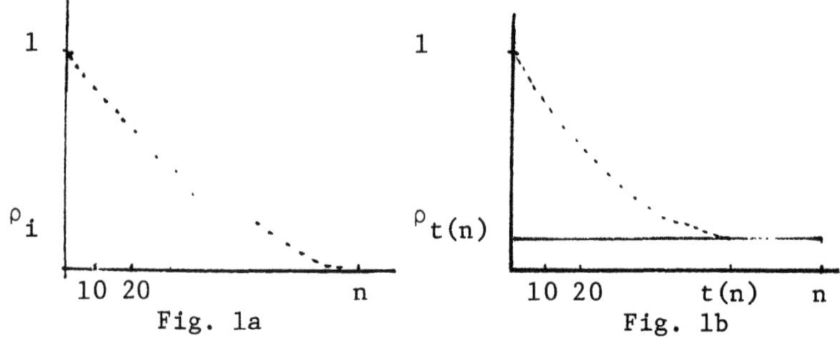

Fig. 1a Fig. 1b

(Note that for convenience of notation we may refer to $\rho_i$ as covariance of $M_n'$. This misuse of notation should not create any confusion since the paper does not ever refer to the actual covariance of the sequence of the maxima. All covariances are those of the sequences that generate the maxima.) The covariance sequence of Figure 1b gives the representation

$$M_n^U = (1-\rho_{t(n)})^{\frac{1}{2}} \mu_n' + \rho_{t(n)}^{\frac{1}{2}} V$$

where $\{\rho_i'\}$ is the covariance of $\mu_n'$ and V is ind. N(0,1). We see that

$$\rho_i' = \frac{\rho_i - \rho_{t(n)}}{1 - \rho_{t(n)}} = \frac{r_i - r_{t(n)}}{1 - r_{t(n)}} \qquad \text{for } 1 \leq i \leq t(n)$$
$$= 0 \qquad \text{for } t(n) < t \leq n \quad . \tag{3.12}$$

The R.H.S. of (3.11) is equal to

$$\int_{-\infty}^{\infty} P(\mu_n' \leq (b_n - \varepsilon\, r_n^{1/2} - \rho_{t(n)}^{1/2} u)(1 - \rho_{t(n)})^{-1/2} \phi(u)du$$
$$\leq \Phi(-\varepsilon\, r_n^{1/2}(2\rho_{t(n)})^{-1/2} + P(\mu_n' \leq u_n'') \tag{3.13}$$

where $u_n'' = (b_n - \frac{\varepsilon}{2} r_n^{1/2})(1 - \rho_{t(n)})^{-1/2}$. Due to (3.10) and definition of $t(n)$, $r_n^{1/2}\rho_{t(n)}^{-1/2} \to \infty$ as $n \to \infty$ and the first term in the R.S.H. of (3.13) converges to zero.

Part III: Remaining Proof:

As commented before, the proof can be completed by a series of subtractions as in part II. However (2.2) now is powerful enough if the sum in the R.H.S. of (2.2) is split in to $q(n)$ ($\to\infty$ as $n\to\infty$) carefully chosen parts. We give the details below.

We first note that $P(M_n^* \leq u_n'') \to 0$ since

$$(u_n'')^2 = b_n^2 - \varepsilon\, b_n r_n^{1/2}(1 + o(1)) \ . \tag{3.14}$$

Using (2.2)

$$|P(\mu_n' < u_n'') - P(M_n^* < u_n'')| \leq (\text{const.})\ n \sum_{i=1}^{t(n)} \rho_k' \exp(-u_n''^2/(1+\rho_k)) \ . \tag{3.15}$$

The proof will be completed by showing that the R.H.S. of (3.15) converges to zero. In the following, we again take notational liberties by excluding constants and factors of $o(1)$ or $O(1)$ from the equalities or inequalities. This would help us concentrate on the mechanics that make the proof work. Substituting the values of $\rho_k'$ and $u_n''$ the R.H.S. of (3.15) is

$$n \sum_{k=1}^{t(n)} (r_k - r_{t(n)}) \exp\{- b_n^2 - (r_k - r_{t(n)}) b_n^2 - \varepsilon\, b_n r_n^{1/2}\} \ . \tag{3.16}$$

We note that the expression in (3.16) is monotone in $r_k$. We will split the sum in to (q(n)+1) parts and substitute the maximum value of $r_k$ for each part thus for $t_0 < t_1 < \dots < t_{q(n)+1} = t(n)$, the expression in (3.16) is at most

$$n \sum_{j=0}^{q(n)} (t_{j+1}-t_j)(r_{t_j}-r_{t(n)})\exp\{-[b_n^2-(r_{t_j}-r_{t(n)})b_n^2 - \varepsilon\, b_n r_n^{\frac{1}{2}}]\} \,. \tag{3.17}$$

Let us define

$$\begin{aligned} t_0 &= [n^\theta] \qquad \text{for} \quad 0 < \theta < (1-r_1)/(1+r_1) \\ t_i &= [\exp\{(1-r_n^{i/2})\ell nn\}] \qquad i = 1,2,\dots,q(n) \end{aligned} \tag{3.18}$$

and

$$t_{q(n)+1} = t(n) \,.$$

Since we have chosen $t(n) = [\exp\{(1-(\ell nn)^{-\frac{1}{2}})\ell nn\}]$ we see that q(n) would be such that $r_n^{(q(n)+1)/2}$ would be smaller than $(\ell nn)^{-\frac{1}{2}}$ for the first time i.e.,

$$r_n^{q(n)/2} > (\ell nn)^{-\frac{1}{2}} \geq r_n^{(q(n)+1)/2}$$

and hence

$$q(n) < \ell n\ell nn/|\ell nr_n| < \ell n\ell nn \quad \text{for large } n \,. \tag{3.19}$$

The term for j = 0 in (3.17) goes to zero exactly as the first term in R.H.S. of (2.5). For $j \geq 1$, we will replace $t_{j+1} - t_j$ by $t_{j+1}$. Also using (3.1 B iii) we have

$$\text{(i)} \quad r_{t_j} - r_{t(n)} < r_{t(n)} \frac{\ell nt(n) - \ell nt_j}{\ell nt(n)}$$

$$< r_{t(n)}\, r_n^{j/2} \,.$$

and

$$\text{(ii)} \quad \frac{r_{t(n)}}{r_n^{\frac{1}{2}}} < r_n^{\frac{1}{2}} \frac{\ell nn}{\ell nn - (\ell nn)^{\frac{1}{2}}} = r_n^{\frac{1}{2}}(1 + (\ell nn)^{-\frac{1}{2}})$$

Hence the $j^{th}$ term in (3.17) is at most

$$n \exp((1-r_n^{\frac{j+1}{2}})\ell nn) r_{t(n)} r_n^{j/2} \exp(-2\ell nn + \ell n\ell nn + 2r_{t(n)} r_n^{j/2} \ell nn + \varepsilon\, r_n^{\frac{1}{2}} (\ell nn)^{\frac{1}{2}}) .$$

$$= (r_{t(n)} r_n^{j/2} \ell nn) \exp(-r_n^{\frac{j+1}{2}} \ell nn [1 - 2 \frac{r_{t(n)}}{r_n^{\frac{1}{2}}} - \frac{\varepsilon}{r_n^{j/2} (\ell nn)^{\frac{1}{2}}} ]) .$$

$$\leq \exp(-(\ell nn)^{\frac{1}{2}}).$$

The last inequality follows by noticing that $r_n^{j/2}(\ell nn)^{\frac{1}{2}} > 1$, $r_{t(n)}/r_n^{\frac{1}{2}} \to 0$ and the terms $r_n^{(j+1)/2}\, \ell nn$ is the exponent is replaced by $(\ell nn)^{\frac{1}{2}}$ when $j = q(n)$ due to the definition of $t(n)$. Since the total number of terms summed is much smaller order than $\exp(-\,(\ell nn)^{\frac{1}{2}})$, the proof is completed.

## Section 4: Examples of the use of TCT

Following we give three examples of strongly dependent cases in which the representation of the type (3.3) leads to correct answer and the subtraction of the first order term $r_n^{\frac{1}{2}}U$ brings the proofs to more conventional forms. We also give one example when the method does not lead to the answers. For other applications see Leadbetter, Lindgren and Rootzen [5].

### Example 1:

Let $\{X_n\}$, $\{r_n\}$ and $\{M_n\}$ be as before. We have looked at the representation

$$X_i \simeq (1 - r_n)^{\frac{1}{2}} Z_i + r_n^{\frac{1}{2}} U \tag{4.1}$$

where $Z_i$ are i.i.d. N(0,1). Thus

$$\bar{X}_n \simeq (1 - r_n)^{\frac{1}{2}} \bar{Z}_n + r_n^{\frac{1}{2}} U . \tag{4.2}$$

We know that $\bar{Z}_n \to 0$ at the rate $n^{-\frac{1}{2}}$. Thus if $r_n \to 0$ at any slower rate, we find $\bar{X}_n \simeq r_n^{\frac{1}{2}} U$ i.e., the binding variable can be thought of as the sample mean.

Also

$$\frac{1}{n-1}\sum_{i=1}^{n}(X_i - \bar{X}_n)^2 \simeq (1 - r_n)^{\frac{1}{2}}\left(\sum_{i=1}^{n}(Z_i - \bar{Z}_n)^2/(n-1)\right)$$
$$\simeq (1 - r_n)^{\frac{1}{2}} . \tag{4.3}$$

The last approximation is due to the fact that $\sum_{i=1}^{n}(Z_i - \bar{Z}_n)^2/(n-1)$ is an unbiased estimate of the population variance viz 1. Hence

$$\max_{1\leq i\leq n}\left(\frac{X_i - \bar{X}_n}{s_n}\right) \simeq \max_{1\leq i\leq n} Z_i = M_n^* \tag{4.4}$$

where $s^2 = \frac{1}{n-1}\sum_{i=1}^{n}(X_i - \bar{X}_n)^2$ is the sample variance. That the above representation is correct and leads to right answers is proved by McCormick in the following theorem. His conditions are more general and include the cases of Theorem 3.1.

Theorem 4.1 (McCormick [7])

For the quantities defired above,

$$\lim_{n\to\infty} P\{c_n(\frac{M_n - \bar{X}_n}{s} - b_n) \leq x\} = \exp(-\exp(-x)) \tag{4.5}$$

for $-\infty < x < \infty$ if

$$((\ell n n)/n)\sum_{k=1}^{n}|r_k - r_n| = o(1) \tag{4.6}$$

Example 2:

Again from (4.1) we could conclude that

$$\frac{2c_n(M_n - \bar{X}_n - (1-r_n)^{\frac{1}{2}}c_n)}{\ell n\ell n n} \simeq (1-r_n)^{\frac{1}{2}}\, 2\left(\frac{M_n^* - c_n)c_n}{\ell n\ell n n}\right.$$
$$\simeq \frac{2(M_n^* - c_n)c_n}{\ell n\ell n n} \tag{4.7}$$

Thus the law of iterated logarithms for maxima in the dependent cases can be concluded.

Theorem 4.2: For the quantities defined above and conditions (B) (i), (ii), (iii) of Theorem 3.1,

$$\liminf_{n\to\infty} \frac{2c_n(M_n - (1 - r_n)^{\frac{1}{2}}c_n - \bar{X}_n)}{\ell n\ell nn} = -1 \quad \text{a.s.}$$
$$\limsup_{n\to\infty} \frac{2c_n(M_n - (1 - r_n)^{\frac{1}{2}}c_n - \bar{X}_n)}{\ell n\ell nn} = +1 \quad \text{a.s.} \tag{4.8}$$

Results (2.8) were first proved by Mittal and Ylvisaker [13] under an additional unnecessary conditions which was later removed by McCormick [6].

Example 3:

Let us denote

$$m_{n,k} = \max_{i\varepsilon G_n} X_i \tag{4.9}$$

where $G_n$ is a subset of $\{1,2,\ldots,n\}$ of size $n/k$. By using 4.1 we see that

$$m_{n,k} \simeq (1 - r_n)^{\frac{1}{2}}\max_{i\varepsilon G_n} Z_i + r_n^{\frac{1}{2}}U \tag{4.10}$$

and

$$M_n \simeq (1 - r_n)^{\frac{1}{2}} \max_{1\leq i\leq n} Z_i + r_n^{\frac{1}{2}}U \quad .$$

We see that as long as $r_n^{\frac{1}{2}}U$ is a negligible term the joint distribution of the maxima of $X_i$ on $G_n$ and $G_n^c$ would be asymptotically independent but would degenerate on the diagonal when $r_n^{\frac{1}{2}}U$ becomes the dominating term. However the distribution of $(M_n - m_{n,k})$ should be practically free of the mixing conditions.

Theorem 4.3: (Mittal [9]) If either $r_n \ell nn \to 0$ or conditions of Theorem 3.1 (A) or (B) hold them

$$\ell t_{n\to\infty} P\{M_n - m_{n,k} \leq x/c_n\} = (1 + (k - 1)e^{-x})^{-1} \tag{4.11}$$

for $0 \leq x < \infty$.

Example 4:

We now give an example where TCT would explain why it is unable to provide answers and a new line of thought is required. Suppose one wants to look at the behavior of the maximum of the normalized sums from the SGP then (4.1) gives

$$\frac{\sum_{i=1}^{j} X_i}{(V(j))^{\frac{1}{2}}} = (1 - r_n)^{\frac{1}{2}} \frac{\sum_{i=1}^{j} Z_i}{(V(j))^{\frac{1}{2}}} + r_n^{\frac{1}{2}}\left(\left(\frac{j}{(V(j))^{\frac{1}{2}}}\right)U\right. . \tag{4.12}$$

Now even if $Z_i$ are originally i.i.d. sequence $V(j)^{-\frac{1}{2}} \sum_{i=1}^{j} Z_i$ is farthest from being asymptotically independent. Here the simple subtraction of the binding variable is not possible.

It is reasonably clear to notice that the behavior of $\mu_n = \max_{1 \leq j \leq n} (V(j))^{-\frac{1}{2}} \sum_{i=1}^{j} X_i$ depends on the covariance sequence $\{r_n\}$ only via $\{\rho_n = \sum_{k=1}^{n} r_k$ ( and hence not directly on the size, smoothness or the rate of decay of $\{r_n\}$). (Notice that the sequence of variance $\{V(j)\}$ depends only on $\{\rho(j)\}$.)

The situation is much more complex here. The behavior of the $\mu_n$ is dictated by the "crowding" of the sequence $\{(V(j)^{-\frac{1}{2}} \sum_{i=1}^{j} X_i\}$ (viz the rate at which the covariance of the two adjacent variables tends to one as $j \to \infty$ ) and the size of the sequence $\{V(j)\}$. For more details see Mittal [10] where we see that arbitrarily slow growths of $\mu_n$ are possible under suitable conditions on $\{r_n\}$.

## References

[1] Berman, S.M. 1964. Limit theorems for the maximum term in stationary sequences. Ann. Math. Stat. 35, pp. 502-516.

[2] Fisher, R.A. and L.H. Tippett 1928. Limiting forms of the frequence distribution of the largest or smallest member of a sample. Proc. Cambridge Phil. Soc. 24, pp. 180-190.

[3] Gnedenko, B.V. 1943. Sur la distribution limite du terme maximum dûne série al éatorie. Ann. Math. 44, pp. 423-453.

[4] Leadbetter, M.R., Lindgren, G., and Rootzen, H. 1978. Conditions for the convergence in distribution of maxima of stationary normal processes. Stoch. Proc. and Appl. 8, pp. 131-139.

[5] Leadbetter, M.R., Lindgren, G., and Rootzen, H. 1983. *Extremes and related properties of random sequences and Processes*. Springer-Verlag series in statistics.

[6] McCormick, W.P. 1980a. An extension to a strong law result of Mittal and Ylvisaker for the maxima of stationary Gaussian processes. Ann. Prob. 8, pp. 498-510.

[7] McCormick, W.P. 1980b. Weak convergence for the maxima of stationary Gaussian processes using random normalization. Ann. Prob. 8, 483-497.

[8] McCormick, W.P. and Mittal, Y. 1976. On weak convergence of the maximum. Tech. Report No. 81. Department of Statistics, Stanford University.

[9] Mittal, Y. 1978. Maxima of partial samples in Gaussian sequences. Ann. Prob. 6, pp. 421-432.

[10] Mittal, Y. 1979a. Extreme value distribution for normalized sums from stationary Gaussian sequences. Stoch. Proc. and Appl. 9, pp. 67-84.

[11] Mittal, Y. 1979b. A new mixing condition for stationary Gaussian processes. Ann. Prob. 7, pp. 724-730.

[12] Mittal, Y. and Ylvisaker, D. 1975. Limit distributions for the maxima of stationary Gaussian processes. Stoch. Proc. and Appl. 3, pp. 1-18.

[13] Mittal, Y. and Ylvisaker, D. 1976. Strong laws for the maxima of stationary Gaussian processes. Ann. Prob. 4, pp. 357-371.

[14] Plackett, R.L. 1954. A reduction formula for normal multivariate integrals. Biometrika 41, pp. 351-360.

[15] Slepian, D. 1962. The one sided barrier problem for Gaussian noise. Bell Syst. Tech. J. 41, pp. 463-501.

# EXTREMES IN HYDROLOGY

Vujica Yevjevich

International Water Resources Institute
School of Engineering and Applied Science
George Washington University
Washington, D.C. 20052, U.S.A.

ABSTRACT

Several variables are needed for proper description of extremes in hydrology, such as peaks or lowest values, volume of excess or deficit, time duration above or below a critical value, etc. Selections of the type of extremes and the underlying hydrologic processes are needed in definition of extremes.

Basic approach to extraction of information on hydrologic extremes is to sample the historic sample. Distributions of extremes are rarely derived directly from properties of underlying processes. The empirical historical approach has resulted in standardized practice.

Four characteristics of hydrologic time series: tendency, intermittency, periodicity and stochasticity, affect the selection of methods for derivation of properties of extremes. Novel approaches to flood extremes are needed in: multivariate treatment of extremes, estimation of values for extremely small probabilities of exceedence (for floods: $10^{-4}$ to $10^{-7}$), optimal extraction of regional information, and relationships of flood parameters to land practice and water resource development.

Estimation of PDF of flood peaks can be improved by using the variance of estimates of flood peaks as weights. The exponential tails of stochastic components of underlying processes predetermine PDF. Average skewness and kurtosis in a region may lead to best PDF fit.



*J. Tiago de Oliveira (ed.), Statistical Extremes and Applications, 197–220.*

## INTRODUCTION

Humanity has always lived with hydrologic extremes, basically and mostly with floods and droughts. Too much or too little water in streams, lakes, springs, soil moisture and aquifers, in comparison with their average water availability, had always put the stress on society in various ways and from time to time. Therefore, it was not unusual for hydrology to have paid so much attention to investigation, for understanding of physical or mathematical properties with descriptions of hydrologic extremes.

The literature on hydrologic extremes is abundant. Thousands of papers, reports, chapters of books and entire books make a large pool of information on the subject. The state-of-the-art presentation would require a long preparation and several hours to cover it properly. In this limited presentation only the highlights are given, considered by the writer of this text as of the most interest to practice and human responses to these extremes.

### Definitions of Extremes of Hydrology

Three steps are useful in defining extremes in hydrology. First, hydrologic extremes occur on both ends of the range of fluctuation of any relevant hydrologic process that defines extremes either as water excess or as water deficit, and in general as extremes in water availability; so this step should decide either to investigate the extreme high or the extreme low conditions of the relevant basic hydrologic process. Second, it is necessary to clearly define what is the hydrologic process (the variable, their points of occurrence in space and their time process) that produces from place to place and from time to time the extreme conditions defined in the first step. This variable and its space-time process is then the extremes-defining or the determinant process. Third, the new, derived variables of this determinant process must be clearly selected, and in such a way that the extremes can be properly described.

As the first example, let us investigate the runoff extremes of a medium-size river. The first step may be the decision to study the flood extremes. The second step defines the runoff process, say as the continuous time series of river flows at a given cross section (Fig. 1), which can be very often approximated by the available daily flows if the instantaneous flood peak data are not available. The third step should single out the flood related variables for the study, such variables as the maximum annual peak ($Q_{max}$), the total excess water volume ($V_{max}$) above the critical flood flows during the passage of a flood wave, and the maximum time duration of flows above the critical flow(s) ($t_{max}$).

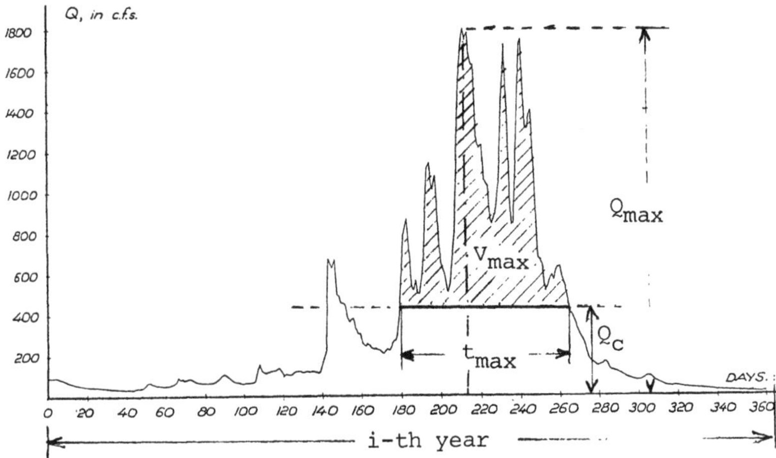


Figure 1. Definition of flood related variables in the i-th year: $Q_{max}$=the maximum instantaneous peak, $V_{max}$=the maximum water volume above the critical flood flow $Q_c$, and $t_{max}$=the maximum duration of flood discharges above the critical flood flow $Q_c$.

These three random variables do not necessarily occur simultaneously at the same flood event within the year; usually they do, but in some cases each of the three may occur during a different flood wave in a year. Each variable has a different practical implication: $Q_{max}$ determines what is flooded during a passage of the flood wave; $V_{max}$ determines the water volume which should be stored in a reservoir or retention basin if flood damages are to be avoided or minimized; and $t_{max}$ determines the time effects on damage (say, such effects as the partial or full loss of flooded crops). Similarly, if the first step decides on the study of low flow extremes, the three random variables of extremes will be: $Q_{min}$, $V_{min}$ and $t_{max}$, with these variables defined in analogous ways as for the high extremes of Fig. 1.

As the second example, let us investigate the sequence of wet and dry precipitation intervals. The first step decision will be to investigate the extremes sequences of dry intervals, or the longest or largest drought in the given sample of size N. The second step will be to select the interval of time series, say the annual, seasonal or monthly series. If the year is that interval, the prolonged sequences of dry years result in droughts. The third step will be to select the drought-defining random variables (Fig. 2), which basically are analogous to those of Fig. 1, namely the longest sequence of dry years (with a proper definition of the dry year), the drought water deficit below a constant value, that is greater than the water volumes of dry years, and the largest one-interval deficit during a sequence of dry years (largest deficit intensity).

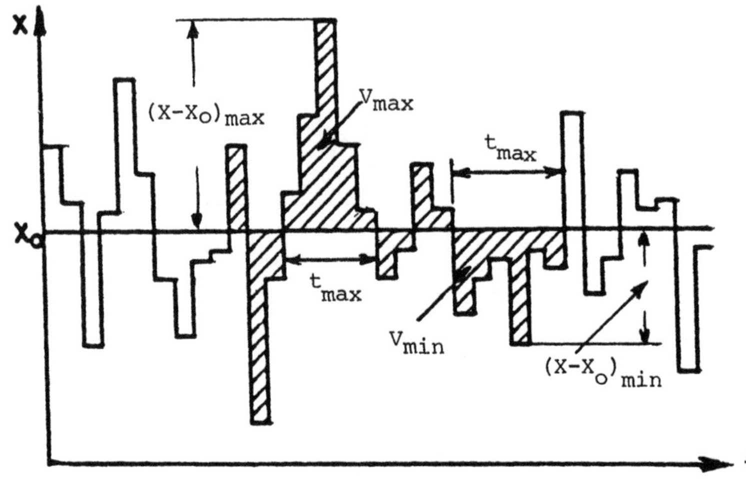


Figure 2. Definition of drought-related variables of annual precipitation series $(X-X_o)_{min}$=the maximum one-year deficit (maximum intensity of the drought); $V_{min}$=the drought deficit (largest negative run-sum in a sample size N); and $t_{max}$=the longest drought duration (longest negative run-length in a sample of size N).

Basically, three random variables in description of hydrologic extremes are: (1) maximum or minimum intensity (which has the same dimension as the time series variable); (2) maximum or minimum volumes (surplus, deficit, or integral or summation of the time series variable over the time of extremes); and (3) maximum duration of extremes (which has the dimension of time).

Many other examples of hydrologic extremes may be cited. Such examples are the extremes of water levels (say levels of rivers, lakes, aquifers). They usually have two random variables of extremes only, namely the maximum ($H_{max}$) or minimum ($H_{min}$) levels for the given sample of size N, and the time duration of extreme levels above or below a critical level $H_c$. The importance of these level extremes is evident if one takes into account, for example, the large effects on many coastal problems of highest or lowest historic levels of the Great Lakes in Canada and USA.

Historic Approach to Extraction of Information

The basic historic approach in estimating the properties of any random variable that describes extremes in hydrology is to sample the sample, namely to select a value of that random variable in each subsample of the sample The smaller the subsample and the larger the historic sample, the larger becomes the size of the derived sample of that random variable. The classical subsample used in practice is the year. However, the seasons of different length have also been used as subsamples, for the basic time series being either continuous or discrete with a small time interval (say, hour or day). Then the probability distributions, or even the space and/or time dependence models, are derived from the estimated frequency distributions, and/or the space/time correlograms or cross-correlograms. Rarely attempts are made to derive the properties of extremes-defining random variables directly from the mathematical models of the underlying hydrologic processes.

The major characteristics of the empirical historic approach in estimating the probability distributions and the other characteristics of extremes may be summarized as follows:

1. The n extreme values, one for each subsample, with $n=N/\Delta t$, where N=the sample size and $\Delta t$=the selected subsample size, are sorted in the ascending order, and plotted on the graph paper either with standard (cartesian) or special coordinate scales (logarithmic, probabilistic, double-logarithmic), as the cumulative frequency distributions.

2. A large number of plotting positions $P(X_m \leqslant x_m)$ for the value of the ascending order m have been suggested, however, with the prevailing plotting position in use in hydrology still being

$$P(X_m \leqslant x_m) = \frac{m}{n+1} \tag{1}$$

3. The plotted cumulative frequency distribution curves have been generally used for these purposes: (3.1) to ascertain their general forms, that lead to hypotheses of the probability distribution functions (PDF) most likely to fit well these cumulative frequency curves; (3.2) to use frequency curves directly for estimating the values of extremes for given probability or for its reciprocal as the average return period; and (3.3) to identify "outliers", namely the extremes of those plotted points, that seem to significantly deviate at either of the two tails from the expected fitted cumulative PDF F(X). The entire set of techniques have been developed for treating these outliers, basically by considering them representative of the much longer sample sizes than is the size N of the historic sample.

4. The following probability distribution functions have been used mainly in hydrology for fitting the cumulative frequency distribution curves of extremes: lognormal, gamma, gamma for lognormal transforms (Log-Pearson Type III), double exponential, and bounded exponential; all the other proposed PDF for hydrologic extremes in literature have been less commonly used.

5. There have been continuous attempts in literature to find theoretical backgrounds for using any of the above PDF or otherwise, with a variety of underlying assumptions. However, the goodness-of-fit tests, or the prevalence of passing these tests in a region for a given type of extreme, have been used as the major criterion in selecting one function over all those being suggested or used.

6. The Log-Pearson Type III function has been selected by the U.S. governmental water-related agencies, basically for two reasons: (a) that a uniformity be applied, namely in fitting the same function within a wide range of agencies; and (b) because this three-parameter function has shown more flexibility to fit a variety of shapes of the resulting cumulative frequency distributions than the other, basically the two-parameter distribution functions.

7. The extrapolation of frequency curves or the fitted PDF beyond the range of the plotted points is performed very cautiously in hydrology. Some agencies even apply the rule that the extrapolation should not go beyond the probability of exceedence or non-exceedence, say of P=0.002, or the corresponding return period of T=500 years (in case of the year as the subsample). For the smaller probability of exceedence, some other, usually non-probability based methods are advocated or used, such as PMF (probable maximum flood) for the case of flood peaks. The problem of extrapolation of PDF of extremes has grown to be a very controversial topic in hydrology.

8. Regional grouping of frequency distributions for extremes is popular though a difficult subject in the analysis of extremes in hydrology.

## Character of Underlying Hydrologic Processes

The major hydrologic time-space processes, for which the study of extremes are important in practice, are basically composed of the four structural properties: tendency, intermittency, periodicity and stochasticity, with the changes in space of parameters which in one way or another describe these properties, also subject to influences of these properties.

Tendency. Tendency results either from the systematic errors (inconsistency in data), or it is a direct consequence of activities by humans on the earth's surface and/or disruptions in nature (nonhomogeneity in data). This is usually a difficult structural property to treat, because a given pattern of tendency is bound to the historic sample and will be completely different in the future samples. Therefore, the basic approaches to treatment have resulted in: (1) identification of trends and the necessary supporting physical or historical evidence for the trends identified; (2) removal of trends in parameters from the historic sample; and (3) extrapolation, or synthesis of future trends in parameters to be superposed to the trendless simulated samples or to mathematical models, as the potential future. Two examples are given here to demonstrate these difficulties in treating tendency in underlying hydrologic processes and, therefore, in the parameters of probability distributions of hydrologic extremes.

Figure 3 gives the maximum annual instantaneous peak discharges of the Danube River in Orshova, Iron Gates, Rumania, for the period 1838-1950. The period 1838-1874 seems to have a constant mean maximum annual peak discharge (basically because of small changes in the upstream flood plains and in the river basin land uses). From 1874-1950 (or a little earlier, say 1945) the mean seems to have started to continuously increase, as the fitted straight line trend demonstrates in Fig. 3.

This trend is definitely a result of building levees for flood protection of fertile flood plains along the Danube and its major tributaries (Drava, Tisza, Sava, etc.). The three-quarters of a century of constructions of levees and pumping stations (therefore, initiated by the then available vapor machines and wood or coal, and then electricity, for pumping water during floods from flood plains behind the levees) have produced a clear trend

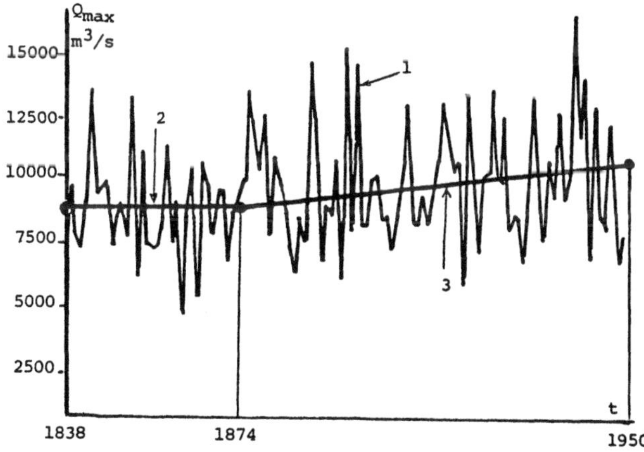


Figure 3. Tendency in the mean of the maximum annual flood peaks of the Danube River at Orshova, Iron Gates, Rumania, in the period 1838-1950: (1) Maximum annual flood peaks (in $m^3/s$), (2) Period 1838-1874 with apparently no trend in the mean, (3) Period 1874-1950 with evidently non-negligible trend in the mean (approximately an increase of 20-25 percent in the mean during this period).

in the mean of the maximum annual flood peaks in the Lower Danube (as well as all over its major course and the courses of its principal tributaries). The trend is reversed after 1940's to 1950's, namely by the construction of reservoirs for hydropower, water supply, irrigation, navigation, flood control and other purposes, with their storage of water in flood seasons "accidentally" or "purposely" decreasing the maximum annual flood peaks. Figure 4 then schematically represents a full cycle of tendency in the mean of the maximum annual flood peak, as the result of a kind of long-range economic cycle. It is, however, difficult to determine whether or when the downward trend will bottom out, since demands continue for reservoir space in the industrial and postindustrial periods.

Similarly as in Figs. 3 and 4 for the mean of the maximum annual flood peaks, the other parameters of flood peaks, particularly their standard deviation, should be investigated for trends (say by using $Q_{max}-\bar{Q}_{max}$ deviations for small segments or subsamples along the sample of historic records). The future tendency then may be best incorporated into the analysis by standardizing the variables of extremes by

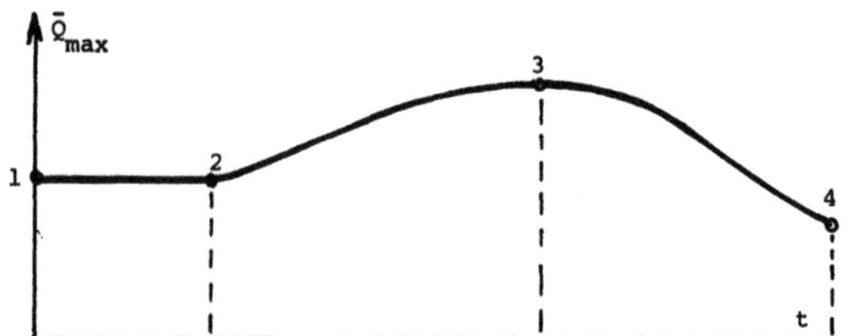


Figure 4. The general trend cycle in the mean of the maximum annual flood peaks of many rivers in the world: 1-2, the period prior to construction of upstream levees; 2-3, the period of construction of upstream levees and the increase downstream of the mean flood peak; and 3-4, the period of intensive construction of general, but particularly of flood control, storage reservoirs.

$$Y_i = \frac{X_i - \bar{X}}{S_x} \tag{2}$$

where $X_i$=the observed values of an extreme, $\bar{X}$=the estimate of its mean (sample mean), and $S_x$=the estimate of its standard deviation. By studying the $Y_i$-variable, often called by hydrologists the frequency factor (Chow 1951), the trends in the mean and standard deviation $\bar{X}_t$ and $S_{x,t}$, can be introduced as their extrapolations into the future, but based on separate studies from the study of the distribution of $Y_i$, with the future extremes evaluated by

$$X_{i,t} = \bar{X}_t + S_{x,t} Y_i \ . \tag{3}$$

The other example of significant trend changes (say as the jumps or slippages in parameters of the probability distribution of extremes) in flood peaks is given by Fig. 5. Three cases of cumulative frequency distribution of maximum annual flood peaks of the Upper Susitna River at the future Watana Dam site, Alaska are studied: (1) Upper graph, with the flood peaks in case of natural flood flows; (2) Middle graph, with the peaks downstream of the reservoir if the future reservoir will be operated only for the hydroelectric power production; and (3) Lowest graph (crossing the middle graph for low flood peaks), with the peaks downstream

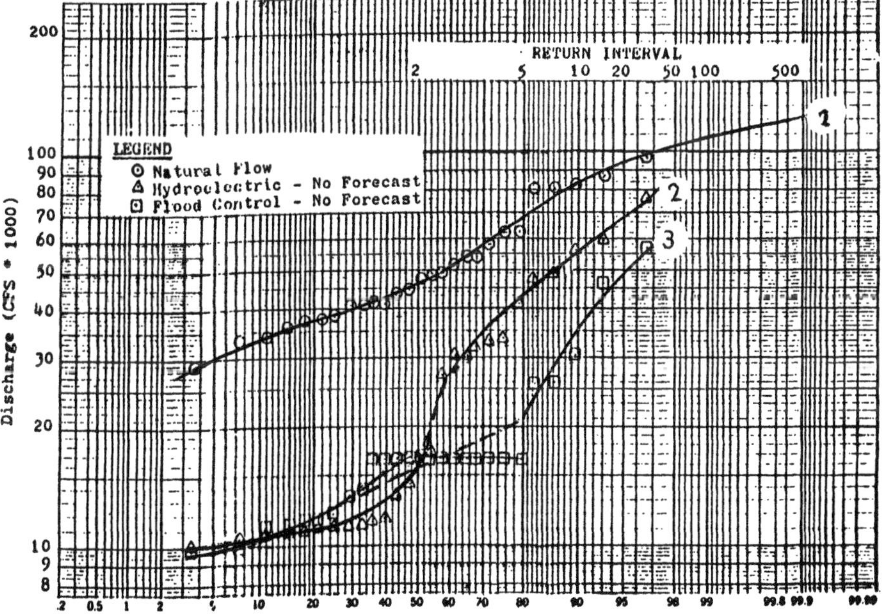


Figure 5. Three cumulative frequency distributions of maximum annual flood peaks at the site of the future Watana Dam, Upper Susitna River, Alaska: (1) Upper graph, for the natural flood peak flows; (2) Middle graph, for the peaks with reservoirs operated only for hydroelectric power production; and (3) Lower graph (crossing middle graph for low peaks) for the special storage space used for flood control (besides the storage use for hydroelectric power production).

of the reservoir if it is operated with a special storage space exclusively reserved for flood control apart from the regulation by the remaining storage capacity for the hydroelectric power production.

Intermittency. Intermittent processes like precipitation, with the Yes-No periodic-stochastic process models determine in the final count the properties of lowest extremes (low precipitation, low flows, droughts). However, little is available in literature on how the properties of lowest extremes are interrelated with properties of intermittency of basic hydrologic processes.

Periodicity. All parameters of basic continuous hydrologic processes seem to exhibit a non-negligible periodicity of the year, especially the mean, variance and autocovariances. The amplitudes of harmonics of the inferred Fourier functions of periodic parameters seem to be the most significant for the mean, variance and autocovariances. They decrease for the autocorrelation, skewness and kurtosis coefficients, basically because of their definitions as the ratios of statistical moments. However, their periodicity is mostly significant because of the non-proportionality of amplitudes of the powers of moments, used in these ratios (Yevjevich and Dyer 1983).

Stochasticity. The stochastic part of hydrologic processes of the TIPS decomposition (TIPS=tendency, intermittency, periodicity and stochasticity) of hydrologic series, basically imposes the probabilistic character to extremes. Ideally, it would be desirable by using the TIPS decomposition of basic hydrologic processes, to reduce the process only to the normal stochastic stationary component. Then the theory of gaussian stationary processes could be used to analytically derive the probabilistic characteristics of extremes.

## FLOOD EXTREMES

In flood extremes, apart from the study of probability distributions of their three random variables of Fig. 1 ($Q_{max}$, $V_{max}$ and $t_{max}$), at least these other complex cases require attention in hydrology, because of the relative lack of appropriate methods of solutions at present:

(1) Joint probability distributions of several extremes, that determine the flood conditions downstream of large lakes and reservoirs. One such case is the tri-variate distribution of rainfall-produced partial flood inflow into these bodies of water, snowmelt-produced partial flood inflow, and the state (level, storage) of the lake or reservoir at the initiation of these partial flood inflows.

(2) Estimation of flood extremes of very small probabilities, especially when stakes are very high in case of their occurrence (large metropolitan areas, atomic power plants, very sensitive and national security industries, etc.), say of the probabilities in

the range

$$P(X_i > x_i) = 10^{-4} - 10^{-7} \quad (4)$$

or of the return periods of 10,000 to 10 million years, still await the reliable solution methods.

(3) Finding the optimal and efficient methods to produce regional probability distributions of flood extremes, with the general standardized variable distributions.

(4) Finding relationships between the major parameters of flood extremes and the changes in land use and water resources developments, would contribute significantly to the treatment of tendency in flood-related parameters.

(5) Assessing the properties of quality of flood waters, such as sediment, debris, mud and, pollutants, with their movement along with the flood waves, in a multi-variate space-time processes approach, is also a relatively unexplored area of flood extremes.

Fit and Extrapolation of Probability Distributions of Flood Peaks

Figure 6 presents 40 flood peak frequency distributions for the 40 simulated samples of size of 25 years, generated from the

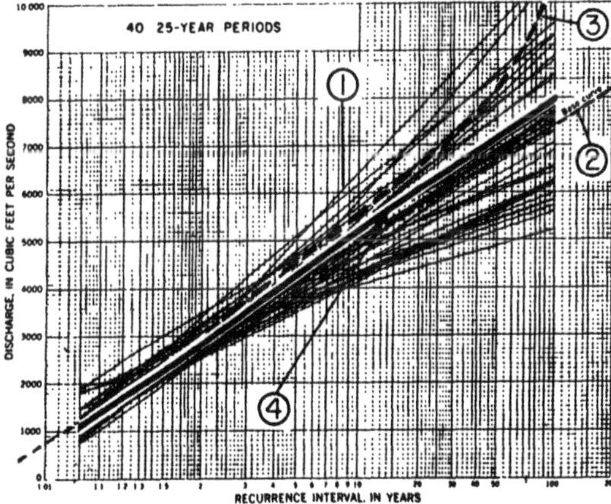


Figure 6. Sampling variance of flood peak discharge for given recurrence intervals, used as weights in fitting the double-exponential probability function to a frequency distribution curve of flood peaks: (1) 40 generated frequency curves of 25-year sample sizes; (2) double-exponential PDF for which the sample and frequency curves are generated; (3) a singled-out frequency distribution curve; and (4) fitted double-exponential function by using the variances of estimates as weights for fitting the frequency curve under (3).

assumed double-exponential probability distribution of maximum annual flood peaks. The variance of the 40 estimates for any given recurrence interval (or small probability of exceedence) increases fast with the increase of this interval. Therefore, in fitting a PDF the more weight should be given to flood peak values of small recurrence than of large recurrence intervals. Regardless of bias (nonreplacement of drawn-out values) in generating samples of the above 40 frequency curves of Fig. 6 (Benson 1960), this example is illustrative. In fitting PDF obtained for each F(Q), the flow series characteristics and the sample size.

Experience shows that estimates of small and medium floods (say, 1-year to 5-year floods) are more accurate than the estimates of large floods (say 50-year to 100-year floods). By giving greater weight to smaller floods than to larger floods in estimating the probability distribution of a flood random variable, a more accurate fit is obtained.

The lognormal, the Pearson Type III applied to logarithms of flood variables, and the double-exponential PDF are the most prevalent for fitting frequency distributions of flood random variables. The use of logarithms is somewhat equivalent to giving greater weight to small floods than to large floods in estimation.

The selection of the best fitting function may be guided not only by the goodness-of-fit tests, but prior to it also by the following three considerations:

1. The stochastic components of hydrologic series (after periodicity in basic parameters are removed) has both tails that are well approximated by the simple exponential functions (Tao et al. 1976). Figure 7 gives an examples of flows of the three rivers (Tioga, Oconto, Current, USA) out of the 17 examples given in that reference. The two straight lines (in graph paper with semi-logarithmic scales) are the confidence limits to the exponential tails (left and right), with the tails themselves given as dotted lines. All of the 17 stations passed the tests of both tails following closely the simple exponential functions. Therefore, all the functions that theoretically represent the distributions of extremes, sampled from the exponential tails of underlying independent stochastic components, are the candidates for fitting the frequency distribution curves of these extremes. In particular, they are the double-exponential and the bounded exponential functions.

2. Another criterion in selecting the probability function is the use of average regional parameters. For instance, in several regions the average unbiased skewness coefficient of flood peak frequency distributions comes out to be close to $\bar{C}_s$=1.3, and the average unbiased kurtosis coefficient close to 5.7, which are the constant skewness and kurtosis of the double exponential func-

tion of Eq.(5). Therefore, the hypothesis becomes very attractive that all flood frequency curves at river points in that region may be conceived as being drawn from this PDF. Because of very large sampling variations of unbiased $C_s$ and k of flood data, the large deviations $C_s$-1.3 and k-5.7 of flood peaks at the individual stations should not imply that it is appropriate to search for the other PDF to fit the cases of these deviations being relatively large, as long as the regional averages are close to 1.3 and 5.7.

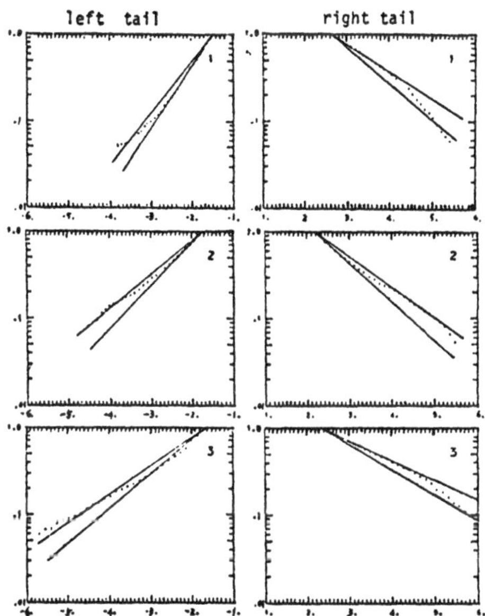


Figure 7. Tests of left and right tails of frequency distributions of independent stochastic components (abscissa) of daily flows of the Tioga River (1), Oconto River (2), and Current River (3), U.S.A., following the simple exponential functions, with tails represented by dotted lines and the two solid lines as the confidence limits.

3. Finding the theoretical background for justifiction of the use of particular population distribution functions for maximum annual flood peaks is also attractive. Example is the use of random number of random variables, when both have the relatively simple PDF, and the simple corresponding PDF for flood peaks.

If the hydrologic series of river flows were independent stationary stochastic process, the sample of the maximum annual flood peaks would contain all the information on flood peaks. However, when this series is highly stochastically dependent ($\rho_1$ ranging from 0.80-0.98) as well as periodic, the extraction of information by using one maximum annual peak only may not be complete. Figure 8 shows the definitions of annual and partial flood peak series (sampling the sample).

The theoretical distribution of all the flood peaks, above a given crossing level (Fig. 8) is a function of this level. The

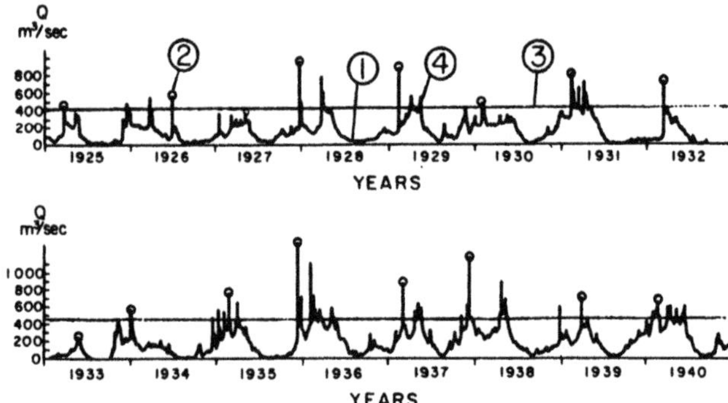


Figure 8. Use of annual and partial series in estimating the annual flood peak distributions: (1) daily flow series; (2) annual flood peaks; (3) crossing level for determining floods in partial series approach; and (4) partial series peaks.

higher the level, the closer becomes the distribution of flood peaks of partial series to a simple exponential probability distribution function. The average number of flood peaks per year in the partial series increases with a decrease of this crossing level. The partial series should give more information on the probability distribution of maximum annual flood peak under some conditions than does the annual flood series approach. Assume that the flood peaks of partial series has a two-parameter $(\alpha,\beta)$-gamma distribution, and that the number of flood peaks per year or season, or any time unit above a given crossing level, is Poissonian distributed with the parameter $\Lambda$. Then the distribution of the annual largest peak may be derived from developments by Todorovic (1970) on the random number of random variables, as:

$$F(x) = e^{-\Lambda} \sum_{k=o}^{\infty} \frac{\Lambda^k}{k!} \left[ 1 - e^{-\beta x} \sum_{i=o}^{\alpha-1} \frac{(\beta x)^i}{i!} \right]^k \tag{7}$$

with x the annual flood peaks resulting from the partial series approach. Equation (7) can also be written as

$$F(x) = \exp \left[ -\Lambda \, e^{-\beta x} \sum_{i=o}^{\alpha-1} \frac{(\beta x)^i}{i!} \right] \tag{8}$$

When the gamma distribution becomes the simple exponential function ($\alpha$=1), Eq.(8) becomes the double exponential function

$$F(x) = \exp\left[-\Lambda e^{-\beta x}\right] \quad , \tag{9}$$

with $\Lambda$ estimated from the Poissonian distribution as the average number of peaks per year, and $\beta$ from the fit of the simple exponential function to partial series of flood peaks above a sufficiently high crossing level for floods.

To apply Eq.(9) two conditions must be fulfilled: (1) that the number of "partial" floods in a year follows a Poissonian distribution, which may not be an exact discrete distribution; and (2) that the crossing level of Fig. 8 is sufficiently high for the partial flood peaks to be exponentially distributed. In general, it was shown to be a very close approximation, as Figs. 9 and 10

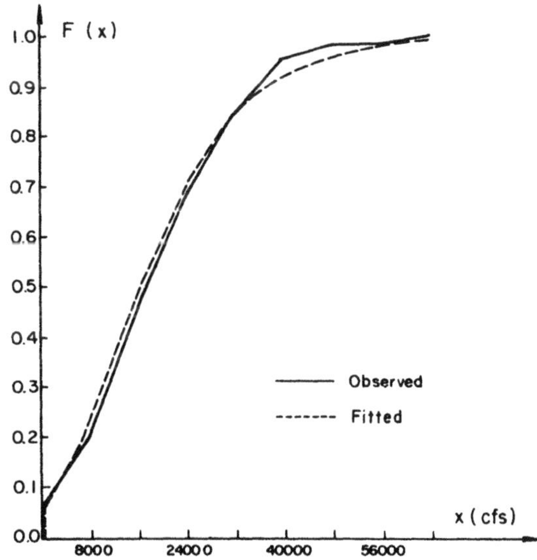


Figure 9. The double-exponential distribution function, and the observed frequency distribution of the largest exceedence, using a one-year time interval for the Greenbrier River at Alderson, USA.

show (Zelenhasic 1970), even when the entire year (and not only the major flood season) is used in the estimation of parameters of Eq.(9).

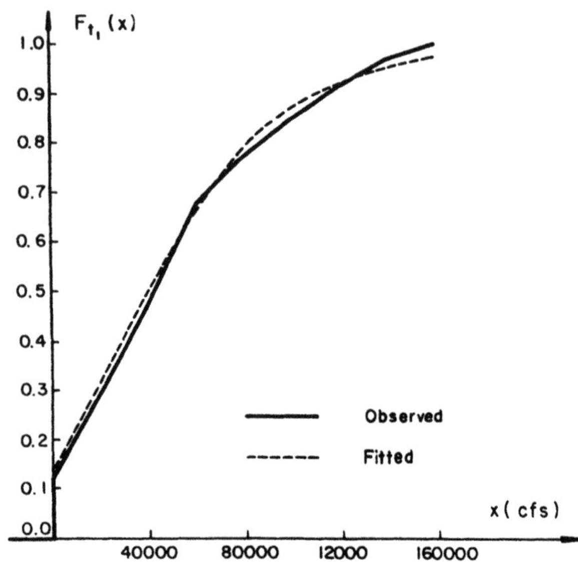


Figure 10. Fitted and observed distributions of the largest exceedence using the one-year time interval for the Susquehanna River at Wilkes-Barre, USA.

Use of Partial Instead of Annual Series

Figure 11 defines the partial series of flood peaks and Fig. 12 gives another example of series peaks for two different

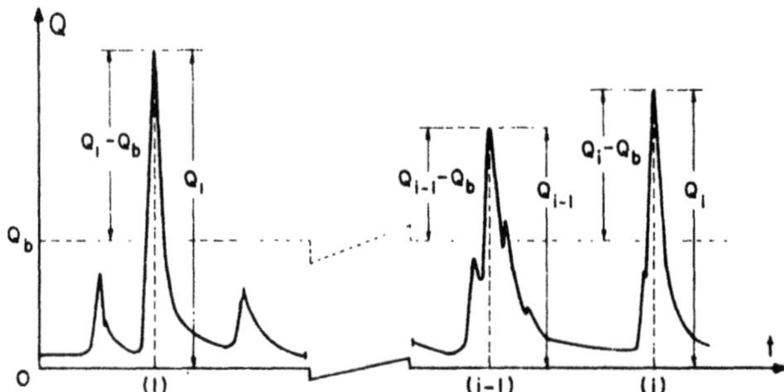


Figure 11. Schematic representation of a stream flow hydrograph for definition of partial series of flood peaks.

crossing levels with two different numbers of peaks. To compare the goodness of fits of flood probability distributions by using the annual and partial series of flood peaks, the variance of es-

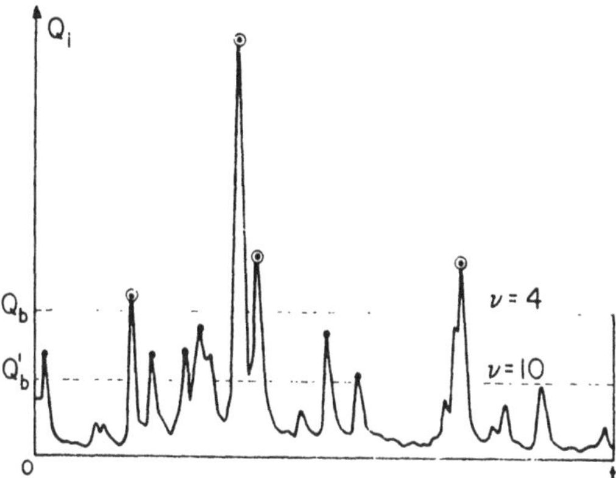


Figure 12. Example of selecting partial series of floods from daily flow hydrograph, with $\nu=4$ peaks for $Q_b$, and $\nu=10$ for $Q'_b$.

timates of flood peaks for given return period is determined by

$$\operatorname{var}\hat{Q}(T) = \frac{1}{m-1} \sum_{i=1}^{m} \left[ \hat{Q}_i(T) - \overline{\hat{Q}_i(T)} \right]^2 \tag{10}$$

where: $\hat{Q}(T)$=the unbiased estimate of highest annual flood peak Q for the return period T, either as $\hat{Q}_a(T)$ for direct estimation from the annual series of peaks, or as $\hat{Q}_p(T)$ for the indirect estimation from the partial series of peaks, and m=the number of generated samples of flows that serve for computing the $\hat{Q}(T)$ values of Eq.(10). It is assumed that the estimate is better when $\operatorname{var}\hat{Q}(T)$ of Eq.(10) is smaller. Therefore,

$$R = \frac{\operatorname{var}\hat{Q}_a(T)}{\operatorname{var}\hat{Q}_p(T)} \tag{11}$$

is the comparison ratio between the two ways of estimation. For $R < 1$, the annual series is a better way of estimation. For $R > 1$, the case is opposite. Equation (11) is then useful for comparison, regrdless of some unspecified doubts in using this ratio R (Taveres and Silva 1983). The ratio R can be obtained from Eq.(10) either experimentally by generating samples (Monte Carlo method), or analytically by finding the two variances of Eq.(11).

In case of the use of Eq.(5), the variance (Yevjevich and Taesombut 1978, after adjustment for the difference in symbols with Eq.(5)), obtained analytically, is

$$\mathrm{var}\ \hat{Q}_a(T) = \frac{1}{\alpha^2 n} \left[1.11 + 0.61\, y^2(T) + 0.52\, y(T)\right] \quad , \tag{12}$$

where

$$y(T) = -\ell n \left[-\ell n\left(1 - \frac{1}{T}\right)\right] = \alpha(Q-\beta) \tag{13}$$

and

$$\hat{Q}_a(T) = \hat{\beta} + y(T)/\hat{\alpha} \quad . \tag{14}$$

In case of Eq.(9), (Yevjevich and Taesombut 1978, after adjustment for difference in symbols with Eq.(9))

$$\mathrm{var}\ \hat{Q}_p(T) = \frac{1}{\lambda\beta^2 n} \{1 + [\ell n\lambda + y(T)]^2\} \quad , \tag{15}$$

with $\lambda$=the average number of partial floods per year, so that the ratio R of Eq.(11) becomes

$$R = \frac{\hat{\lambda}\hat{\beta}^2}{\hat{\alpha}^2} \cdot \frac{[1.11 + 0.52\, y(T) + 0.61\, y^2(T)]}{\{1 + [\ell n\hat{\lambda} + y(T)]^2\}} \tag{16}$$

Figure 13 presents the three variable relationship: R against y(T) or T for given $\lambda$, with $\hat{\alpha}=\hat{\beta}$. Two conclusions can be drawn from Fig. 13 and Eq.(16):

(1) The simplified theoretical approach shows that the partial series in general has more information for the highest annual flood peaks than the annual series as soon as $\lambda$ reaches the highest value of about 1.5 (or one and one-half flood peaks per year, on the average); and

(2) The smaller the flood return period T, in this case, the more information is contained in the partial series on the highest annual peak in comparison with the annual series, for all the other conditions being the same.

In general, there should be a corresponding reduction in the ratio R of Fig. 13 as $\lambda$ increases, because with a decrease of the truncation level $Q_b$ of Figs. 11 and 12, the average number $\lambda$ of partial flood peaks per year increases, but also the autocorrelation of the sequence of partial flood peaks increases (meaning a decrease of information).

Figure 14 gives the example for the generated samples of size N=50 of daily flows of the Powell River (USA), as the relationship of the ratio R of Eq.(11) versus y(T) or T for given $\lambda$, showing that for about $\lambda$=2.2 and T > 50, the partial series has more information on annual flood peaks than the annual series.

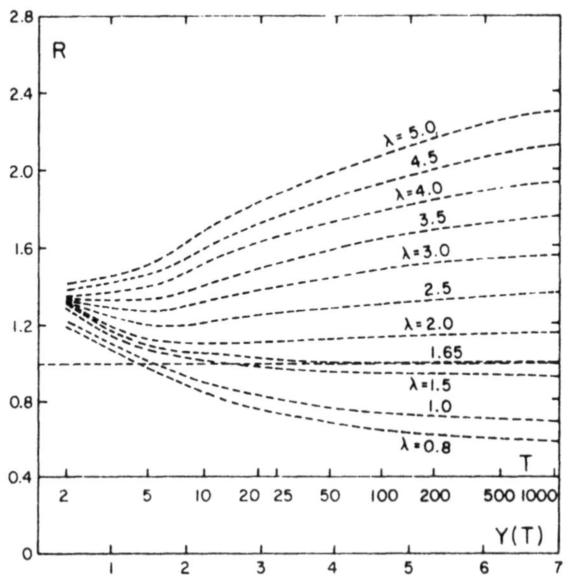


Figure 13. Relationship between the ratio R of sampling variances $\hat{Q}(T)_a$ and $\hat{Q}(T)_p$, as the index of information in estimates of highest annual flood peaks, based on theoretical approach, versus the return period T, for given $\lambda$.

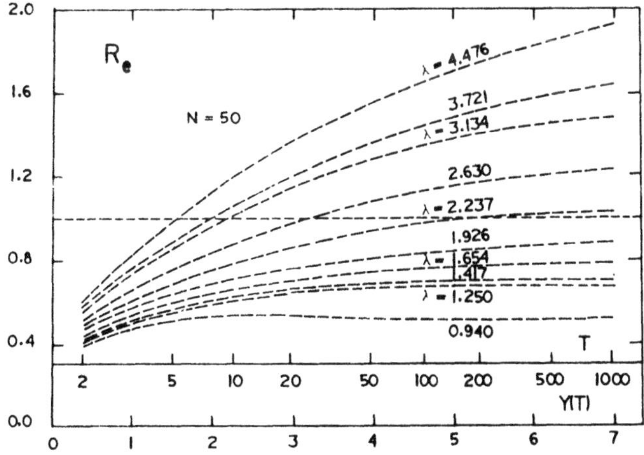


Figure 14. Relationship of ratio R to return period for $\lambda$=0.940-4.476 with $Q_b$=4500-9500 cfs, for N=50, for generated daily flow samples of the Powell River, USA (after Yevjevich and Taesombut 1978).

### Distributions of Flood Volumes and Flood Durations

The experience shows that frequency distributions of flood volumes are often well fitted either by the two-parameter lognormal or the two-parameter gamma PDF, similarly as for the monthly or annual total water volumes. The distribution of durations of flood discharges above a critical flood discharge often have shown to be the two-parameter gamma for continuous times (or eventually the geometric discrete distribution for duration counted as the discrete number of basic time intervals).

## LOW FLOW AND DROUGHT EXTREMES

### Low Flow Distributions

The three basic low-flow describing random variables are: $Q_{min}$=the instantaneous lowest value; $V_d$=the deficit of water during the low-flow season below a given constant (expected draft) value; and $t_{max}$=the duration of this deficit (low flows).

The experience has shown that the most attractive PDF for fitting the frequency distributions of $Q_{min}$ is the bounded (limited) exponential distribution of the smallest value

$$F(x) = \exp\left[-\left(\frac{x-\gamma}{\beta-\gamma}\right)^{\alpha}\right] \tag{17}$$

introduced by R.A. Fisher and L.H.C. Tippett, and highly advocated by E.J. Gumbel, with $\alpha > 0$, $\beta > 0$, and $\gamma \geqslant 0$, and the transformatio

$$Y = \left(\frac{x-\gamma}{\beta-\gamma}\right)^{\alpha} \tag{18}$$

that leads to

$$F(x) = e^{-y} \tag{19}$$

The distribution of the water deficits during the low-flow season seem also to be best fit by either the two-parameter lognormal or the two-parameter gamma PDF, as the case is with many other volume related random variables of hydrology and water resources. Similarly, the duration of long low-flow seasons may be fitted well by the two-parameter gamma distribution or the geometric distribution, respectively for the continuous or discrete case of measuring the time duration.

## Point Droughts

The crossing theory or the theory of runs seems to be the most appropriate probabilistic approach to the study of point droughts, as the range theory seems to be the most pertinent approach in the study of water storage capacity and regulation of water flows. The theory of runs helps produce the distributions of runs (length, sum), with the negative run associated either with the length of run (negative run-length) or with the sum of values in the run (negative run-sum), as the duration and volume descriptions of droughts, respectively.

Mostly, the hydrologic extremes in droughts are related to the maximum negative run-length in the sample of size N (longest drought), maximum negative run-sum in the sample of size N (largest drought by its deficit of water), or the maximum instantaneous or time-interval deficit intensity of all the runs in a sample.

For independent series of annual precipitation or annual runoff, the probability of a negative run-length of n time intervals is

$$f(n) = p\,q^{n-1} \tag{20}$$

where q=probability of non-exceedence of the crossing level that defines run (droughts), and p=1-q. When the annual series become time dependent, as the case is for most annual runoff series, the computation of probabilities of runs of a given length becomes somewhat more complicated (Saldarriaga and Yevjevich, 1970).

The asymptotic expected value of longest negative run-length (longest drought) in the sample of size N is given by Cramer and Leadbetter (1967) for independent series by

$$E(L_m) = -\frac{\log N}{\log(1-q)} + 0(1) \tag{21}$$

with $q=P(X \leqslant x_0)$, $x_0$=the crossing level, and 0(1)=the error of the order of one.

Most of distributions of longest negative run-lengths (duration of longest drought), but especially of largest negative run-sums (largest deficit of the drought) in the sample of size N, for non-normal time-dependent random variables, of varying crossing level of $q=P(X \leqslant x_0)$, are obtained by the Monte Carlo method in generating a large set of samples of a given size, and the corresponding cumulative frequency distributions of these variables, because no exact distributions are available in the literature. As examples, Fig. 15 gives the frequency distributions of the

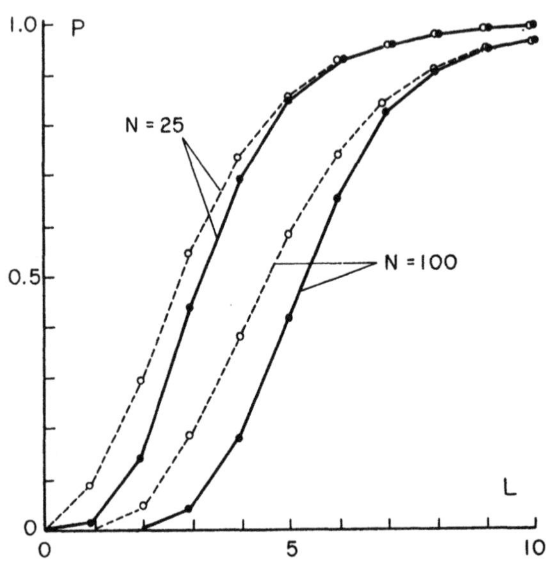


Figure 15. Distributions of the longest negative run-length for q=0.50, ρ=0.0, and γ=00, and two samples N=25 and N=200, (solid lines), and of the negative run-length for the largest negative run-sum (dashed lines); (after Millan and Yevjevich 1971).

longest negative run-length of normal independent process ($\gamma$=0, $\rho_1$=0, q=0.50, and N=25 and 100), while Fig. 16 gives the frequency distributions of the largest negative run-sums (maximum drought deficit) for the same samples and other properties. The dashed lines of Figs. 15 and 16 give the frequency distributions of the negative run-length for the largest negative run-sum (Fig. 15) or of the negative run-sum for the longest negative run-length (Fig. 16), because the longest negative run-legnth of a sample does not necessarily coincide with the run giving the largest negative run-sum, or vice versa (for details, see Millan and Yevjevich, 1971). For the state-of-the-art on point droughts, see Guerrero-Salazar and Yevjevich, 1975.

Regional Droughts

Droughts are regional phenomena. Therefore, the spread of water deficits or shortages in both time and space are important. Therefore, the areal spread of droughts as well as their time duration along with their severities as water deficits, are very relevant topics of hydrologic extremes. For some aspects of regional droughts, see Tase (1976) and the book "Coping with Droughts" (edited by Yevjevich, da Cunha and Vlachos, 1983) which gives not only a summary of the state of knowledge on regional droughts,

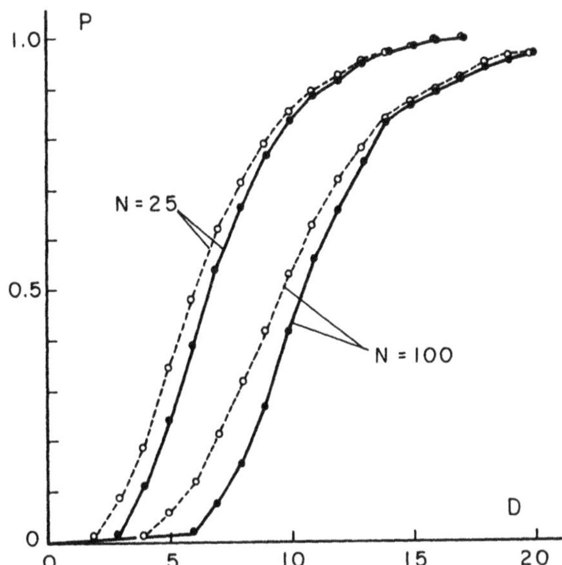


Figure 16. Distributions of the largest negative run-sum for q=0.50, ρ=0.00, γ=0.00 and two samples N=25 and N=100, (solid lines), and of the negative run-sum for the longest negative run-length (dashed lines); (after Millan and Yevjevich 1971).

but several case studies from around the world.

## ACKNOWLEDGMENT

The research leading to this paper is sponsored by the U.S. National Science Foundation, Grant No. CME-79-16817. This support is respectfully acknowledged.

## REFERENCES

Benson, M., 1960, "Flood Frequency Analysis", U.S. Geological Survey, Water Supply Paper 1543-A.

Chow, V.T., 1951, "A General Formula for Hydrologic Frequency Analysis", Trans. Am. Geophys. Union, Vol. 32, p. 231-237.

Guerrero-Salazar P. and V. Yevjevich, 1975, "Analysis of Drought Characteristics by the Theory of Runs", Colorado State University Hydrology Papers, No. 80, September, Fort Collins, Colorado.

Millan, Jaime, and Vujica Yevjevich, 1971, "Probabilities of Observed Droughts", Colorado State University Hydrology Papers, No. 50, June, Fort Collins, Colorado.

Saldarriaga, J., and V. Yevjevich, 1970, "Application of Run-Lengths to Hydrologic Series", Colorado State University Hydrology Papers, No. 80, September, Fort Collins, Colorado.

Taesombut, V., and V. Yevjevich, 1978, "Use of Partial Flood Series for Estimating Distribution of Maximum Annual Flood Peaks", Colorado State University Hydrology Papers, No. 97, October, Fort Collins, Colorado.

Tao, P.C., V. Yevjevich and N. Kottegoda, 1976, "Distributions of Hydrologic Independent Stochastic Components", Colorado State University Hydrology Papers, No. 82, January, Fort Collins, Colorado.

Tase, Norio, 1976, "Area-Deficit-Intensity Characteristics of Droughts", Colorado State University Hydrology Papers, No. 87, November, Fort Collins, Colorado.

Tavares, L.V. and J.E. Da Silva, 1983, "Partial Duration Series Methods Revisited", Journal of Hydrology, Vol. 64, No. 1/4, July, pp. 1-14.

Todorovic, P., 1970, "On Some Problems Involving Random Number of Random Variables", Annals of Math. Statistics, V. 43, No. 3.

Yevjevich, V. and V. Taesombut, 1978, "Information on Flood Peaks in Daily Flow Series", Proc. Int. Symp. on Risk and Reliability in Water Resources, University of Waterloo, Waterloo, Canada.

Yevjevich, V. and T.G.J. Dyer, 1983, "Basic Structure of Daily Precipitation Series", Journal of Hydrology, Vol. 64, No. 1/4, pp. 49-68.

Yevjevich, V., L. da Cunha and E. Vlachos, Editors, 1983, "Coping with Droughts", Book published by Water Resources Publications, POB 2841, Littleton, Colorado, USA.

Zelehasic, Emir, 1970, "Theoretical Probability Distributions for Flood Peaks", Colorado State University Hydrology Papers, No. 42, November, Fort Collins, Colorado.

# APPLICATION OF EXTREME VALUES IN STRUCTURAL ENGINEERING

G.I. Schuëller

Institut für Mechanik, Universität Innsbruck (Austria)

Basically, the design process in structural engineering is based on extreme value considerations. In previous times, in which statistical properties of structural loading as well as structural resistances were not recognized, the design procedures have been based on the selection of largest, i.e. "maximum" values of loads and smallest i.e. "minimum" resistances purely on an empirical basis. In modern developments of the theory of structural safety and reliability, loads as well as resistances are dealt with in a scientific, i.e. rational way which implies their modeling by extreme value distributions. This contribution contains various examples of application of extreme value models particularly in the areas of wind-, earthquake- and ocean engineering.

## 1. INTRODUCTION

Although the first attempt to introduce probabilistic concepts to structural problems goes back as far as 1926 (1) it was A.M. Freudenthal (2-5) and Freudenthal and Shinozuka (6,7), who developed what is now called the classical structural reliability theory. The basic principles of this theory will be outlined briefly in the following section; for a treatise, however, it is referred to Schuëller (8) and Ang and Tang (9). It is interesting to note that from the very beginning the discussion centered on the justification for applying certain probability models to certain classes of problems. Needless to say that an enormous amount of data is required to be able to give preference to one particular type of distribution over an other one, if it is referred to the entire statistical population. In



J. Tiago de Oliveira (ed.), Statistical Extremes and Applications, 221–234.

structural engineering this aspect is most important as mostly extrapolations have to be made far beyond the range over which data, i.e. observations are available. Therefore, in addition to statistical fit procedures and additional criterion has to be provided to select a particular type of distribution. For instance this might be that the probability model used is required to be physically germane to the problem under consideration. An excellent example of physical relevance within the framework of structural reliability analysis is the application of models which lead to extreme value distributions. The concept of structural safety is based on the fact that the "maximum" load experienced during the design life of the structure does not exceed its "minimum" resistance. This leads directly to the consideration of the largest values of samples of load observations and the smallest values of samples of strength observarions (see Figure 1).

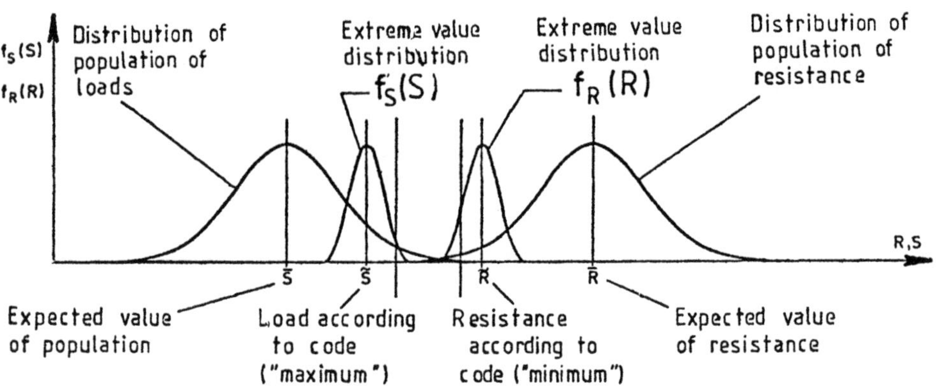


Figure 1. Schematic Sketch of Extreme Value Distributions of S and R in Relation to the Distributions of Population (1o).

The definition of a random variable as an extreme of a large sample of an unknown distribution of the population makes the prediction, particularly in the range of interest, much more reliable as if it were based on statistical fit procedures of the initial population. In the latter case the bulk of data which influences the statistical estimation of the parameters of the distribution most significantly is located in a range far removed from the actual range of interest.

## 2. BASIC PRINCIPLES OF STRUCTURAL RELIABILITY

The fundamental concepts of structural reliability theory are based on the derivation of mathematical models from principles of mechanics and experimental data which relate structural resistance and load variables. This relation might be defined by a limit state- or failure surface of the following form:

$$g(\underline{X})=g(X_1, X_2, \ldots, X_n)=0 \tag{1}$$

where the X represent the random variables of structural resistance and loading. The limit state is reached, i.e. failure occurs when the condition $g(\underline{X}\leqslant 0)$ is met. Limit states are a matter of definition. They might be defined as ultimate load failure, uncerviceability, etc.. Depending on the respective problem the limit states may be considered individually or combined. The probability that a particular limit state will be reached, i.e. the probability of failure may be expressed as (7):

$$p_f=\int_D f_{\underline{X}}(\underline{\eta})\,d\underline{\eta} \tag{2}$$

where $f_{\underline{X}}(\underline{\eta})$ is the joint density function of $X_1, X_2 \ldots, X_n$, i.e.

$$f_{\underline{X}}(\underline{\eta})=f_{X_1,X_2\ldots,X_n}(\eta_1,\eta_2,\ldots,\eta_n),$$

D the domain in the n-dimensional space in which the failure condition is satisfied and $d\underline{\eta}=d\eta_1, d\eta_2 \ldots, d\eta_n$. In the case of two stochastically independent random variables modeling the load S and the resistance R the above equation can be expressed in terms of a convolution integral:

$$p_f=P(R\leqslant S)=\int_{-\infty}^{\infty} F_R(x)\,f_S(x)\,dx \tag{3}$$

where $F_R(x)$ is the cumulative distribution function (CDF) for R and $f_S(x)$ the probability density function (PDF) for S. To illustrate the problem a schematic sketch is shown in Figure 2.

It is obvious that the direct solution of the equ.(2) is quite difficult if not to say in most cases intractable and this for mainly two reasons: first, in general the scarce data base is insufficient to estimate the parameters of the joint density function and second, there are great numerical difficulties to perform the multidimensional integration over the domain D, particularly for irregular shapes of D. In order to circumvent this problem, first order approximations which are based on the second-statistical moments (FOSM) have been developed to various degrees of sophistication. Although the procedure can not bypass the data problem it can be, however, simplified by introducing correlations between the various variables involved. The method which has been introduced in its basic form, i.e. for two vari-

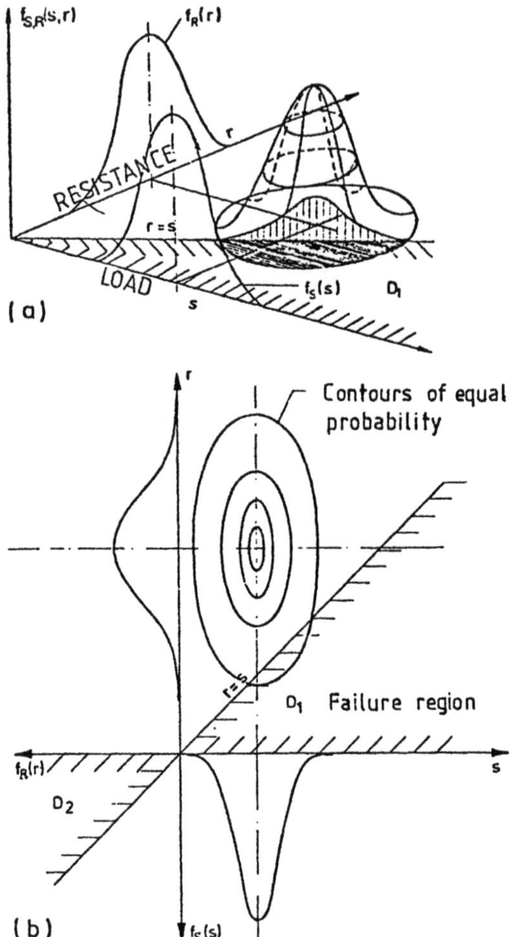


Figure 2. Schematic Sketch for Calculation of Failure Probability $p_f$ for Case of Two Random Variables (8).

ables, by Freudenthal (4) utilizes the standardized form of the random variables, i.e.

$$x_i = \frac{X_i - E(X_i)}{\sigma_i} \tag{4}$$

The limit state equation is then

$$g_1(\underline{x})=g_1(x_1,x_2,\ldots,x_n)\leqslant 0 \tag{5}$$

In this context it should be noted that the expansion of the procedure to n dimensions has been carried out first by Hasofer and Lind (11).

The problem is now to find the so called design point $x_*$. (See Figure 3)

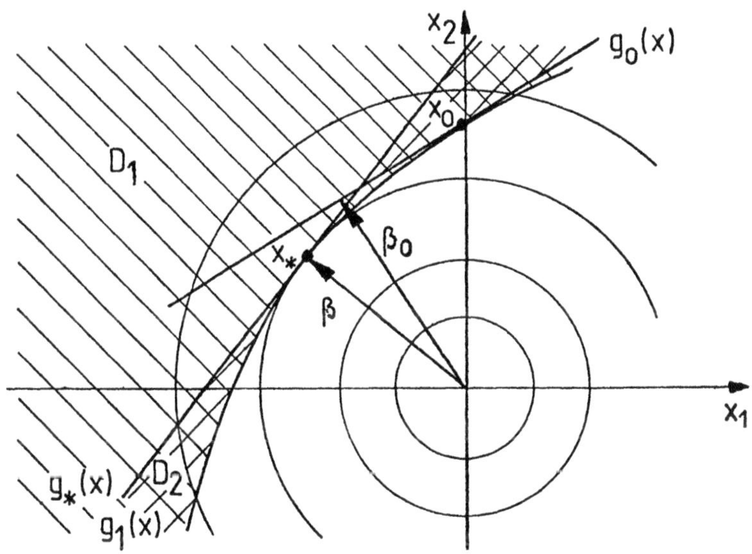


Figure 3. Schematic Sketch of Design Point (for Uncorrelated Variables).

It can be found by existing optimization procedures, such as, for example Lagrangian multiplier (Shinozuka (12)) or any appropriate library routine available in most computer centers (Dolinski (13)). It has been shown by Freudenthal and Shinozuka (4,6), that for correlated or non correlated Gaussian variables the design point is also the point of maximum likelihood. For non Gaussian variables the following transformation may be introduced:

$$\underline{u}_i=\Phi^{-1}[F_i(x_i)] \tag{6}$$

in which $F_i(\cdot)$ is the CDF of $x_i$. The transformation of the vector $\underline{x}$ into $\underline{u}$ requires also the transformation of the limit state function $g_i(\underline{x})$

$$x_i = F_i^{-1}[\Phi(u_i)] \tag{7}$$

and

$$g_i(\underline{x}) = g_i\{F^{-1}[\Phi(\underline{u})]\} = h_i(\underline{u}) \leq 0 \tag{8}$$

Therefore, in general terms the method would also work for extreme value distributions. In practical application, however, due to the transformation, the limit state function becomes quite complicated. In this case, however, being still the shortest distance to the origin, the design point looses its maximum likelihood property. Finally, for the calculation of the limit state, i.e. failure probabilities the following integral has to be evaluated - preferably by Monte Carlo (MC) simulation procedures -

$$p_f = \int_D f_{\underline{x}}(\underline{\eta})\,d\underline{\eta} \tag{9}$$

where D is the domain, where $g_1(\underline{x}) \leq 0$ and $f_x(\eta) = f_{x_1, x_2 \ldots, x_n}(\eta_1, \eta_2 \ldots, \eta_1)$ is the joint density function of $\underline{x}$. The above equation is similar to equ. (2). For MC evaluation for example the following expression is suggested by Shinozuka (12):

$$p_f = \frac{A}{N} \sum_{i=1}^{N} \frac{\delta_i}{(2\pi)^{n/2}} \exp\left(-\frac{1}{2} \underline{\eta}_i^T \underline{\eta}_i\right) \tag{1o}$$

with $\delta_i = \begin{cases} 0 \text{ if } g(\underline{\eta}_i) > 0 \\ 1 \text{ if } g(\underline{\eta}_i) \leq 0, \end{cases}$

A is the area of the domain of integration and n the number of simulated samples. By doing this, the advantage of importance sampling becomes quite obvious. Needless to say that like any approximate method FOSM has just a limited possibility of application as well as accuracy. Aside some comparisons of selected problems with exact methods only recently a first attempt has been made by Stix and Schueller (14) to illuminate these aspects (see also Breitung (18)).

The failure probability as defined by equ. (2) and (3), however, is not yet a direct measure of the safety of a structure. This measure may be expressed as the reliability which is a time varying quantity, i.e.

$$\begin{aligned} L_T(t) &= 1 - F_T(t) \\ &= \sum_{x=0}^{n} P(\cdot)\,p_x(t) \end{aligned} \tag{11}$$

where $F_T(t)$ is the failure probability now expressed as a function of time, T the design life of the structure, $P(\cdot)$ the survival probability of a structure under x load applications, i.e. $(1-p_f)^x$ and $p_x(t)$ the probability of occurrence of the load events E, such as severe storms, earthquakes, etc.. Environmental hazards, for example may be modeled best by some type of a Poisson process (Russell and Schuëller (15,16)). The load intensity may also be modeled by distributions of annual maximum loads. A model for reliability assessment is shown in Figure 4.

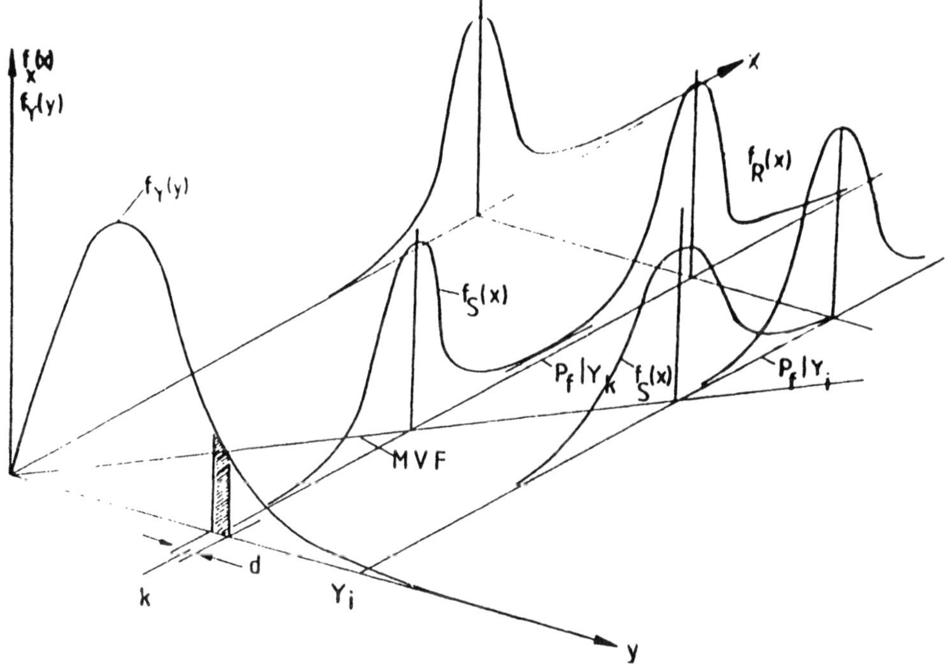


Figure 4. Schematic Sketch of Reliability Concept

Within this model the $f_Y(y)$ is defined as the PDF of the load in terms of extreme wind velocities (pressures), waves or earthquake accelerations etc.. The distribution $f_S(x)$ is the conditional PDF of the load effect given a particular realization of load intensity and $f_R(x)$ the PDF of the structural resistance. Depending on the particular problem, type and parameters of this distribution may be determined by structural - either in the static or dynamic - analysis. The conditional probability of survival is then

$$L|E = \int_0^\infty (1-p_f|Y) f_Y(y)\,dy \tag{12}$$

and the survival probability

$$L_T(t) = \sum_{x=o}^{n} L|E \cdot p_x(t) \tag{13}$$

## 3. APPLICATIONS

### 3.1 General

It should be mentioned at this point that this paper concentrates on aspects of applications of extreme value distributions to problems in structural engineering rather than on theoretical development. For the latter purpose it is referred to other contributions to this study and other pertinent contributions (i.e. (17,1o)).

In brief, the asymptotic distributions are based on the following considerations. In a sample of the size n (where n is large) - which represents observations of a random variable X described by a distribution function F(x) - the probability that all n (independent) observations are smaller than $x_n$, i.e. that $x_n$ is the largest of the n observations, is defined by

$$F_n(X_n) = [F(x)]^n \tag{14}$$

and therefore

$$f_n(X_n) = n[F(x)]^{n-1} f(x) \tag{15}$$

In other words, equ.(14) and (15) represent the CDF and PDF of the largest of n independent observations. The distribution function of the smallest of n independent observations is therefore

$$F_1(X_1) = 1-[1-F(x)]^n \tag{16}$$

and respective PDF

$$f_1(X_1) = n[1-F(x)]^{n-1} f(x) \tag{17}$$

Equ. (14) to (16) represent the basic equations of the theory of extreme value distributions. If the distributions of the population F(x) are known, the extreme value distributions may be derived (generally by numerical integration).
The asymptotic distributions which are valid for large values of n

$$\lim_{n \to \infty} F_n(X_n) = \lim_{n \to \infty} [F(x)]^n \tag{18}$$

and

$$\lim_{n \to \infty} [1-F_1(X_1)] = \lim_{n \to \infty} [1-F(x)]^n \qquad (19)$$

lead directly to undefined expressions which may be avoided by studying the asymptotic behavior with increasing n.
Now if for the case of the largest values as form of the basic function $F(x)=1-e^{-\alpha_n x}$ is assumed the distribution of

$$F_n(X_n) = [1-e^{-\alpha_n x}]^n \qquad (2o)$$

is to be sought.
This leads to the so called "First Asymptotic Distribution" of the largest values, which in its standardized form can be written

$$F_n(y) = \exp[-e^{-y}] \qquad (21)$$

and

$$f_n(y) = \alpha_n \exp[-ye^{-y}] \qquad (22)$$

For the First Asymptotic Distribtuion of the smallest values one obtaines similarly

$$F_1(y) = 1-e^{-e^y} \qquad (23)$$

and

$$f_1(y) = \alpha_1 c^{y-e^y} \qquad (24)$$

where $y=\alpha(x-u)$. The First Asymptotic is also called the Gumbel distribution (17).

Now if for the basic function, i.e. distribution of the population the form

$$F(x) = 1-x^{-\beta} \qquad (25)$$

is assumed, the following distribution for the largest values is obtained:

$$F_n(x_n) = \exp[-\frac{u}{x}]^{\beta} \qquad 0 \leqslant x \infty \qquad (26)$$

This distribution is also called the Second Asymptotic or Frêchet distribution. It may also be derived from the First Asymptotic by the following transformation: $y=\beta \ln(x/n)$.
The distribution of the smallest values is easily obtained by considering properties of symmetry, i.e.

$$F_1(-x_1) = 1-F_n(x_n) \qquad (27)$$

Therefore the Second Asymptotic distribution of the smallest values is

$$F_1(x_1) = 1-\exp[-\frac{u}{x}]^{\beta} \tag{28}$$

Again, the First Asymptotic is based on a exponentially decreasing basic function, i.e. distribution of the population, while the Second Asymptotic is based on a polynomial type. For cases where the basic functions are limited a Third Asymptotic may be obtained. For the largest values the functional form is

$$F_n(x_n) = \exp\left[-\left(\frac{\omega-x}{\omega-n}\right)^{\gamma}\right] \qquad -\infty<x\leq\omega \tag{29}$$

and for the smallest values

$$F_1(x_1) = 1-\exp\left[-\left(\frac{x-\omega}{u-\omega}\right)^{\gamma}\right] \qquad x\geq\omega \tag{3o}$$

Equ. (3o) is also called the Weibull distribution.
Generally, in engineering applications only three types of Asymptotic distributions are used, i.e. the Gumbel- and Frêchet distributions (equ. (21),(26)) and the Weibull distribution for the smallest values (equ. (3o)).

## 3.2 Wind Exposed Structures

Wind engineering is a field where the application of extreme value distributions is of particular interest. Knowledge on expected extreme wind velocities comprise the basic information for analysis of structures under wind loading. In this context generally data of annual maximum wind velocities -either referring to gusts, hourly - or 1o min. average velocities - are utilized. This implies that the annual maximum value is representative for all possible wind velocities occurring during that year. It has been pointed out before, under these circumstances only the class of extreme value distributions apply. Due to the scarcity of data however it is difficult to make a statistical choice as to which type of extreme value distribution fit the data best. Tiago de Oliveira (19) has shown that for this purpose a minimum of 5o data points would be required. In Fig. 5 the fit of both the Gumbel- and the Frêchet distribution are fitted to 17 years of data only in few cases reliable data are available for a longer period of time additional arguments are given for preference of a particular type of distribution. It is claimed, for example, that the distribution of the population of wind velocities is Gaussian. This, of course supports the assumption of the Gumbel distributed annual maximum wind velocities. It should however be mentioned, that other arguments might support the assumption of the Frêchet distribution as well (2o).

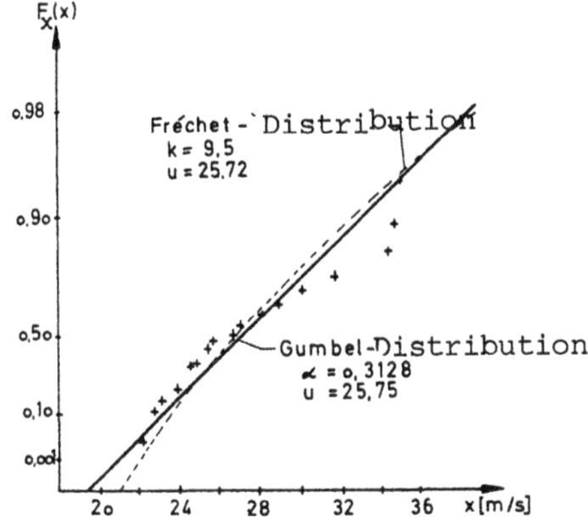


Figure 5. Wind velocity data - Annual Maximum Values of 17 Years of Record (Gumbel Paper)

## 3.3 Offshore Structures

It is generally recognized that the most significant factor for the risk of failure of platforms is due to storm generated water waves (21). Again, the extreme wave heights are the governing parameter for the load distribution and consequently for the reliability analysis (22). For this reason extensive research has been carried out to develop models for decision on particular types of distributions (23-28). Fig. 6 shows for example the fit

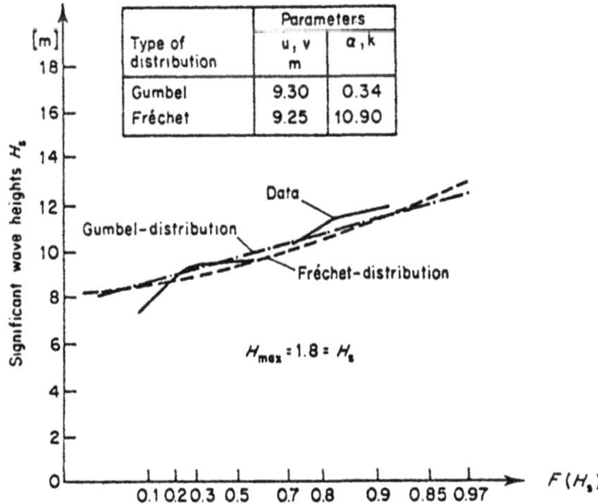


| Type of distribution | Parameters u, v m | α, k |
|---|---|---|
| Gumbel | 9.30 | 0.34 |
| Fréchet | 9.25 | 10.90 |

Figure 6. Fit of Asymptotic Distributions to Maximum Significant Wave Heights ($H_s$) Resulting from 1o Storm Events observed within 5o Years (16).

of the Gumbel- and the Frêchet distribution to hurricane wave data, which is also not too conclusive. Again, the argument of Gaussian sea state is also used to support the Gumbel distribution for long term wave forecast.

### 3.4 Earthquake Hazard

Similar to the two types of natural hazards discussed in the previous section, i.e. wind- and wave loading, earthquake induced loading - in terms of ground acceleration and velocities - is also to be modeled by extreme values. Due to lack of data no statistical choice with respect to the types of asymptotic distribution can be made. Gumbel-, Frêchet as well as Weibull distributions are used to model earthquake risk (Kanda (3o)). More than the other two this type of natural hazard is an open area for statistical research.

## 4. CONCLUSIONS

It has been shown, that in structural reliability extreme value distributions play an important if not a dominating role. Risk of structural failure may only be estimated rationally by dealing with the extremes of load and resistances. For the purpose of extrapolation beyond the range of observations, i.e. in the regions of the tails of the distributions, it is most important to provide sound statistical estimation procedures. Models for statistical choice between various types of distributions should be further developed.

## REFERENCES

[1] Mayer, M.: Die Sicherheit der Bauwerke, J. Springer Verlag, Berlin, 1926.
[2] Freudenthal, A.M.: The Safety of Structures, Transactions, ASCE, Vol. 112, 1947.
[3] Freudenthal, A.M.: Reflections on Standard Specifications for Structural Design, Transactions, ASCE, Vol. 113, 1948, pp.269 - 292.
[4] Freudenthal, A.M.: Safety and the Probability of Structural Failure, Transactions, ASCE, Vol. 121, Proc. Paper 2843, 1956, pp. 1337-1397.
[5] Freudenthal, A.M.: Safety, Reliability and Structural Design, J Struct. Div., ASCE, Vol. 87, No. ST3, 1961.
[6] Freudenthal, A.M. and Shinozuka, M.: Structural Safety Under Conditions of Ultimate Load Failure and Fatigue, WADD Techn. Report 61-177, Aeronautical Systems Div., Wright-Patterson Air Force Base Ohio, October 1961.
[7] Freudenthal, A.M., Garrelts, J.M. and Shinozuka, M.: The

Analysis of Structural Safety, J. Struct. Div., Proc. ASCE, Vol. 92, No. ST1, Feb., 1966, pp. 267-325.

[ 8] Schüeller, G.I.: Einführung in die Sicherheit und Zuverlässigkeit von Tragwerken, W. Ernst u. Sohn Verlag, Düsseldorf, 1981.

[ 9] Ang, A.H-S. and Tang, W.H.: Probability Concepts in Engineering in Planning and Design, Vol. II, John Wiley, New York, 1984.

[1o] Freudenthal, A.M. and Schüeller, G.I.: Risikoanalyse von Ingenieurtragwerken, Reports, Konstr. Ingenieurbau, Ed.: Zerna, Report No. 25, Vulkan Verlag, Essen, Aug. 1976, pp. 7-95.

[11] Hasofer, A.M. and Lind N.C.: Exact and Invariant Second Moment Code Format, Journ. Engr. Mech. Div., ASCE, Vol. 1oo, No. EM1, Proc. Paper 1o376, Feb., 1974, pp. 111-121.

[12] Shinozuka, M.: Basic Analysis of Structural Safety, ASCE J. Struct. Eng., Vol. 1o9, No. 3, March 1983, pp. 721-74o.

[13] Dolinski, K.: First-Order Second-Moment Approximation in Reliability of Structural Systems: Critical Review and Alternative Approach, J. Struct. Safety, Vol. 1, No 3, April, 1983, pp. 211-231.

[14] Stix, R. und Schüeller, G.I.: Problemstellung bei der Berechnung der Versagenswahrscheinlichkeit, Institut für Mechanik, Univ. Innsbruck, Internal Working Report No. 3, 1983.

[15] Russell, L.R. and Schüeller, G.I.: Probabilistic Models for Texas Gulf Coast Hurricane Occurrences, J. Petroleum Techn., March, 1974, pp. 279 - 200.

[16] Schüeller, G.I.: External Hazards, Chapt. 3.1. of "High Risk Safety Technology", A.E. Green (Ed.), John Wiley and Sons, London, 1982, pp. 483-496.

[17] Gumbel, E.J.: Statistics of Extremes, Columbia Univ. Press, New York, 1958.

[18] Breitung, K.: Asymptotic Approximations for Multinormal Domain and Surface Integrals, Proc. 4th Int. Conf. Appl. Statist. & Probability in Soil & Struct. Engr., G. Augusti et al. (Ed.) Pitagora Editrice, Bologna, 1983.

[19] Tiago de Oliveira, J.: Statistical Choice of Extreme Models, NATO-ASI, Vimeiro, Portugal, Aug., 1983.

[2o] Schüeller, G.I. and Panggabean, H.: Probabilistic Determination of Design Wind Velocity in Germany, Proc. Instn. Civ. Engrs., Part 2, 1976, 61, Dec., pp. 673-683.

[21] Schüeller, G.I.: Risk Criteria for the Design of Fixed Platforms, Proc., 2nd. Int. IAHR Symp. on Stoch. Hydraulics, Lund, Sweden, Water Res. Publ., Fort Collings, Col., USA, pp. 349-367.

[22] Schüeller, G.I. and Choi, H.S.: Offshore Platform Risk Based on a Reliability Function Model, Proc. Offshore Techn. Conf., Houston, May, 1977, pp. 473-479.

[23] Flatseth, J.H. and Pedersen, B.: Distribution of Wave Height

[23] in different Ocean Areas around the world, Det Norske Veritas, Res. Dep. Rep. No. 7o-7-S, 197o.

[24] Petrauskas, C. and Aagaard, P.M.: Extrapolation of Historical Storm Data for Estimation of Design-Wave Heights, Soc. of Petr. Eng. J., March, 1971, pp. 23-37.

[25] Jahns, H.O. and Wheeler, J.D.:Long-Term Wave Probabilities Based on Hindcasting of Severe Storms, Preprints, Offshore Techn. Conf., Paper No. 159o, May, 1972.

[26] Battjes, J.A.: Long-Term Wave Height Distribution at Seven Stations around British Isles, Deutsche Hydr. Zeitschrift, Jahrgang 25, Heft 4, 1972.

[27] Thom, H.C.S.: Extreme Wave Height Distributions over Oceans, J. Waterw. Harb. Coast. Engr., Proc. ASCE, Vol. 99, Aug. 1973, pp. 355-374.

[28] Borgman, L.E.: Probabilities for the Highest Wave in Hurricane, J. Waterways, Harbors and Coastal Engr., Proc. ASCE, Vol. 99, No. WW2, May, 1973, pp. 185-2o7.

[29] Bea, R.G.: Gulf of Mexico Hurricane Wave Heights, Prepr. Offshore Techn. Conf., Vol. II, Houston, May, 1974, pp. 791-81o.

[3o] Kanda, J.: A New Extreme Value Distribution with Lower and Upper Limits for Earthquake Motions and Wind Speeds, Theoretical and Applied Mechanics, Vol. 31, 1981, pp. 351-36o.

# EXTREMES IN METEOROLOGY

R. Sneyers

Royal Meteorological Institute, Brussels

ABSTRACT.

Statistics of meteorological extremes are generally needed for economic purposes.

In this paper a review is made successively on the following topics :
1) generalities on the statistical analysis of time series of extremes,
2) distribution of meteorological extremes and extreme distributions in meteorology,
3) evaluation of extremes in the case of short series of observations :
   a) statistical analysis of large values,
   b) extremes in autocorrelated series.

Examples are given with Belgian climatological series.

## 1. INTRODUCTION.

Extremes may be considered in meteorology at different time levels. There are daily, monthly, annual and for given periods absolute extremes. Daily extremes are used to characterize the weather of 24-hour periods. The average of the daily maximum and minimum is f.i. an excellent tool for calculating fuel consumption through degree days. Monthly and annual extremes are generally bounded to unusual nice or bad weather. Therefore they may be put into connection f.i. with good performance of travel industry or of certain branches of commerce or on the contrary they may be the cause of severe damages to agriculture, to the environment or even to buildings. In each case it turns out into economic advantages or losses and raises the question of the advisability of investments for productiveness or for protection.

In the case of protection two attitudes are possible in this problem : either fix an upper limit to the economic investment necessary for protection against unusual weather conditions or avoid the damage at any price.



*J. Tiago de Oliveira (ed.), Statistical Extremes and Applications, 235–252.*

In each case the solution of the problem lies in the knowledge of the probabilities associated with extreme weather conditions that is with extreme values. Examples are easy to give : maximum temperature and functioning of airplane engines, minimum temperature and frost damages, maximum rainfall and floods, droughts and crop damages, maximum windspeed and security of buildings, maximum snow cover and means of communication, persistence of snow cover and practice of wintersports, etc.

For the answer of all those questions long series of observations are generally needed. Fortunately, meteorological extremes have been observed since relatively a long time owing to the interest of the meteorologists for such series. Moreover, it may be shown that for establishing the distribution of extremes, it is also possible to use relatively short series.

This being said, the paper will be devided into three parts :
1) generalities on statistical analysis of time series of extremes;
2) distributions of meteorological extremes and extreme distributions in meteorology;
3) evaluation of extremes in the case of short series of observations :
 a) statistical analysis of large values,
 b) extremes in autocorrelated series.

Moreover, examples will be given originating generally from Belgian climatological series.

## 2. STATISTICAL ANALYSIS OF TIME SERIES OF EXTREMES.

To make statistical prediction possible two kinds of properties are necessary : stability of the series of observations or even better independence of the series and goodness of fit of the adjusted distribution to the series of observations.

### 2.1 Independence.

Independence may be tested by examining the series for trend and for serial correlation which are respectively long range and short range alternatives to independence.

#### 2.1.1. Test against trend.

The trend test which has our favour is the one of Mann (cf [1] p. 11). If $x_i$, $i=1, 2, ..., n$ is the time series and if for each element $x_i$, $n_i$ is the number of the preceding $x_j$ $(i>j)$ such as $x_i>x_j$, the test statistic is then given by the relation :

$$t=\Sigma n_i \tag{1}$$

Under the hypothesis of independence, its distribution function is asymptotically a normal distribution with mean and variance :

$$E(t)=n(n-1)/4 \quad \text{and} \quad \operatorname{var} t=n(n-1)(2n+5)/72 \tag{2}$$

The test is non parametric and has the advantage to be able to be performed progressively from the first element to the following ones or from any element of the series to the last one.

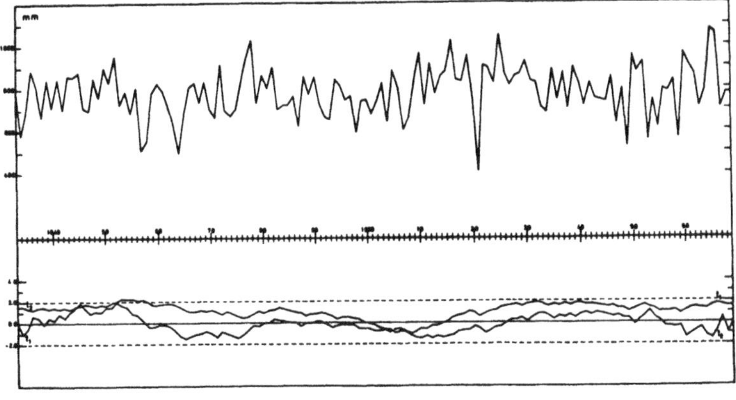

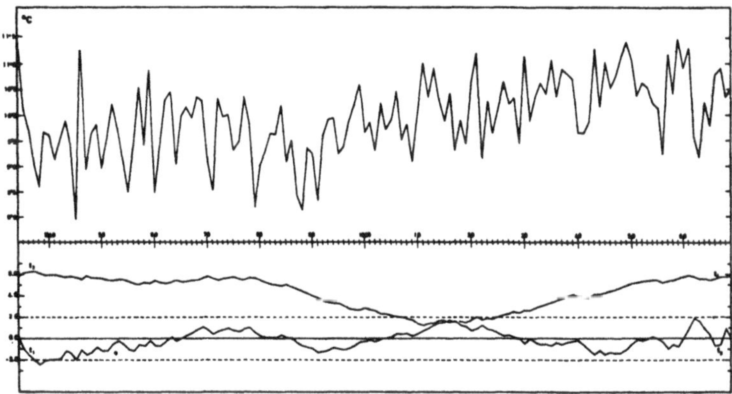

**Fig. 1. Annual rainfall and annual averages of the air-temperature at Brussels-Uccle from 1833 to 1969. Progressive values of the statistics $u(t)$ and $u'(t)$.**

Fig. 1 gives an example of the application of the trend test to the annual rainfall at Uccle from 1833 to 1969 and to the annual average of the temperature for the same period. The rainfall is found to be stable, the temperature not.

2.1.2 Test against serial correlation.

Though the absence of serial correlation is not essential, it is often necessary for making testing and estimation possible when independence is an underlying hypothesis, what is frequently the case. In particular, the distribution of the test statistic of the trend test given above is only true if the series is uncorrelated. Therefore it is advisable to test the series for serial correlation before testing it for trend.

The test for serial correlation that we use is the one of Wald-Wolfowitz. With a change of origin of the series $x_i$ such that

**Table 1. Ranks $y_i$ of the annual averages of the windspeed at Uccle from 1933 to 1969 arranged in ascending order. Auxiliary values $y'_i$, $y'_i - y'_{i+1}$, $n_i$, $n'_i$ needed for the computation of the statistics $r$, $r_s$ and $t$. Progressive values of the statistic $u(t)$ of the trend test.**

| $i$ | $y_i$ | $y'_i$ | $y'_i - y'_{i+1}$ | $y_i - i$ | $n_i$ | $n'_i$ | $t$ | $u(t)$ | $i'$ |
|---|---|---|---|---|---|---|---|---|---|
| 1 | 20 | 1 | −11 | 19 | 0 | 19 | 0 | 0 | 37 |
| 2 | 31 | 12 | −6 | 29 | 1 | 29 | 1 | 1 | 36 |
| 3 | 37 | 18 | 20 | 34 | 2 | 34 | 3 | 1,57 | 35 |
| 4 | 17 | −2 | 10 | 13 | 0 | 16 | 3 | 0 | 34 |
| 5 | 7 | −12 | −19 | 2 | 0 | 6 | 3 | −0,98 | 33 |
| 6 | 26 | 7 | 17 | 20 | 3 | 22 | 6 | −0,56 | 32 |
| 7 | 9 | −10 | −20 | 2 | 1 | 7 | 7 | −1,05 | 31 |
| 8 | 29 | 10 | 27 | 21 | 5 | 23 | 12 | −0,49 | 30 |
| 9 | 2 | −17 | −14 | −7 | 0 | 1 | 12 | −1,25 | 29 |
| 10 | 16 | −3 | −11 | 6 | 3 | 12 | 15 | −1,34 | 28 |
| 11 | 27 | 8 | −7 | 16 | 7 | 19 | 22 | −0,86 | 27 |
| 12 | 34 | 15 | 10 | 22 | 10 | 23 | 32 | −0,14 | 26 |
| 13 | 24 | 5 | −9 | 11 | 6 | 17 | 38 | −0,12 | 25 |
| 14 | 33 | 14 | 12 | 19 | 11 | 21 | 49 | 0,38 | 24 |
| 15 | 21 | 2 | −9 | 6 | 6 | 14 | 55 | 0,25 | 23 |
| 16 | 30 | 11 | 20 | 14 | 11 | 18 | 66 | 0,51 | 22 |
| 17 | 10 | −9 | −22 | −7 | 3 | 6 | 69 | 0,08 | 21 |
| 18 | 32 | 13 | −3 | 14 | 14 | 17 | 83 | 0,49 | 20 |
| 19 | 35 | 16 | 23 | 16 | 17 | 17 | 100 | 1,01 | 19 |
| 20 | 12 | −7 | 8 | −8 | 4 | 7 | 104 | 0,58 | 18 |
| 21 | 4 | −15 | −32 | −17 | 1 | 2 | 105 | 0,00 | 17 |
| 22 | 36 | 17 | 25 | 14 | 20 | 15 | 125 | 0,54 | 16 |
| 23 | 11 | −8 | −7 | −12 | 5 | 5 | 130 | 0,19 | 15 |
| 24 | 18 | −1 | −10 | −6 | 9 | 8 | 139 | 0,05 | 14 |
| 25 | 28 | 9 | 15 | 3 | 15 | 12 | 154 | 0,13 | 13 |
| 26 | 13 | −6 | −6 | −13 | 7 | 5 | 161 | −0,07 | 12 |
| 27 | 19 | 0 | 5 | −8 | 11 | 7 | 172 | −0,14 | 11 |
| 28 | 14 | −5 | 6 | −14 | 8 | 5 | 180 | −0,36 | 10 |
| 29 | 8 | −11 | −17 | −21 | 3 | 4 | 183 | −0,75 | 9 |
| 30 | 25 | 6 | 20 | −5 | 17 | 7 | 200 | −0,63 | 8 |
| 31 | 5 | −14 | 4 | −26 | 2 | 2 | 202 | −1,04 | 7 |
| 32 | 1 | −18 | −22 | −31 | 0 | 0 | 202 | −1,49 | 6 |
| 33 | 23 | 4 | 8 | −10 | 18 | 4 | 220 | −1,36 | 5 |
| 34 | 15 | −4 | −7 | −19 | 12 | 2 | 232 | −1,44 | 4 |
| 35 | 22 | 3 | 16 | −13 | 19 | 2 | 251 | −1,32 | 3 |
| 36 | 6 | −13 | 3 | −30 | 4 | 1 | 255 | −1,63 | 2 |
| 37 | 3 | −16 | −17 | −34 | 2 | 0 | 257 | −1,99 | 1 |

$$\Sigma x_i = 0 \tag{1}$$

and putting :

$$x_{n+1} = x_1 \tag{2}$$

the test statistic becomes :

$$R = \sum_{i=1}^{n} x_i x_{i+1} \tag{3}$$

Under the hypothesis of independence it has asymptotically a normal distribution with mean and variance :

$$E(R) = -S_2/(n-1) \quad \text{and} \quad \text{var } R = S_2^2/(n-1) \tag{4}$$

where $S_2 = \Sigma x_i^2$.

If we change the test statistic into :

$$r = R/S_2 \tag{5}$$

the mean and the variance of the distribution become :

$$E(r) = -1/(n-1) \quad \text{and} \quad \text{var } r = 1/(n-1) \tag{6}$$

and the test becomes identical with the one of R.L. Anderson.

The test of Wald-Wolfowitz is non parametric while the one of R.L. Anderson has been derived under the assumption of normality. This shows that the latter test is robust.

One way of arrangement of both tests is to replace each $x_i$ of the series by its rank $y_i$ in the series arranged in ascending order. Both tests may then be applied on the $y_i$ which allows some check procedures when the computations are made by hand.

As an example we have given in table 1 the trend and serial correlation analysis of the series of mean annual windspeed at Uccle (Brussels). For the serial correlation we find :

$$R = -117 \;,\; S_2 = 4.218 \;,\; r = -0{,}02774 \;,\; u(r) = \frac{r - E(r)}{\sqrt{\text{var } r}} = 0{,}00023$$

which shows that the observed value of the statistic is near its median. The hypothesis of absence of serial correlation may thus be accepted.

For the trend test, the test statistic computed progressively shows decreasing negative values from the $26^{\text{th}}$ on. The last value is significant at the 0,05 level. The series may thus be considered as being significantly decreasing and is consequently not stable.

2.2 Goodness of fit.

When a distribution function has been adjusted to a series of observations, good statistical prediction may only be expected if the goodness of fit of the distribution to the observations has been tested.

In the case of the normal distribution (cf. [1] p. 74) there are two kinds of tests of goodness of fit : those based on the reduced central moments of higher order : coefficients of skewness or of kurtosis and the other based on the comparison of linear estimates of the standard deviation to its quadratic estimate. When the alternative hypothesis is not fixed the latter kind of tests has a slightly better power relatively to a set of alternatives than the former. For these tests, if $x'_m$, $m=1, 2, \ldots, n$ is the ranked series in ascending order, the test statistic is of the form :

$$X = \Sigma a_m (x'_{n+1-m} - x'_m)/s \quad , \quad m = 1, 2, \ldots, (n-1)/2 \text{ or } n/2 \tag{1}$$

where $s^2 = \sum_1^n (x'_m - \bar{x})^2 / n$, $\bar{x} = \sum_1^n x'_m / n$ and $a_m$ are special values.

For $n$ up to 50, Shapiro and Wilk have given the values of $a_m$ and the test is a one sided one based on $X^2$. For $n \geqslant 50$ a simplified form for $a_m$ has been used by D'Agostino and the test is two sided.

For other distributions one solution of the problem of goodness of fit is to use the Kolmogorov-test. If $F(x)$ is the distribution function of the series of observations the test is based on the fact that there exists a constant $D_\alpha$ such as the relation :

$$\frac{m}{n} - D_\alpha < F(x'_m) < D_\alpha - \frac{m-1}{n} \tag{2}$$

**Table 2. Annual maximum dry and rainy spells. Observed distribution $F_d(i)$ of the dry spells and $F_r(i)$ of the rainy spells (1901-69); corresponding theoretical distributions $\phi_d(i)$ and $\phi_r(i)$**

| $i$ | $F_d(i)$ | $\phi_d(i)$ | $F_r(i)$ | $\phi_r(i)$ |
|---|---|---|---|---|
| 8 | 2 | 4,04 | 1 | 0,49 |
| 9 | 7 | 8,69 | 1 | 1,67 |
| 10 | 13 | 15,05 | 3 | 4,14 |
| 11 | 25 | 22,44 | 8 | 8,14 |
| 12 | 34 | 30,03 | 11 | 13,53 |
| 13 | 39 | 37,21 | 20 | 19,85 |
| 14 | 44 | 43,57 | 23 | 26,54 |
| 15 | 48 | 48,97 | 28 | 33,10 |
| 16 | 54 | 53,40 | 34 | 39,20 |
| 17 | 55 | 56,96 | 40 | 44,59 |
| 18 | 57 | 59,77 | 41 | 48,75 |
| 19 | 58 | 61,95 | 45 | 53,16 |
| 20 | 62 | 63,63 | 50 | 56,37 |
| 21 | 64 | 64,93 | 56 | 58,97 |
| 22 | 64 | 65,91 | 60 | 61,08 |
| 23 | 65 | 66,66 | 64 | 62,77 |
| 24 | 67 | 67,23 | 65 | 64,10 |
| 25 | 67 | 67,66 | 66 | 65,21 |
| 26 | 67 | 67,99 | 67 | 65,98 |
| 27 | 67 | 68,23 | 67 | 66,64 |
| 28 | 67 | 68,42 | 67 | 67,13 |
| 29 | 68 | 68,56 | 68 | 67,55 |
| 30 | 69 | 68,67 | 68 | 67,86 |
| 31 | | 68,75 | 69 | 68,10 |

is verified with probability $(1 - \alpha)$ for any value of $m$.

The disadvantage of the Kolmogorov-test is that it assigns the same interval-width to all the values of $F(x'_m)$ though it is known that the distribution function of $F(x'_m)$ has smaller variances at the extremes than in the neighbourhood of the median. Therefore it may be advisable to use the distribution function of the extremes to test the goodness of fit of the adjusted distribution function. In particular the distribution function of the lowest and of the largest value $x'_1$ and $x'_n$ are respectively :

$$1 - [\,1 - F(x)]^n \quad \text{and} \quad [F(x)]^n \tag{3}$$

functions which may be used to compute the probability corresponding to $x'_1$ and to $x'_n$.

Moreover, these distributions being asymptotically independent, a joint test may be applied through the use of the Fisher-test on joint probabilities.

## 3. EXTREME DISTRIBUTIONS IN METEOROLOGY.

The topics that we will consider successively are : the exact distributions of extremes, the Fisher-Tippett asymptotes, special extreme distributions, extremes adjusted by ordinary distributions and the use of the Fisher-Tippett asymptotes to adjust skew distributions.

### 3.1 Exact distributions of extremes.

If the $N$ variates $x_1, x_2, ..., x_N$ are independently distributed following the distribution functions :

$$F_1(x_1), F_2(x_2), ..., F_N(x_N) \tag{1}$$

the distribution function of the largest value $x$ of a sample of $N$ observed values of $x_1, x_2, ..., x_N$ is :

$$\phi(x) = F_1(x) \,.\, F_2(x) \,.\, ... \,.\, F_N(x) \tag{2}$$

This principle may of course be applied as well to continuous variates as to discrete variates.

As an example, we have considered the spells of dry and rainy days at Uccle (Brussels). In order to determine the distribution of the annual maximum of dry or wet spells we have considered for each month the negative binomial distribution adjusted to the observed dry or wet spells for each month for the period (1901-30)(cf. [2]). It is found that the distribution of the dry spells is the same for each of the twelve months but differs from month to month for the wet spells. Application of formula (2) leads to the results given in table 2 where the observed maximum spells of 69 years are compared with the expected distribution of the maximum spells derived from the period 1901-30. The fit is found to be fairly good. It should be noted that for the dry spells the exact distribution of the annual maximum becomes :

$$\phi(x) = [F(x)]^N \tag{3}$$

with $N = 12$, since the distribution function of the monthly spells is the same for the twelve months.

### 3.2 The Fisher-Tippett asymptotes.

When in formula 3.1(3) $N$ is large, the distribution function $\phi(x)$ may be approximated following the case by asymptotic distributions given by Fisher and Tippett in 1928.

The three cases are (cf [3] p.164) :
1) $x$ is unlimited at both sides;
2) $x$ has a lower bound;
3) $x$ has an upper bound.

The first case is realized with initial distributions of the exponential type. The asymptote is then the double exponential distribution :

$$\phi(x) = \exp[-e^{-y}] \quad \text{with} \quad y = (x - \mu)/\sigma \quad \text{and} \quad \sigma > 0 \tag{1}$$

The second case occurs for distributions of the Cauchy type. The asymptote is then the three parameter distribution connected with the three parameter Weibull distribution :

$$\phi(x) = \exp\left[-\left(\frac{\alpha-\beta}{x-\beta}\right)^{\gamma}\right] \quad \text{with} \quad \alpha>\beta>0 \;, \gamma>0 \;\text{ and }\; x\geqslant\beta \tag{2}$$

Moreover, with $y = \log(x-\beta)$, it takes the form of the first asymptote.

The third case, named the bounded case, leads to the distribution function :

$$\phi(x) = \exp\left[-\left(\frac{\omega-x}{\omega-\epsilon}\right)^{\delta}\right] \quad \text{with} \quad \epsilon<\omega \;,\; \delta>0 \;\text{ and }\; x\leqslant\omega \tag{3}$$

For practical applications the two first asymptotes are the most frequently used. Moreover, when $\beta = 0$ , the adjustement of the second type reduces to an adjustement of the first type by substituting $\log x$ to $x$.

For the first asymptote, three methods of adjustement are usually used : the method of moments, the method of least squares and the maximum likelihood method.

The first method is based on the fact that we have :

$$E(y) = 0{,}577216 \ldots \text{ (Euler's constant)} \quad \text{and} \quad \operatorname{var} y = \pi^2/6 \tag{4}$$

For the second method, from (1) we have :

$$-\log(-\log\phi) = y \tag{5}$$

and with attributing to the ranked values $x'_m$ of the observations $x_i$ the probabilities $\phi_m = m/(n+1)$, from (5) we may estimate $\mu/\sigma$ and $1/\sigma$ by least squares.

The third method has been derived from the maximum likehood equations by Kimball (cf. [1] p. 54). The equation giving the value of the scale parameter is :

$$n\sigma = \Sigma x'_m (1 + \log F_m) \tag{6}$$

where the $x'_m$ are the ranked observations in ascending order and where :

$$\log F_m = -\left(\frac{1}{m} + \frac{1}{m+1} + \ldots + \frac{1}{n}\right) \tag{7}$$

Moreover, the estimate of $\sigma$ has to be corrected for the bias with the factor $b_n$ given by the relation for $n > 11$ :

$$\log(b_n - 1) = -0{,}975652 \log(n-1) + 1{,}043532 - 1{,}950309/\log(n-1) + 3{,}574231/\log^2(n-1) \tag{8}$$

the log's being naperian ones.

For the location parameter $\mu$, the estimate is then :

$$\mu = \bar{x} - E(y) \,.\, \sigma \quad \text{with} \quad \bar{x} = \Sigma x_i/n \tag{9}$$

The advantage of these estimates is that they are apparently asymptotically efficient.

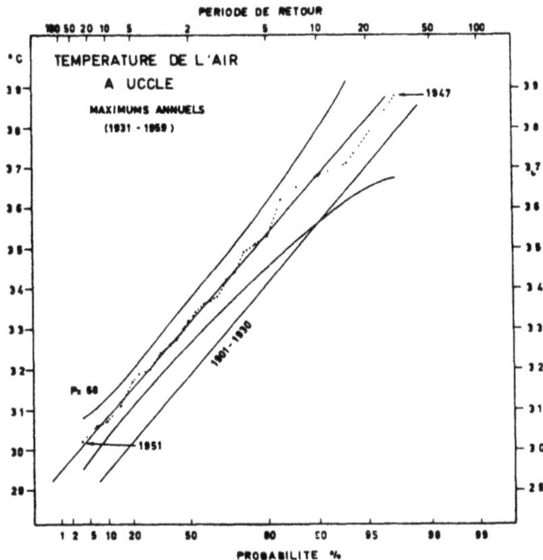


**Fig. 2 Annual maximum of the air temperature at Brussels-Uccle, (1931-1959). Observed values and adjusted curve. Adjusted curve for the period 1901-1930. Confidence bounds at the 68% level for the order statistics.**

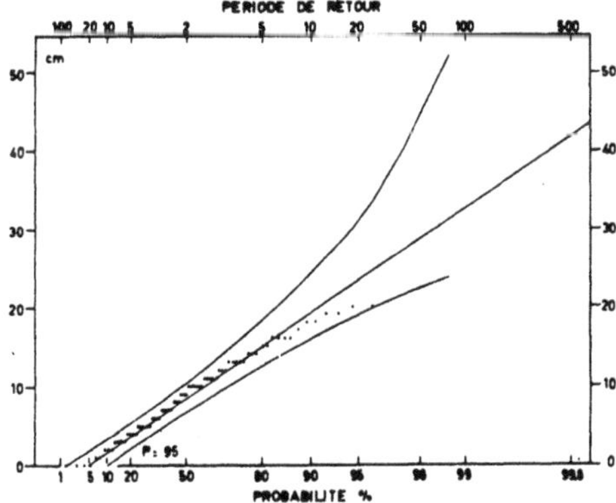


**Fig. 3. Annual maximum snow-depth at Brussels-Uccle in cm. Adjusted double exponential distribution, observed values and confidence bounds of the order statistics at the 95% level.**

As an example of the first asymtote we have considered the annual maximum of the temperature at Uccle.

In Fig. 2 (cf [4] p. 13) we give the observed values for the period 1931-1959 plotted on double exponential probability paper, the adjusted straight line to these observations as well as the straight line adjusted to the observations of the period 1901-30. The plot shows a good fit for the period 1931-1959. Moreover, the difference between the two straight lines shows that there was a trend in the series with higher temperatures during the period 1931-1959 than during the former.

Another example is given in fig. 3 (cf [5] p. 517) with the maximum depth of the snow cover during the winter at Uccle. Here too, there is a good agreement between the observed values and the adjusted curve.

For the second asymptote a first example is given with the maximum spell $L$ of consecutive days with a snowcover at Uccle. It is found that $\log(L+10)$ follows a double exponential law and fig. 4 (cf [6] p. 9) shows a good fit of the distribution to the observations. The constant 10 was found empirically. It should be however amphasized that for the smallest value $L'_1$ of the sample of 75 observations of $L$ the distribution function $F(L+10)$ leads to the result :

$$1 - [1 - F(L'_1 + 10)]^{75} \cong 0{,}50 \quad (10)$$

which shows that the choice of the constant 10 was a good one.

Another example is given with the annual maximum $R$ of the daily rainfall at Uccle (table 3). Here, the distribution of $\log R$ has a double exponential distribution.

The relation with the reduced variate $y$ of the double exponential distribution is :

$$\log R = 5{,}612 + 0{,}2421\, y \quad (11)$$

For the lowest and the largest value of the series of observations we have :

$$1 - [1 - F(187)]^{60} = 0{,}38 \quad \text{and} \quad [F(723)]^{60} = 0{,}34 \quad (12)$$

which shows a good agreement of the distribution with the observations.

3.3 Special extreme distributions

When the extremes come from two populations of extremes and when $F_1(x)$ , $F_2(x)$ are the distribution functions of each population, the distribution function $\phi(x)$ of the largest value is then :

$$\phi(x) = F_1(x) \,.\, F_2(x) \quad (1)$$

Moreover if large values of the first population are small values of the second one, on probability paper where we assume that $F_1(x)$ and $F_2(x)$ are represented by straight lines, $\phi(x)$ will have a representation being asymptotically a straight line for small values and for large values. Consequently the simplest way of representing such a curve will be a branch of a hyperbola.

A first example of such an adjustment is given by the monthly maximum of the rainfall intensity in 1 min at Uccle. If the rainfall intensity $t$ is given with taking the mean of the monthly

**Table 3. Annual maximum of the daily rainfall at Uccle.**
**Random simple sample of 60 values arranged in ascending order (0,1 mm).**

| $m$ | $x$ | $m$ | $x$ | $m$ | $x$ |
|---|---|---|---|---|---|
| 1 | 187 | 21 | 277 | 41 | 343 |
| 2 | 198 | 22 | 279 | 42 | 354 |
| 3 | 198 | 23 | 282 | 43 | 356 |
| 4 | 203 | 24 | 284 | 44 | 368 |
| 5 | 204 | 25 | 285 | 45 | 372 |
| 6 | 214 | 26 | 290 | 46 | 375 |
| 7 | 217 | 27 | 298 | 47 | 378 |
| 8 | 222 | 28 | 301 | 48 | 392 |
| 9 | 232 | 29 | 306 | 49 | 398 |
| 10 | 238 | 30 | 325 | 50 | 405 |
| 11 | 240 | 31 | 329 | 51 | 406 |
| 12 | 243 | 32 | 333 | 52 | 406 |
| 13 | 249 | 33 | 334 | 53 | 412 |
| 14 | 255 | 34 | 336 | 54 | 415 |
| 15 | 265 | 35 | 338 | 55 | 468 |
| 16 | 265 | 36 | 338 | 56 | 480 |
| 17 | 265 | 37 | 340 | 57 | 507 |
| 18 | 266 | 38 | 340 | 58 | 511 |
| 19 | 271 | 39 | 342 | 59 | 600 |
| 20 | 272 | 40 | 343 | 60 | 723 |

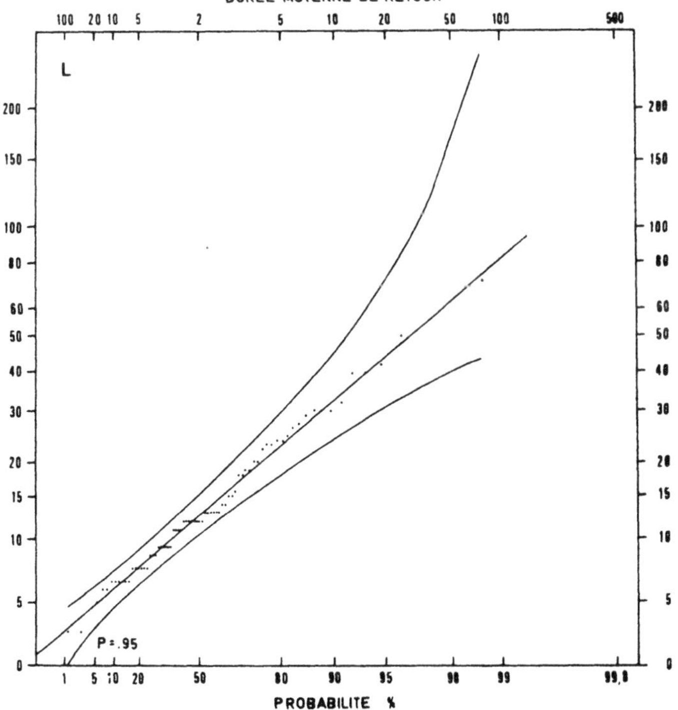


**Fig. 4. Annual maximum of the spells $L$ of days with snow lying at Uccle. Adjusted double exponential distribution to log $(L + 10)$, observed values and confidence bounds of the order statistics at the 95% level.**

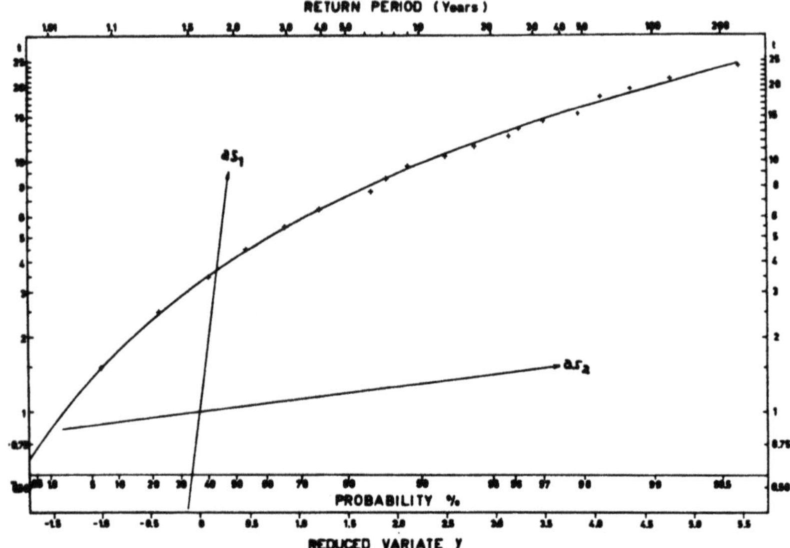


Fig. 5. Monthly maximum of the rainfall intensity in 1 min, at Uccle. Adjusted curve and observed values plotted on double exponential probability paper. $t$ is given in fifths of the mean monthly maximum.

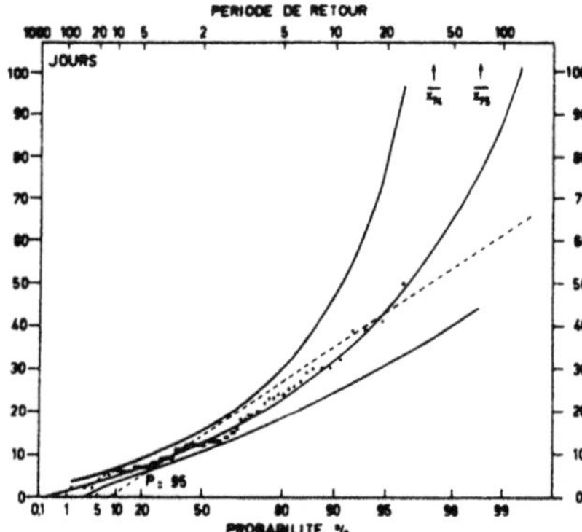


Fig. 6. Maximum spell of consecutive days with snow lying, at Uccle. Adjusted double exponential distribution (interrupted line), modified double exponential distribution (continuous line), oberved values and for the modified distribution confidence bounds for the order statistics at the 95% level.

maximum intensity as a unit, it is found that we may consider the distribution of $t$ as being the same for the twelve months. Moreover with $x = \log t$,

$$y = ax + b + \sqrt{c^2(x-d)^2 + e^2} \quad \text{with} \quad a > |c| \tag{2}$$

leads to a double exponential distribution.

The graphical representation of this distribution on double exponential probability paper is given in fig. 5 (cf [7] p. 68). The method of adjustment used was fitting first a linear function of $x$ to $y$ and secondly fitting by a graphical method a branch of a hyperbola to the discrepancies of the observed $x$ with the corresponding theoretical values. It should be emphasized that in this graphical representation of (2) the independent variate is plotted on the vertical axis.

Another example of special extreme distribution is a particular case of the preceding one. In a first study the distribution of the maximum spell of consecutive days with snow lying at Uccle was found to be not a double exponential distribution, but the transform :

$$y = ax + b + c/(x+d) \quad \text{with} \quad a > 0 \quad \text{and} \quad c < 0 \tag{3}$$

led to a double exponential distribution.

This means that on the double exponential probability paper the curve (3) has an horizontal asymptote.

The result of such an adjustment is given in fig. 6 (cf [5] p. 516) and seems to be very good. The method used was estimating by a graphical method some special fractiles after which the equation of the curve was computed.

3.4 Ordinary extreme distributions.

The basic assumptions leading to the Fisher-Tippett asymptotes are not always fulfilled in meteorology.

Indeed, long series of observations are not necessarily fitted by one of these asymptotes. Moreover, even if it may be tempting to use transforms as was made with the special distributions it should always be put in mind that the increase of the number of parameters under estimation increases the error of estimation at the same time. Therefore, when possible it is always advisable to make adjustments with distributions having the smallest number of parameter.

A first example of an extreme distribution following not a Fisher-Tippett asymptote is given by the annual maximum of the temperature at De Bilt (Netherlands). Fig. 7 (cf [8] p. 53) shows the graphical plot on normal probability paper of the annual maximum of temperature for 133 years. The plot shows an excellent fit of the normal distribution.

Another example is given by the monthly maximum of the rainfall in 1 min to 24 hour at Uccle as well as the monthly maximum of the rainfall duration [9]. It has been found that the log-normal distribution function gives a very good fit to the observations so that estimations of the maximum rainfall could be given for any duration between 1 min and 24 hour. For example, fig. 5 has been replotted on log-normal probability paper in fig. 8. It shows that the observed maximum rainfall in 1 min is very well fitted by a log-normal distribution.

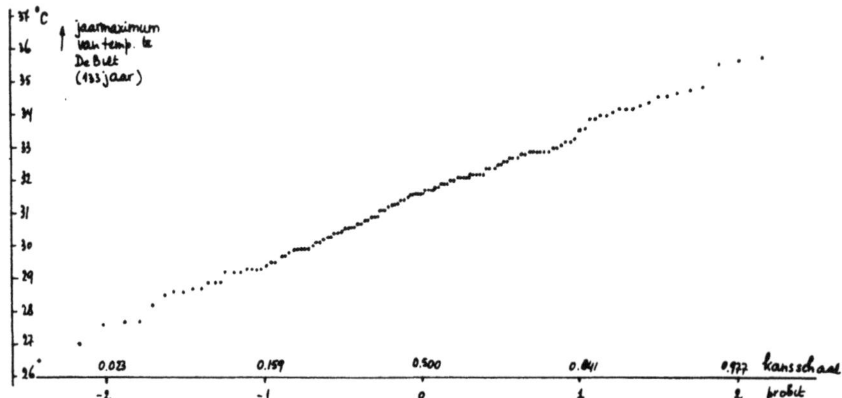

Fig. 7. Annual maximum of the air temperature at De Bilt (Netherlands) (1849-1981). Observed values plotted on normal probability paper.

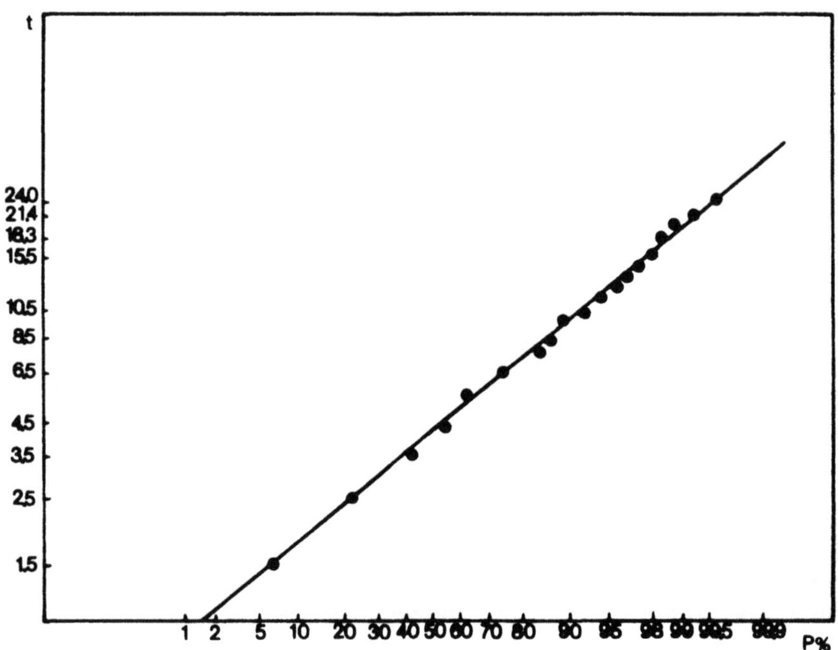


Fig. 8. Monthly maximum of the rainfall in 1 min, at Uccle. Plot of the logarithm of the observations on normal probability paper.

**Table 4. Fractiles with mean return period $T$ of the seasonal and annual duration of sunshine at Uccle.**

| T | H | P | $P_v$ | E | A | Année |
|---|---|---|---|---|---|---|
| 100 | 95 | 329 | 403 | 390 | 207 | 1296 |
| 50 | 102 | 344 | 426 | 409 | 219 | 1315 |
| 30 | 107 | 357 | 442 | 425 | 228 | 1333 |
| 25 | 110 | 362 | 451 | 432 | 232 | 1339 |
| 10 | 123 | 390 | 490 | 468 | 254 | 1380 |
| 6 | 132 | 410 | 516 | 493 | 270 | 1411 |
| 2 | 165 | 475 | 596 | 577 | 321 | 1527 |
| 6 | 202 | 547 | 676 | 670 | 379 | 1690 |
| 10 | 215 | 572 | 703 | 703 | 399 | 1757 |
| 25 | 236 | 610 | 741 | 752 | 431 | 1873 |
| 30 | 240 | 617 | 751 | 762 | 436 | 1895 |
| 50 | 250 | 636 | 767 | 786 | 452 | 1958 |
| 100 | 263 | 660 | 789 | 817 | 471 | 2044 |

### 3.5 Fisher-Tippett type I distributions for ordinary variates.

The Fisher-Tippett type I distribution is a skew distribution and as such may eventually be fitted to skewed distributions. In particular, the sunshine duration at Uccle has monthly and seasonal totals distributed according gamma distributions but the annual total has been found to be well fitted by a double exponential distribution [10]. Table 4 gives different fractiles which have been derived through gamma distributions for the seasonal sunshine duration and through the double exponential distribution for the year (hours).

## 4. EXTREMES IN SHORT SERIES.

The statistical analysis of extreme values makes it generally necessary to have at disposal long series of observations since in the case of annual extremes only one value is obtained each year. This drawback may be avoided in two cases : 1) by the consideration of large values; 2) by modelling the time series of daily values.

### 4.1 Use of large values.

If we consider the $k$ largest values accompanying the largest value and if these values are independently distributed, the distribution function $F$ of the large values may be put into relation with the distribution function $\phi$ of the largest value following the equation :

$$F^k = \phi \tag{1}$$

Hence, if we have for a small number $N$ of years the $k$ largest values, the sample of $kN$ large values may be sufficiently large for making an accurate adjustment of the distribution $F$ and with (1), of $\phi$.

As an example (cf [11]) we have again considered the annual maximum of the daily rainfall at Uccle. In table 5 we have grouped for five years the ten largest values of the daily rainfall, which makes a sample of 5 x 10 = 50 values. Five distribution functions have been fitted to this sample : the normal, the log-normal, the gamma, the double exponential and the logarithmic double exponential function.

Table 5.

Daily rainfall at Uccle.
Large values for the years 1976 to 1980 in chronological order (mm).

| | | | | | | | | | | |
|---|---|---|---|---|---|---|---|---|---|---|
| 1976 | 10,2 | 16,4 | 34,0 | 25,3 | 11,8 | 11,3 | 16,5 | 12,2 | 19,5 | 20,8 |
| 77 | 15,0 | 26,7 | 16,4 | 15,4 | 23,8 | 15,2 | 24,4 | 16,1 | 14,8 | 14,0 |
| 78 | 17,5 | 47,3 | 19,6 | 16,9 | 12,4 | 19,9 | 19,4 | 13,3 | 15,0 | 16,1 |
| 79 | 13,3 | 17,8 | 14,3 | 36,0 | 19,5 | 19,9 | 14,5 | 22,8 | 13,3 | 14,2 |
| 80 | 18,1 | 24,1 | 17,2 | 20,2 | 22,3 | 39,1 | 25,1 | 17,0 | 16,8 | 18,3 |

Table 6.

Probabilities associated to the smallest and to the largest daily rainfall at Uccle

| Distribution | gamma | double exponential | logarithmic double exponential |
|---|---|---|---|
| $P_1$ | 0,073 | 0,25 | 0,79 |
| $P_2$ | 0,980 | 0,92 | 0,51 |

Table 7.

Annual maximum of the daily rainfall at Uccle.
Fractiles with return period T (mm).

| T | 2 | 6 | 10 | 25 | 30 | 50 | 100 |
|---|---|---|---|---|---|---|---|
| 1976-80 | 31,1 | 43,8 | 50,5 | 64,4 | 67,5 | 77,1 | 92,2 |
| 1901-72 | 31,2 | 44,7 | 51,8 | 66,8 | 70,2 | 80,6 | 97,3 |

The Shapiro-Wilk test for normality has rejected both normal distributions. For the other distributions the goodness of fit has been tested through the probabilities given to the extremes of the sample. Table 6 shows that the best fit is given by the logarithmic double exponential distribution confirming in this manner previous results. In table 7 we have given the fractiles for usual return periods and have compared them with those given by the analysis of the series of extremes of the period 1901-71. The differences have been found to be not significant at the 0,05 level.

4.2 Modelling time series

For a random sampl[illegible] $x_1$, $x_2$, ..., $x_n$ of size $n$ with distribution function $F(x)$, the distribution function $\phi(x)$ of the [illegible] $x$ is :

**Table 8. Maximum values of the daily average of the atmospheric pressure at Uccle in April, in 0,1 mm of normal mercury. Excess $x'_i$ on 7000, for thirty consecutive years, arranged in ascending order. Corresponding values $F(x'_i)$ of the distribution function of the pressures and values $[F(x'_i)]^{6.33}$ of the distribution function of the largest value of the month. Limits $F_{0,025}$ and $F_{0,975}$ of the two-sided confidence interval at the level 0,95 of the probabilities associated to the maximum value $x'_i$.**

| $i$ | $x'_i$ | $F(x'_i)$ | $[F(x'_i)]^{6,33}$ | $F_{0,025}$ | $F_{0,975}$ |
|---|---|---|---|---|---|
| 1 | 552 | 0,655 | 0,07 | 0,001 | 0,12 |
| 2 | 564 | 709 | 11 | 01 | 17 |
| 3 | 571 | 742 | 15 | 02 | 22 |
| 4 | 583 | 788 | 22 | 04 | 27 |
| 5 | 601 | 851 | 36 | 06 | 31 |
| 6 | 603 | 858 | 38 | 08 | 35 |
| 7 | 605 | 862 | 39 | 10 | 39 |
| 8 | 608 | 871 | 42 | 12 | 42 |
| 9 | 608 | 871 | 42 | 15 | 46 |
| 10 | 615 | 891 | 48 | 17 | 49 |
| 11 | 618 | 898 | 51 | 20 | 53 |
| 12 | 618 | 898 | 51 | 23 | 56 |
| 13 | 619 | 900 | 51 | 25 | 59 |
| 14 | 619 | 900 | 51 | 28 | 63 |
| 15 | 627 | 915 | 57 | 31 | 66 |
| 16 | 627 | 915 | 57 | 34 | 69 |
| 17 | 632 | 927 | 62 | 37 | 72 |
| 18 | 632 | 927 | 62 | 41 | 75 |
| 19 | 635 | 932 | 64 | 44 | 77 |
| 20 | 636 | 933 | 65 | 47 | 80 |
| 21 | 638 | 937 | 66 | 51 | 83 |
| 22 | 640 | 941 | 68 | 54 | 85 |
| 23 | 651 | 955 | 75 | 58 | 88 |
| 24 | 652 | 956 | 75 | 62 | 90 |
| 25 | 654 | 959 | 77 | 65 | 92 |
| 26 | 654 | 959 | 77 | 69 | 94 |
| 27 | 658 | 963 | 79 | 74 | 96 |
| 28 | 665 | 971 | 83 | 78 | 98 |
| 29 | 689 | 986 | 92 | 83 | 99 |
| 30 | 703 | 992 | 948 | 88 | 999 |

$$\phi(x) = [F(x)]^n \qquad (1)$$

This formula may however not be applied to series of daily observations because such observations are generally autocorrelated.

Nevertheless if we define the equivalent number of repetitions $\omega$ by the relation :

$$\omega = (n \operatorname{var} \overline{x}_n)/\operatorname{var} x \qquad (2)$$

where $\overline{x}_n = \Sigma x_i / n$ , the quantity :

$$n' = n/\omega \qquad (3)$$

may be considered as the equivalent size of the series and the relation (1) may be replaced by the approximate relation :

$$\phi(x) = [F(x)]^{n'} \qquad (4)$$

In the case where the series $x_i$ is autoregressed, $\omega$ is a function of the coefficients of the autoregression and thus $n'$ may be computed straightforward.

As an example (cf [1] p. 169) we have considered the series of daily averages of the atmospheric pressure at Uccle. With two years of observations it has been found that the series followed an autoregressive scheme of the second order. Moreover, the seasonal variation of the mean being very small, the autoregression scheme may be applied to every month. Taking then the special case of April, with $\omega=4{,}74$ computed from the autoregression, we have :

$$n'=30/4{,}74=6{,}33 \tag{5}$$

value which has been used in (4) to determine the distribution function of the monthly maximum of the atmospheric pressure observed in April during thirty years.

The computation of $\phi(x)$ for each of the thirty ranked observations is given in table 8. It has been done under the assumption of normally distributed daily averages. The mean of the distribution is the mean $m$ of the autoregressive scheme and the variance var $x$ has been derived from (2) after having computed var $\bar{x}_{30}$ from a series of 144 monthly averages. The comparison with the confidence intervals at the 0,95 probability level shows a general agreement for the computed probabilities except for observations n$^{r}$ 5, 6 and 7.

## REFERENCES

[1] Sneyers, R. (1975), Sur l'analyse statistique des séries d'observations. Organisation météorologique mondiale, Note technique n° 143, Genève.

[2] Sneyers, R. (1975), Über die Anwendung von Markoff'schen Ketten in der Klimatologie, Arbeiten aus der Zentralanstalt für Meteorologie und Geodynamik, Heft 31, Wien, pp. 12/1-12/10.

[3] Gumbel, E.J. (1958), Statistics of extremes, Columbia University Press, New York.

[4] Sneyers, R. (1960), Le hasard en météorologie, Institut royal météorologique de Belgique, Contributions n° 63.

[5] Sneyers, R. (1964), La statistique de l'enneigement du sol en Belgique, Arch. Met., Geoph. und Biokl., B, 13, pp. 503-520.

[6] Sneyers, R. (1967), Les propriétés statistiques de l'enneigement du sol en Belgique, Institut royal météorologique de Belgique, Pub. A, n° 63.

[7] Sneyers, R. (1961), On a special distribution of maximum values, Monthly Weather Review, 88, pp. 66-69.

[8] Van Montfort, M.A.J. (1982), Modellen voor maximum en minima, schattingen en betrouwbaarheidsintervallen, keuze tussen modellen, Agricultural University Wageningen, Netherlands, Department of Mathematics, Statistics Division, Technical Note 82-02.

[9] Sneyers, R. (1979), L'intensité et la durée maximales des précipitations en Belgique, Institut royal météorologique de Belgique, Pub. B, n° 99.

[10] Sneyers, R. et Vandiepenbeeck, M. (1982), La durée d'insolation à Uccle et en Belgique, Institut royal météorologique de Belgique, Pub. B, n° 118.

[11] Sneyers, R. et Vandiepenbeeck, M. (1983), On the use of large values for the determination of the distribution of maximum values, to be published in Arch. Met. Geoph. Biokl, Ser. B.

# EXTREME VALUES IN INSURANCE MATHEMATICS

Jozef L. Teugels

Katholieke Universiteit Leuven

ABSTRACT : We survey the existing actuarial literature for information on large claims. We postulate a specific form for the claim size distribution and investigate its consequences. Special attention is given to reinsurance.

## INTRODUCTION

More and more the actuary is facing situations where large claims upset the statistical procedures on which he used to base decisions concerning premiums, type of reinsurance, retention, etc.

Let us assume that we are looking at a homogeneous risk portfolio, that claims come in according to a Poisson process $\{N(t);t \geqslant 0\}$ and that, independently, the amounts $Y_1,Y_2,\ldots$ come from a claim size distribution F of $Y \geqslant 0$. General references are [3,4,9,16].

## 1. EXTREME VALUES

On other occasions we had ample opportunity to discuss the concept of an extreme value or of a *large claim* both from the theoretical [29] as from the practical point [30] of view. For an alternative approach, see [32].

Practitioners generally agree that large claims do exist in certain portfolios but individual definitions differ according to experience and/or the type of reinsurance used in the portfolio by the respective companies.



*J. Tiago de Oliveira (ed.), Statistical Extremes and Applications, 253–259.*

## 2. MATHEMATICAL FORMULATION

2.a. In [3] the use of so-called *shadow claims* is discussed. Here

$$F(x) = pF_1(x) + qF_2(x)$$

where $0 \leqslant p \leqslant 1$, $p+q = 1$, $F_1$ is a "decent" distribution while $F_2$ is "dangerous" although q is small.
The above model shows the intricacies resulting from large claims : the number of large claims (used to estimate q) is very small, while the sizes (used to estimate $F_2$) have all the defects attributed to outliers.
We focus attention on $F_2$ by assuming p=0.

2.b. Theoretically it seems appropriate to assume that

$$1 - F(x) = x^{-\alpha}L(x) \qquad (*)$$

for some slowly varying L and $\alpha \geqslant 0$. We call F of *Pareto-type* with *index* $\alpha$.
The next section contains some justification for (*) and some of its consequences. Section 5 deals with the statistical aspect of estimating $\alpha$.

## 3. CONSEQUENCES OF THE ASSUMPTION (*).

That (*) reflects, at least to some extent, what one might call appropriate behaviour has been discussed in [29, 30]. Here are some of the relevant issues.

3.a. Condition (*) implies that $F \in \mathcal{S}$ , the class of *subexponential distributions*. As such, if $S_n = Y_1 + \ldots + Y_n$ and $M_n = \max(Y_1,\ldots,Y_n)$ then for $x \to \infty$

$$P[S_n > x] \sim P[M_n > x]$$

which means that $S_n$ is large because of $M_n$.

3.b. Let $R(t) = u + ct - \sum_1^{N(t)} Y_i$ be the *reserve* left at time t where u is the initial capital and ct the income from premiums. Define $T_u = \inf\{t:R(t) < 0\}$. Then the *infinite horizon ruin problem* consists in the evaluation of $P[T_u < \infty]$. From (*) and [14, 33] we see that for $u \to \infty$

$$1 - P[T_u < \infty] \sim Cu^{-\alpha+1}L(u)$$

for some constant C.

3.c. If $\alpha < 1$ then (*) is equivalent to the weak convergence of $S_n/M_n$ to a non-degenerate limit [8,11]. Even $E(S_n/M_n) \to c \geq 1$ is equivalent [7]. This shows that individual claims can partially control the entire portfolio.

3.d. If $\alpha < 1$ then $EY = \infty$ : a premium calculation of the form

$$P(t) = (EY)(1 + \delta)t$$

with safety loading $\delta > 0$ cannot be postulated. All principles involving Var Y are obsolete if $\alpha < 2$.

## 4. REINSURANCE

As explained in [30] the insurer can safeguard himself against large risks by taking some kind of reinsurance.

4.a. *Proportional reinsurance* is given by $p \sum_{i=1}^{N(t)} Y_i$; from the point of view of large claims this type of reinsurance is not very satisfactory as all small claims are still in use and the largest claims are only partly covered.

4.b. Another kind is the *excess-loss*, $\sum_{i=1}^{N(t)} (Y_i - L)^+$ reinsurance with *retention* L; this reinsurance form considers the claims individually and not as part of the portfolio.

4.c. In *stop-loss* reinsurance, $\{\sum_{i=1}^{N(t)} Y_i - C\}^+$ is reinsured at retention C, to be determined.

4.d. The *r largest claims* or $\sum_{k=1}^{r} Y^*_{N(t)-k+1}$ reinsurance form covers too much; this is not so with

4.e. the *Ecomor* treaty

$$\sum_{k=1}^{r} Y^*_{N(t)-k+1} - r\, Y^*_{n(t)-r} \tag{1}$$

where the claim $Y^*_{N(t)-r}$ acts as a random retention.

In the actuarial literature a number of papers deal with the quantities associated with the above reinsurance policies. See [1,2,5,6,10,15,22,23,24,25].

In practice both form and level of reinsurance are mostly decided by the board of governors; the resulting premiums are negotiated by brookers. Completely unsatisfactory, the lack of a solid mathematical theory may be stressed.

## 5. ESTIMATION OF $\alpha$

Let us finally turn to the important statistical problem of estimating $\alpha$ in (*).

5.a. Using *one order statistic*, one can rely on a result by Smirnov [27] to derive that under (*)

$$(\log n)^{-1} \log Y^*_{n-k+1} \xrightarrow{P} \frac{1}{\alpha}$$

for every fixed k.
Using a variable $k_n \to \infty$ this can be refined. Take for example $F(x) = \exp(-x^{-\alpha})$; then [13]

$$\sqrt{k_n}\,\{\alpha \log Y^*_{n-k_n+1} - \log \frac{n}{k_n}\} \xrightarrow{\mathcal{D}} \mathcal{N}(0,1)$$

for $k_n = o(n^{2/3})$.

5.b. Using *two order statistics* a result of Rossberg [26] yields that under (*)

$$P\left\{\log \frac{Y^*_{n-k-m+1}}{Y^*_{n-k+1}} \leq \frac{1}{\alpha} \log w\right\} \xrightarrow{\mathcal{D}} \frac{1}{B(m,k)} \int_0^w x^{k-1}(1-x)^{m-1}dx$$

as $n \to \infty$ for fixed k and m.
We can let m vary with n. If $L(x) \to c > 0$ then for $m_n = o(n)$ [13]

$$P\left\{\alpha \log \frac{Y^*_{n-m_n-k+1}}{Y^*_{n-k+1}} + \log m_n \leq v\right\} \xrightarrow{\mathcal{D}} \frac{1}{\Gamma(k)} \int_0^{e^v} e^{-x}x^{k-1}dx .$$

For k=1 this reduces to a result of de Haan - Resnick [12].

5.c. Instead of extending this procedure, let us use the *extreme part of the sample*. Recall that (*) with $\alpha > 0$ implies that as $x \to \infty$

$$\frac{1}{1-F(x)} \int_x^\infty \frac{1-F(y)}{y}\,dy \to \frac{1}{\alpha} .$$

Hence it seems natural to estimate $\frac{1}{\alpha}$ by

$$\frac{1}{1-F_n(a_n)} \int_{a_n}^\infty \frac{1}{y}\{1 - F_n(y)\}dy$$

where $F_n$ is the empirical distribution and $a_n \to \infty$. For $a_n = 0(n^\delta)$ $(\delta > 0)$ Haeusler - Schneemeier have obtained a strong law and a normal limit [17]. For $a_n = Y^*_{n-k}$ one obtains an estimator due to Hill [19], i.e.

$$H_r^{(n)} = \frac{1}{r} \sum_{k=1}^{r} \log Y^*_{n-k+1} - \log Y^*_{n-r}$$

and discussed by Hall [18].
Notice the similarity with (1). Moreover under (*), $H_r^{(n)} \xrightarrow{\mathcal{D}} Z_r$, a gamma-distributed variable. Taking also $r \to \infty$, degenerate and normal laws can be obtained. For r=n and $Y_0^* \equiv 1$ we find the estimator suggested in [20].

5.d. The above estimators can often be used to construct asymptotic confidence intervals. They do not depend on L in (*). It is clear that, whenever information on L is available, more refined results should be obtainable. For example rates of convergence results like in Smith [28] and in C.W. Andersen and I. Weissmann in this volume are of great importance.

## CONCLUSIONS

The relevance of (*) in practical applications has been established in a few publications [21]. Recently acquired data reinforce this feeling while the theoretical consequences lead to plausible conclusions.

Even if (*) is accepted as a satisfactory assumption a lot of problems remain. We just mention a few
- other procedures to get information on $\alpha$; the use of record values has been illustrated in [31];
- bayesian methods might prove to be very helpful;
- the influence of (*) on the different types of reinsurance has to be studied in great detail; useful formulas for retentions and corresponding premiums have to be developed.

## REFERENCES

[1] AMMETER, H. Note concerning the distribution function of the total loss excluding the largest individual claims. Astin Bulletin, 2, 1963, 32-43.

[2] AMMETER, H. The rating of "largest claim" reinsurance covers. Quart. Letter Allg. Reinsurance Co. Jubilee, 1964, 2.

[3] BEARD, R.E., PENTIKAINEN, T., PESONEN, E. Risk Theory, Methuen, London, 1977.

[4] BEEKMAN, J.A. Two Stochastic Processes. J. Wiley, N.Y., 1974.

[5] BENKTANDER, G. Largest claims reinsurance (LCR) A quick method to calculate LCR-risk rates from excess of loss risk rates. Astin Bulletin, 10, 1978, 54-58.

[6] BERLINER, B. Correlations between excess of loss reinsurance covers and reinsurance of the n largest claims. Astin Bulletin, 6, 1972, 260-275.

[7] BINGHAM, H.H., TEUGELS, J.L. Conditions implying domains of attraction. Proc. Sixth Prob. Th., Brasov, 1981, 23-34.

[8] BREIMAN, L. On some limit theorems similar to the arc sine law, Th. Prob. Appl., 10, 1965, 323-333.

[9] BUHLMAN, H. Mathematical methods in risk theory. Springer, Berlin, 1970.

[10] CIMINELLI, E. On the distribution of the highest claims and its application to the automobile insurance liability. Tr. 20th Int. Congress Act., 1976, 501-517.

[11] DARLING, D.A. The influence of the maximum term in the addition of independent random variables. Trans. Amer. Math. Soc., 73, 1952, 95-107.

[12] DE HAAN, L., RESNICK, S. A simple asymptotic estimate for the index of a stable distribution. J. Roy Statist. Soc. B, 42, 1980, 83-88.

[13] DE MEYER, A., TEUGELS, J.L. Limit theorems for Pareto-type distributions, Trans. S. Banach Int. Inst. Warsaw, 1983, to appear.

[14] EMBRECHTS, P., VERAVERBEKE, N. Estimates for the probability of ruin with special emphasis on the possibility of large claims. Insurance : Math. Econ. 1, 1981, 55-72.

[15] FRANCKX, E. Sur la fonction de distribution du sinistre le plus élevé. Astin Bulletin, 2, 1963, 415-424.

[16] GERBER, H. An introduction to mathematical risk theory. Univ. Pennsylvania, 1979.

[17] HAEUSLER, E., SCHNEEMEIER, W. Estimating the index in Pareto-type distributions, preliminary report, 1983.

[18] HALL, P. On de Haan and Resnick's estimates of the index

of a stable law, J.R. Statist. Soc. B, 44, 1982, 37-42.

[19] HILL, B.M. A simple general approach to inference about the tail of a distribution. Ann. Statistics, 3, 1975, 1163-1174

[20] ILIESCU, D.V., VODA, V. Some notes on Pareto distributions. Revue Roum. Math. P. Appl., 24, 1979, 597-609.

[21] JUNG, J., BENCKERT, L.G. Statistical models of claim distributions in fire insurance, Astin Bulletin, 8, 1974, 1-25.

[22] KREMER, E. Rating of largest claims and ECOMOR reinsurance treaties for large portfolios, Astin Bulletin, 13, 1981, 47-56.

[23] KUPPER, J. Contributions to the theory of the largest claim cover. Astin Bulletin, 6, 1971, 34-46.

[24] KUPPER, J. Kapazität und Höchschadenversicherung. Mitt. Ver. Schw. Versich. Math., 72, 249-258.

[25] RAMACHANDRAN, G. Extreme value theory and fire insurance, Trans. 20th Int. Congress Act., 1976, 695-706.

[26] ROSSBERG, H.J. Über die Verteilungsfunktionen der Differenzen und Quotienten von Ranggrössen, Math. Nachr., 21, 1960, 3-79.

[27] SMIRNOV, N.V. Limit distributions for the terms of a variational series. Transl. Amer. Math. Soc., 11, 1962, 82-143.

[28] SMITH, R.L. Uniform rates of convergence in extreme value theory. Adv. Appl. Probability, 14, 1982, 600-622.

[29] TEUGELS, J.L. Remarks on large claims. Bull. Inst. Stat. Inst., 49, 1981, 1490-1500.

[30] TEUGELS, J.L. Large claims in insurance mathematics. Astin Bulletin, 15, 1983, 81-88.

[31] TEUGELS, J.L. On successive record values in a sequence of independent identically distributed random variables, 1984, this volume.

[32] TIAGO DE OLIVEIRA, J. Statistical methodology for large claims, Astin Bulletin, 9, 1977, 1-9.

[33] VON BAHR, B. Asymptotic ruin probabilities when exponential moments do not exist, Scand. Actuarial J., 1975, 6-10.

# USE AND STRUCTURE OF SLEPIAN MODEL PROCESSES FOR PREDICTION AND DETECTION IN CROSSING AND EXTREME VALUE THEORY*

Georg Lindgren

Department of Mathematical Statistics
University of Lund
Box 725
S-220 07 Lund, Sweden

SUMMARY. A Slepian model is a random function representation of the conditional behaviour of a Gaussian process after events defined by its level or curve crossings. It contains one regression term with random (non-Gaussian) parameters, describing initial values of derivatives etc. at the crossing, and one (Gaussian) residual process. Its explicit structure makes it well suited for probabilistic manipulations, finite approximations, and asymptotic expansions.

Part of the paper deals with the model structure for univariate processes and with generalizations to vector processes conditioned on crossings of smooth boundaries, and to multiparameter fields, conditioned on local extremes or level curve conditions.

The usefulness of the Slepian model is illustrated by examples dealing with optimal level crossing prediction in discrete and continuous time, non-linear jump phenomena in vehicle dynamics, click noise in FM radio, and wave-characteristic distributions in random waves with application to fatigue.

## 1. INTRODUCTION

The object of study in this paper is the *Slepian model process*, so named after D. Slepian, who introduced it (1963) to represent a normal process after zerocrossings. It describes in functional form the stochastic behaviour of a normal (stationary or non-stationary) process near a general level or curve crossing. It contains one regression term with random non-normal parameters, describing in-



*J. Tiago de Oliveira (ed.), Statistical Extremes and Applications, 261–284.*

itial values, such as derivatives etc., at the crossing, and one normal residual process. Its explicit structure makes it well suited for probabilistic manipulations, finite approximations, and asymptotic expansions.

The paper will present the general model structure for univariate processes, and generalizations to vector processes conditioned on crossings of smooth boundaries, and to multiparameter fields, conditioned on local extremes or level curve conditions. The Slepian model is also used to derive optimal level crossing predictors for continuous time and discrete time processes, and to study wave-characteristic distributions in random waves with application to fatigue.

To further motivate the study, and point at the types of conclusions which can be drawn, we shall briefly mention two non-trivial examples, the first of which has its roots in a classical application of crossing theory in telecommunication.

EXAMPLE 1.1 (Clicks in FM radio) In FM radio a signal $S(t)$ is transmitted as an argument modulated cosine wave $A\cos(\omega_0 t+S(t))$, which is corrupted by additive noise $\xi(t)$ during transmission. In the receiver which ideally should reconstruct $S(t)$, the noise causes more or less serious disturbances in the output signal. One type of such disturbances is the so called FM click, noticed as a short spike with high amplitude in the audible signal.

The clicks can be described statistically by means of crossings in a certain bivariate noise process $(\xi_1(t), \xi_2(t))$, which is a function of the original noise process $\xi(t)$; see [13]. A click occurs any time $\xi_1(t)$ crosses the zero level, under the extra condition that $\xi_2(t)$ takes a value greater than the carrier amplitude A. By modelling both $\xi_1(t)$ and $\xi_2(t)$ near such conditioned crossing points by Slepian model processes one obtains simple functional expressions for the random click shape and amplitude.

In particular, if signal is absent, i.e. $S(t) \equiv 0$, the clicks are approximately of the form $A\Psi(At)$ for large A, where $\Psi$ is a standardized click form given by the random function

$$\Psi(t) = \frac{\zeta_2(\xi+t^2/2)}{(-\xi+t^2/2+\zeta_1 t)^2+\zeta_2^2 t^2}$$

Here $\xi, \zeta_1, \zeta_2$ are independent random variables with exponential, normal, and Rayleigh distributions, respectively, taking different (independent) values at each new click. Figure 1, taken from [8], shows examples of typical click forms for a few outcomes of the

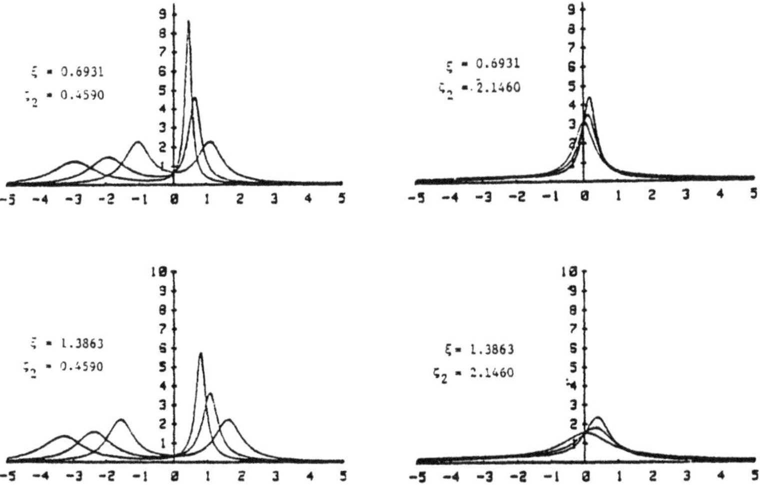

Figure 1: Normalized click shape for various outcomes of random coefficients; $\zeta_1 = 0$ (symmetric curves), $\zeta_1 = 0.6745$, and $\zeta_1 = 1.216$ (most unsymmetric curves) in each diagram.

random coefficients.

Similar explicit, but more complicated Slepian models can be derived for clicks in the presence of a signal $S(t)$; see [9]. □

EXAMPLE 2 (Jumps and bumps on random roads) The vertical movements of a car travelling on a rough road can be modelled as a stationary process. The event that a wheel jumps and bumps, i.e. leaves the ground for a short while, can be characterized as a level crossing in the process that describes the normal forces between the wheel and the ground. The conditional properties of this force and of other related processes, such as the road and wheel elevation after a jump, can be described explicitly in terms of Slepian model processes.

Figure 2 shows, for a simplified case, the mean values of two such Slepian models, here describing the road and wheel after a jump at time 0. In particular, the curves clearly show the speed dependence, both with regards to jump length, and to the height over the zero level at which jumps are likely to occur. Further they give some indication about the angle under which the wheel hits the ground after a jump; for details on the models, see [7].

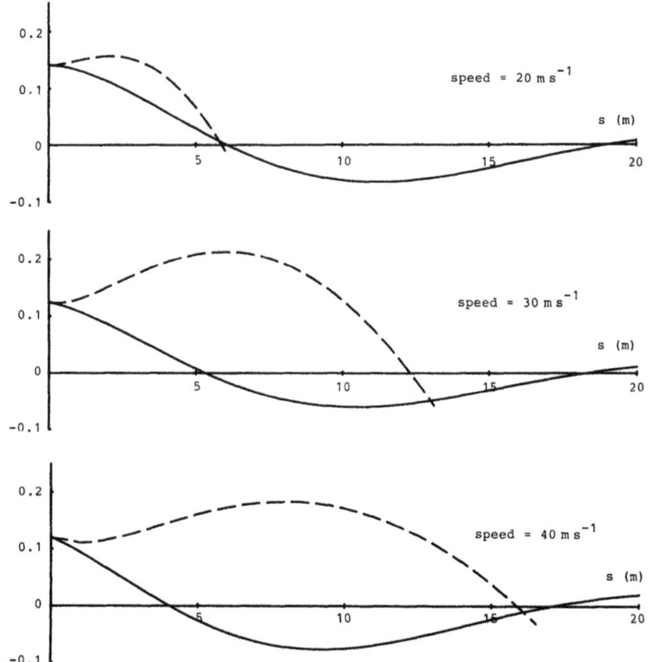


Figure 2: Mean values for road surface (——) and wheel elevation (- - -) after a jump at $s = 0$. □

## 2. GENERAL STRUCTURE OF THE UNIVARIATE SLEPIAN MODEL

Let $\{\xi(t),\ t \in R\}$ be a stationary normal process with mean zero and unit variance, and with covariance function satisfying

$$r(t) = 1 - \lambda_2 t^2/2 + o(t^2) \quad \text{as} \quad t \to 0 , \tag{2.1}$$

and with continuously differentiable sample paths. A sufficient condition for differentiability is that the process is separable and that

$$-r''(t) = \lambda_2 + o(|\log|t||^{-a}) \quad \text{as} \quad t \to 0 , \tag{2.2}$$

for some $a > 1$, a condition we shall assume satisfied. When needed we also assume $\lambda_4 = r^{IV}(0) < \infty$, and that $r^{IV}(t)$ satisfies a condition like (2.2), in which case $\xi(t)$ is twice continuously differentiable. Further, we assume that $\xi(0)$, $\xi'(0)$, $\xi(s_1)$, ..., $\xi(s_n)$ have a non-singular distribution for all distinct non-zero $s_1$, ... $s_n$.

Since $\lambda_2 = -r''(0) < \infty$, the number of upcrossings of a level $u$ in any bounded interval is finite, and the mean number $\mu$ of upcrossings per time unit is given by Rice's formula

$$\mu = \frac{1}{2\pi} \lambda_2^{1/2} e^{-u^2/2} .$$

Let $\{t_k\}$ be the locations of the upcrossings, indexed so that $\ldots t_{-1} < t_0 \leq 0 < t_1 \ldots$ . In order to retain information about $\{\xi(t)\}$ near its upcrossings we attach to each $t_k$ a *mark* $\eta_k$, defined as the sample function $\xi(\cdot)$ translated back the distance $t_k$,

$$\eta_k(t) = \xi(t_k+t) .$$

In particular, $\eta_k(0) = u$, upcrossing.

A realization of $\{\xi(t)\}$ generates a realization of the sequence of marks $\{\eta_1(t)\}$, $\{\eta_2(t)\}$, ... . For each such realization one can form the *empirical distribution* of the observed values $\eta_k(s)$, $k = 1, 2, \ldots$, i.e. of $\xi(t_k+s)$ a fixed time $s$ after the upcrossings. For increasing observation intervals these empirical distributions converge to the distributions of a certain stochastic processes, which we shall call the *Slepian model process*. It has the form

$$\xi_u(t) = ur(t) - \zeta r'(t)/\lambda_2 + \kappa(t) ,$$

where $\zeta$ is a Rayleigh distributed random variable, and $\kappa$ is a zero-mean, non-stationary normal process. The following derivation of the Slepian process follows mainly the presentation in [2], Chapter 10.

Write, for any $T > 0$,

$$N_u(T) = \#\{t_k \in (0, T]\} ,$$

let $s = (s_1, \ldots, s_n)^T$, $y = (y_1, \ldots, y_n)^T$ be vectors of real numbers, and define

$$\tilde{N}_u(T) = \#\{t_k \in (0, T]\} ; \xi(t_k+s_j) \leq y_j , j = 1, \ldots, n\} .$$

Then $F_s^T(y) = \tilde{N}_u(T)/N_u(T)$ is the empirical distribution of $\xi(\cdot)$ at times $s_1, \ldots, s_n$ after its u-crossings in $(0, T]$.

THEOREM 2.1 If the process $\{\xi(t)\}$ is ergodic with $E(N_u(1)) < \infty$ then, with probability one,

$$F_s^T(y) \to F_s(y) = \frac{E(\tilde{N}_u(1))}{E(N_u(1))}, \quad \text{as} \quad T \to \infty. \tag{2.3}$$

PROOF This is a simple consequence of the ergodic property $N_u(T)/T \to E(N_u(1))$, $\tilde{N}_u(T)/T \to \tilde{E}(N_u(1))$. □

Up to this point we have not been specific about the basic underlying probability space. Now let $C_1$ be the space of continuously differentiable functions defined on the real line, and denote a typical element in $C_1$ by $\xi$, with value $\xi(t)$ at time t. We can then think of the probability measure P for our original process $\{\xi(t)\}$ as defined on $C_1$ or, more precisely, on the smallest $\sigma$-field $F$ of subsets of $C_1$ which makes all the projections $\xi \rightsquigarrow \xi(t)$ measurable.

As a generalization of the point process $\tilde{N}_u$ studied above, define, for each $E \in F$, a new point process $\tilde{N}_{u,E}$ by deleting from $N_u$ all points $t_k$ for which $\xi(t_k+\cdot) \notin E$, i.e. for which the function $\eta_k(\cdot) = \xi(t_k+\cdot)$ does not belong to E. Then, motivated by (2.3), we define a second probability measure $P^u$ on $C_1$ by

$$P^u(E) = \frac{E(\tilde{N}_{u,E}(1))}{E(N_u(1))} = \frac{E(\#\{t_k \in (0,1];\xi(t_k+\cdot) \in E\}}{E\#\{t_k \in (0,1]\})} .$$

The measure $P^u$ is the *Palm distribution* or the *ergodic distribution* of $\xi$ after a u-upcrossing. By additivity of expectations $P^u$ is in fact a probability measure, and the finite-dimensional distributions of $\xi(s_j)$ under $P^u$ are clearly exactly the functions $F_s$ appearing in (2.3). Thus

$$P^u(\xi(s_j) \le y_j,\ j = 1, \ldots, n) = F_s(y) =$$

$$= \lim_{T\to\infty} \frac{\#\{t_k \in (0,T];\xi(t_k+s_j) \le y_j, j=1,\ldots,n\}}{\#\{t_k \in (0,T]\}} .$$

Obviously, the result of Theorem 2.1 about the empirical distributions of $\eta_k(t)$ holds also with $\tilde{N}_u$ replaced by $\tilde{N}_{u,E}$ for arbitrary $E \in F$, i.e. with P-probability one,

$$\frac{\tilde{N}_{u,E}(T)}{N_u(T)} \to \frac{E(\tilde{N}_{u,E}(1))}{E(N_u(1))} = P^u(E) .$$

This leads to the convergence of the empirical distributions of many other interesting functionals, such as the excursion time, i.e. the time from an upcrossing to the next downcrossing of the same level, and the maximum in intervals of fixed length following an upcrossing; for more examples, see [5].

The measure $P^u$ formalizes the notion of a conditional distribution of the process given that it has a u-upcrossing at time 0, and it describes the long run properties of $\xi(t_k+t)$ as a function of $t$, when $t_k$ runs through the set of all u-upcrossings. One can therefore interpret the $P^u$-distribution of $\xi(t)$ as the "conditional distribution of the original process at time $t$ after an arbitrary u-upcrossing". In particular

$$P^u(\xi(t) \text{ has an upcrossing of } u \text{ for } t = 0) = 1 .$$

Furthermore, all $\eta_k(t) = \xi(t_k+t)$ have the same $P^u$-distribution for $k = 0, 1, \ldots,$ being "equally arbitrary".

We now turn to a study in more detail of the properties of any of the marks under the Palm distribution, and to the special representation of $\eta_0(t) = \xi(t)$ in terms of the Slepian model process. The following theorem uses the definition of the Palm distribution and forms the basis for the Slepian representation.

THEOREM 2.2 Let $\mu = E(N_u(1)) = (2\pi)^{-1}\lambda_2^{1/2}\exp(-u^2/2)$. Then, for $t \neq 0$,

$$P^u(\xi(t) \leq y) = \int_{x=-\infty}^{y} \{\mu^{-1} \int_{z=0}^{\infty} z f(u, z) f(x|u, z) dz\} dx , \tag{2.4}$$

where $f(u, z)$ is the joint P-density of $\xi(0)$ and $\xi'(0)$, and $f(x|u, z)$ is the conditional P-density of $\xi(t)$ given $\xi(0) = u$, $\xi'(0) = z$. Thus, the $P^u$-distribution of $\xi(t)$ is absolutely continuous with density

$$\mu^{-1} \int_{z=0}^{\infty} z f(u, z) f(x|u, z) dz . \tag{2.5}$$

The n-dimensional $P^u$-distribution of $\xi(s_1), \ldots, \xi(s_n)$ is obtained by replacing $f(x|u, z)$ by $f(x_1, \ldots, x_n|u, z)$, the conditional P-density of $\xi(s_1), \ldots, \xi(s_n)$ given $\xi(0) = u$, $\xi'(0) = z$.

PROOF This follows simply from the theory of marked crossings. For example, with $\tilde{N}_u(1) = \#\{t_k \in (0, T]; \xi(t_k+t) \leq y\}$,

$$E(\tilde{N}_u(1)) = \int_{z=0}^{\infty} z f(u, z) P(\xi(t) \leq y \mid \xi(0) = u, \xi'(0) = z) dz =$$

$$= \int_{x=-\infty}^{y} \int_{z=0}^{\infty} z f(u, z) f(x \mid u, z) dz dx ,$$

proving (2.4). The multivariate version is quite analogous. □

The joint density of $\xi(s_1), \ldots, \xi(s_n)$ under $P^u$,

$$\mu^{-1} \int_{z=0}^{\infty} z f(u, z) f(x_1, \ldots, x_n \mid u, z) dz , \tag{2.6}$$

is obviously a mixture of (conditional) normal densities, with mixing function $\mu^{-1} z f(u, z)$. Since $\xi(0)$ and $\xi'(0)$ are independent and normal with mean zero and $V(\xi'(0)) = \lambda_2$, we have $f(u, z) = (\mu/\lambda_2)\exp(-z^2/2\lambda_2)$, and thus (2.6) can be written

$$\int_{z=0}^{\infty} \frac{z}{\lambda_2} \exp(-z^2/2\lambda_2) f(x_1, \ldots, x_n \mid u, z) dz . \tag{2.7}$$

Since the covariance matrix of $\xi(0), \xi'(0), \xi(s_1), \ldots, \xi(s_n)$ is

$$\left(\begin{array}{c|c|cccc} 1 & 0 & r(s_1) & \cdots & & r(s_n) \\ \hline 0 & \lambda_2 & -r'(s_1) & \cdots & & -r'(s_n) \\ \hline r(s_1) & -r'(s_1) & 1 & \cdots & & r(s_1-s_n) \\ \vdots & \vdots & \vdots & & & \vdots \\ r(s_n) & -r'(s_n) & r(s_n-s_1) & \cdots & & 1 \end{array}\right) = \begin{pmatrix} 1 & 0 & \Sigma_{13} \\ 0 & \lambda_2 & \Sigma_{23} \\ \Sigma_{31} & \Sigma_{32} & \Sigma_{33} \end{pmatrix}$$

it follows from standard properties of conditional normal densities that $f(x_1, \ldots, x_n \mid u, z)$ is n-variate normal with mean and covariances given by $u\Sigma_{31} + z\lambda_2^{-1}\Sigma_{32}$ and $\Sigma_{33} - \Sigma_{31}\Sigma_{31} - \lambda_2^{-1}\Sigma_{32}\Sigma_{23}$, so that

$$E(\xi(s_i) \mid \xi(0) = u, \xi'(0) = z) = ur(s_i) - zr'(s_i)/\lambda_2 \tag{2.8}$$

$$\mathrm{Cov}(\xi(s_i),\ \xi(s_j)\,|\,\xi(0) = u,\ \xi'(0) = z) = \tag{2.9}$$

$$= r(s_i - s_j) - r(s_i)r(s_j) - r'(s_i)r'(s_j)/\lambda_2.$$

Thus, the density (2.7) is a mixture of normal densities with the same covariances (2.9) but with different means (2.8), mixed in proportion to the Rayleigh density

$$\frac{z}{\lambda_2}\exp(-z^2/2\lambda_2)\ ,\ z > 0\ . \tag{2.10}$$

We are now ready to introduce the Slepian model process. Let $\zeta$ have a Rayleigh distribution with density (2.10), and let $\{\kappa(t),\ t \in R\}$ be a non-stationary normal process, independent of $\zeta$, with mean zero and covariance function

$$r_\kappa(s,\ t) = r(s-t) - r(s)r(t) - r'(s)r'(t)/\lambda_2\ .$$

That this really is a covariance function follows from (2.9), and hence, on some probability space, which need not be specified further, there exist $\zeta$ and $\kappa$ with these properties. The process

$$\xi_u(t) = ur(t) - \zeta r'(t)/\lambda_2 + \kappa(t) \tag{2.11}$$

is the Slepian model process for $\xi(t)$ after u-upcrossings. Obviously, conditional on $\zeta = z$, it is a normal process with mean and covariances given by the right hand sides of (2.8) and (2.9), respectively, and so its finite-dimensional distributions, and hence its entire distribution, are given by the densities (2.7), as stated in the following theorem.

THEOREM 2.3 The Palm distribution of the mark $\eta_0(t) = \xi(t)$, (and thus of all marks $\eta_k(t)$), is equal to the distribution of the Slepian model

$$\xi_u(t) = ur(t) - \zeta r'(t)/\lambda_2 + \kappa(t)\ ,$$

i.e. $P^u(\xi(s_j) \in B_j,\ j = 1, \ldots, n) = P(\xi_u(s_j) \in B_j,\ j = 1, \ldots, n)$ for any Borel sets $B_1, \ldots, B_n$. The process $\{\xi_u(t)\}$ therefore describes $\xi(t_k+t)$ around u-upcrossings. □

One should note that the level u enters into $\xi_u(t)$ only via the function $ur(t)$, while $\zeta$ and $\kappa(t)$ can be the same for all u. This is a major advantage of the Slepian model, and it simplifies considerably comparison of crossing properties for different levels. It also makes it possible to derive strong limit theorems as $u \to \infty$, which are then, by means of Theorems 2.3 and 2.1, translated into limit theorems for the empirical distribution of functionals such as the maximum in a specified interval following the upcrossing.

It should also be noted that if $r(t)$ satisfies (2.2), $\kappa(t)$ is continuously differentiable with $E(\kappa(t)) = E(\kappa'(t)) = 0$, $V(\kappa(0)) = V(\kappa'(0)) = 0$, so that $\kappa(0) = \kappa'(0) = 0$, with probability one. Since $\lambda_2 = -r''(0)$ one has $\xi_u'(0) = \zeta$, so that $\zeta$ represents the distribution of the derivative of $\xi(\cdot)$ at its u-upcrossings.

COROLLARY 2.4 The long run distribution of $\xi'(t_k)$ at the points of u-upcrossing is given by the Rayleigh density $(z/\lambda_2)\cdot\exp(-z^2/2\lambda_2)$, $z > 0$.

We end this section with two remarks with extensions and consequences of the model process.

REMARK 2.5 (Local maxima — wave-length and amplitude distributions) Let $\{t_k\}$ be the local maxima of a twice continuously differentiable normal process $\xi(t)$, i.e. the downcrossing zeros of $\xi'(t)$. By conditioning both on the height $\xi(t_k) = u$, say, and the second derivative $\xi''(t_k) = z$ at the maximum one can construct a Slepian model for the Palm distribution of $\xi(t_k+t)$.

The regression term is

$$E(\xi(t)|\xi(0) = u,\ \xi'(0) = 0,\ \xi''(0) = z) = \tag{2.12}$$

$$= u\,\frac{\lambda_4 r(t)+\lambda_2 r''(t)}{\lambda_4-\lambda_2^2} + z\,\frac{\lambda_2 r(t)+r''(t)}{\lambda_4-\lambda_2^2} = uA(t) + zC(t)\ ,$$

say, and the conditional covariance function $\mathrm{Cov}(\xi(s),\ \xi(t)|\xi(0),\ \xi'(0),\ \xi''(0))$ is

$$r_\Delta(s,\ t) = r(s-t) - \{\lambda_2(\lambda_4-\lambda_2^2)\}^{-1}\{\lambda_2\lambda_4 r(s)r(t)+\lambda_2^2 r(s)r''(t)+$$

$$+(\lambda_4-\lambda_2^2)r'(s)r(t)+\lambda_2^2 r''(s)r(t)+\lambda_2 r''(s)r''(t)\}\ . \tag{2.13}$$

Further, the Palm distribution of $\xi(t_k)$ and $\xi''(t_k)$ at local maxima has the bivariate density

$$f(u,\ z) = \{2\pi\lambda_4^2(\lambda_4-\lambda_2^2)\}^{-1/2}|z|\exp\{-\frac{1}{2(\lambda_4-\lambda_2^2)}(z^2+2\lambda_2 uz+\lambda_4 u^2)\} =$$

$$= f_{max}(u)q_u(z) \ , \ u \in R, \ z < 0 \ , \tag{2.14}$$

where

$$f_{max}(u) = \varepsilon\phi(u/\varepsilon) + (1-\varepsilon^2)^{1/2}ue^{-u^2/2}\Phi(u(1-\varepsilon^2)^{1/2}/\varepsilon) \ , \tag{2.15}$$

($\varepsilon^2 = 1-\lambda_4/\lambda_2^2$), is the Rice distribution of the maximum height, and

$$q_u(z) = c_u|z|\exp(-(z+\lambda_2 u)^2/2(\lambda_4-\lambda_2^2)) \ , \ z < 0 \ , \tag{2.16}$$

the conditional density of $\xi''(0)$ given $\xi(0) = u$, $\xi'(0) = 0$; see [3]. The Slepian model for $\xi(t_k+t)$ is therefore

$$\xi_{max}(t) = \xi_o A(t) + \zeta C(t) + \Delta(t) \ ,$$

where the distribution of $\xi_o$, $\zeta$ is given by (2.15) and (2.16), and $\Delta$ is non-stationary and normal with mean zero and covariance function (2.13).

As a typical use of a Slepian model we shall now use $\xi_{max}(t)$ to discuss the statistical properties of wave-length and amplitude in a stationary normal process, i.e. the time and height difference between a local maximum and the following minimum. As an example we shall then use only the regression term $\xi_{reg}(t) = \xi_o A(t) + \zeta C(t)$, since for moderate values of $t$ this contains most of the information about $\xi_{max}(t)$. Thus let $T$, $H$ denote the wave-length and amplitude of the first wave in $\xi_{reg}(t)$, i.e.

$$T = \inf\{t > 0; \ \xi_{reg}(t) \ \text{is a local mimimum}\} \ ,$$

$$H = \xi_{reg}(0) - \xi_{reg}(T) \ .$$

Since $\xi_{reg}(t) = \zeta(\xi_o A(t)/\zeta + C(t))$ it is clear that $T$ is a function only of $\xi_o/\zeta$. In fact, if there is no $t_o > 0$ such that $r'(t_o) = r'''(t_o) = 0$, it is even a one-to-one function of $\xi_o/\zeta$. Since if for some $c_1 \neq c_2$, $t_o > 0$,

$$c_1 A'(t_o) + C'(t_o) = c_2 A'(t_o) + C'(t_o) = 0$$

then $A'(t_o) = C'(t_o) = 0$, and consequently $r'(t_o) = r'''(t_o) = 0$.

If $T$ is a one-to-one function of $\xi_o/\zeta$ there are functions

$p$ and $q$ such that

$$\xi_o/\zeta = p(T) \ , \ \zeta = Hq(T) \tag{2.17}$$

and

$$T = p^{-1}(\xi_o/\zeta) \ ,$$

$$H = \xi_{reg}(0) - \xi_{reg}(T) = \zeta(\frac{\xi_o}{\zeta}(A(0)-A(T))+(C(0)-C(T))) =$$

$$= \zeta(p(T)(1-A(T))-C(T)) \ . \tag{2.18}$$

Since $\xi'_{reg}(T) = 0 = \zeta(\xi_o A'(T)/\zeta + C'(T))$ one obtains

$$p(t) = -C'(t)/A'(t) \ ,$$

and thus, by (2.18) and (2.17),

$$q(t) = -\frac{A'(t)}{C'(t)(1-A(t))+A'(t)C(t)} \ .$$

Now it follows simply from (2.17) that $T$, $H$ have joint density

$$f_{T,H}(t, h) = h|p'(t)|q^2(t)f_{\xi_o,\zeta}(hp(t)q(t), hq(t)) \ ,$$

where the density of $\xi_o$, $\zeta$ is given by (2.14). We get the following explicit density for wave-length and amplitude in the regression part of the Slepian model:

$$f_{T,H}(t, h) = ch^2|q^3(t)p'(t)| \cdot$$

$$\cdot \exp\{-\frac{h^2q^2(t)}{2\varepsilon^2}\frac{T_m^4}{\pi^4}\{(\frac{\pi^2}{T_m^2}p(t)+1)^2+\frac{\varepsilon^2}{1-\varepsilon^2}\}\}$$

where $T_m = \pi\sqrt{\lambda_2/\lambda_4}$ is the mean wave-length, and $c$ is a normalizing constant. Since $p(t)$ and $q(t)$ are given explicitly in terms of $A(t)$ and $C(t)$ the density $f_{T,H}$ can be simply calculated for any covariance function. It is the only known closed form approximation for the joint distribution of wave-length and amplitude in the original stationary process $\xi(t)$ that is in agreement with observed data, and depends on the full covariance function; see [11], from which the following level curve plot of $f_{T,H}(t, h)$ is taken.

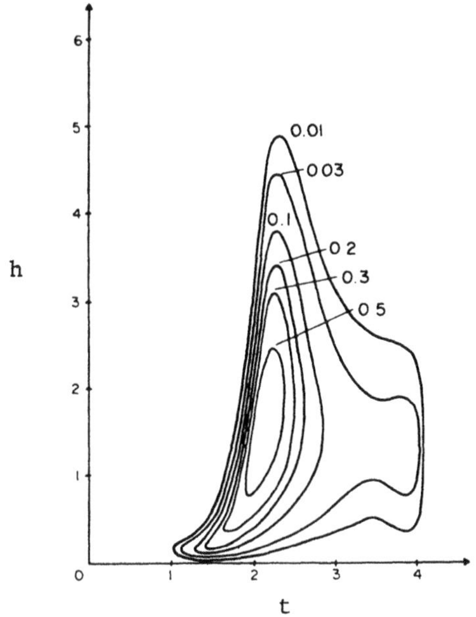


Figure 3: Approximate density $f_{T,H}(t, h)$ of wavelength and amplitude in low frequency white noise; $\varepsilon = 0.667$.

## 3. OPTIMAL LEVEL CROSSING PREDICTION

Suppose we want to predict (in advance!) when a stationary normal process $\eta(t)$ exceeds a predetermined level $u$, and that the prediction may be based on a univariate or multivariate process $\xi(t)$, jointly normal and stationary with $\eta(t)$. For example, we may take $\xi(t) = (\eta(t), \eta'(t))^T$, or even $\xi(t) = (\eta(s), s \leq t)$, but it is also possible to include in $\xi(t)$ other processes correlated with $\eta(t)$. In a practical example, extreme flood events may be predicted from current water levels and meteorological variables such as wind and barometric pressure.

Denote by $X$ with $\sigma$-field $\mathcal{F}$, the state space of $\xi(t)$. Let $\{t_k\}$ be the times of upcrossings of the level $u$ by $\eta(t)$, and let $C_t$ denote the event that $\eta(\cdot)$ has a u-upcrossing at $t$, occasionally writing $P(\cdot|C_t)$ for the Palm probability given an upcrossing at $t$. By a *crossing predictor with warning time* $h$ we mean any set $\Gamma_h \subseteq X$, such that any time $\xi(t)$ belongs to $\Gamma_h$ we consider it likely that $C_{t+h}$ will occur, and set the *alarm* for and upcrossing at $t + h$.

The following probabilities and conditional (Palm) probabilities describe the relevant properties of $\Gamma_h$ as an alarm region,

$$\alpha_h = P(\xi(t) \in \Gamma_h) ,$$

$$\gamma_h = P(\xi(t) \in \Gamma_h | C_{t+h}) .$$

We call $\alpha_h$ the *size of the alarm region* or "the stationary probability of alarm", while $\gamma_h$ is the *detection probability with warning time* $h$.

It is important that in case of an upcrossing at time $t_k$ the conditional probability $\gamma_h = P(\xi(t_k-h) \in \Gamma_h | C_{t_k})$ is as large as possible for each fixed warning time $h$. Since $\gamma_h$ is increasing in $\Gamma_h$, i.e. $\Gamma_h \subset \Gamma_h'$ implies $\gamma_h \leq \gamma_h'$, it is clear that one should only compare alarm policies with the same $\alpha_h$-function. We therefore make the following definition.

By an *optimal alarm policy of size* $\{\alpha_h\}$ is meant a family of sets $\{\Gamma_h\}$, $\Gamma_h \in F$, such that

$$P(\xi(t) \in \Gamma_h) = \alpha_h , \quad h > 0 ,$$

$$P(\xi(t) \in \Gamma_h | C_{t+h}) = \sup_{\Gamma \in F} P(\xi(t) \in \Gamma | C_{t+h}) ,$$

where the sup is taken over $\Gamma$ such that $P(\xi(t) \in \Gamma) \leq \alpha_h$. An optimal alarm policy could be described as the alarm policy which, for given detection function $\gamma_h$, spends the shortest possible time in alarm state. This is of course not the same as having the smallest number of false alarms.

In practice the total alarm time $\alpha_h$ must be balanced against what can be obtained in detection probability. In order to get a reasonable $\gamma_h$ for some particular $h$, e.g. for the minimum *useful* warning time, it may be necessary to accept a long total alarm time for that h-value.

To obtain an optimal level crossing predictor or alarm policy, we shall now use an analogy with "most powerful tests" and consider the likelihood ratio between the unconditional distribution of $\xi(t)$ and its conditional distribution given a u-upcrossing in $\eta(\cdot)$ at time $t + h$. This is done by means of the Palm distribution $P^u(\xi(h) \in \cdot)$ and its explicit Slepian representation $\xi_u(t)$, so that

$$\gamma_h = P^u(\xi(h) \in \Gamma_h) = P(\xi_u(h) \in \Gamma_h) .$$

In general the Slepian model for $\xi(t_k+t)$ given a u-upcrossing

$\eta(t_k)$ is

$$\xi_u(t) = m_{u,\zeta}(t) + \kappa(t) ,$$

where

$$m_{u,z}(t) = E(\xi(t)\,|\,\eta(0) = u,\ \eta'(0) = z) ,$$

and $\zeta$ is a Rayleigh distributed random variable, and $\kappa(t)$ a non-stationary normal residual.

We now specialize to the finite dimensional case, $\xi(t) \in R^p$, and write

$$\Sigma_k = \begin{pmatrix} 1 & 0 & \Sigma_{13}(k) \\ 0 & \lambda_2 & \Sigma_{23}(k) \\ \Sigma_{31}(k) & \Sigma_{32}(k) & \Sigma_{33} \end{pmatrix}$$

for the covariance matrix of $\eta(0)$, $\eta'(0)$, $\xi(k)$. Suppose $\Sigma_{33}$ is non-singular and that no nonzero linear function of $\eta(0)$ and $\eta'(0)$ can be predicted perfectly by $\xi(k)$. Then $\Sigma_k$ is invertible and the distribution of $\kappa(k)$ is non-singular normal with covariance matrix

$$\Lambda_k = r_\kappa(k,\ k) = \Sigma_{33} - \Sigma_{31}(k)\Sigma_{13}(k) - \lambda_2^{-1}\Sigma_{32}(k)\Sigma_{23}(k) .$$

Further write, in analogy with previous notation,

$$m_{u,z}(k) = u\Sigma_{31}(k) + z\Sigma_{32}(k)/\lambda_2 = uA(k) + zB(k) .$$

THEOREM 3.1 If $\xi(-h) \in R^p$, $\eta(0)$, $\eta'(0)$ have a non-singular distribution the optimal alarm region is defined by (dropping $k = -h$),

$$\Gamma_h = \{x \in R^p;\ u^2 - \left(\frac{u - A^T\Sigma_{33}^{-1}x}{\sigma_{\eta\cdot\xi}}\right)^2 + 2\log\Psi(\sigma_{\eta'\cdot\xi\eta}B^T\Lambda^{-1}(x-uA)) \geq K\} , \tag{3.1}$$

where $\sigma^2_{\eta\cdot\xi} = V(\eta(0)\,|\,\xi(-h)) = 1 - A^T\Sigma_{33}^{-1}A$, $\sigma^2_{\eta'\cdot\xi\eta} = V(\eta'(0)\,|\,\xi(-h),\ \eta(0)) = 1/\lambda_2^{-1}+B^T\Lambda^{-1}B)$, and $\Psi(x) = \phi(x) + x\Phi(x)$, $\phi$ and $\Phi$ being the standard normal density and distributions functions. Note that

$A^T\Sigma_{33}^{-1}x = E(\eta(0)|\xi(-h)=x)$, $\sigma^2_{\eta'\cdot\xi\eta} B^T\Lambda^{-1}(x-uA) = E(\eta'(0)|\xi(-h) = x, \eta(0) = u)$.

PROOF The optimal alarm region is defined by means of the likelihood ratio $p(x) = dP^u/dP(x)$ between the conditional (Palm) distribution of $\xi(-h)$ given an upcrossing $\eta(0) = u$, and its unconditional normal distribution, and it was shown by de Maré (1980) that $\Gamma_h = \{x \in R^p;\ p(x) \geq K\}$, where $K$ is such that $P(\xi(-h) \in \Gamma_h) = \alpha_h$; cf. the Neyman-Pearson Lemma and most powerful tests.

Here $dP$ is normal with mean zero and covariance matrix $\Sigma_{33}$, while $dP^u$ has density

$$\int_{z=0}^{\infty} \frac{z}{\lambda_2} e^{-z^2/2\lambda_2} \frac{1}{(2\pi)^{p/2}\sqrt{\det \Lambda}} \exp(-\tfrac{1}{2}(x-uA-zB)^T\Lambda^{-1}(x-uA-zB))dz. \tag{3.2}$$

By integration and some matrix manipulation this can be seen to be equal to a constant times

$$\exp(-\tfrac{1}{2}(x^T\Sigma_{33}^{-1}x-u^2+\left(\frac{u-A^T\Sigma_{33}^{-1}x}{\sigma_{\eta\cdot\xi}}\right)^2)\Psi(\sigma_{\eta'\cdot\xi\eta}B^T\Lambda^{-1}(x-uA)),$$

and taking logarithms yields the result. □

The form (3.1) of $\Gamma_h$ allows comparison with alarm regions based on the mean square predictor $\hat{\eta}(t+h) = E(\eta(t+h)|\xi(t)=x) = A^T\Sigma_{33}^{-1}x$. The optimal alarm is set when $\hat{\eta}(t+h)$ is near $u$, but slightly adjusted by the $\Psi$-term, according to the expected change rate $E(\eta'(t+h)|\xi(t)=x, \eta(t+h)=u)$.

Fore a more direct proof of (3.1), write (3.2) in a form analogous to (2.6) as

$$\mu^{-1}\int_{z=0}^{\infty} z f_{\eta,\eta'}(u,z)f_{\xi|\eta=u,\eta'=z}(x)dz =$$

$$= \mu^{-1}f_\xi(x)f_{\eta|\xi=x}(u)\int_{z=0}^{\infty} z f_{\eta'|\xi=x,\eta=u}(z)dz =$$

$$= \mu^{-1}f_\xi(x)f_{\eta|\xi=x}(u)E(\eta'(0)^+|\xi(-h)=x,\eta(0)=u),$$

givning $p(x) = \mu^{-1}\mu_h(x)$, where $\mu_h(x)$ is the conditional upcrossing intensity in $\eta(t+h)$ given $\xi(t)=x$. This gives the optimal alarm region

$$\Gamma_h = \{x \in R^p;\ \mu_h(x) \geq K\} \tag{3.3}$$

REMARK 3.2 (Optimal level crossing prediction in discrete time.) If $\{\eta_n\}$ and $\{\xi_n\}$ are jointly stationary and normal, $\xi_n \in R^p$, one can find an optimal alarm region $\Gamma_h$ for detecting a level crossing $\eta_{n-1}<u<\eta_n$, by comparing the unconditional density of $\xi_{-h}$ with the conditional density given $\eta_{-1}<u<\eta_0$. In this case the optimal region is simply those outcomes of $\xi_{-h}$ with the highest conditional probability of upcrossing, i.e. of $\eta_{-1}<u<\eta_0$:

$$\Gamma_h = \{x \in R^p;\ \Phi(u; m_{-h}(x), \Delta_{-h}) \geq K\}$$

where $\Phi(u; m_k(x), \Delta_k) = P(\eta_{-1}<u<\eta_0 | \xi_k=x)$. □

## 4. EXTENSIONS TO MULTIVARIATE PROCESSES AND FIELDS

We now turn to Slepian models for more complicated types of crossing problems, in multivariate processes and in homogeneous fields with multidimensional parameter. First let the process $\xi(t) \in R^p$ be multivariate and consider, instead of a simple level $u$, a smooth boundary $\partial\Gamma$ in $R^p$. We saw in Chapter 3, e.g. Theorem 3.1, how such a boundary naturally appeared when predicting level crossings.

To evaluate in more detail the properties of an alarm region like the one in Theorem 3.1, and compare it with other simpler region, e.g. with linear boundary, one needs a description both of the alarm process $\xi(s)$ near upcrossings in the alarmed process $\eta(t)$, $t \approx s+h$, and of $\eta(t+h)$ near alarms at time $t$, i.e. when $\xi(t)$ enters the alarm region.

Obviously, both the above case are obtained by specializing the model for p-variate process near its crossings of a curved boundary in $R^p$, as the following analysis shows.

Let $\xi(t) = (\xi_1(t), \ldots, \xi_p(t))^T \in R^p$ be a stationary, zero-mean normal vector process with covariance function $r_\xi(t) = E(\xi(s+t)\cdot\xi(s)^T)$, and assume every component is $\xi(t)$ is continuously differentiable. Further, let $\partial\Gamma$ be a smooth surface in $R^p$ defined by a real-valued function $\gamma(x)$ of $x = (x_1, \ldots, x_p)^T$, such that $x \in \partial\Gamma \Leftrightarrow \gamma(x) = 0$, and assume $\gamma$ is continuously differentiable near $\partial\Gamma$. Write $\nu_x$ for the unit normal to the surface $\gamma(x) = C$, i.e. $\nu_x = \dot{\gamma}(x)/|\dot{\gamma}(x)|$, where $\dot{\gamma}(x) = (\partial\gamma/\partial x_1, \ldots, \partial\gamma/\partial x_p)^T$ is supposed to be non-zero near $\partial\Gamma$.

The process $\xi(t)$ passes through the surface $\partial\Gamma$ any time the univariate process $\gamma(\xi(t))$ has a zero crossing, and we denote by $t_k$ the times of the crossings in the positive direction. As in the univariate case we shall study the marks $\eta_k(t) = \xi(t_k+t)$, describing $\xi(\cdot)$ near the crossings and derive the Slepian model for their Palm distribution.

In the model process $\xi_u(t)$ defined by (2.11) an important role is played by $\xi_u(0)$ (= the constant level $u$), and by the Rayleigh distributed derivate $\xi'_u(0) = \zeta$, describing the long run variation in $\xi(t_k)$ and $\xi'(t_k)$. A similar role will here be played by variables which describe the coordinates $\xi_j(t_k)$ and their derivative $\xi'_j(t_k)$ at the points of crossings. These distributions can be expressed as a surface density over $\partial\Gamma$ and a regular density over $R^p$ in the following way.

Denote by $f_{\xi(t)}(x)$ the density of $\xi(t)$ at the point $x \in \partial\Gamma$, representing a mass distribution over $\partial\Gamma$. The derivative $\xi'(t)$ expresses the direction of a flow through $\partial\Gamma$, and the total flow at the point $\xi(t) = x$ is equal to the scalar product $\nu_x^T \cdot \xi'(t)$ between the unit normal and $\xi'(t)$. The net flow in the positive direction is the positive part $(\nu_x^T \cdot \xi'(t))^+$. The total flow at time $t$ through any region $A \subset \partial\Gamma$ is then equal to the surface integral

$$\int_{x \in A} (\nu_x^T \cdot \xi'(t))^+ f_{\xi(t)}(x) ds(x),$$

($ds(x)$ denoting the surface element). Taking expectations and integrating $t$ over $[0,1]$ one obtains the mean flow through $A$ as

$$\int_{t=0}^{1} \int_{x \in A} E((\nu_x^T \cdot \xi'(t))^+ | \xi(t)=x) \, f_{\xi(t)}(x) \, ds(x) \, dt. \tag{4.1}$$

This is Belayev's (1968) formula for the mean number of exits, interpreted as a mean flow. For a generalization, see Lindgren (1980).

If the process is stationary the mean number of positive crossings per time unit through $A$ is simply

$$\mu_\xi(A) = \int_{x \in A} E((\nu_x^T \cdot \xi'(0))^+ | \xi(0)=x) \, f_{\xi(0)}(x) \, ds(x) =$$

$$= \int_{z \in R^p} \int_{x \in A} (\nu_x^T \cdot z)^+ \, p(ds|x) \, f_{\xi(0)}(x) \, ds(x), \tag{4.2}$$

where $p(dz|x)$ is the conditional distribution of $\xi'(0) \in R^p$ given $\xi(0) = x \in R^p$.

The following theorem is the multivariate analogue of Corollary 2.4, and it is proved in [6], where the precise conditions are specified. (To retain notational analogy with the univariate case we condition on $\xi(0) = u = (u_1, \ldots, u_p)^T$.)

THEOREM 4.2 If $\{\xi(t), t \in R\}$ is p-variate, stationary and ergodic, and satisfies certain regularity conditions, the long run distribution of $\xi(t_k)$ and $\xi'(t_k)$ at the points $t_k$ of positive crossings of the boundary $\partial\Gamma$ is given by

$$\mu_\xi(\partial\Gamma)^{-1} (\nu_u^T \cdot z)^+ p(dz|u) f_{\xi(0)}(u) ds(u), \quad u \in \partial\Gamma, \ z \in R^p. \quad (4.3)$$

□

We can now state the multivariate version of Theorem 2.2 about the Palm distribution $P^u$ of $\xi(t_k+t)$ near positive crossings of $\partial\Gamma$.

THEOREM 4.3 Let $\mu = \mu_\xi(\partial\Gamma)$ be given by (4.2). Then, for $t \neq 0$, the density of the Palm distribution $P^u$ for $\xi(t_k+t)$ near positive crossings through $\partial\Gamma$ is

$$\mu^{-1} \int_{u \in \partial\Gamma} \int_{z \in R^p} (\nu_u^T \cdot z)^+ p(dz|u) f_{\xi(0)}(u) f(x|u,z) ds(u), \quad (4.4)$$

where $f(x|u,z)$ is the conditional P-density of $\xi(t)$ given $\xi(0) = u$, $\xi'(0) = z$. □

For a normal process it is a simple matter to express the conditional normal distribution $f(x|u,z)$ in terms of conditional means and covariances. Let, as above $r_\xi(t)$ be the covariance function matrix of $\xi(t)$, with $E(\xi(t)\cdot\xi'(0)^T) = -r_\xi'(t) = r_\xi'(t)^T$, $E(\xi'(t)\cdot\xi'(0)^T) = -r_\xi''(t)$, so that the covariance matrix of $\xi(0)$, $\xi'(0)$, $\xi(t)$ (of type $3p \times 3p$) is

$$\begin{pmatrix} r_\xi(0) & r_\xi'(0)^T & r_\xi(t)^T \\ r_\xi'(0) & -r_\xi''(0) & -r_\xi'(t)^T \\ r_\xi(t) & -r_\xi'(t) & r_\xi(0) \end{pmatrix}.$$

If we denote by $r_{\xi,\xi'}(0)$ the covariance matrix for $\xi(0)$ and $\xi'(0)$, the conditional distribution of $\xi(t)$ given $\xi(0) = u$, $\xi'(0) = z$ is normal with mean

$$E(\xi(t)|\xi(0) = u, \xi'(0)=z) = m_{u,z}(t) =$$

$$= (r_\xi(t), -r_\xi'(t)) r_{\xi,\xi'}(0)^- \binom{u}{z} = \alpha(t)\cdot u + \beta(t)\cdot z,$$

(where $\alpha(t)$ and $\beta(t)$ are functions of type $p \times p$, depending on $r_\xi(t)$ and $r_\xi'(t)$), and with covariances

$$r_\kappa(s,\ t) = \mathrm{Cov}(\xi(s),\ \xi(t)\,|\,\xi(0),\ \xi'(0)) =$$

$$= r_\xi(s-t) - (r_\xi(s),\ -r_\xi'(s))\,r_{\xi,\xi'}(0)^{-}\begin{pmatrix} r_\xi(t)^T \\ -r_\xi'(t)^T \end{pmatrix}. \quad (4.5)$$

Here $r_{\xi,\xi'}(0)^{-}$ denotes the generalized inverse of $r_{\xi,\xi'}(0)$, which even in simple applications need not be invertible.

Now, considering the density (4.4) as a density of a non-stationary normal process with a random mean function $m_{u,z}(t)$, where $u$ and $z$ are allowed to vary with the density (4.3), we immediately obtain the following generalization of the Slepian model after exits through $\partial\Gamma$.

THEOREM 4.4 The Palm distribution of the mark $\eta_0(t) = \xi(t)$, and thus of the process $\xi(t_k+t)$ after positive crossings through $\partial\Gamma$, is equal to the distribution of the Slepian model

$$\xi_{\partial\Gamma}(t) = m_{\xi,\zeta}(t) + \kappa(t) = \alpha(t)\cdot\xi + \beta(t)\cdot\zeta + \kappa(t)\ , \quad (4.6)$$

where $\xi$, $\zeta$ are p-variate random variables, distributed over $\partial\Gamma$ and $R^p$, respectively, with joint distribution

$$f_{\xi,\zeta}(u,\ dz)ds(u) = \mu_\xi(\partial\Gamma)^{-1}(\nu_u^T\cdot z)^+ p(dz|u) f_{\xi(0)}(u)ds(u)\ ,$$

and $\kappa$ is a non-stationary normal process, independent of $\xi$, $\zeta$, with covariance function $r_\kappa(s,\ t)$ given by (4.5).

The model process has exactly the same structure as in the one-dimensional case, with $\xi$ and $\zeta$ playing the roles of the values of the random $\xi(t_k)$ and $\xi'(t_k)$ at the crossings. □

EXAMPLE 4.5 Let $\xi(t) = (\xi_1(t),\ \ldots,\ \xi_p(t))^T$ have independent standardized normal components and consider the Slepian model for $\xi(t)$ after exits through the sphere $\partial\Gamma = \{x \in R^p;\ |x| = (\Sigma_1^p x_i^2)^{1/2} = r\}$. In this case $\xi(t)$ and $\xi'(t)$ are independent with $V(\xi_i'(t)) = \lambda_{2i}$, say, so that

$$f_{\xi(0)}(u) = (2\pi)^{-p/2}\exp(-|u|^2/2)\ ,$$

$$p(dz\mid u) = \prod_{i=1}^{p}(2\pi\lambda_{2i})^{-1/2}\exp(-z_i^2/2\lambda_{2i})dz\ .$$

Further, since $\nu_u = r^{-1}u$ on $\partial\Gamma$ and the outcrossing intensity

$$\mu_\xi(\partial\Gamma) = C'(\lambda)r^{p-1}\exp(-r^2/2) ,$$

(where $C'(\lambda)$ depends only on $\lambda_{2i}$, $i = 1, \ldots, p$), we get the density for $\xi$ and $\zeta$, describing the location and direction at crossings, as

$$C''(\lambda)r^{-p+1}(r^{-1}u^T\cdot z)^+\exp(-\|z\|_y^2/2) , \tag{4.7}$$

with $\|z\|_\lambda^2 = \Sigma_1^p z_i^2/\lambda_{2i}$, and $C''(\lambda)$ depends on $\lambda_{2i}$, $i = 1, \ldots, p$.

The functions $\alpha(t)$ and $\beta(t)$ in $m_{\xi,\zeta}(t)$ are quite simple. Writing $\Lambda = \text{diag}(\lambda_{2i})$, we have $r_\xi(t) = \text{diag}(r_i(t))$, and

$$r_{\xi,\xi'}(0) = \begin{pmatrix} I & 0 \\ 0 & \Lambda \end{pmatrix} ,$$

so that the $i^{th}$ element of $\alpha(t)\cdot u + \beta(t)\cdot z$ is

$$u_i r_i(t) - z_i r_i'(t)/\lambda_{2i} ,$$

and $r_\kappa(s, t)$ is diagonal with diagonal elements

$$r_i(s, t) = r_i(s-t) - r_i(s)r_i(t) - r_i'(s)r_i'(t)/\lambda_{2i}$$

as in the one-dimensional case. The p-variate model process is therefore $\xi_{\partial\Gamma}(t) = (\xi_{\partial\Gamma}^{(1)}(t), \ldots, \xi_{\partial\Gamma}^{(p)}(t))^T$ where

$$\xi_{\partial\Gamma}^{(i)}(t) = \xi_i r_i(t) - \zeta_i r_i'(t)/\lambda_{2i} + \kappa_i(t)$$

with independent $\kappa_i$-processes and dependent $\xi_1, \ldots, \xi_p, \zeta_1, \ldots, \zeta_p$ with density (4.7). □

EXAMPLE 4.6 (Level crossing prediction, continued) Suppose, as a special case, that we want to predict level crossings in a normal process $\eta(\cdot)$ making use of $\eta(t)$ and $\eta'(t)$, and that we want to predict crossings $h$ time units ahead. Then take

$$\xi(t) = (\eta(t), \eta'(t), \eta(t+h))^T$$

and consider alarms when $(\eta(t), \eta'(t))^T$ enters an alarm region $\tilde{\Gamma}$ through the boundary $\partial\tilde{\Gamma}$. With $\partial\Gamma = \partial\tilde{\Gamma} \times R$, the conditional behaviour of $\xi(t)$ after alarms can then be modelled in terms of a Slepian model after crossings of $\partial\Gamma$.

The tri-variate model $\xi_{\partial\Gamma}(t)$ carries information both about the predicted variable $\eta(t_k+t+h)$ and about the predictor $(\eta(t_k+t), \eta'(t_k+t))^T$ near alarms at time $t_k$, and we consider here as an example the model for $\eta(t_k+t+h)$, i.e. $\xi_{\partial\Gamma}^{(3)}(t)$.

Writing $\nu_{\tilde{u}}$ for the unit normal to $\partial\tilde{\Gamma}$ in $R^2$ at $\tilde{u} = (u_1, u_2)^T$, and with $\tilde{z} = (z_1, z_2)^T$, we have that $\nu_u^T\cdot z = \nu_{\tilde{u}}^T\cdot\tilde{z}$. This implies that there is no information to gain by keeping the variables $\xi_3$ and $\zeta_3$ in the model process. They describe $\eta(t_k+h)$ and $\eta'(t_k+h)$, and, borrowing a term from statistical theory, $\xi_1$ $(= \eta(t_k))$, $\xi_2 = \zeta_1$ $(= \eta'(t_k))$, and $\zeta_2$ $(= \eta''(t_k))$ are "sufficient" for the description of $\eta(t_k+t+h)$ for all $t$.

The simplified Slepian model for $\eta(t_k+t+h)$ is obtained directly from Theorem 4.1 by conditioning on suitable derivatives, as follows. The conditional distribution of $\eta(t+h)$ given $\eta(0) = x_1$, $\eta'(0) = x_2$, $\eta''(0) = z_2$ is normal with mean

$$m_{x_1,x_2,z_2}(t) = x_1A(t+h) + x_2B(t+h) + z_2C(t+h) , \tag{4.8}$$

with $A(t)$ and $C(t)$ as in (2.12), and $B(t) = -r'(t)/\lambda_2$, (giving the regression part), and with covariance function equal to $r_\Delta(s+h, t+h)$ as defined by (2.13) (giving the covariance function for the residual process).

Further, the conditional distribution of $\eta''(0)$ given $\eta(0) = x_1$, $\eta'(0) = x_2$, is normal with mean $-x_1\lambda_2$ and variance $\lambda_4-\lambda_2^2$. Hence, by (4.3), the distribution of the initial values $\xi_1$, $\xi_2$ $(= \zeta_1)$, $\zeta_2$ at the boundary point $\tilde{u} = (u_1, u_2)^T \in \partial\tilde{\Gamma}$ and $z_2$ is, writing $\tilde{z} = (u_2, z_2)^T$,

$$C(\nu_u^T\cdot z)^+\exp\left\{-\frac{(z_2+\lambda_2u_1)^2}{2(\lambda_4-\lambda_2^2)} - \frac{u_1^2}{2} - \frac{u_2^2}{2\lambda_2}\right\} . \tag{4.9}$$

Thus, the Slepian model for $\eta(t_k+t+h)$ is

$$\eta_{\partial\Gamma}(t) = \xi_1A(t+h) + \xi_2B(t+h) + \zeta_2C(t+h) + \kappa(t) ,$$

where $\xi_1$, $\xi_2$, $\zeta_2$ have density (4.9), and the non-stationary normal process $\kappa(t)$ has covariance function $r_\kappa(s, t) = r_\Delta(s+h, t+h)$ defined by (2.13). By examining this model it is possible to calculate, at least approximately, important quantities such as

$$P(\eta(t_k+t+h) \text{ has a u-upcrossing for some } t \in I \mid \text{alarm at } t_k)$$

for any interval $I$. □

EXAMPLE 4.7 (Slepian models for random fields) If $\{\xi(t), t \in R^p\}$ is a homogeneous normal field, a Slepian model can easily be constructed for the behaviour near points characterized by some type of level crossing, such as a local maximum or a specific point at a contour curve.

Recently, in [15] Slepian models near upcrossings in any fixed direction were constructed. The field $\xi(t)$ is said to have a u-upcrossing in the $t_p$-direction at $t^*$ if $\xi(t_1^*, \ldots, t_{p-1}^*, t) = u$, upcrossing, as a function of $t$ for $t = t_p^*$, and if further $\xi(t_1, \ldots, t_{p-1}, t_p^*)$ has a local maximum for $t_j = t_j^*$, $j = 1, \ldots, p-1$, considered as a function of $t_1, \ldots, t_{p-1}$.

The Slepian model for $\xi(t^*+t)$ near a u-upcrossing in the $t_p$-direction at $t^*$ is

$$\xi_u(t) = ur(t) - \zeta a(t) - \zeta_u^T \cdot b(t) + \kappa(t) ,$$

where $a(t)$ and $b(t)$ (of type $p(p-1)/2 \times 1$) are functions, $\zeta$ is a Rayleigh variable, describing $\partial\xi/\partial t_p$, and $\zeta_u$ is a $p(p-1)/2$-variate random variable modelling the curvature in the remaining variables, i.e. equal to $\partial^2\xi/\partial t_i \partial t_j$, $1 \le i, j \le p-1$, both $\partial\xi/\partial t_p$ and $\partial^2\xi/\partial t_i \partial t_j$ taken at the point of maximum. The non-stationary normal process $\kappa(t)$ has a similary structure as in previous cases.

A similar model

$$\xi_{max}(t) = \xi_o r(t) - \zeta_u^T \cdot c(t) + \Delta(t)$$

can be derived for $\xi(t)$ near local maxima of fixed ($\xi_o = u$) or random height; see [4]. Here $\zeta_u$ is a $p(p+1)/2$-variate random variable, describing $\partial^2\xi/\partial t_i \partial t_j$, $1 \le i, j \le p$ at the points of maximum. □

* Part of this work has previously appeared in *Probability and mathematical statistics*; Essays in honour of Carl-Gustav Esseen; see [10].

## REFERENCES

[1] Belayev, Uy. K. (196[illegible] On the number of exits across a boundary of a region by a [illegible]ctor stochastic process. *Theor. Probability Appl.* 13, 320-324.

[2] Leadbetter, M.R., Lindgren, G. & Rootzén, H. (1983) *Extremes and related properties of random sequences and processes*. Springer Verlag, New York.
[3] Lindgren, G. (1970) Some properties of a normal process near a local maximum. *Ann. Math. Statist.* 41, 1870-1883.
[4] Lindgren, G. (1972) Local maxima of Gaussian fields. *Ark. Mat.* 10, 195-218.
[5] Lindgren, G. (1977) Functional limits of empirical distributions in crossing theory. *Stochastic Processes Appl.* 5, 143-149.
[6] Lindgren, G. (1980) Model processes in nonlinear prediction with application to detection and alarm. *Ann. Probability* 8, 775--792.
[7] Lindgren, G. (1981) Jumps and bumps on random roads. *J. Sound and Vibration* 78, 383-395.
[8] Lindgren, G. (1983a) On the shape and duration of FM-clicks. *IEEE Trans. Information Theory* -29, 536-543.
[9] Lindgren, G. (1983b) Shape and duration of clicks in modulated FM transmission. Techn. Report 1983:1, Dept. of Mathematical Statistics, University of Lund.
[10] Lindgren, G. (1983c) Use and structure of Slepian model processes in crossing theory. In *Probability and mathematical statistics; Essays in honour of Carl-Gustav Esseen,* (A. Gut & L. Holst, Eds.). Department of mathematics, Uppsala.
[11] Lindgren, G. & Rychlik, I. (1982) Wave characteristic distributions for Gaussian waves — wave-length, amplitude and steepness. *Ocean Engng.* 9, 411-432.
[12] Maré, J. de (1980) Optimal prediction of catastrophes with applications to Gaussian processes. *Ann. Probability* 8, 841-850.
[13] Rice, S.O. (1963) Noise in FM receivers. In *Time Series Analysis* (M. Rosenblatt, Ed. ) 395-422. Wiley, New York.
[14] Slepian, D. (1963) On the zeros of Gaussian noise. In *Time Series Analysis* (M. Rosenblatt, Ed.) 104-115. Wiley, New York.
[15] Wilson, R.J. & Adler, R.J. (1982) The structure of Gaussian fields near a level crossing. *Adv. Appl. Probability* 14, 543-565.

# SPLINE AND ISOTONIC ESTIMATION OF THE PARETO FUNCTION

James Pickands III

University Of Pennsylvania

Summary:

The Pareto function is defined to be $d(1 - F(x))/f(x)dx$. It is nearly constant in the upper tail of every continuous "textbook" distribution. We discuss non-parametric estimation of it by a spline method. Under order restrictions we consider isotonic estimation.

## 1. Introduction

Suppose that $\{X_i, i = 1, 2, \ldots n\}$ are mutually independent with common CDF (cumulative distribution function) $F(x) \equiv p\{X \leq x\}$. Recently there has been a lot of research on estimation of the density function $f(x) \equiv F'(x)$ and on estimation of the hazard function $h(x) \equiv f(x)/(1 - F(x))$. We discuss, here, the estimation of what we call the Pareto function $\phi(x)$ where

$$(1.1) \qquad \phi(x) \equiv d(1 - F(x))/f(x)dx.$$

This is intuitively less appealing than the density function or the hazard function and yet it turns out that it has properties which make it more appropriate. An important property is the fact that for most "textbook" continuous distributions,

$$(1.2) \qquad \phi(x) \to c \in (-\infty, \infty)$$

as $x \to x_\infty$ where

$$x_\infty \equiv \text{lub } x | F(x) < 1.$$



*J. Tiago de Oliveira (ed.), Statistical Extremes and Applications, 285–296.*

We call $x_\infty$ the "upper limit".

We discuss, here, the non-parametric estimation of the Pareto function $\phi(x)$, given by (1.1). We use two general methods applied extensively to the density function f(x) and the hazard function h(x). The first is the "spline" method. For a survey see the recent paper by Wegman and Wright (1983). For an application see, for example, Wahba and Wendelberger (1980). Our estimator, $\tilde{\phi}(x)$, is constant between "knots" and is such that the density function f(x) is continuous everywhere. The knots are the ascending order statistics $X_{k:n}$ or a subset of them.

The second general method is "isotonic" estimation. See the book by Barlow et. al. (1972) for a survey on isotonic estimation. This method is applied when order restrictions are assumed. We would use this method when it is assumed that $\phi(x)$ is nondecreasing or nonincreasing.

In the von-Mises parametrization, the extreme value distributions are $\Lambda(x|a, b, c)$ where

$$- \log \Lambda(x|a, b, c) \equiv (1 + cy)^{-1/c},$$

$c \neq 0$ and

$$- \log \Lambda(x|a, b, 0) \equiv e^{-y}$$

where

$$y = (x - b)/a.$$

In order that F lie in the domain of attraction of $\Lambda(x|a, b, c)$ it is sufficient that (1.2) hold. See de Haan (1970) pages 108-113. The author has shown (1983) that (1.2) is necessary as well as sufficient in order that F lie twice differentiably in the domain of attraction of $\Lambda(x|a, b, c)$. The author has shown (1975) that

$$\lim_{x \to x_\infty} p\{X > x + y\, m(x) | X > x\}$$

$$= 1 - F(y|a, c),\ a \in (0, \infty),\ c \in (-\infty, \infty),$$

where $F(y|a, c)$ is the generalized Pareto distribution function and m(x) is the conditional median of X - x given that X > x. We considered an adaptive sample method for choosing x and proved convergence. The generalized Pareto family is defined in Section 2 and we show, there, that it is characterized by a constant Pareto function.

In Section 2 we examine the Pareto function $\phi(x)$, relating it the cumulative distribution function $F(x) \equiv p\{X \leq x\}$ and the cumulative hazard function $H(x) \equiv - \log(1 - F(x))$ and their derivatives. We show, there, that the distribution function F(x) is a generalized Pareto one if and only if $\phi(x)$ is constant. In Section 3, we show that an at least equally simple representation exists for $\phi(x)$ in terms of the inverse cumulative hazard function or ICHF. Using the constant Pareto function characterization we find the ICHFs for the generalized Pareto family in terms of the function expp x which we introduce there. In Section 4 we consider the generalized Pareto family and introduce the function logg x. In Section 5 we show algebraicly how to fit a spline function, as described above, between some of the order statistics. We introduce, there, the function $\text{expp}^{-1}\, x$, the inverse function for expp x. In Section 6 we find the log-likelihood function for a sample. In Section 7 we consider the spline method of estimation of $\phi(x)$ and raise some questions about the choice of an estimator. In Section 8 we examine estimation under order restrictions. We include the algorithms for isotonic estimation, that is for estimation assuming that $\phi(x)$ is monotone nonincreasing or nondecreasing as $x \to x_\infty \equiv \text{lub}\, x | F(x) < 1$. Finally, in Section 9, we consider the analogy in which 1-F is replaced by -log F and the generalized Pareto family is replaced by the generalized extreme value family. Proofs will be published elsewhere.

## 2. Preliminaries

We assume that X is a non-negative random variable with CDF (cumulative distribution function) $F(x) \equiv p\{X \leq x\}$. The probability function $P(x) = 1 - F(x)$. We assume that $F(x)$ is twice differentiable and we call the first two derivatives $f(x)$ (the density function) and $f'(x)$. The "hazard function" $h(x) \equiv f(x)/(1 - F(x)) \equiv H'(x)$, the derivative of the "cumulative hazard function" or CHF: $H(x) \equiv -\log(1 - F(x))$. Of considerable importance in extreme value theory is what we call the Pareto function

$$\phi(x) \equiv d(1/h(x))/dx = d(1 - F(x))/f(x)dx.$$

The generalized Pareto distribution function is $F(x) \equiv F(x|a, b, c)$ where

$$H(x) \equiv -\log(1 - F(x)) \equiv \int_0^{x/a} du/(1 + cu)_+ \tag{2.1}$$

with $x_+ \equiv x \vee 0$, $a \in (0, \infty)$ and $c \in (-\infty, \infty)$. It follows that

$$1/h(x) \equiv 1/H'(x) \equiv a + cx$$

and the Pareto function

$$\phi(x) \equiv d(1/h(x))/dx = c,$$

constant for all $x \in (0, x_\infty)$. By looking at these equations in the reverse order we can see that the converse is true. That is, we have just proved the following:

Lemma 2.1: A CDF $F(x)$ is a generalized Pareto CDF with shape parameter $c \in (-\infty, \infty)$ if and only if $\phi(x) \equiv c$, for all $x \in (x_{-\infty}, x_\infty)$. Here $x_{-\infty} \equiv \text{glb}\{x|F(x) > 0\} \in (-\infty, \infty)$.

We assume from here on, without loss of generality, that the lower limit $x_{-\infty} \equiv 0$. To what extent does the Pareto function $\phi(x)$ determine the CDF $F(x)$? We assume that $F(0) = 0$ which implies that $H(0) = 0$. This is a permissable assumption from a practical point of view. Notice that

$$H(x) = - \log (1 - F(x)) = \int_0^x h(u)du.$$

Thus $F(x)$ is uniquely determined by $1/h(x)$. So the CDFs $F(x)$ are in 1 to 1 relation to $\{h(0), \phi(x) \equiv d(1/h(x))/dx\}$ provided that the latter exists. In other words, for every such distribution, the "scale factor" $h(0)$ and the $\phi(x)$ are uniquely determined. To avoid confusion, note that $h(0)$ is coupled with $\phi(x)$, not with $F(x)$ in the 1-1 relation.

## 3. The inverse cumulative hazard function

It is well known that $F(X) \sim U(0, 1)$ if $F$ is continuous, as we assume. That is, $F(X)$ has the standard uniform distribution. It follows that $H(X) \equiv - \log(1 - F(X)) \sim e(1)$. That is it has the standard negative exponential distribution with $-\log p\{Y > y\} = y_+$. So we can write

$$X \equiv H^{-1} (Z),\ Z \sim e(1),$$

where $H^{-1}(z)$ is the unique (by continuity) inverse function for $H$. We call $H^{-1}$ the inverse cumulative hazard function or ICHF.

Lemma 3.1: The function $\phi(x)$ can be written

$$\phi \equiv d(\log(H^{-1})'(z))/dz \equiv (H^{-1})''(z)/(H^{-1})'(z)$$

where $z \equiv H(x)$.

Proof. Notice that

$$dz/dx \equiv H'(x) \equiv h(x) \equiv f(x)/(1 - F(x))$$

$$\equiv 1/(dx/dz) \equiv 1/(H^{-1})'(z).$$

The Pareto function is $\phi(x)$ where

$$\phi(x) \equiv d(1/h(x))/dx = d(H^{-1})'(z)/dx = d(H^{-1})'\ (z)/dH^{-1}(z).$$

Dividing both sides by dz we get that

$$\phi = (H^{-1})''(z)/(H^{-1})'(z) = (\log(H^{-1})')'(z).$$

The lemma is proved.

Suppose $\phi = c \in (-\infty, \infty)$ is constant. Then

$$\log(H^{-1})'(z) \equiv a + cz$$

for some $a \in (-\infty, \infty)$. So

$$(H^{-1})'(z) \equiv e^{a} \exp cz.$$

It follows that

$$H^{-1}(z) \equiv e^{a} \int_0^z e^{cu}\, du \equiv e^{a}(e^{cz} - 1)/c$$

if $c \neq 0$. Then

$$H^{-1}(z) \equiv e^{a} z \operatorname{expp} cz \tag{3.2}$$

where

$$\operatorname{expp} x \equiv (e^{x} - 1)/x \equiv \int_0^x e^{u}\, du/x \tag{3.3}$$

when $x \neq 0$. Suppose $c = 0$. Then eq. (3.2) is still true if

$$\operatorname{expp} 0 \equiv 1 \tag{3.4}$$

We define it to be so and note that by L'Hopital's rule

$$\lim_{x\to 0} \operatorname{expp} x \equiv 1.$$

In fact expp is not only continuous at $x = 0$ but analytic there. Now we can assert that (3.2) is true for all $c \in (-\infty, \infty)$.

## 4. The generalized Pareto family

A CDF $F(x)$ is a member of the generalized Pareto family if and only if

$$H(x|a, c) \equiv -\log(1 - F(x|a, c)) = \int_0^{x/a} du/(1 + cu)_{+}.$$

Notice that the upper limit $x_\infty (\equiv \text{lub } x|F(x) < 1)$ is $\infty$

if $c \geq 0$. If $c < 0$, $x_\infty = a/|c|$. We can write

(4.1) $H(x|a, c) \equiv c^{-1} \log(1 + cx/a) = (x/a) \operatorname{logg} cx/a$

where

(4.2) $\operatorname{logg} x \equiv x^{-1} \log(1 + x) = x^{-1} \int_0^x du/(1 + u)_+$

if $x \neq 0$. By L'Hopital's rule,

$$\lim_{c \to 0} c^{-1} \log(1 + cx/a) = x/a \equiv H(x|a, 0).$$

Thus $F(x|a, 0)$ is the negative exponential distribution with mean $a$. So (4.1) is true even for $c = 0$ if

(4.3) $\operatorname{logg} 0 \equiv 1 \equiv \lim_{x \to 0} \operatorname{logg} x$

as we define it to be. By L'Hopital's rule, the function logg x is not only continuous at 0. It is analytic there.

For interesting recent applications see the papers by Davison (1983) and Smith (1983).

## 5. Spline fitting

Let $z_k$, $k = 1, 2, \ldots n$ be an increasing sequence of numbers in $(0, \infty)$. Let $x_k$ be a similar sequence and let

$(H^{-1})'(0) \in (0, \infty)$ be specified. There is a unique inverse cumulative hazard function $H^{-1}(z)$ on $z \in (0, x_n)$, with

$H^{-1}(z_k) \equiv x_k$, for all k, which is such that $(H^{-1})'(z)$ is continuous everywhere and $\phi$ is constant on each interval of the form $(z_{k-1}, z_k)$. By Lemma 3.1, the Pareto function

$$\phi \equiv (\log(H^{-1})')'(z).$$

Let $x_0 \equiv z_0 \equiv 0$, let $a_k \equiv (H^{-1})'(z_k)$ and let $c_k$ be the constant

value of $\phi(z)$ on $(z_k, z_{k+1})$. Recall that $a_0$ is prespecified.

Now, for $z \in (0, z_1)$,

(5.1) $H^{-1}(z) \equiv H^{-1}(0) + \int_0^z (H^{-1})'(u)du = \int_0^z (H^{-1})'(u)du.$

But

$$\log (H^{-1})'(z) \equiv \log(H^{-1})'(0) + c_0 z = \log a_0 + c_0 z$$

and so

$$(5.2) \quad (H^{-1})'(z) \equiv a_0 e^{c_0 z}.$$

By (5.1) and (5.2),

$$H^{-1}(z) \equiv a_0 \int_0^z e^{c_0 u} du = a_0 \; z \text{ expp } c_0 z$$

where expp x is given by (3.3) and (3.4). By our continuity assumption

$$H^{-1}(z_1) \equiv x_1 \equiv a_0 z_1 \text{ expp } c_0 z_1.$$

So

$$(5.3) \quad c_0 \equiv z_1^{-1} \text{ expp}^{-1} \; x_1/a_0 z_1$$

where $\text{expp}^{-1}$ x is the inverse function for expp x. More generally suppose that $z \in (z_k, z_{k+1})$. Then

$$(5.4) \quad H^{-1}(z) \equiv H^{-1}(z_k) + \int_0^{z-z_k} (H^{-1})'(z_k + u)du$$

now

$$\log \, (H^{-1})'(z_k + u) \equiv \log(H^{-1})'(z_k) + c_k u$$

and so

$$(5.5) \quad (H^{-1})'(z_k + u) \equiv (H^{-1})'(z_k)e^{c_k u}.$$

By (5.4) and (5.5),

$$(5.6) \; H^{-1}(z) \equiv x_k + a_k \int_0^{z-z_k} e^{c_k u} \, du$$

$$= x_k + a_k(z - z_k) \text{ expp } c_k(z - z_k).$$

By continuity

$$H^{-1}(z_{k+1}) \equiv x_{k+1} = x_k + a_k(z_{k+1} - z_k) \text{ expp } c_k(z_{k+1} - z_k)$$

and so

$$(5.7) \quad c_k \equiv (z_{k+1} - z_k)^{-1} \text{ expp}^{-1} \; (x_{k+1} - x_k)/a_k(z_{k+1} - z_k).$$

Now, by (5.2),

$$(5.8) \quad \log a_k/a_{k-1} \equiv \log a_k - \log a_{k-1}$$

$$= \log (H^{-1})'(z_k) - \log (H^{-1})'(z_{k-1}) = c_{k-1}(z_k - z_{k-1}).$$

Given $a_o$ and the sequences $\{z_k\}$ and $\{x_k\}$, for $z \varepsilon (z_k, z_{k-1})$, $H^{-1}(z)$ is given by (5.6) where the sequences $\{a_k\}$ and $\{c_k\}$ are generated by (5.8) and (5.7). We could extend $H^{-1}(z)$ for $z > x_n$ as follows. We extend the last constant value of $\phi$ which is $c_{n-1}$. For $z > z_n$,

$$\log(H^{-1})'(z) \equiv \log(H^{-1})'(z_n) + c_{n-1} (z - z_n),$$

$$(H^{-1})'(z) \equiv (H^{-1})'(z_n)e^{c_{n-1}(z - z_n)}$$

and so

$$H^{-1}(z) \equiv H^{-1}(z_n) + \int_{z_n}^{z} (H^{-1})'(u)du$$

$$= x_n + (H^{-1})'(z_n) \int_0^{z-z_n} e^{c_{n-1}u} du$$

$$= x_n + a_n(z - z_n) \text{ expp } c_{n-1}(z - z_n).$$

## 6. The log-likelihood

Notice that $f(x) \equiv h(x)(1 - F(x)) = h(x)e^{-H(x)}$. The log likelihood, then, is log L where

$$(6.1) \quad \log L \equiv \sum_{k=1}^{n} \log f(X_{k:n}) = \sum_{k=1}^{n} \log h(X_{k:n}) - \sum_{k=1}^{n} H(X_{k:n})$$

where $X_{k:n}$, $k = 1, 2, \ldots n$ are the ascending order statistics. We assume, here, that the density is continuous and that $\phi$ is constant between order statistics. Let $c_k$ be the constant value of $\phi(x)$ in $(X_{k:n}, x_{k+1:n})$. We are still assuming that $H(0) = 0$. Now

$$h(0) \equiv f(0)/(1 - F(0)) = f(0) = \lim_{x \to 0} H(x)/x \equiv \lim_{z \to 0} z/H^{-1}(z)$$

$$= 1/(H^{-1})'(0) = 1/a_0.$$

For $x \leq X_{1:n}$,

$$1/h(x) = a_0 + c_0 x$$

and

$$h(x) = a_0^{-1}(1 + c_0x/a_0)^{-1}$$

and so

$$H(x) = a_0^{-1} \int_0^x dt/(1 + c_0t/a_0) \equiv c_0^{-1} \log(1 + c_0x/a_0)$$
$$\equiv (x/a_0) \operatorname{logg} c_0x/a_0$$

where logg x is given by (4.2) and (4.3). It follows that

$$(6.2) \qquad 1/h(X_{1:n}) = a_0 + c_0 X_{1:n}$$

and

$$(6.3) \qquad H(X_{1:n}) \equiv (X_{1:n}/a_0) \operatorname{logg} c_0X_{1:n}/a_0.$$

Let $x \in (X_{k:n}, X_{k+1:n}]$. Now

$$1/h(x) = (1/h(X_{k:n})) + c_k(x - X_{k:n})$$

So

$$h(x) \equiv h(X_{k:n})(1 + c_kh(X_{k:n})(x - X_{k:n}))^{-1}$$

and so

$$H(x) = H(X_{k:n}) + h(X_{k:n}) \int_0^{x-X_{k:n}} dt/(1 - c_kh(X_{k:n})t) = H(X_{k:n})$$
$$+ c_k^{-1} \log(1 + c_kh(X_{k:n})(x - X_{k:n})) = H(X_{k:n})$$
$$+ [h(X_{k:n})(x - X_{k:n})] \operatorname{logg} c_kh(X_{k:n})(x - X_{k:n}).$$

It follows that

$$(6.4) \qquad 1/h(X_{k+1:n}) = (1/h(X_{k:n})) + c_k(X_{k+1:n} - X_{k:n})$$

and that

$$(6.5) \qquad H(X_{k+1:n}) = H(X_{k:n})$$
$$+ [h(X_{k:n})(X_{k+1:n} - X_{k:n})] \operatorname{logg} c_kh(X_{k:n})(X_{k+1:n} - X_{k:n}).$$

The log-likelihood is log L where log L is given by (6.1) with (6.2), (6.3), (6.4) and (6.5).

## 7. Estimation

Given the order statistics $X_{k:n}$, $k = 1, 2, \ldots n$, we can estimate $\phi(x)$ as follows. Let $\ell = 1, 2, \ldots, m \leq n$ and let $k(\ell)$ be a nondecreasing sequence of integers such that $k(0) \equiv 0$, $k(m) \equiv n$ and $X_{0:n} \equiv 0$. By the method of Section 5 we fit a CHF $H(x)$ so that the density is continuous everywhere, $\phi(x)$ is constant on $(X_{k(\ell-1):n}, X_{k(\ell):n}]$, $\ell = 1, 2, \ldots m$ and

$$X_{k(\ell):n} \equiv H^{-1}(-\log(1 - k(\ell)/(n+1))).$$

Suppose that $\phi(x)$ has a nowhere dense discontinuity set. For uniform consistency ($\sup_{x\varepsilon(0,x^\infty)} |\tilde{\phi}(x) - \phi(x)| \to 0$ i.p. as $n \to \infty$), it is necessary and sufficient that

$$\min_{\ell=1}^{m} (k(\ell) - k(\ell-1)) \to \infty$$

and

$$n^{-1} \max_{\ell=1}^{m} (k(\ell) - k(\ell-1)) \to 0$$

as $n \to \infty$. The efficiencies of estimators of this kind remain to be investigated.

Another approach is to estimate $\phi(x)$ using a roughness penalty. For an example of this approach see Good and Gaskins (1980) for example. Any estimator of our type has a sample Pareto function $\tilde{\phi}(x)$ which is constant between order statistics and so the maximum log-likelihood can only increase as the $k(\ell)$ sequence becomes more finely grained. We recommend that a roughness penalty be substracted. Specifically we recommend that the objective function (to be maximized) be $\log L - \alpha m$, $\alpha \varepsilon (0, \infty)$. For reasons not discussed here we prefer a functional of this form, in particular with $\alpha \equiv 1$. That is we seek to maximize $\log L-m$. In any of these procedures one must initially specify $h(0) \equiv 1/H'(0) \equiv 1/a_0$ which we may call the "initial scale factor." Appropriate density estimators are not hard to find.

## 8. Isotonic estimation

Suppose we wish to estimate the $\phi(x)$ on the assumption that it is nondecreasing (or nonincreasing) as x increases. Following Barlow et. al. (1972) we proceed as follows. First we use the method at the beginning of the preceding section with $m = n$ and $k(\ell) \equiv \ell$. That is we fit $\phi(x)$ in such a way that it is constant between successive pairs of order statistics. When successive interval constant values of $\phi$ are in the wrong order the intervals are combined and a new iteration is performed. We continue this procedure until all of the interval constant values are in the correct order. The number of iterations is finite.

## 9. Discussion

The author has verified that the preceding development has an analogy where 1-F, in the definition of $\phi(x)$, is replaced by -log F. Notice that $(-\log F)/(1-F) \to 1$ as $F \to 1$. The generalized Pareto distribution is, then, replaced by the generalized extreme value distribution, which is characterized by the constancy of this function. We would call it the "Gumbel function". For the latter there is no necessary lower limit and so the estimation of the density at 0 would be replaced by, say, the estimator of density at the sample median.

## REFERENCES

Barlow, R.E., Bartholomew, D.J., Bremner, J.M. and Brunk, H.D., (1972) Statistical Inference Under Order Restrictions. Wiley, New York.

Davison, A.C., (1984). Modeling excesses over high thresholds with an application. Proceedings of the NATO Advanced Study Institute on Statistical Extremes and Applications. This volume.

Galambos J. (1978). The Asymtptotic Theory of Extreme Order Statistics. John Wiley and Sons Inc., New York.

Good, I.J. and Gaskins, R.A. (1980). Density estimation and bump hunting by penalized maximum likelihood method exemplified by scattering and meteorite data (invited paper with discussion), J. Amer. Statist. Assoc., 75, 42-73.

de Haan L. (1970). On Regular Variation and its Application to the Weak Convergence of Sample Extremes. Mathematical Centre Tracts, Mathematisch Centrum, Amsterdam.

Pickands, J. (1975). Statistical inference using extreme order statistics. Ann. Statist., 3, 119-131.

Pickands, J. (1983). The continuous and differentiable domains of attraction of the extreme value distributions. Submitted.

Smith, R.L., (1984). Threshold methods for sample extremes. Proceedings of the NATO Advanced Study Institute on Statistical Extremes and Applications. This volume.

Wahba, G. and Wendelberger, J. (1980). Some new mathematical methods for variational object analysis using splines and cross-validation. Monthly Weather Review, 108, 36-57.

Wegman, E.J. and Wright L.W. (1983). Splines in Statistics. J. Amer. Statist. Assoc. 78, 351-365.

# EXTREMAL PROCESSES

Laurens de Haan

Erasmus University Rotterdam

Our approach to extremal processes will be via the convergence of the point process generated by the i.i.d. observations to a Poisson point process as originally proposed by J. Pickands.

Let $X_1, X_2, \ldots$ be i.i.d. random variables from a distribution in the domain of attraction of some extreme-value distribution. Consider for $n = 1, 2, \ldots$ the simple point processes $P_n$ in $\mathbb{R}_+ \times \mathbb{R}_+$ with points $\{(k/n, (X_k - b_n)/a_n)\}_{k=1}^{\infty}$ with $a_n > 0$ and $b_n$ the attraction coefficients for convergence of the sequence $\max_{1 \leq i \leq n} X_i$. We shall show that, as $n \to \infty$ these points converge to a Poisson point process P' on $\mathbb{R}_+ \times \mathbb{R}$ with intensity measure $dt \times d\upsilon$ with $\upsilon(x,\infty) = -\log G(x)$ for $0 < G(x) < 1$ where G is the limiting extreme-value distribution function. According to Skorokhods theorem there must be a probability space and a sequence of random elements $\tilde{P}_n$ defined on that space such that $\tilde{P}_n \overset{d}{=} P_n$ and $\tilde{P}_n \to P$ almost surely. We are going to exhibit such an almost sure construction. In fact the $\tilde{P}_n$ will be (random) functions of P.



*J. Tiago de Oliveira (ed.), Statistical Extremes and Applications, 297–309.*

In order to keep the presentation as simple as possible we shall first consider lower order statistics instead of upper order statistics and confine ourselves to the uniform distribution on [0,1].

The proofs are new but there is only one new result about extremal processes (property 7). With no claim of completeness I mention the names of Dwass, Lamperti, Pickands, Robbins and Siegmund, further Resnick and Deheuvels.

## 1. The basic point process convergence.

Let P be a Poisson point process on $\mathbb{R}_+ \times \mathbb{R}_+$ with Lebesgue measure as intensity measure. Let $P_n$ be the simple point process on $\mathbb{R}_+ \times \mathbb{R}_+$ with points $\{(k/n, nX_k)\}_{k=1}^{\infty}$ where $X_1$, $X_2$, ... are i.i.d. uniform [0,1] random variables. We are going to construct a sequence of point processes $\tilde{P}_n$ on $\mathbb{R}_+ \times \mathbb{R}_+$ which are functions of P and such that $\tilde{P}_n \overset{d}{=} P_n$ for all n.

Denote the restriction of P to $\mathbb{R}_+ \times [0,n]$ by $Q_n$. The first coordinates of the points of $Q_n$ form a homogeneous Poisson point process on the line. So these first coordinates just form a sequence $0 < T_1^{(n)} < T_2^{(n)} < \ldots$ and the random variables

$T_1^{(n)}$, $T_2^{(n)} - T_1^{(n)}$, $T_3^{(n)} - T_2^{(n)}$, ... are i.i.d. random variables

with $P\{T_1^{(n)} \leq x\} = 1 - \exp -(nx)$ for $x > 0$.

We can thus enumerate and exhibit the points of $Q_n$ as $\{(T_k^{(n)}, Y_k^{(n)})\}_{k=1}^{\infty}$. Note that since two-dimensional Lebesgue measure is a product measure, the random variables $n^{-1} Y_k^{(n)}$ ($k = 1, 2, 3, \ldots$) are i.i.d. with uniform distribution. Hence the point process $\tilde{P}_n$ whose points are $\{(k/n, Y_k^{(n)})\}_{k=1}^{\infty}$ has the same distribution as $P_n$.

In order to get $\tilde{P}_n$ from $Q_n$ we thus have to shift each point $(T_k^{(n)}, Y_k^{(n)})$ over a random distance $k/n - T_k^{(n)}$ horizontally. This shift will be the random mapping producing $\tilde{P}_n$ from $Q_n$.

We claim that, as n increases, the sequence of point processes $\tilde{P}_n$ converges to P a.s. It is sufficient to show that the shift mappings, introduced above, converge to the identity mapping as n increases.

We have to show (note that it is sufficient to prove convergence of the processes on any strip $(0,M) \times \mathbb{R}$)

$$\sup_{0 \leq k/n \leq M} |k/n - T_k^{(n)}| \to 0 \text{ a.s. } (n \to \infty) \text{ for } M > 0.$$

This is equivalent to

$$\sup_{0 \leq x \leq M} \left| \frac{[nx]}{n} - T_{[nx]}^{(n)} \right| \to 0 \text{ a.s.},$$

to

$$\sup_{0 \leq x \leq M} |x - T_{[nx]}^{(n)}| \to 0 \text{ a.s. } (n \to \infty) \text{ for } M > 0$$

and to (note that uniform convergence is automatic since $T_{[nx]}^{(n)}$ is non-decreasing in x and the limit function is continuous)

$$(1) \qquad \lim_{n \to \infty} T_{[nx]}^{(n)} = x \text{ a.s.}$$

for $0 \leq x \leq M$ (since convergence for all rationals entails convergence everywhere). Now the inverse function of $T_{[nx]}^{(n)}$ is the function $x \to n^{-1}\max\{k | T_k^{(n)} \leq x\}$ (recall that the $T_k^{(n)}$ form an increasing sequence in k). Since the lefthand side of (1) is a monotone function for all n, the pointwise convergence in (1) is equivalent to the pointwise convergence of the inverse functions of $T^{(n)}_{[nx]}$ to the inverse function of the righthand side of (1), i.e. (1) is equivalent to

$$(2) \qquad \lim_{n\to\infty} n^{-1}\max\{k \mid T_k^{(n)} \leq x\} = x \text{ a.s.}$$

for $0 \leq x \leq M$.

This is easy to prove once we realize how the $T_k^{(n)}$ are formed. The $T_k^{(n)}$ are projections of points of the original Poisson point process on the first coordinate axis. Thus, if $N_k(x)$ is the number of points of P in the interval $(0,x] \times (k-1, k]$, then $N_1(x)$, $N_2(x)$, $N_3(x)$, .. are i.i.d. Poisson (x) and

$$\max\{k \mid T_k^{(n)} \leq x\} = N_1(x) + \ldots + N_n(x).$$

Hence (2) is an easy consequence of the strong law of large numbers.

We have proved:

Theorem 1.

Let $P_n$ be the simple Poisson point process on $\mathbb{R}_+ \times \mathbb{R}_+$ with points $\{(k/n, nX_k)\}_{k=1}^{\infty}$ where $X_1$, $X_2$, ... are i.i.d. uniform [0,1] random variables. There exists a sequence of point processes $\tilde{P}_n$ defined on one probability space with $\tilde{P}_n \stackrel{d}{=} P_n$ for all n and $\tilde{P}_n \to P$ a.s. $(n\to\infty)$ where P is a homogeneous Poisson point process on $\mathbb{R}_+ \times \mathbb{R}_+$.

## 2. The lower extremal process based on the exponential distribution.

Let $Q$ be a simple point process on $\mathbb{R}_+ \times \mathbb{R}_+$. Define $m_Q(t)$ as the infimum of the second coordinates of those points of $Q$ whose first coordinate is not greater than $t$, with the understanding that $m_Q(t) = \infty$ if there are no points whose first coordinates are not greater than $t$. This function is a continuous function of $Q$ for all $t > 0$. It follows from theorem 1 that

(3) $$\lim_{n\to\infty} m_{\tilde{P}_n}(t) = m_P(t) \text{ a.s.}$$

for all $t > 0$. This implies that

$$\lim_{n\to\infty} m_{\tilde{P}_n}(t) = m_P(t)$$

for all continuity points of $m_P(t)$ with $a \leq t \leq b$ a.s. for all $0 \leq a \leq b < \infty$.

For non-decreasing functions $\psi_1$ and $\psi_2$ on an interval $I$ let $L_I(\psi_1, \psi_2)$ be their Lévy-distance.

Theorem 2 (cf. Deheuvels ZfW 58, 1981, 1-6).

(4) $$\lim_{n\to\infty} L_{(0,\infty)}(m_{\tilde{P}_n}, m_P) = 0 \text{ a.s.}$$

Proof.

Note that $m_{\tilde{P}_n}(t)$ and $m_P(t)$ are non-increasing in $t$.

According to our construction in the previous section we have for any rational $r > 0$ excluding a nullset $N_r$

$$\sup_{x\leq r} |T^{(n)}_{[nx]} - x| \to 0 \ (n\to\infty).$$

Hence, excluding the nullset $\bigcup_{r\in Q} N_r$, we have

$$\sup_{x\leq y} |T^{(n)}_{[nx]} - x| \to 0 \ (n\to\infty) \text{ for any real } y > 0.$$

Now with probability one $m(t) \to \infty$ or 0 according to $t \to 0$ or $\infty$, hence for fixed $\varepsilon > 0$ there is a (random) $M_\varepsilon$ such that $m(M_\varepsilon) < 2^{-\frac{1}{2}}\varepsilon$ a.s. We just showed that there exists $n_0(\varepsilon)$ such that for $n \geq n_0$

$$\sup_{x\leq M_\varepsilon} |T^{(n)}_{[nx]} - x| < 2^{-\frac{1}{2}}\varepsilon \text{ and (by (3)) } m_{\tilde{P}_n}(M_\varepsilon) < 2^{-\frac{1}{2}}\varepsilon.$$

Now the Lévy-distance between $m_{\tilde{P}_n}(t)$ and $m_P(t)$ on $(0,\infty)$ is at most (recall how $\tilde{P}_n$ was constructed from P)

$$2^{\frac{1}{2}}.\max\left(m(M_\varepsilon),\ m_{\tilde{P}_n}(M_\varepsilon),\ \sup_{x\leq M_\varepsilon} |T^{(n)}_{[nx]} - x|\right).$$

This proves the theorem.

Note that $m_{\tilde{P}_n}(t) \overset{d}{=} m_{P_n}(t) = \min\{nX_k | k \leq nt\}$. In particular for $x = 1$ we have $m_{P_n}(1) = n.\min(X_1, X_2, \ldots, X_n)$. The above assertion is thus a "process" version of the convergence of minima from a uniform distribution to the exponential distribution.

The process $m_P(t)$, which we abbreviate now as $m(t)$, being the limit of the process of partial minima of i.i.d. uniform observations, is an example of an extremal process.
The structure of extremal processes from other distributions is similar as we shall see. We now list a number of properties of the process $m(t)$.

1. $m(t)$ is non-increasing. In fact $m(t)$ is constant except for jumps which are isolated except near $t = 0$. Further

$$\lim_{t \downarrow 0} m(t) = \infty \text{ and } \lim_{t \to \infty} m(t) = 0 \text{ a.s.}$$

2. $m(t)$ is a Markov-process with stationary transition probabilities. In fact for $0 < t_1 < t_2 < \ldots < t_n$ we have

$$\left(m(t_1), m(t_2), \ldots, m(t_n)\right) \stackrel{d}{=} \left(m(t_1), \min\{m(t_1), d(t_1, t_2)\},\right.$$

$$\left., \ldots, \min\{m(t_1), d(t_1, t_2), \ldots, d(t_{n-1}, t_n)\}\right)$$

where $m(t_1)$, $d(t_1, t_2)$, ..., $d(t_{n-1}, t_n)$ are independent random variables and
$m(t_1) \in \exp(t_1)$, $d(t_i, t_{i+1}) \in \exp(t_{i+1} - t_i)$ for
$i = 1, 2, \ldots, n-1$.

3. The sample paths of $m(t)$ can be described as follows: for any $t > 0$, if the process is in state a at time t, it remains there for a random length of time having exponential distribution with parameter a and then jumps to a point uniformly distributed in $(0,a)$.

4. The inverse function of the trajectory of $m(t)$ as a random process has the same distribution as the process $m(t)$ itself.

5. The process $\{m(-t)\}_{t<0}$ has independent increments.

6. The epochs of the jumps of the process $m(t)$ form a Poisson point process with intensity measure $dt/t$ on $(0,\infty)$.

7. If $\{(S_k, Y_k)\}_{k=1}^{\infty}$ is a enumeration of the points of a Poisson point process on $(0,\infty) \times (0,\infty)$ with intensity measure $e^{-tx}dtdx$, then

$$\{m(t)\}_t \stackrel{d}{=} \{\sum_{S_k>t} Y_k\}_t.$$

Proof.
All statements follow readily from the construction of the process. Note that most statements are analogous of similar statements for $m_{P_n}(t)$.

1. Follows from 3. It is obvious that $m(t) \to \infty$ or 0 in probability according to $t \to 0$ or $\infty$. This is then also true with probability one since $m(t)$ is a.s. a monotone function.

2. Let $\{(T_k, X_k)\}_{k=1}^{\infty}$ be an enumeration of the points of P. Then $m(t):= \min\{X_k | T_k < t\}$. The statement is now obvious with $d(t_i, t_{i+1}):= \min\{X_k | t_i < T_k \leq t_{i+1}\}$.

3. If $m(t) = x$, then $m(t+h) = x$ for some $h > 0$ if and only if the interval $(t,t + h] \times [0,x]$ contains no points of P. This means that the probability that the process remains in the state x for at least h time units, is $e^{-xh}$; this holding time thus has an exponential (x) distribution.
The second coordinate of the point of P in $(t,\infty) \times [0,x]$ whose first coordinate (which we call $T_1^{[0,x]}$) is smallest, is uniformly distributed over $[0,x]$ since the intensity measure is a product measure and more specifically Lebesgue-measure. Now obviously $T_1^{[0,x]}$ is the time epoch at which the process first jumps after time t and the second coordinate of the point of P with first coordinate $T_1^{[0,x]}$ is the state to which the process jumps. This proves the second statement of 3.

4. If we interchange the two coordinates of all the points of the point process P we get a point process of the same structure by the symmetry of Lebesgue measure.

5. For $x, y > 0$ and $0 < t_1 < t_2$ by 2

$$P\{m(t_1) - m(t_2) > x \text{ and } m(t_2) > y\} = P\{m(t_1) - d(t_1, t_2) > x$$

$$\text{and } d(t_1, t_2) > y\} = \int_y^\infty P\{m(t_1) > x+u\}\, dP\{d(t_1, t_2) \le u\} =$$

$$= t_2^{-1}(t_2 - t_1) \exp -(t_1 x + t_2 y).$$

Note that $P\{m(t_1) - m(t_2) = 0\} = t_2^{-1} t_1$.

6. By 4 it is sufficient to show that the set of values of $m(t)$ forms a Poisson point process of the same structure. From 3 we get for two consecutive values $X < Y$ of the process

$$P\{Y^{-1}X \le x | Y\} = x \quad (0<x<1)$$

hence

$$p\{-\log X + \log Y > x | \log Y\} = e^{-x} \quad (x>0).$$

So by the Markov property the logarithms of the values of the process $m(t)$ form a homogeneous Poisson process.

7. Consider the Poisson point process Q with points $\{(S_k, \tilde{Y}_k)\}_{k=1}^\infty$ where $\tilde{Y}_k := S_k Y_k$ for $k = 1, 2, \ldots$ Its intensity measure is $t^{-1}dte^{-x}dx$. We have to show that the process $\mu(t) := \sum_{S_k > t} \tilde{Y}_k / s_k$ for $t > 0$ has the same structure as the process $\{m(t)\}_t$.

Clearly, if $\mu(t) = a$, for some $s < t$ the probability that $\mu(s) = a$ equals the probability of having no point in the point process $\{S_k, \tilde{Y}_k)\}$ in the time interval $[s,t)$ i.e. $\exp -(\log t - \log s) = t^{-1}s$. Also, given there is a jump in the process $\{\mu(-t)\}_{t<0}$ at time $t = -s$, the height of the jump has an exponential distribution with parameter s (use the fact that the intensity measure is product measure). The process thus develops in the same way as the extremal process $\{m(t)\}$ according to 3 and 4 (i.e. the properties of 3 are checked for the inverse function of the process according to 4).

## 3. Other extremal processes.

We shall extend the results of the previous sections to results for upper order statistics and at the same time extend the results to arbitrary max-stable probability distributions.
Let $X_1, X_2, \ldots$ be i.i.d. random variables with distribution function F. Let F be in the domain of attraction of G, one of the extreme-value distributions i.e. for some positive function a

$$(5)\qquad \lim_{t\to\infty} \frac{U(tx) - U(t)}{a(t)} = \lim_{n\to\infty} \frac{U(nx) - U(n)}{a(n)} = \left(\frac{1}{-\log G}\right)^{\leftarrow}(x) =: H(x)$$

for $x > 0$ where U is defined as an inverse function by
$U := \left(\frac{1}{1-F}\right)^{\leftarrow}$.
Define the Poisson point processes P and $Q_n$ as in section 1.
Let $\{(T_k^{(n)}, Y_k^{(n)})\}_{k=1}^{\infty}$ be the points of $Q_n$ where $T_1^{(n)} \leq T_2^{(n)} \leq \ldots$ as in section 1. We now form a point process $\tilde{P}_n$ from $Q_n$ as follows: the points of $\tilde{P}_n$ are

$$\{(k/n,\ a_n^{-1}(U(1/Y_k^{(n)}) - b_n))\}_{k=1}^{\infty}$$

with $b_n = U(n)$ and $a_n = a(n)$. The random shift in the first coordinate $T_k^{(n)} \to k/n$ is as in section 1, the transformation of the second coordinate is not random. Obviously the point process $\tilde{P}_n$ has the same distribution as the point process $P_n$ with points

$$\{(k/n,\ a_n^{-1}(X_k - b_n))\}_{k=1}^{\infty}.$$

From the results of section 1 and (5) the point processes $\tilde{P}_n$ converge almost surely to the Poisson point process that is obtained from P by applying the transformation H to the second coordinator of its points.

We have proved

Theorem 2.

Let $P_n$ be the simple point process on $\mathbb{R}_+ \times \mathbb{R}$ with points $\{(k/n,\ a_n^{-1}(X_k - b_n))\}_{k=1}^{\infty}$ where $X_1, X_2, \ldots$ are i.i.d. random variables with distribution function F. Suppose F is the domain of attraction of the extreme-value distribution G.
Let $b_n := H(n)$, $a_n := H(ne) - H(n)$ with $H := \left(\frac{1}{1-F}\right)^{\leftarrow}$. There exists a sequence of point processes $\tilde{P}_n$ defined on one probability space with $\tilde{P}_n \overset{d}{=} P_n$ $(n = 1, 2, \ldots)$ and $\tilde{P}_n$ converges almost surely to a Poisson point process on $\mathbb{R}_+ \times \mathbb{R}$ with intensity measure $dt \times d\upsilon$ where $\upsilon(x,\infty) = -\log G(x)$ for those x for which $0 < G(x) < 1$ and $\upsilon$ places no mass elsewhere.

It follows from the theorem that (with $m_Q(t)$ and $m(t)$ as in section 2 but $\tilde{P}_n$ and P as in this section)

$$\lim_{n\to\infty} m_{\tilde{P}_n}(t) = m_P(t) = H(1/m(t)) =: M(t)$$

for all continuity points of $m_P(t)$ with $a \leq t \leq b$ a.s. for all $0 \leqq a \leqq b < \infty$. The properties of the process $H(1/m(t))$ which is the limit in distribution of the sequence of processes

$m_{P_n}(t) = \max\{(X_k - b_n)/a_n | k \leq nt\}$, follow directly from those of the process m(t) listed in section 2.

We formulate the translation of some of the properties of M(t) above for the general limiting extremal process for maxima:

1. M(t) is non-decreasing. In fact M(t) is constant except for jumps which are isolated except near t = 0. Further

$$\lim_{\downarrow 0} M(t) = \lim_{u\downarrow 0} H(u) \text{ and } \lim_{t\to\infty} M(t) = \lim_{u\to\infty} H(u) \text{ a.s.}$$

2. M(t) is a Markov-process with stationary transition probabilities. In fact for $0 < t_1 < t_2 < \ldots < t_n$ we have

$$(M(t_1), M(t_2), \ldots, M(t_n)) \stackrel{d}{=} (M(t_1), \max\{M(t_1), D(t_1, t_2)\}, \ldots, \max\{M(t_1), D(t_1, t_2), \ldots, D(t_{n-1}, t_n)\})$$

   where $M(t_1), D(t_1, t_2), \ldots, D(t_{n-1}, t_n)$ are independent random variables and $M(t_1) \in G^{t_1}$ (i.e. $M(t_1)$ has probability distribution function $G^{t_1}(x)$ where G is the limiting extreme-value distribution) and $D(t_i, t_{i+1}) \in G^{t_{i+1}-t_i}$ for $i = 1, 2, \ldots, n-1$.

3. I formulate this property only for the double-exponential distribution $\Lambda$. The sample paths of M(t) with underlying distribution $\Lambda$ are described as follows: for any $t > 0$, if the process is in state a at time t, it remains there for a random length of time having exponential distribution with parameter a; the height of the jump that occurs then, has an exponential distribution (i.e. the consecutive jumps are i.i.d. exponential).

5. The inverse function of the trajectories of $M(t)$ as a random process has independent increments.

6. The epochs of the jumps of the process $M(t)$ form a Poisson point process with intensity measure $dt/t$ on $(0,\infty)$.

One can generalize the definition of an extremal process beyond the limiting distributions G. Let F be any probability distribution function. If one replaces G everywhere in property 2 by F, then this gives the definition of a consistent family of marginal distributions, hence of a stochastic process, in fact this process has the same distribution as $(1/-\log F)^{\leftarrow}(1/m(t))$ and the properties can be read off easily from this representation.
This is possible since for any pdf $F(x)$ and any $t \in \mathbb{R}_+$ the function $F^t(x)$ is again the distribution function of a probability distribution. This is no longer true for a (say) two-dimensional distribution function $F(x, y)$. To see this take e.g. a random vector $(X, Y)$ with $X = -Y$.
So the higher-dimensional version of an extremal process (the definition of which is obvious from property 2) is only possible if the underlying distribution function $F(x, y)$ is max-infinitely divisible i.e. if $F^t(x,y)$ is the distribution function of a probability distribution for all $t > 0$. It is known that a pdf F is infinitely divisible if and only if $-\log F(x, y)$ is the distribution function of a measure. It is easy to see that any 2-dimensional extreme-value distribution is max-infinitely divisible, so it generates an extremal process. Obviously any one-dimensional pdf is max-infinitely divisible.

# EXTREME VALUES FOR SEQUENCES OF STABLE RANDOM VARIABLES

Michael B. Marcus*

Texas A&M University

Let $\xi = (\xi_1, \ldots, \xi_n)$ be an $R^n$-valued p-stable sequence. We obtain bounds on the expectation and distribution of $\|\xi\| = \sup_{1 \leq i \leq n} |\xi_i|$ both in terms of the spectrum of $\xi$ and the quantity $\inf_{1 \leq j \neq k \leq n} (E|\xi_j - \xi_k|^r)^{1/r}$, $r < p$. These results are compared with their well known counterparts for Gaussian processes (the case $p = 2$).

## 1. INTRODUCTION

Let $X = (X_1, \ldots, X_n)$ be a Gaussian process on $R^n$ with mean zero. A great deal is known about the properties of $\|X\| := \sup_{1 \leq i \leq n} |X_i|$. For example for $0 < r < \infty$

$$\sqrt{\frac{2}{\pi}} \leq \frac{(E\|X\|^r)^{1/r}}{\sup_{1 \leq i \leq n} (E(X_i)^2)^{1/2}} \leq c_r (\log n)^{1/2} \qquad (1.1)$$

where $c_r > 0$ is a constant depending only on $r$. Also for $0 < r < \infty$,

*This research was supported in part by a grant from the National Science Foundation.



*J. Tiago de Oliveira (ed.), Statistical Extremes and Applications, 311–324.*

$$E\|X\| \geq d_r \inf_{1\leq j\neq k\leq n} (E|X_j - X_k|^r)^{1/r}(\log n)^{1/2} \tag{1.2}$$

where $d_r$ is a constant depending only on r. The inequalities in (1.1) are completely elementary. The result in (1.2) is a direct consequence of Slepian's Lemma. The purpose of this paper is to find analogues of (1.1) and (1.2) for p-stable processes on $R^n$ for $0 < p < 2$. Doing this will give us an opportunity to review some new useful ways by which p-stable processes can be represented. (Throughout this paper read $\log n$ to be $\max(1, \log n)$.)

## 2. STABLE SEQUENCES

A sequence $\xi = (\xi_1,\ldots,\xi_n)$ is called a real valued symmetric p-stable sequence, $0 < p \leq 2$ if for all sequences of real numbers $\{\alpha_i\}_{i=1}^n$

$$E \exp i \sum_{i=1}^n \alpha_i\xi_i = \exp[ -\int_{R^n} \sum_{i=1}^n |\alpha_i\beta_i|^p m(d\beta)] \tag{2.1}$$

where $\beta = (\beta_1,\ldots,\beta_n) \in R^n$ and $m$ is a finite positive measure on $R^n$. Note that (2.1) implies

$$\sum_{i=1}^n \alpha_i\xi_i = ( \int_{R^n} \sum_{i=1}^n |\alpha_i\beta_i|^p m(d\beta))^{1/p}\theta_{p,1} \tag{2.1a}$$

where " $\mathcal{D}$ " denotes equal in distribution and

$$E \exp i\lambda\theta_{p,1} = \exp[-|\lambda|^p], \quad \lambda \text{ real.} \tag{2.2}$$

I.e., all real linear functionals of $\xi$ are stable with an appropriate scale parameter determined by m. (Relation (2.1) defines stable measures on any linear space. This is well known; see [4] for further discussion and references.)

Let $\|\xi\| = \sup_{1\leq i\leq n} |\xi_i|$. $R^n$ equipped with this norm is a Banach space denoted by $\ell_n^\infty$. Let $|m| = m(R^n)$ and define

$$\sigma(m;p) = \left( \int_{R^n} \|\beta\|^p m(d\beta) \right)^{1/p}. \tag{2.2a}$$

We shall refer to $\sigma(m;p)$ as the spectrum of the process $\xi$. ($|m|$ is often called the spectrum of $\xi$ because $\beta$ is generally taken to lie on the unit ball of $\ell_n^\infty$. The reader will see that it is useful not to impose this condition.)

We will now give a very useful representation for stable processes due to Le Page, Woodroofe and Zinn [1] (see also Proposition 1.5, [4]) as Lemma 2.1.

Lemma 2.1: Let $Y$ be a positive real valued random variable satisfying $P(Y > \lambda) = e^{-\lambda}$. Let $\{Y_k\}_{k=1}^\infty$ be i.i.d. copies of $Y$ and let $\Gamma_j = Y_1 + \dots + Y_j$. Let $\mu$ be an $R^n$ valued random variable distributed according to $\frac{m}{|m|}$ and let $\{\mu_j\}_{j=1}^\infty$ be i.i.d copies of $\mu$. Let $\{\varepsilon_j\}_{j=1}^\infty$ be a Rademacher sequence independent of $\{\mu_j\}_{j=1}^\infty$, i.e. $P(\varepsilon_1 = 1) = P(\varepsilon_1 = -1) = 1/2$ and $\{\varepsilon_j\}_{j=1}^\infty$ are i.i.d. Then the p-stable process $\xi$ defined by (2.1) can be represented by

$$\xi = c(p)|m|^{1/p} \sum_{j=1}^\infty \varepsilon_j (\Gamma_j)^{-1/p} \mu_j \tag{2.3}$$

where $c(p) = \left( \int_0^\infty \frac{\sin v}{v^p} dv \right)^{1/p}$.

Alternately, we define the measure $\nu$ on the unit ball of $\ell_n^\infty$ by

$$\nu(A) = (\sigma(m;p))^{-p} \int_{\{\beta: \frac{\beta}{\|\beta\|} \varepsilon A\}} \|\beta\|^p \, m(d\beta),$$

for measurable subsets $A$ of the unit ball. Let $v$ be an $R^n$ valued random variable distributed according to $\nu$ and let $\{v_j\}_{j=\infty}^\infty$ be i.i.d. copies of $v$. The p-stable process $\xi$ defined by (2.1) can also be represented by

$$\xi = c(p)\sigma(m;p) \sum_{j=1}^\infty \varepsilon_j (\Gamma_j)^{-1/p} v_j \tag{2.4}$$

where, $\{v_j\}_{j=1}^{\infty}$ takes values on the unit ball of $\ell_n^{\infty}$ and $\{\varepsilon_j\}_{j=1}^{\infty}$ is independent of $\{v_j\}_{j=1}^{\infty}$.

Proof: For a proof of (2.3) see (1.34)' in [4]. The representation (2.4) follows immediately from (2.3) if one recognizes that the right side of (2.1) can be written as

$$\exp \left[ - \int_{R^n} \left| \sum_{i=1}^{n} \alpha_i \frac{\beta_i}{\|\beta\|} \right|^p \|\beta\|^p \, m(d\beta) \right]$$

(Actually the series in (2.3) and (2.4) both converge a.s. This follows by fixing $\{\Gamma_j\}_{j=1}^{\infty}$ and using the Three Series Theorem on the resulting marginal sums of independent random variables.)

We now come to our first result on the size of $\|\xi\|$.

Theorem 2.1: Let $\xi = (\xi_1, \ldots, \xi_n)$ be a symmetric p-stable sequence as defined above. Then for $1 < p < 2$ and $r < p$

$$c_{p,r} \le \frac{\left(E\|\xi\|^r\right)^{1/r}}{\sigma(m,p)} \le c'_{p,r} (\log n)^{1/q}, \tag{2.5}$$

where $\frac{1}{p} + \frac{1}{q} = 1$. For $p = 1$ and $r < p$

$$c_{1,r} \le \frac{\left(E\|\xi\|^r\right)^{1/r}}{\sigma(m,1)} \le c'_{1,r} \, L_2 n \tag{2.6}$$

where $L_2 n = \max(1, \log\log n)$. For $0 < p < 1$ and $r < p$

$$c_{p,r} \le \frac{\left(E\|\xi\|^r\right)^{1/r}}{\sigma(m;p)} \le c'_{p,r} \, . \tag{2.7}$$

In (2.5), (2.6) and (2.7), $0 < c_{p,r}$, $c'_{p,r} < \infty$ are constants depending only on p and r.

Proof: The lower bounds are easy. Write $\xi$ as in (2.4). By Levy's inequality (see Lemma 4.1, Chapter II [3]) and the fact that $\|v_1\| = 1$ we have

$$P\left[\left\| \sum_{j=1}^{\infty} \varepsilon_j (\Gamma_j)^{-1/p} v_j \right\| > \lambda\right] \ge \frac{1}{2} P\left[\|\Gamma_1^{-1/p} v_1\| > \frac{\lambda}{2}\right]$$

$$= \frac{1}{2} P\left[ |\Gamma_1|^{-1/p} > \frac{\lambda}{2} \right] .$$

Therefore

$$E\|\sum_{j=1}^{\infty} \varepsilon_j(\Gamma_j)^{-1/p} v_j\|^r \geq 2^{r-1} E|\Gamma_1|^{-r/p}$$

and we get the lower bounds in (2.5), (2.6) and (2.7) with

$$c_{p,r} = c(p)2^{1-\frac{1}{r}}(E|\Gamma_1|^{-r/p})^{1/r}.$$

We now give some facts needed in proving the upper bounds. First we observe that for $r < p$

$$(E\|\sum_{j=1}^{\infty} \varepsilon_j(\Gamma_j)^{-1/p} v_j\|^r)^{1/r}$$

$$\leq 2^{1/r}(E \sup_j (\frac{j}{\Gamma_j})^{r/p})^{1/r} (E\|\sum_{j=1}^{\infty} \varepsilon_j \frac{v_j}{j^{1/p}}\|^r)^{1/r}. \quad (2.8)$$

This follows from the contraction principle for $r \geq 1$, (without the term $2^{1/r}$ on the right). For $r < 1$ one can obtain (2.8) by using a version of the contraction principle for probabilities (see Lemma 1.3, (iii) [5]). Also note that by Lemma 1.2 [4]

$$E \sup_j (\frac{j}{\Gamma_j})^{r/p} \leq \frac{cr}{p-r} \quad (2.9)$$

for some constant $c$ independent of $p$. Next let $\{a_j\}_{j=1}^{\infty}$ be a sequence of real numbers with $|a_j| \leq 1$. For $1 < p < 2$ (with $\frac{1}{p} + \frac{1}{q} = 1$) there exist constants $0 < d_1(p), d_2(p) < \infty$ such that for $\lambda \geq \lambda_0(p)$

$$P[|\sum_{j=1}^{\infty} \varepsilon_j \frac{a_j}{j^{1/p}}| > \lambda] \leq \exp[-d_2(p)\lambda^q], \quad (2.10)$$

and

$$P[|\sum_{j=1}^{\infty} \frac{\varepsilon_j}{j^{1/p}}| > \lambda] \geq \exp[-d_1(p)\lambda^q] . \quad (2.10a)$$

Also for $\lambda \geq \lambda_0(1)$

$$P[|\sum_{j=1}^{\infty} \varepsilon_j \frac{a_j}{j}| > \lambda] \leq \exp[- d_2(1)e^{\lambda}]. \quad (2.11)$$

and

$$P\left[\left|\sum_{j=1}^{\infty} \frac{\varepsilon_j}{j}\right| > \lambda\right] \geq \exp[-d_1(1)e^{\lambda}] \tag{2.11a}$$

For proofs of (2.10), (2.10a), (2.11) and (2.11a) see either [2] or Lemma 3.1 [4].

We now complete the proof of Theorem 2.1. We represent $\xi$ by (2.4) and use (2.8) to obtain

$$(E\|\xi\|^r)^{1/r} \leq$$

$$c(p)\sigma(m;p)2^{1/r}\left(E \sup_j \left(\frac{j}{\Gamma_j}\right)^{r/p}\right)^{1/r} \left(E\left\|\sum_{j=1}^{\infty} \varepsilon_j \frac{v_j}{j^{1/p}}\right\|^r\right)^{1r} . \tag{2.12}$$

We write $v_j = (v_{j1},\ldots,v_{ji},\ldots,v_{jn})$. Clearly,

$$P\left(\left\|\sum_{j=1}^{\infty} \varepsilon_j \frac{v_j}{j^{1/p}}\right\| > \lambda\right) \leq \sum_{j=1}^{n} P\left(\left|\sum_{j=1}^{\infty} \varepsilon_j \frac{v_{ji}}{j^{1/p}}\right| > \lambda\right).$$

Therefore for $1 < p < 2$, by (2.10) we have

$$P\left(\left\|\sum_{j=1}^{\infty} \varepsilon_j \frac{v_j}{j^{1/p}}\right\| > \lambda\right) \leq n \exp[-d_2(p)\lambda^q] \tag{2.13}$$

and for $p = 1$, we use (2.11) to obtain

$$P\left(\left\|\sum_{j=1}^{\infty} \varepsilon_j \frac{v_j}{j}\right\| > \lambda\right) \leq n \exp[-d_2(1)e^{\lambda}], \tag{2.14}$$

for $\lambda \geq \lambda_0(p)$ sufficiently large. Using (2.9) in (2.12) along with either (2.13) or (2.14) we get the upper bounds in (2.5) and (2.6). When $p < 1$

$$\left\|\sum_{j=1}^{\infty} \varepsilon_j \frac{v_j}{j^{1/p}}\right\| \leq \sum_{j=1}^{\infty} \frac{1}{j^{1/p}} < \infty.$$

This gives us the upper bound in (2.7) and completes the proof of the theorem.

We now give some examples of p-stable sequences.

1.) The canonical i.i.d. sequence $\theta_p = (\theta_{p,1};\ldots;\theta_{p,n})$, $0 < p < 2$, where $\{\theta_{p,i}\}_{i=1}^{n}$ are i.i.d. copies of $\theta_{p,1}$ given in (2.2). The measure $m$ which defines this sequence by means of (2.1) assigns mass $1$ to each of the $n$ canonical unit vectors $(1,0,\ldots,0),\ldots,(0,0,\ldots,0,1)$ in $R^n$. Thus in this case $\sigma(m,p) = n^{1/p}$. Also in this case it is completely elementary to show that

$$\left(E\|\theta_p\|^r\right)^{1/r} \sim n^{1/p} , \tag{2.15}$$

(where we write $a \sim b$ if constants $0 < c_1 \leq c_2 < \infty$ such that $c_1 b \leq a \leq c_2 b$.) Note that by (2.1a) for $r < p$ and $1 \leq j \neq k \leq n$

$$\left(E|\theta_{p,j} - \theta_{p,k}|^r\right)^{1/r} = 2^{1/p}\left(E|\theta_{p,1}|^r\right)^{1/r} . \tag{2.16}$$

2.) For $p < t \leq 2$ let $\theta_t = (\theta_{t,1};\ldots;\theta_{t,n})$ be the canonical i.i.d. sequence of t-stable random variables, i.e. the distribution of $\theta_{t,1}$ is given by (2.2). Let $m$ be the measure induced on $R^n$ by $(\theta_{t,1};\ldots;\theta_{t,n})$ and let $\xi_t = (\xi_{t,1};\ldots;\xi_{t,n})$ be the p-stable sequence defined by (2.1) with measure $m_t$. In this case, for $t < 2$

$$\sigma(m_t;p) = \left(E\|\theta_t\|^p\right)^{1/p} \sim n^{1/t}, \tag{2.17}$$

and for $t = 2$

$$\sigma(m_2;2) = \left(E\|\theta_2\|^p\right)^{1/p} \sim (\log n)^{1/2} \tag{2.18}$$

(These estimates are completely elementary since in these cases we are dealing with sequences of independent stable or Gaussian random variables.) We now show that for $t < 2$, $r < p$

$$\left(E\|\xi_t\|^r\right)^{1/r} \sim n^{1/t} \tag{2.19}$$

and for $t = 2$, $r < p$

$$\left(E\|\xi_2\|^r\right)^{1/r} \sim (\log n)^{1/2} . \tag{2.20}$$

Lower bounds in (2.19) and (2.20) are given by Theorem 2.1 along with (2.17) and (2.18). To obtain the upper bounds we represent $\xi_t$ by (2.3). Using (2.8) and the fact that $|m| = 1$ we have

$$\left(E\|\xi_t\|^r\right)^{1/r} = c(p)\left(E\|\sum_{j=1}^{\infty} (\Gamma_j)^{-1/p}(\theta_t)_j\|^r\right)^{1/r} \tag{2.21}$$

$$\leq c(p)2^{1/r}\left(E \sup_j\left(\frac{j}{\Gamma_j}\right)^{r/p}\right)^{1/r}\left(E\|\sum_{j=1}^{\infty} \frac{(\theta_t)_j}{j^{1/p}}\|^r\right)^{1/r},$$

where $\{(\theta_t)_j\}_{j=1}^{\infty}$ are i.i.d. copies of $\theta_t$ and we use the fact that $\{(\theta_t)_j\}_{j=1}^{\infty} \overset{\mathcal{D}}{=} \{\varepsilon_j(\theta_t)_j\}_{j=1}^{\infty}$. We write

$$(\theta_t)_j = ((\theta_{t,1})_j; \ldots; (\theta_{t,n})_j)$$

and observe that

$$\sum_{j=1}^{\infty} \frac{(\theta_t)_j}{j^{1/p}} = \Big(\sum_{j=1}^{\infty} \frac{(\theta_{t,1})_j}{j^{1/p}}; \ldots; \sum_{j=1}^{\infty} \frac{(\theta_{t,n})_j}{j^{1/p}}\Big)$$

$$\overset{\mathcal{D}}{=} \Big(\sum_{j=1}^{\infty} \frac{1}{j^{t/p}}\Big)^{1/t} \theta_t$$

because for each $i$, $\{(\theta_{t,i})_j\}_{j=1}^{\infty}$ is an i.i.d. t-stable sequence. Thus

$$\big(E\|\xi_t\|^r\big)^{1/r} \leq c(p)2^{1/r}\big(E \sup_j\big(\tfrac{j}{\Gamma_j}\big)^{r/p}\big)^{1/r}\Big(\sum_{j=1}^{\infty} j^{-t/p}\Big)^{1/t}\big(E\|\theta_t\|^r\big)^{1/r}.$$

Since, (as in (2.17) and (2.18)), $(E\|\theta_t\|^r)^{1/r} \sim n^{1/t}$, $t < 2$ and $(E\|\theta_2\|^r)^{1/r} \sim (\log n)^{1/2}$ we have completed the proof of (2.19) and (2.20). As in example 1 by (2.1a), for $1 \leq j \neq k \leq n$,

$$\big(E|\xi_{t,j} - \xi_{t,k}|^r\big)^{1/r} = \big(E|\theta_{t,j} - \theta_{t,k}|^p\big)^{1/p}\big(E|\theta_{p,1}|^r\big)^{1/r} \tag{2.21a}$$

$$= 2^{1/q}(E|\theta_{t,1}|^p)^{1/p}(E|\theta_{p,1}|^r)^{1/r} .$$

3.) Let $\{r_j\}_{j=1}^{n}$ be a Rademacher sequence and let $m_\varepsilon$ be the measure induced on $R^n$ by $\varepsilon = \{r_1,\ldots,r_n\}$. For fixed $0 < p < 2$ we denote by $\xi_\varepsilon$ the p-stable sequence defined by (2.1) with measure $m$. Clearly,

$$\sigma(m_\varepsilon, p) = 1 .$$

We will show that for $r < p$, $1 < p < 2$

$$(E\|\xi_\varepsilon\|^r)^{1/r} \sim (\log n)^{1/q} \tag{2.22}$$

and for $p = 1$, $r < 1$

$$(E\|\xi_\varepsilon\|^r)^{1/r} \sim L_2 n \ . \tag{2.23}$$

The upper bound in (2.22) follows from (2.5) and (2.23) from (2.6). For the lower bound we proceed as in (2.8). Using Lemma 1.3 (iii) [5] we get

$$\Big(E\|\sum_{j=1}^{\infty} (\Gamma_j)^{-1/p}\varepsilon_j\|^r\Big)^{1/r} \geq 2^{-1/r}\Big(E \inf_j\big(\frac{j}{\Gamma_j}\big)^{r/p}\Big)^{1/r}\Big(E\|\sum_{j=1}^{\infty} \frac{\varepsilon_j}{j^{1/p}}\|^r\Big)^{1/r} \tag{2.24}$$

where $\{\varepsilon_j\}_{j=1}^{\infty}$ is an i.i.d. sequence with $\varepsilon_1 = \varepsilon$. By Jensen's inequality

$$\Big(E \inf_j\big(\frac{j}{\Gamma_j}\big)^{r/p}\Big)^{1/r} \geq \frac{1}{\Big(E \sup_j\big(\frac{\Gamma_j}{j}\big)^{r/p}\Big)^{1/r}} > 0 \tag{2.25}$$

since, following the ideas of Lemma 1.2 [4], one can show that

$$\Big(E \sup_j\big(\frac{\Gamma_j}{j}\big)^{r/p}\Big)^{1/r} < \infty \ .$$

Denoting $\varepsilon_j = (r_{1,j}; \ldots; r_{n,j})$ we see that

$$\sum_{j=1}^{\infty} \frac{\varepsilon_j}{j^{1/p}} = \Big(\sum_{j=1}^{\infty} \frac{r_{1,j}}{j^{1/p}}, \ \ldots, \ \sum_{j=1}^{\infty} \frac{r_{n,j}}{j^{1/p}}\Big)$$

where each of the identically distributed random variables has a distribution which satisfies (2.10a) for $1 < p < 2$ or (2.11a) when $p = 1$. It follows that

$$\Big(E\|\sum_{j=1}^{\infty} \frac{\varepsilon_j}{j^{1/p}}\|^r\Big)^{1/r} \geq k_{p,r}(\log n)^{1/q} \tag{2.26}$$

from some constant $k_{p,r}$ depending only on $p$ and $r$;

$$\Big(E\|\sum_{j=1}^{\infty} \frac{\varepsilon_j}{j}\|^r\Big)^{1/2} \geq k_{1,r}L_2 n \tag{2.27}$$

for some constant $k_{1,r}$ depending only on $r$. Using (2.25) and (2.26) or (2.27) in (2.24) we complete the demonstration of (2.22) and (2.23). Note that by (2.1a) for fixed $0 < p < 2$

$$\begin{aligned}\big(E|(\xi_\varepsilon)_j - (\xi_\varepsilon)_k|^r\big)^{1/r} &= \big(E|r_j - r_k|^p\big)^{1/p}\big(E|\theta_{p,1}|^r\big)^{1/r} \\ &= 2^{1/q}(E|\theta_{p,1}|^r)^{1/r}\end{aligned} \tag{2.28}$$

(where $(\xi_\varepsilon)_j$, $(\xi_\varepsilon)_k$ represent the j-th and k-th component of $\xi_\varepsilon$, $1 \leq j \neq k \leq n$.)

These examples show that both the lower bound and the upper bound in Theorem 2.1 can be achieved.

Theorem 2.1 shows that given $\sigma(m;p)$ the fluctuation in the value of $(E\|\xi\|^r)^{1/r}$ over all possible p-stable sequences on $R^n$ is rather limited. Comparing (2.5) with (1.1), since $q > 2$, one might say that it is even more restricted than the fluctuation of Gaussian processes on $R^n$.

Let us now consider (1.2). This is a very interesting result because it states that whatever the Gaussian processes $X$ on $R^n$, if the components have a certain minimal separation in the $L^r$ metric, then $E\|X\|$ has a certain minimal rate of growth as a function of $n$. An analogous result for p-stable sequences on $R^n$ was obtained in Corollary 2.7, [4] and is given here as Theorem 2.2.

<u>Theorem 2.2</u>: Let $\xi = (\xi_1,\dots,\xi_n)$ be a symmetric p-stable sequence as defined above. Then for $1 < p < 2$, $r < p$ and $\frac{1}{p} + \frac{1}{q} = 1$,

$$E\|\xi\| \geq d_{r,p} \inf_{1\leq j\neq k\leq n} (E|\xi_j - \xi_k|^r)^{1/r}(\log n)^{1/q} \tag{2.29}$$

where $d_{r,p}$ is a constant depending only on $r$ and $p$.

<u>Proof</u>: In [4] this result is obtained as a corollary of a more general theorem. The proof is somewhat easier in the case considered in this paper so we will sketch the proof here. The proof makes use of (1.2), which, as we remarked above, is a consequence of Slepian's lemma. The representation of $\xi$ in (2.3) is equal in distribution to

$$\xi(\omega,\omega_1) = c(p)|m|^{1/p} \sum_{j=1}^{\infty} (\Gamma_j)^{-1/p} g_j\mu_j \tag{2.30}$$

where $\{g_j\}_{j=1}^\infty$ are i.i.d normal random variables with mean zero and $(E|g_1|^p)^{1/p} = 1$. Furthermore, we take $\{\Gamma_j\}_{j=1}^\infty$ and $\{\mu_j\}_{j=1}^\infty$ to be defined on the probability space $\Omega$, $(\omega \in \Omega)$ and $\{g_j\}_{j=1}^\infty$

on $\Omega_1$, $(\omega_1 \in \Omega_1)$ so that $\xi(\omega,\omega_1)$ is an $R^n$-valued random variable on $\Omega \times \Omega_1$.

We will show that (2.3) and (2.30) are equal in distribution by showing that they have the same characteristic functionals. Let $\nu$ be the probablity law of $g_1$ and define

$$\eta(d\beta) = \int_{-\infty}^{\infty} m\left(\frac{d\beta}{y}\right)\nu(dy).$$

Note that $g_1\mu_1$ is the random variable with distribution $\frac{\eta}{|\eta|}$ and also that $|\eta| = |m|$. The exponent on the right in (2.1) for the measure $\eta$ is

$$\int_{R^n} \left|\sum_{i=1}^{n} \alpha_i\beta_i\right|^p \eta(d\beta) = \int_{-\infty}^{\infty} \int_{R^n} \left|\sum_{i=1}^{n} \alpha_i\beta_i y\right|^p m(d\beta)\nu(dy)$$

$$= \int_{R^n} \left|\sum_{i=1}^{n} \alpha_i\beta_i\right|^p m(d\beta) \int_{-\infty}^{\infty} |y|^p \nu(dy)$$

Thus, since the last integral is equal to 1, (2.3) and (2.30) have the same characteristic functionals. We also use the fact that $\{g_j\}_{j=1}^{\infty} \overset{\mathcal{D}}{=} \{\varepsilon_j g_j\}_{j=1}^{\infty}$.

By (2.1a)

$$(E|\xi_j - \xi_k|^r)^{1/r} = \left(\int_{R^n} |\beta_j - \beta_k|^p m(d\beta)\right)^{1/p} (E|\theta_{p,1}|^r)^{1/r} \tag{2.31}$$

We define

$$d(j,k) = \left(\int_{R^n} |\beta_j - \beta_k|^p m(d\beta)\right)^{1/p}.$$

Note that for fixed $\omega \in \Omega$, $\xi(\omega,\omega_1)$ is an $R^n$-valued Gaussian process on the probability space $\Omega_1$. The customary $L^2(\Omega_1)$ metric for this process is

$$d_\omega(j,k) = (E_{\omega_1}|(\xi(\omega,\omega_1))_j - (\xi(\omega,\omega_1))_k|^2)^{1/2}$$

$$= c(p)|m|^{1/p}\left(\sum_{i=1}^{\infty} (\Gamma_i)^{-2/p}\left|(\mu_i)_j - (\mu_i)_k\right|^2\right)^{1/2},$$

where $(\xi)_j$ denotes the jth component of $\xi$, etc. We have the following significant relationship between $d_\omega(j,k)$ and $d(j,k)$:

$$E\exp[i\lambda(\xi_j - \xi_k)] = E_\omega \exp[-\lambda^2 b(p) d_\omega^2(j,k)] \tag{2.32}$$

$$= \exp[-|\lambda|^p d^p(j,k)]$$

where $b(p)$ is a constant depending only on $p$ and $E_\omega$ denotes exectation with respect to $\Omega$. The first equality results from taking expectation with respect to the marginal Gaussian process. Equality between the first and third terms is just (2.1). The equality between the second and third term in (2.32) enables us to use an exponential Chebyshev inequality to obtain

$$P[\omega \in \Omega: d_\omega(j,k) \leq \varepsilon d(j,k)] \tag{2.33}$$

$$\leq \exp[-\frac{2}{\alpha}(\frac{pb(p)}{2})^{\alpha/2}(\frac{1}{\varepsilon})^\alpha]$$

where $\frac{1}{\alpha} = \frac{1}{p} - \frac{1}{2}$. If we let $\varepsilon = (d(p)\log n)^{-1/\alpha}$ for a constant $d(p)$, sufficiently large, depending only on $p$, we can make the right side of (2.33) less than $n^{-4}$. Then, by the Borel-Cantelli lemma we have that for each $\omega \in \Omega'$, $\Omega' \subset \Omega$, $\mathrm{Prob}(\Omega') = 1$, there exists an $n_0(\omega)$ such that

$$d_\omega(j,k) \geq \frac{d(j,k)}{(d(p)\log n)^{\frac{1}{p}-\frac{1}{2}}} \tag{2.34}$$

for all $1 \leq j \neq k \leq n$ and $n \geq n_0(\omega)$.

Applying (1.2) (with $r = 2$) to the marginal Gaussian process $\xi(\omega,\omega_1)$ we get

$$E_{\omega_1}\|\xi(\omega,\omega_1)\| \geq b_p \inf_{1\leq j\neq k\leq n} d_\omega(j,k)(\log n)^{1/2}$$

for a constant $b_p$ depending only on $p$. Let us now take $n'$ sufficiently large so that (2.34) is satisfied on a set $\omega \in \Omega''$, $\mathrm{Prob}(\Omega'') \geq 1/2$. Then for $n \geq n'$

$$E\|\xi\| = E_\omega E_{\omega_1}\|\xi(\omega,\omega_1)\| \geq \frac{b_p}{2} \inf_{1\leq j\neq k\leq n} d(j,k)(\log n)^{1/q}. \tag{2.35}$$

The inequality (2.9) follows immediately from (2.31) and (2.35) by taking $d_{r,p}$ sufficiently small, (since we know from (2.5) that $E\|\xi\| > 0$).

In Examples 1.) - 3.) above the term $\inf_{1\le j\ne k\le n} (E|\xi_j - \xi_k|^r)^{1/r}$ is some small constant (see (2.16), (2.21a) and (2.28)). From (2.26) we see that the rate of growth of (2.29) can not be increased (modulo some constant). On the other hand (2.29) is very weak when considering the processes in Examples 1.) and 2.). When $p < 1$, Theorem 2.1 applied to Example 3.) shows that $E\|\xi_\varepsilon\|$ can be bounded in $n$. However when $p = 1$ one would guess that an analogue of (2.9) would be valid with $(\log n)^{1/q}$ replaced by $L_2 n$. Whether or not this is the case is an open question.

In conclusion we will mention a general result on stable processes based on a method of Yurinski, that was obtained recently with J. Zinn and E. Giné. It is valid for all stable processes with values in a linear measurable space with a quasi-norm. In the context of this paper and for $p > 1$ it is

$$P[\|\xi\| > E\|\xi\| + \lambda] \le c_p \sigma^p(m,p)\lambda^{-p} \tag{2.36}$$

where $c_p$ is a constant depending only on $p$ and $\sigma(m,p)$ is given in (2.2a). It is well known that

$$\lim_{\lambda\to\infty} \lambda^p P[\|\xi\| > \lambda] = c'_p \sigma^p(m,p)$$

for some constant $c'_p$ depending only on $p$. The relevance of (2.36) is that $E\|\xi\|$ can be much larger than $\sigma(m,p)$, as Example 3.) shows. A proof of (2.36) and further discussion on these points will be published shortly.

## REFERENCES

1. Le Page, R., Woodroofe, M. and Zinn, J., Convergence to a stable distribution via order statistics, Ann. of Probability, 9, (1981), 624-632.

2. Marcus, M. B. Tail probability estimates for certain Rademacher sums, Colloquium Mathematicum, 36 (1976), 153-155.

3. Marcus, M. B. and Pisier, G., Random Fourier sries with applications to harmonic analysis, Ann. Math. Studies, Vol. 101 (1981), Princeton Univ. Press, Princeton, N. J.

4. Marcus, M. B. and Pisier, G., Characterizations of almost surely continuous p-stable random Fourier series and strongly stationary processes, Acta Matematica, to appear.

5. Marcus, M. B. and Zinn, J., The bounded law of the iterated logarithm for the weighted empirical process in the non-i.i.d. case, Ann. of Probability, to appear.

# LARGE DEVIATIONS OF EXTREMES

C.W. Anderson

University of Sheffield

## 1. INTRODUCTION

The aim of this talk is to discuss large deviation limit results for extremes. In particular I will (i) describe how interest in such results is motivated by statistical applications; (ii) review and offer some extensions of current results, and relate them to other rate-of-convergence theorems; and (iii) discuss some possible extensions and open questions.

## 2. THE KEY APPROXIMATION

Much of the classical approach to the statistical analysis of large observations is based on the approximation, suggested by various probability limit theorems,

$$P\left(\frac{M_n - b_n}{a_n} \leq x\right) \simeq G(x) \qquad (K)$$

where $M_n$ is a random variable representing the maximum of n individual observations, $a_n$ and $b_n$ are constants, and G is one of the extreme value distributions. An alternative form is :

$$P\left(M_n \leq y\right) \simeq G\left(\frac{y - b_n}{a_n}\right) \qquad (K)$$

I will refer to these as the 'Key Approximation'.

The statistical importance of the approximation is that it suggests the fittings of an extreme value distribution to the observations of $M_n$ and the basing of inferences, for example



J. Tiago de Oliveira (ed.), Statistical Extremes and Applications, 325–340.

about future large values, on the resulting estimated distribution of the observation. An apparent advantage of such a procedure is that it appears to avoid the need for too specific assumptions about the distribution of the observations, $\{X_i\}_1^n$ say, of which $M_n$ is the largest: the limit theorems on which the Key Approximation rests are known to hold under quite mild conditions on the tail of the distribution function F of X (and dependence of the $\{X_i\}$) and so it does not matter if we do not know the tail behaviour of F, so long as we are willing to assume it satisfies the rather broad domain of attraction conditions. In this way the approach seems appealingly robust.

Possible errors are either (i) statistical, or (ii) due to the approximation. We concentrate here on the latter.

The argument for the approximation (K) of the true distribution of $M_n$ by an extreme value distribution is analagous to that provided by the Central Limit Theorem for the use of the Normal distribution in mainstream Statistics. However (K) is not supported by such strong rate-of-convergence ramifications as the Central Limit Theorem: some results described briefly in the next section show that in many cases the difference between the two sides in (K) goes to zero very slowly.

Another aspect of the approximation arises from the fact that very often we wish to estimate values x of the variable X which will be exceeded with very low probability over a specified length of time. A way of formulating such problems is in terms of the estimation of high quantiles of $M_n$, and for these (K) in the form

$$P\left(M_n > y\right) \simeq 1 - G\left(\frac{y-b}{a}\right)$$

or equivalently, $(\bar{K})$

$$P\left(\frac{M_n - b}{a} > x\right) \simeq 1 - G(x)$$

suggests use of high quantiles of the fitted G as estimates. Now, both sides in $(\bar{K})$ are small for the x or y of interest, and so their difference is bound to be small too. Of more importance is the relative error

$$P\left(\frac{M_n - b_n}{a_n} > x\right) / \left(1-G(x)\right), \qquad (R)$$

which, ideally, one would hope would be close to 1 when both terms are individually small; that is, when x is large. The question of whether, and when, this hope is justified motivates the study of the large deviations limit theorems which are the main point of this talk. Similar large deviations limit theorems occupy an important place in central limit theory: I would like to suggest, however, that the role of such results in extreme value theory is even more central because of their link with the statistical problem above. Section 4 reviews known results of this kind and sketches a few extensions. Section 5 contains a variety of comments relating to the preceding sections.

## 3. ABSOLUTE ERRORS

Several workers (for example Galambos (1978), Reiss (1981), Davis (1982), Gomes (1978), Smith (1982), Cohen (1982), Anderson (1971)) have discussed aspects of the difference between the two sides in (K). For our present purpose results formulated in terms of as broad conditions on F as possible are of interest, and the most comprehensive of this kind are due to Smith and Cohen. They show, under conditions sketched below, that

$$P\left(\frac{M_n - b_n}{a_n} < x\right) - G(x) = \gamma(x)G'(x)\delta_n + o(\delta_n) \qquad (3.1)$$

uniformly in x as $n \to \infty$, where $\gamma(x)$ is a function depending to some extent on F, and $\delta_n \to 0$ at a rate depending heavily on F.

Smith considered the cases $G = \Phi_\alpha$ and $\Psi_\alpha$. For F to belong to the domain of attraction of $\Phi_\alpha$ it is necessary and sufficient that

$$L(t) = -t^{\alpha}\log F(t) \qquad (3.2)$$

should be slowly-varying (SV) at $\infty$. Smith showed that if slow-variation is strengthened to

$$\frac{L(tx)}{L(t)} - 1 \sim \upsilon(x)g(t) \text{ , } x > 0, \; t \to \infty \qquad (3.3)$$

where g is regularly-varying with index $< 0$ then (3.1) holds, for suitable $a_n$ and $b_n$, with $\delta_n = g(a_n)$ and $\gamma(x) = x\upsilon(x)/\alpha$. Note that (3.3) is a condition on the rate of slow variation of L (a 'slow variation-with-remainder' condition in Goldie's (1980) terminology).

There is a similar result for $\Psi_\alpha$.

Cohen discusses $G = \Lambda$. He considers two classes of distributions in the domain of attraction of $\Lambda$. The first, class $\underset{\sim}{E}$, consists of distributions of random variables X such that $e^X$ is in the domain of attraction of $\Phi_\alpha$. Results for these distributions can be carried over from Smith's work. The second class, N, is similar to the following. Take as starting point Balkema and de Haan's (1972) characterization of F in the domain of attraction of $\Lambda$:

$$- \log F(x) = c(x) \exp\left\{- \int_{-\infty}^{x} (1/\phi(t))dt\right\} \tag{3.4}$$

where $c(x) \to 1$, $\phi > 0$ and $\phi' \to 0$ as $x \to x_0 = \sup\{x: F(x)<1\}$. If $\phi(x)$ is of constant sign for large x, $c(x)-c \sim s\ \phi'(x)$ as $x \to x_0$, for some s and

| | | |
|---|---|---|
| $\phi'(x)$ | when $x_0 = \infty$ | regularly varying |
| $\phi'(x_0 - x^{-1})$ | when $x_0 < \infty$ | at $x_0$ |

then (3.1) holds with $\delta_n = \phi'(b_n)$ and $\gamma(x) = -x^2/2$, for suitable $a_n$ and $b_n$. (This does not coincide exactly with Cohen's formulation, but it uses his ideas and is more convenient for comparisons later). The regular variation of $\phi'$ is used to show (in the $x_0 = \infty$ case)

$$\phi'(x + y\,\phi(x)) \sim \phi'(x), \quad (x \to \infty) \tag{3.5}$$

uniformly in $|y| < -K \log|\phi'(x)|$ for any $K>1$, a condition we record for future comparison.

## 4. RELATIVE ERRORS: LARGE DEVIATIONS LIMITS

The description of results about relative errors is broken up into three parts, the first discussing the precise formulation of the question of interest, the second giving results for $\Phi_\alpha$, and the third results for $\Lambda$.

### 4.1 Formulation

We are interested in conditions under which

$$P\left(\frac{M_n - b_n}{a_n} > x\right)/(1-G(x)) \to 1 \text{ as } n \to \infty \tag{4.1}$$

when x can grow with n. A way of formulating the question precisely is to ask how quickly a sequence $\{x_n\}$ can increase with n and yet (4.1) remain true for all $x = O(x_n)$ : $\{x_n\}$ then defines a zone in which the approximation of small probabilities for $M_n$ by G is adequate. More explicitly, we seek conditions on $\{x_n\}$ and F so that

$$\lim_{n\to\infty} P(\frac{M_n - b_n}{a_n} > y_n)/(1-G\ (y_n)) = 1 \qquad (4.2)$$

for every sequence $\{y_n\}$ such that $y_n = O(x_n)$; or, equivalently, so that

$$\lim_{n\to\infty} \sup_{x \leq Ax_n} \left| P(\frac{M_n - b_n}{a_n} > x)/(1-G(x)) -1 \right| = 0 \quad (4.3)$$

for each positive A. Since we are interested more in broad behaviour than in local fluctuations, it is reasonable to take $\{x_n\}$ to be non-decreasing.

4.2 Large Deviations for $\Phi_\alpha$

4.2.1 Super-Slow Variation. When F belongs to the domain of attraction of $\Phi_\alpha$ it is well known that the normalizing constants may be taken to be $b_n=0$ and $a_n = \alpha_n = \inf\ \{t\ :\ -\ \log F(t) \leq 1/n\}$.

With these constants (4.2) becomes $1 - F^n(\alpha_n y_n) \sim 1 - \Phi_\alpha(y_n)$, which, from the fact that $1-t \sim -\log t$ as $t\to 1$, is equivalent to

$$-\log F(\alpha_n y_n)/ -\log F(\alpha_n) \sim y_n^{-\alpha} ,$$

or, from (3.2),

$$\lim_{n\to\infty} L(\alpha_n y_n)\ /\ L(\alpha_n) = 1.$$

Thus, for (4.3) to hold it suffices that the slow variation condition $\lim\limits_{x\to\infty} L(xy)\ /\ L(x) = 1$ for each $y>0$ should be strengthened to $\lim\limits_{x\to\infty} L(xy(x)/L(x)) =1$ for each function y such that $y(x) = o(\xi(x))$, where $\xi$ is a non-decreasing function for which $\xi(a_n) = x_n$. Equivalently

$$\lim_{x\to\infty} L\left(x\ \xi^{\delta}(x)\right)/\ L(x) = 1 \qquad (4.4)$$

uniformly in $\delta\epsilon[0,1]$. In fact it turns out (Anderson (1978)) that (4.4) is necessary as well as sufficient for (4.3), even when $a_n$ and $b_n$ are arbitrary normalizing constants giving convergence to $\Phi_\alpha$ for fixed x. Condition (4.4) is a restriction on the slowness of variation of L and functions satisfying it are said to be $\xi$-super-slowly varying ($\xi$-SSV). The condition provides an internal measure of slowness of variation, in contrast to (3.3), which measures it externally. A relationship between the two types of condition will emerge in the next section.

4.2.2 Conditions for $\xi$ - Super-Slow Variation. How restrictive on a slowly-varying function is the $\xi$-SSV property? To answer this question it is convenient to make the transformation $L^*(t)=\log L(e^t)$, $\xi^*(t)=\log \xi(e^t)$ and similarly for other functions. The $\xi$-SSV condition (4.4) translates into

$$\lim_{t\to\infty} L^*(t+\delta\xi^*(t)) - L^*(t) = 0 \qquad (4.5)$$

uniformly in $\delta\varepsilon[0,1]$.

The Karamata representation for SV L is (in * form)

$$L^*(t) = c(t) + \int_1^t \varepsilon(u)du \qquad (4.6)$$

where $c(t)$ tends to a finite limit and $\varepsilon(t)$ to 0 as $t \to \infty$. For (4.5) to hold it is necessary and sufficient that

$$\int_t^{t+\delta\xi^*(t)} \varepsilon(u)du = \int_0^\delta \xi^*(t)\,\varepsilon(t+w\xi^*(t))dw \to 0$$

uniformly in $\delta\varepsilon[0,1]$, for which it is clearly sufficient that

$$\lim_{t\to\infty} \xi^*(t)\,\varepsilon(t+\omega\xi^*(t)) = 0$$

uniformly over $[0,1]$, and for this in turn, since $\xi^*$ is non-decreasing, it is necessary and sufficient that

$$\lim_{t\to\infty} \xi^*(t)\,\varepsilon(t) = 0 \qquad (4.7)$$

Thus if $\varepsilon(u) = o(1/\xi^*(u))$ in (4.6) then L is $\xi$-SSV.

What is not so immediate, but true, is that under the following restriction on the rate of growth of $\xi^*$:

$$\xi^*(t+\xi^*(t)) = O(\xi^*(t)), \qquad (4.8)$$

(4.7) is in fact necessary as well as sufficient for $\xi$-super-slow variation of L (Anderson (1978)). The proof of this works by showing the existence of a transformation s turning $L^*(t+\delta\xi^*(t))-L^*(t)$ into $L^*(s(x+\beta))-L^*(s(x))$ for a bounded $\beta$. The $\xi$-SSV property of $L^*$ becomes thus the SV property of $L^*(s)$, for which (4.6) can be invoked, and reversal of the transformation leads to (4.7).

We note that the function $\varepsilon$ of the Karamata representation may be taken proportional to $L^*(t+\alpha)-L^*(t)$ for any $\alpha > 0$ (Seneta (1976) p.13), so that, under (4.8), $\xi$-SSV is equivalent to

$$\lim_{t\to\infty} \xi(t)(L^*(t+\alpha)-L^*(t)) = 0,$$

which is (the * version of) a o-version of (3.3); it is also the condition for $\xi^*$-slow variation as defined by Ash, Erdos and Rubel (1974). Condition (4.8) is an O-version of Beurling slow variation: see Goldie (1980).

The foregoing shows that the case when (4.8) holds is fairly well understood: it is worth noting too that (4.8) covers a wide range of functions $\xi$, including all power functions. We now turn however to the case when $\xi$ grows faster than (4.8) allows.

A first observation is that (4.7) is still sufficient for $\xi$-SSV, but may be too strong. If $\xi^*$ grows rapidly enough for $\int_1^\infty du/\xi^*(u)<\infty$ and (4.7) is true then L converges to a finite constant, and so is $\xi$-SSV for any $\xi$, a property of L that we will call universal-super-slow variation (U-SSV). There are functions which are not U-SSV but are $\xi$-SSV for a $\xi$ with $\int_1^\infty du/\xi^*(u)<\infty$: an example is

$$L^*(t) = \log_3 t, \quad \xi^*(t) = t^\beta, \quad \beta>1.$$

Thus (4.7) is stronger than necessary when (4.8) no longer holds.

Suppose $\xi^*(x) = x\phi(x)$ where $\phi$ is non-decreasing. The $\xi$-SSV condition (4.5) becomes then

$$\lim_{t\to\infty} L^*(t(1+\delta\phi(t)))-L^*(t) = 0$$

uniformly in $\delta\varepsilon[0,1]$, and it is not difficult to see that this implies

$$\lim_{t\to\infty} L^*(\exp(t+\delta\phi^*(t)))-L^*(\exp(t)) = 0$$

uniformly in $\delta\varepsilon[0,1]$, where $\phi^*(t) = \log\phi(e^t)$. If $\phi^*$ satisfies (4.8) then the representation previously derived may be used for $L^*(\exp(.))$, and this yields for $L^*$ the conclusion $\varepsilon(t) = o(1/t \log \phi(t))$, which is readily seen also to be sufficient for $\xi$-SSV. Further, a closer look at the proof of the earlier representation shows that in fact monotonicity of $\phi$ is not needed here.

The argument above can be repeated to bring into the discussion more and more rapidly-increasing $\xi^*$. Define a transformation T on real functions f by $Tf = f^*-I$, where I is the identity function. Thus T is the transformation taking $\xi^*$ into $\phi^*$. Then:

Theorem 1: If $\xi_r = T^r\xi^*$ satisfies (4.8) for some integer $r \geq 1$ then L is $\xi$-SSV if and only if it has a Karamata representation (4.6) in which

$$\varepsilon(t) = o\left(t \log t \log_2 t \ldots \log_{r-1} t \ \xi_r(\log_r t)\right)^{-1}.$$

Example If $\xi^*(x) = x^{1+(\log x)^{\beta}}$ where $\beta>0$ then $\xi_2(t) = \beta t$ and so $\varepsilon(t) = o(t \log t \log_2 t)^{-1}$. Evidently $L^*(t) = \log_4 t$ has such a representation.

The theorem does not cover all $\xi$. For example, if $\xi^*(x) = e^x$ then $\xi_r(x) \sim e^x$ for every r, so (4.8) is never satisfied. I do not know whether there is a $\xi$-SSV L which is not U-SSV for such a rapidly-increasing $\xi$, but I conjecture that there is. (On the other hand it must be so lethargic in its variation that it would be scarcely distinguishable from a constant.)

4.2.3 Uniformity. For slowly-varying functions, and for several variants of them, pointwise convergence in the defining relationship implies uniformity (see, for example, de Haan (1970), Ash, Erdos and Reubel (1974), Seneta (1976), Bingham and Goldie (1982).) It is natural to ask, therefore, whether the requirement of uniformity in the definition of $\xi$-super-slow variation can be dispensed with. The following theorem asserts that it can if (4.8) holds.

Theorem 2: Let $\xi$ be a non-decreasing function satisfying (4.8) and suppose that L is (the * version of) a measurable SV function satisfying

$$\lim_{t\to\infty} L(t+\delta\xi(t))-L(t) = 0$$

for each $\delta$ [0,1]. Then convergence is uniform over [0,1].

The proof is an adaptation of that of Korevaar, van Aardenne-Ehrenfest and de Bruijn (1949).

4.2.4 Further Terms. What other behaviour is possible when $\{x_n\}$ grows too quickly for $\left(1-F^n(\alpha_n x_n)\right)/\left(1-\Phi_\alpha(x_n)\right)$ to converge to 1? Does the ratio diverge, for example, in a specific way? The next Theorem uses some of the large deviations results for $\Lambda$ in section 4.3 to throw some light on such questions.

Theorem 3: Suppose that F has a representation via (3.2) and (4.6) in which $\varepsilon(u)>0$ eventually and $\phi(u) = 1/\varepsilon(u)$ is 'uniformly Beurling slowly varying' (UBSV) in the sense $\phi(u+w\phi(u)) \sim \phi(u)$ as $u \to \infty$, locally uniformly. If

$$\log x_n = 0\, g(\log \alpha_n)\phi(\log \alpha_n)$$

for a function g such that $\phi$ is $g^2$ - BSV in the sense of (4.15), then

$$1-F^n(\alpha_n x_n) \sim x_n^{\varepsilon(\log \alpha_n)}(1-\Phi_\alpha(x_n)) \quad n \to \infty.$$

4.3 Large Deviations for $\Lambda$

4.3.1 de Haan and Hordijk's Condition. First we look at a slightly generalised version of de Haan and Hordijk's (1972) sufficient condition for (4.2), supposing $x_o = \infty$.

From $1-t \sim -\log t$ as $t \to 1$ (4.2) is seen to be equivalent to

$$-\log F(b_n+a_n y_n)/-\log F(b_n) \sim e^{-y_n} \tag{4.12}$$

for $y_n = O(x_n)$, and this, by the Balkema -de Haan representation (3.4), is the same as

$$\lim_{n\to\infty} \int_{b_n}^{b_n+a_n y_n} \frac{du}{\phi(u)} - y_n = 0, \quad y_n = O(x_n).$$

Equivalently, when $a_n = \phi(b_n)$,

$$\lim_{n\to\infty} \int_o^1 y_n \left\{ \frac{\phi(b_n)}{\phi(b_n+wy_n\phi(b_n))} - 1 \right\} dw = 0 \tag{4.13}$$

for $y_n = O(x_n)$. Let $\xi$ be a non-decreasing function with $\xi(b_n)=x_n$. Then an alternative way of stating (4.13) is

$$\lim \int_o^\delta \xi(x) \left\{ \frac{\phi(x)}{\phi(x+w\xi(x)\phi(x))} -1 \right\} dw = 0 \tag{4.14}$$

uniformly in $\delta$ in finite intervals of $[0,\infty)$, where the limit is as $x\to\infty$ through $\{b_n\}$.

A sufficient condition for (4.14) is evidently

$$\lim_{x\to\infty} \xi(x) \left\{\frac{\phi(x)}{\phi(x+w\xi(x)\phi(x))} -1 \right\} = 0 \tag{4.15}$$

uniformly in finite intervals in $[0,\infty)$. This is a rate condition for Beurling slow variation (4.10) incorporating both internal and external measures of slowness. I will call it $\xi^2$- Beurling slow variation ($\xi^2$-BSV).

In turn a sufficient condition for (4.15) is de Haan & Hordijk's condition

$$\lim_{x\to\infty} \xi^2(x)\ \phi'(x) = 0. \tag{4.16}$$

In summary then:

| | |
|---|---|
| (4.14) is necessary and sufficient | for (4.2) |
| (4.15) is sufficient | with |
| (4.16) is sufficient | $a_n = \phi(b_n)$. |

4.3.2 <u>A Connexion with Super-Slow Variation.</u> The discussion in the previous section was in terms of $K = -\log(-\log F)$, i.e. $K=\Lambda^{-1}F$, which, apart from a term converging to zero (and therefore playing a minor role in questions about large deviations) is equal to $\int_\infty^x dt/\phi(t)$.
An alternative approach is via the inverse function

$$J=K^{-1}=F^{-1}\Lambda,$$

where $F^{-1}(y) = \inf\{x : F(x) \geq y\}$. Though the two approaches are ultimately equivalent the second links with the class $\Pi$ of de Haan (1970) and appears to lead to conditions that (because they are expressed in terms of slow variation itself rather than Beurling slow variation) are more directly comparable with those for $\Phi_\alpha$.

For each sequence $\{y_n\}$ appearing in (4.2) define a new sequence $\{\eta_n\}$ by

$$F^n(b_n + a_n y_n) = \Lambda(\eta_n). \tag{4.17}$$

In fact $\eta_n = K(b_n + a_n y_n) - K(b_n) + o(1)$ and $y_n + o(1) = (J(\eta_n + \log n) - J(\log n))/a_n$. Then (4.2) holds if and only if $1 - \Lambda(\eta_n) \sim 1 - \Lambda(y_n)$ for $y_n = O(x_n)$, that is, if and only if

$$\lim_{n\to\infty} y_n - \eta_n = 0 \text{ for } y_n = O(x_n). \tag{4.18}$$

Notice that, when $\eta_n$ is expressed in terms of K, (4.18) becomes the same as (4.12). On the other hand substitution of $y_n$ gives as necessary and sufficient for (4.2)

$$\lim_{n\to\infty} \frac{J(\log n + \eta_n) - J(\log n)}{a_n} - \eta_n = 0.$$

for $y_n = O(x_n)$, where $\eta_n$ is defined by (4.17).

The implicit character of $\eta_n$ here makes this condition unappealing: it would be more satisfactory if we could replace $\eta_n$ by $y_n$ itself. The next theorem shows that for the sufficiency half at least we can.

Theorem 4 A sufficient condition for (4.2) is that

$$\lim_{n\to\infty} \frac{J(\log n + y_n) - J(\log n)}{a_n} - y_n = 0 \qquad (4.19)$$

for each $y_n = O(x_n)$.

Proof: Write $\Delta_n = \eta_n - y_n$ where $\eta_n$ is as in (4.17), so that $(J(\log n + y_n + \Delta_n) - J(\log n))/a_n = y_n + o(1)$. If $\Delta_n = O(x_n)$ then (4.19) shows immediately that $\Delta_n \to 0$, which is (4.18), as required. In fact $\Delta_n = O(1)$, for if not there is an $A > 0$ with $\Delta_n > A$ frequently, and so by monotonicity of J,

$$\frac{J(\log n + y_n + \Delta_n) - J(\log n)}{a_n} - y_n > \frac{J(\log n + y_n + A) - J(\log n)}{a_n} - y_n$$

frequently. Since the left side tends to 0 and the right to A we have a contradiction.

Remark Is (4.19) necessary? I do not know. Whether it is or not seems to be bound up with the possibility or removing the restriction on the limiting process in (4.14).

Note that (4.19) is a strengthening of a condition

$$\lim_{n\to\infty} \frac{J(\log n + y) - J(\log n)}{a_n} = y$$

for each fixed y, or, as it turns out, equivalently,

$$\lim_{x\to\infty} \frac{J(x+y) - J(x)}{\alpha(x)} = y \qquad (4.20)$$

for some function $\alpha(x)$ and each fixed y, which are necessary and sufficient for membership of the domain of attraction of $\Lambda$. (The second has, I think, been given by Pickands).

In fact (4.20) is just an additive-argument version of the defining condition for de Haan's class $\Pi$ of functions (non-strictly monotonic version, de Haan (1970), Bingham and Goldie (1982)p.175). The representation for such functions given by de Haan (1970 p.41) yields

$$J(x) = c + \gamma(x) + \int_0^x \gamma(u)$$

where $\gamma$ is an additive-argument SV function: $\lim_{u\to\infty} \gamma(u+w)/\gamma(u) = 1$ for each w. With the natural choice of $a_n = \gamma(\log n)$ therefore (4.19) may be re-expressed as

$$\lim_{n\to\infty} \frac{\gamma(\log n + y_n)}{\gamma(\log n)} -1 + \int_0^1 y_n \left\{\frac{\gamma(\log n + wy_n)}{\gamma(\log n)} -1 \right\} dw$$

for each $y_n = O(x_n)$. A sufficient condition for this is

$$\lim_{n\to\infty} x_n \left\{ \frac{\gamma(\log n + wx_n)}{\gamma(\log n)} -1 \right\} = 0 \qquad (4.21)$$

uniformly over positive finite intervals of w. Note the similarity with (4.15) : (4.21) is a hybrid rate condition for slow variation, whereas (4.15) is one for Beurling slow variation.

If we write $g = \log \gamma$ we see that (4.21) is

$$\lim_{n\to\infty} x_n (g(\log n + wx_n) - g(\log n)) = 0 \qquad (4.22)$$

uniformly over finite positive intervals of w, a form which emphasises its SSV/SV-with-remainder character. The parallelism with SSV theory suggests:

Theorem 5 Let $\zeta$ be any non-decreasing function such that $\zeta(\log n) = x_n$, and let $\varepsilon_g$ denote the integrand in the Karamata representation (4.6) of g.

(i) If $\lim_{x\to\infty} \zeta^2(x)\ \varepsilon_g(x) = 0$ (4.23)

then

$$\lim_{x\to\infty} \zeta(x)\ (g(x + w\zeta(x)) - g(x)) = 0 \qquad (4.24)$$

uniformly over finite positive intervals of w.

(ii) If $\zeta(x + \zeta^2(x)) = O(\zeta(x))$ then (4.23) is necessary as well as sufficient for (4.24).

## 5. DISCUSSION

There is much work still to be done in large deviation theory before it reaches a definitive state, but two broad conclusions seem to be indicated by what is known so far:

(a) the size of the zone $O(x_n)$ in which the large deviations property holds is governed by aspects of the tail behaviour of F (e.g. (4.7) (4.15), (4.23)) which are similar in character to those (e.g. (3.3), (3.5)) which determine the size of absolute errors in (3.1).

(b) examples in de Haan & Hordijk (1972) and Anderson (1978) suggest that the $x_n$ above grows satisfactorily fast under mild conditions on F when F belongs to the domain of attraction of $\Phi_\alpha$ and presumably also to the subset E of the domain of attraction of $\Lambda$, but that it does not for other F in the domain of attraction of $\Lambda$.

Further checking of (b) is needed, but if the conclusions is confirmed then (a) and (b) would tend to show that the message from large deviations theory about the robustness of the classical statistical use of the extreme value distributions is the same as that from the study of absolute errors (Anderson(1976)): that the apparent robustness of the procedure to properties of the tail of F is illusory. In other words use of the Key Approximation entails the tacit assumption (for n not too enormous) that F belongs to a rather restricted subset of distributions whose tail behaviour is already close to that of G.

In the light of this we might be prompted to ask what is the aim of limit theorems generally, as far as statistical procedures for extremes are concerned? Why <u>linear</u> normalization? There are very general limit theorems with other normalizations and rapid convergence: for example

$$\lim_{n\to\infty} P\left(n\bar{F}\left(M_n\right)<x\right) = 1-e^{-x}$$

whenever F is continuous and $\bar{F} = 1-F$. (In fact if we allow ourselves so much scope for 'normalization' we can easily engineer exact equality for all n, to the exponential or any other continuous distribution.) But such results are of little use statistically when F is unknown since they do not tell us which distribution to try to fit to data: the above suggests only

$$P\left(M_n<y\right) \simeq e^{-n\bar{F}(y)}.$$

The statistical utility of the classical Fisher-Tippett-Gnedenko limit theorems, with their linear normalization, is precisely that they suggest the fitting of a distribution known up to at most three parameters. The drawback, however, as we have seen, is that the approximation can be poor.

It is tempting to speculate about some ideal alternative, intermediate betwen the two above. In this we would find transformations $\phi_n$ such that

$$\lim_{n\to\infty} P\left(\phi_n\left(M_n\right) \leqslant x\right) = H(x)$$

for some approximating H, where

(a) H is of some simple form depending at most only on broad properties of F;

(b) $\phi_n$ is simple in some sense, and does not require detailed knowledge of F; and

(c) convergence is fast.

It seems unrealistic to expect such a scheme to exist. The idea of it, though, does suggest that in analyzing extreme data it would be sensible and useful to consider transformation of the observations, in the hope of improving the fit of the standard extreme value distributions; that is, for some reasonably rich family of transformations t (including, perhaps, powers, logs, exponentials), to use

$$P\left(t\left(M_n\right) \leqslant y\right) \approx G\left(\frac{y-b}{a}\right)$$

in place of the Key Approximation. I suspect many people do this already.

Finally I list some more detailed comments about section 4.

1. The main problems in section 4.2 for $G = \Phi_\alpha$ are (a) removal of the restrictions on the rate of growth of $\xi^*$ in the characterization of $\xi$-SSV functions, and in the discussion of uniformity; and (b) the development of further results on further terms.

2. In section 4.3 with $G = \Lambda$ a major problem is to find simple necessary conditions. Related to this is the elucidation of when the restriction on the limiting process in (4.14) can be removed. Further exploration of the link with SSV functions is desirable, and study of the relationship between rate conditions for SV and BSV functions.

3. The transformation used in 4.2.2 to convert a $\xi$-SSV function into a SV function can be viewed as setting up a correspondence between F and a new distribution function $F_T$ with the property that large deviations behaviour of F corresponds to behaviour of $F_T^n(\alpha_n x)$ at finite x. This is reminiscent of the Esscher transformation in central limit theory, in which large deviations behaviour of one distribution is related to standard convergence to normality of another (see e.g. Feller (1966)). A natural question is whether we can find a similar correspondence in other cases in large deviations for extremes.

## REFERENCES


Anderson C.W. (1971) Contributions to the Asymptotic Theory of Extreme Values. Ph.D. Thesis, University of London.

Anderson C.W. (1976) Extreme value theory and its approximations. Proc.Symp.Reliability Technology. Bradford, U.K. Atomic Energy Authority.

Anderson C.W. (1978) Super-slowly varying functions in extreme value theory. J.R.Statist.Soc.B, 40, 197-202.

Ash, J.M., Erdos, P. and Rubel, L.A. (1974) Very slowly varying functions. Aequationes Math. 10, 1-9.

Balkema, A.A. and de Haan, L. (1972) On R. von Mises' condition for the domain of attraction of $\exp(-e^{-x})$. Ann.Math.Statist. 43, 1352-1354.

Bingham, N.H. and Goldie, C.M. (1982) Extensions of regular variation, I: Uniformity and quantifiers. Proc.London Math.Soc. 44, 473-496.

Cohen, J.P. (1982) Convergence rates for the ultimate and penultimate approximations in extreme-value theory. Adv.Appl.Prob. 14, 833-854.

Davis, R.A. (1982) The rate of convergence in distribution of the maxima. Statistica Neerlandica 36, 31-35.

Feller, W. (1966) An Introduction to Probability Theory and its Application, Vol.2, Wiley, New York.

Galambos, J. (1978) The Asymptotic Theory of Extreme Order Statistics. Wiley, New York.

Goldie, C.M. (1980) Slow variation with remainder. Unpublished manuscript. University of Sussex.

Gomes, M.I. (1978) Some probabilistic and statistical problems in extreme value theory. Ph.D. Thesis, University of Sheffield.

Haan, L. de (1970) On Regular Variation and its Application to the Weak Convergence of Sample Extremes. Mathematical Centre Tracts 32, Amsterdam.

Haan, L. de and Hordijk, A. (1972) The rate of growth of sample maxima. Ann.Math.Statist. 43, 1185-1196.

Korevaar, J. Van Aardenne-Ehrenfest, T., and de Bruijn, N.G. (1949) A note on slowly oscillating functions. Nieuw.Arch.Wisk. 23, 77-86.

Pickands, J. (1975) Statistical inference using extreme order statistics Ann.Statist. 3, 119-131.

Reiss, R-D. (1981) Uniform approximation to distributions of extreme order statistics. Adv. Appl. Prob. 13, 533-547.

Seneta, E. (1976) Regularly Varying Functions. Springer-Verlag Lecture Notes on Mathematics 508, Berlin.

Smith, R.L. (1982) Uniform rates of convergence in extreme-value theory. Adv.Appl.Prob. 14, 600-622.

# UNIFORM RATES OF CONVERGENCE TO EXTREME VALUE DISTRIBUTIONS

A.A. Balkema, L. de Haan, S. Resnick

This paper is intended as an introduction to the problem of finding reasonable rates of convergence to be presented at the Workshop on Rates of convergence at the Vimeiro meeting.

In 1936 von Mises gave simple sufficient conditions for the existence of a limit distribution for the sequence of partial maxima, properly normalized, of an iid sequence. These conditions are formulated in terms of the tail behaviour of the underlying distribution function $F$ and its derivatives $F'$ and $F''$.

We concentrate our attention on the case that the limit is the double exponential distribution $\Lambda(x) = \exp -e^{-x}$. Let there exist norming constants $a_n$, positive, and $b_n$, real, so that $H_n(x) := F^n(a_n x+b_n) \to \Lambda(x)$ weakly for $n \to \infty$. We are interested in inequalities of the form

$$\text{(0)} \qquad \|H_n - \Lambda\| := \sup|H_n(x) - \Lambda(x)| \leq \varepsilon \quad \text{for} \quad n \geq m.$$

For convenience we allow $n$ and $m$ to vary continuously. We define $\varepsilon_m^+$ to be the least value of $\varepsilon$ for which the inequality (0) holds. Obviously $\varepsilon_m^+$ depends on the norming constants $a_n$ and $b_n$. In this paper we shall only consider two cases. These are defined in the von Mises theorem below.



J. Tiago de Oliveira (ed.), Statistical Extremes and Applications, 341–346.

Theorem (von Mises) Let $F$ be a df with tail $R = 1 - F$. If

(1) $F'(x)/R(x) \to 1$ for $x \to \infty$,

then $G_n(x) := F^n(x+b_n) \to \Lambda(x)$ weakly, $n \to \infty$, where we choose $b_n$ so that $G_n(0) = \Lambda(0)$; if

(2) $(R(x)/F'(x))' \to 0$ for $x \to F_+$,

then $H_n(x) := F^n(a_n x+b_n) \to \Lambda(x)$ weakly for $n \to \infty$ where we choose $a_n > 0$ and $b_n$ real so that $H_n(0) = \Lambda(0) = H_n'(0) = \Lambda'(0)$.

Here $F_+ := \sup\{x \mid F(x) < 1\}$ denotes the right endpoint of $F$. We refer to (1) and to (2) as the first and second von Mises conditions. For (2) we assume that the density $F'$ is strictly positive on a left neighbourhood of $F_+$ and absolutely continuous. In extreme value theory it is reasonable to assume that the tail of the df is smooth. In practice one has to make inferences about the tail on the basis of observations over a finite range. This is only possible if there exists a simple analytical expression for the distribution function.

There is no real loss of generality in assuming that the limit distribution is $\Lambda$. If $X$ has df $F_o$ in the domain of attraction of $\Phi_1(x) = \exp -1/x$, $x > 0$, and if $F_o$ satisfies the corresponding von Mises condition $xF_o'(x)/R_o(x) \to 1$, then $\log X$ has df $F(x) = F_o(e^x)$ which satisfies (1). Define $G_n$ and $b_n$ as in the von Mises theorem. Then $G_{on}(x) := F_o^n(x.\exp b_n)$ satisfies

$$||G_{on} - \Phi_1|| = ||G_n - \Lambda||.$$

Similar remarks apply to the other possible limit distributions.

The von Mises condition (2) does not determine the complete domain of attraction of $\Lambda$, even for analytic dfs. However, the condition is simple and it comes close to being necessary (see Balkema & de Haan, 1976). It is still widely used to-day.

We define $m_\varepsilon^+$ for $\varepsilon$ positive to be the least value of $m$ for which the inequality (0) holds. Clearly $\varepsilon_m^+$ and $m_\varepsilon^+$ are non-increasing and right continuous, and $m_\varepsilon^+$ is the inverse of $\varepsilon_m^+$.

If $F$ satisfies the von Mises condition (1) then $F'$ is strictly positive on a neighbourhood of $\infty$, hence $F$ is continuous and strictly increasing, and so is $b_n$ as a function of $n$ for $n \geq n_o$. Hence the optimal value $\varepsilon_m^+$ in (0) is well-defined and continuous for $m > n_o$. The same remark applies if condition (2) holds.

In order to measure the accuracy of an upper bound $\varepsilon_n$ for $\varepsilon_n^+$, introduce the discrepancy

$$\delta_n := \log(\varepsilon_n/\varepsilon_n^+).$$

In many cases convergence in extreme value theory is slow: $\varepsilon_n^+$ is of the order of $1/(\log n)$ or of $1/(\log n)^2$ for $n \to \infty$. This has a surprising consequence: If $\varepsilon_n = a/\log n$ and $\delta_n = C\varepsilon_n$, then the discrepancy is small: $\varepsilon_n - C\varepsilon_n^2 < \varepsilon_n^+ < \varepsilon_n$, but the discrepancy $d_\varepsilon := \log(n_\varepsilon/n_\varepsilon^+)$ for the corresponding upper bound $n_\varepsilon$ tends to a positive constant! Example: If $\varepsilon_n = 1/\log n$, $\varepsilon_n^+ = 1/\log 3n$, then $\delta_n = \log(\log 3n/\log n) < \log 3/\log n$, and $d_\varepsilon = \log 3$.

One would like to have upper bounds for $\varepsilon_n^+$ (and for $n_\varepsilon^+$) which are simple, accurate, valid for a wide range of values of $m$ (or $\varepsilon$) and for a large class of distribution functions. It would also be pleasant to have bounds on the discrepancy and to be able to decrease the discrepancy by a series expansion.

These desiderata are incompatible! We aim at results which hold for $\varepsilon \leq 0.1$. In the case of one norming constant there exist reasonable bounds for such large values of $\varepsilon$. For two norming constants and for asymptotic expansions one needs the restriction $\varepsilon \leq 0.01$ to obtain interesting results. This is essential to the problem. It is due to the fact that the expression $|G_n - \Lambda|$ which we try to maximize, has two local maxima, one on the negative, and one on the positive axis. If the function $F'/R$ in the first von Mises condition tends rapidly to 1 the supremum in (0) will be realized in the local maximum on the negative axis. If the function changes slowly, the supremum is realized in the local maximum on the positive axis. (A similar phenomenon holds for the second von Mises condition.) If $F$ has a tail $R$ which not only is asymptotic to the exponential tail $e^{-cx}$ but for which $e^{cx}R(x)$ tends to 1 exponentially fast with exponent exceeding 0.328.. then the supremum will be realized on the negative axis for all $n$, leaving aside initial disturbances, and we speak of rapid convergence: $\varepsilon_n$ is of the order of a negative power of $n$. In most other cases of interest the rate of convergence in (0) will be less and for $n$ sufficiently large it is the local maximum on the positive axis which determines the value of the supremum $\|G_n - \Lambda\|$ in (0). From a practical point of view this is pleasant since we need only consider the behaviour of the tail of $F$ to the right of $b_n$ in order to compute this supremum if $n$ exceeds some constant $n_o$ depending on the distribution $F$. Unfortunately, this value of $n_o$, or rather our estimate of it, is usually quite large.

We conclude this introduction with two examples to show what happens under the best possible circumstances: if $F \equiv \Lambda$ on some neighbourhood of $\infty$.

Example (Influence of the left tail) Assume $F(x) = \Lambda(x)$ for $x \geq x_o$. Since $\Lambda^n(x+\log n) = \Lambda(x)$ we have $a_n = 1$, $b_n = \log n$ for $n \geq n_o := \exp x_o$. Hence

$$\|H_n - \Lambda\| \leq \Lambda(x_o - \log n) = \exp -n/n_o \quad \text{for} \quad n \geq n_o.$$

The bound $\exp -n/n_o$ is quite good if $x_o$ is minimal. Suppose $F(x_1) = \Lambda(x_2)$ with $x_1 \neq x_2$. Choose $x_3 < \max(x_1,x_2) \leq x_o$.

Then for $n \geq n_o$ with $n_i = \exp x_i$, $i = 0,\ldots,3$, we find

$$\|H_n - \Lambda\| \geq |F^n(x_1) - \Lambda^n(x_1)| = |\exp -n/n_2 - \exp -n/n_1|$$
$$\geq \exp -n/n_3$$

for $n$ sufficiently large. Since we may choose $x$ as close to $x_o$ as we wish, we find

$$\delta_n = -n/n_o + n/n_3 = o(n) \qquad \text{for} \quad n \to \infty.$$

The next example describes a df $F$ which is close to the limit $\Lambda$. The sequence $\|H_n - \Lambda\|$ displays the best possible asymptotic behaviour, but the difference $\|H_n - \Lambda\|$ is large for $n \approx 10^8$ and hence $\varepsilon_n^+$ is large for $n \leq 10^8$.

<u>Example</u> Assume $F$ is continuous, coincides with $\Lambda$ outside the interval $[20, 25]$ and is linear on this interval. Then $F$ is close to $\Lambda$ and $\|G_n - \Lambda\| \leq 10^{-4}$ for $n \leq 10^4$. For $n = [\exp 20]$

$$\|G_n - \Lambda\| \geq \Lambda(3) - H_n(3) > 0.25$$

since $1 - \Lambda(21) \leq 0.4(1-\Lambda(20)) \leq 1 - F(23)$ implies $G_n(23-\log n) = F^n(23) \leq \Lambda^n(21) = \Lambda(21-\log n)$ for $\log n < 20$. Thus $\varepsilon_n^+ > 0.25$ for $n \leq 10^8$.

A small disturbance in the tail causes havoc to the rate of convergence. Note that the asymptotic behaviour is good. By the example above

$$\varepsilon_n^+ \leq C_1 . \exp -n/n_o \qquad n \geq N_1 .$$

One can compute explicit bounds for the constants $C_1$ and $N_1$. Indeed $C_1 = 2$ and $N_1 = 0$ will do.

Does the sequence $H_n$ in this example behave well or badly?

The reader is warned that this is the best behaved example he will encounter.

Indeed in our example $\varepsilon_n^+$ tends to zero exponentially fast. We shall prove that faster than exponential convergence is impossible unless $\varepsilon_n \equiv 0$ identically. This settles a question raised by J. Galambos at the conference.

<u>Proposition</u> Suppose $F$ is not of double exponential type. Then there exists a constant $C > 0$ such that

$$\sup_{x \in \mathbb{R}} |F^n(A_n x + B_n) - \Lambda(x)| \geq C^n \qquad n = 1,2,\ldots$$

for all possible norming sequences $A_n > 0$, $B_n$.

<u>Proof</u> Let $U$ have df $\Lambda$. There exists a unique right continuous non-decreasing function $f$ such that $f(U)$ has df $F$. Then $h_n(U) = (f(U+\log n) - B_n)/A_n$ has df $H_n(x) = F_n(A_n x+B_n)$ and the sup above, say $D_n$, equals $\sup_{u \in \mathbb{R}} |\Lambda(h_n(u)) - \Lambda(u)|$.

By assumption $f$ is not linear: there exist points $u_1 < u_2 < u_3$ so that the horizontal distance between $(u_2,y_2)$ and the line through $(u_1,y_1)$ and $(u_3,y_3)$ is $2d > 0$ where $y_i = f(u_i)$ for $i = 1,2,3$. Hence for $A > 0$, $B \in \mathbb{R}$ one of the points $u_i$, say $\xi$, satisfies $|(f(\xi) - B)/A - \xi| \geq d$.

In particular $|h_n(\xi_n) - \xi_n| \geq \dot{d}$ for one of the three points $u_i - \log n$. Now given $d > 0$ there exists a constant $\xi_o$ depending on $d$ such that $|\Lambda(\eta) - \Lambda(\xi)| \geq \Lambda(\xi)/2$ for $\xi \leq \xi_o$ if $|\eta - \xi| \geq d$. Hence if $\log n \geq u_3 - \xi_o$ then

$$D_n \geq |\Lambda(h_n(\xi_n))-\Lambda(\xi_n)| \geq \tfrac{1}{2}\Lambda(\xi_n) \geq 2^{-n}\Lambda(u_1-\log n)$$
$$= (\Lambda(u_1)/2)^n.$$

# RATES OF CONVERGENCE IN EXTREME VALUE THEORY

Janos Galambos

Temple University, Philadelphia, PA

## ABSTRACT

If $H_n(x)$ is the exact distribution function of the maximum in a sample of size n and H(x) is a limiting distribution, one can consider $|H_n(a_n+b_nx)-H(x)|$ for fixed x, or its supremum in x on a finite interval, or on the whole real line. The paper discusses these three possibilities. The practical side of these estimates is discussed, and several open questions are raised.

## DISCUSSION

For easier reference, we reintroduce some notations of previous contributions by the author to these same Proceedings.

For a sequence $X_1, X_2, \ldots, X_n$ of random variables, we put $Z_n = \max(X_1, X_2, \ldots, X_n)$ and

$$H_n(z) = P(Z_n < z). \tag{1}$$

When the random variables $X_j$ are independent with common distribution function F(z), we speak of the classical model $\{X, F(z), n\}$. In this case,

$$H_n(z) = F^n(z). \tag{2}$$



J. Tiago de Oliveira (ed.), Statistical Extremes and Applications, 347–352.

The aim of the asymptotic theory of extremes is to find a function such that, with suitable constants $a_n$ and $b_n > 0$,

$$H_n(a_n+b_nx) \sim H(x). \tag{3}$$

In such approximation, one would like to know the accuracy of (3) for given n. Such accuracy estimates can be sought for different reasons, and the reason decides the type of estimates to be developed. We classify the reasons as follows:

(i) It could be purely mathematical curiosity, in which case pointwise convergence, uniform convergence over an interval, and uniform convergence on the whole real line are all of interest.

(ii) One may face a set of observations $X_1,X_2,\ldots,X_n$ for a given n from a given population and, for computational reasons, the approximation (2) and (3) is to be utilized, in which case an estimate of accuracy is needed at one or at a few points.

(iii) One would like to use the asymptotic theory of extremes without reference to the original population distribution; in this case speed of convergence results are needed as an assurance that the asymptotic theory is practical, but the actual estimates are never used (and they cannot be used, because the population distribution is totally ignored).

The present note is devoted to the analysis of the available results on estimating the speed of convergence in (3) for the classical models, and to the formulations of some problems whose solutions would be significant contributions to extreme value theory.

From now on, we assume (2). For given $F,n,a_n,b_n > 0$ and H, we put

$$d_n(F,H,x) = d(F,H,n,x;a_n,b_n) = |F^n(a_n+b_nx)-H(x)|. \tag{4}$$

The aim is to estimate $d_n(F,H,x)$. We put our discussion into numbered remarks and problems.

Remark 1. Note that the dependence of $d(F,H,n,x;a_n,b_n)$ on $a_n$ and $b_n$ is very significant. The clearest example for this is the case of $F(x) = H(x) = \exp(-e^{-x})$, when

$$d(H,H,n,x;\log n,1) = 0 \text{ for all } x \text{ and } n,$$

and this same difference is different from zero for any other choices of $a_n$ and $b_n$, although it converges to 0 for a large set of pairs $(a_n,b_n)$. In fact, given an arbitrary function $g(n) > 0$ as $n \to +\infty$, $a_n$ and $b_n$ can be chosen so that, for all $n \geq n_0$,

$$e^{-x}g(n) < d(H,H,n,x;a_n,b_n) < 2e^{-x}g(n).$$

Therefore, the suppression of $a_n$ and $b_n$ in $d_n(F,H,x)$ is rarely justified. Its meaning is that, given $a_n$ and $b_n$, an estimate is developed.

This now leads to our first problem.

Problem 1. When H is an extreme value distribution, meaning that it is an asymptotic distribution of properly normalized maximum in a classical model, then $a_n$ and $b_n$ can always be chosen so that $d_n(H,H,x) = 0$ for all n and x. However, if F is not an extreme value distribution, then whatever be $a_n$ and $b_n$, $d_n(F,H,x)$ cannot be zero for all n and x. The question therefore arises whether there is a function $g(n)$ to every extreme value distribution $H(x)$ such that $g(n) > 0$, $g(n) \to 0$ as $n \to +\infty$, and if, for some $a_n$ and $b_n > 0$, and for all x,

$$d_n(F,H,x) < g(n) \text{ for all } n \geq n_0,$$

it then follows that F = H. In other words, can extreme value distributions be approached arbitrarily closely, or are they "isolated" in the sense that they cannot be approached beyond a certain bound?

Remark 2. In (4), it is not required a priori that $a_n$ and $b_n$ be computed by a given rule, and neither is it a requirement that H(x) is an extreme value distribution function. We, of course, want to make $d_n(F,H,x)$ to be small (to converge to zero as $n \to +\infty$), and when we deviate from an extreme value distribution in the choice of H, we want to achieve an even better approximation to $F^n(a_n+b_nx)$ than asymptotic extreme value theory would provide. This is the aim of the studies of penultimate limiting forms for extremes by Gomes (1978, Chapter 5) and (1981), which is further extended in Gomes (1983), and in Cohen (1982 a,b). Therefore, when a general estimate is available for (4), it then includes estimates for the rate of approximation by penultimate forms.

Our second problem is in this connection.

Problem 2. The difference between the penultimate and ultimate approximations in extreme value theory is somewhat striking, but deeper analysis suggests that it might be a part of a more general prospective theory. Since $F^n(a_n+b_nx)$, for large n, is unimodal, and the shape paramater of the Weibull family of distributions makes it possible "to imitate" most (if not all) unimodal distributions, the penultimate behavior of extremes can perhaps be explained by developing a theory of approximations to unimodal distributions by a single family of distribution possessing shape parameter (Weibull, lognormal, etc). Is not it true, for example, that, at least in some cases, the lognormal distributions would provide better aprroximations to $F^n(a_n+b_nx)$ than the ultimate limiting distribution function?

Remark 3. The aims, and thus the conclusions, in cases (i) - (iii) can be different, or even outright opposite to each other. For example, in mathematics, 100/n converges to 0 faster than 1/2(loglog n)/n, say, but in practice, since the sample size n is rarely in the millions, the latter is far better than the first one for estimating the difference in (4). More seriously, the shocking (and beautiful) result of P. Hall (1979) (which is extended to all upper extremes by Reiss (1981)) says that, for the standard normal distribution function F, and with $H(x) = \exp(-e^{-x})$, the best estimate for the supremum in x of $d_n(F,H,x)$ is

$$c/\log n < \sup_x d(F,H,n,x;a_n,b_n) < 3/\log n,$$

which, from the mathematician's point of view, implies that the convergence in asymptotic extreme value theory is so slow that it is almost meaningless (the logarithm of 3.2 million is about 15). But the applied scientists have "known" that, with $n \geq 100$, the asymptotic theory is quite good even for the normal populations, (see Gumbel (1958)), which requires an explanation. It turns out that if, instead of the supremum in x, an estimate is developed for a fixed x, then the magnitude of $d_n(F,H,x)$

is d/n, where $d = d_n(x)$ depends on $a_n$ and $b_n$, but, by proper choice, d can be controlled to be "small" for all n (Galambos (1978), Section 2.10). The same "order of magnitude" is obtained as x goes over a finite interval $a < x < b$. Because of the well known properties of the tails of the normal distribution, with very large $|a|$ and b, all observations from a normal population will be between a and b, and this is why in practice better approximations are oberrved than what Hall's result would give. There is, therefore, no contradiction between theory and practice.

We can conclude: the asymptotic extreme value theory is quite accurate for the empirically determined sample size $n > 100$, which is mathematically justified by the work of Galambos (1978), Section 2.10, if the range of x in (4) is limited to a finite interval (which is just a matter of view-point in many instances). For actual computations (for given n, F, H, x, $a_n$ and $b_n$), the formulas of Galambos (1978, Section 2.10), provide the most practical values (see also Dziubdziela (1978), Gomes (1979), and, for the normal distribution, there is a more precise result in Nair (1981)). For mathematical arguments, the results of P. Hall (1979) and (1980), W.J. Hall and J.A. Wellner (1979), Reiss (1981), Davies (1982) and Smith (1982) can be utilized, whenever the population distribution is covered by the prospective reference.

Problem 3. As was pointed out in the previous contributions of the author to these same Proceedings, the way in which asymptotic extreme value theory is applied is as follows. Since the theory says that, with some constants $a_n$ and $b_n > 0$, (2) and (3) apply, H(x) is assumed to be the exact distribution of $Z_n$, whose parameters are estimated from observations on $Z_n$ but by methods available for H(x). Therefore, the problem of speed of convergence in (4) becomes the problem of estimating $d_n(F,H,x)$ when the suppressed constants $a_n$ and $b_n$ are random variables, namely, some statistics of observations on $Z_n$(such as point estimators of the parameters of H even though the observations follow $F^n$ with some F). In this direction, there is not a single investigation.

References.

Cohen, J.P. (1982a). The penultimate form of approximation to normal extremes. Adv. Appl. Prob. 14, 324-339.
Cohen, J.P. (1982b). Convergence rates for the ultimate and penultimate approximations in extreme-value theory. Adv. Appl. Prob. 14, 833-854.
Davies, R.A. (1982). The rate of convergence in distribution of the maxima. Statist. Neerlandica 36, 31-35.
Dziubdziela, W. (1978). On convergence rates in the limit laws for extreme order statistics. Transactions of the Seventh Prague Conference, Vol. B, 119-127.
Galambos, J. (1978). The asymptotic theory of extreme order statistics. Wiley, New York.
Gomes, M.I. (1978). Some probabilistic and statistical problems in extreme value theory. Thesis for Ph.D., Univ. of Sheffield.
Gomes, M.I. (1979). Rates of convergence in extreme value theory. Rev. Univ. Santander 2, 1021-1023.
Gomes, M.I. (1981). Closeness of penultimate approximations in extreme value theory. 14th European Meeting of Stat., Wroclaw.
Gomes, M.I. (1983). Penultimate limiting forms in extreme value theory, Ann. Inst. Statist. Math. (to appear).
Gumbel, E.J. (1958). Statistics of extremes. Columbia University Press, New York.
Hall, P. (1979). On the rate of convergence of normal extremes. J. Appl. Prob. 16, 433-439.
Hall, P. (1980). Estimating probabilities for normal extremes. Adv. Appl. Prob. 12, 491-500.
Hall, W.J. and J.A. Wellner (1979). The rate of convergence in law of the maximum of an exponential sample. Statist. Neerlandica 33, 151-154.
Nair, K.A. (1981). Asymptotic distribution and moments of normal extremes. Ann. Prob. 9, 150-153.
Reiss, R.-D. (1981). Uniform approximation to distributions of extreme order statistics. Adv. Appl. Prob. 13, 533-547.
Smith, R.L. (1982). Uniform rates of convergence in extreme value theory. Adv. Appl. Prob. 14, 600-622.

# CONCOMITANTS IN A MULTIDIMENSIONAL EXTREME MODEL

M. Ivette Gomes

Departamento de Estatística e
Centro de Estatística e Aplicações
da Universidade de Lisboa

ABSTRACT. Ordering of multivariate data is referred, with special emphasis to induced ordering, where the order statistics of a marginal sample together with their concomitants are considered.

For a particular multidimensional extreme model, useful in statistical analysis of extremes, the importance of concomitants is discussed. Computational problems related to the best linear estimators of unknown parameters based on the multivariate sample of order statistics of largest values and concomitants are put forward. A brief account of asymptotic theory of extreme concomitants in this particular model is given.

## 1. INTRODUCTION

We are here mainly interested in an <u>extremal multidimensional</u> model [Gomes, 1978, Weissman, 1978]: $\overline{(\underset{\sim}{Z}_1, \underset{\sim}{Z}_2, \ldots, \underset{\sim}{Z}_n)}$ is a multivariate sample of size $n$, where $\underset{\sim}{Z}_j$, $1 \leq j \leq n$, are independent $i_j$-vectors, $1 \leq j \leq n$, the joint probability density function(p.d.f.) of $\underset{\sim}{Z}_j = (Z_{1,j}, \ldots, Z_{i_j,j})$ being

$$h_{k,i_j}\left( (z_{1,j}-\lambda)/\delta, \ldots, (z_{i_j,j}-\lambda)/\delta\right)/\delta^{i_j}$$

where $\lambda$ and $\delta$ are unknown location and scale parameters



J. Tiago de Oliveira (ed.), Statistical Extremes and Applications, 353–364.

respectively. The joint p.d.f. $h_{k,i_j}$ - the extremal $i_j$-dimensional p.d.f. -, is the joint p.d.f. corresponding to the non-degenerate joint limiting distribution function (d.f.), whenever it exists of the $i_j$ top order statistics (o.s.), suitably normalized, of a random independent sample of size $m$, as $m \to \infty$.

We thus have

$$h_{k,i_j}(z_1,\dots,z_{i_j}) = g_k(z_{i_j}) \prod_{m=1}^{i_j-1} \left[g_k(z_m)/G_k(z_m)\right] \qquad \text{if } z_1>z_2>\dots>z_{i_j} \tag{1.1}$$

where $G_k(z) = \exp(-(1+kz)^{-1/k})$, $1+kz>0$, $k\neq 0$, $G_0(z)=\exp(-\exp(-x))$ $x\in \mathbb{R}$, the standard von Mises-Jenkinson d.f., is a stable d.f. for maxima, $g_k(z)$ denoting its derivative.

For univariate samples the concept of o.s. plays an important role in statistical inference and is defined clearly and unambiguously. For multivariate samples we find in the literature different generalizations of the concept of order - reduced ordering, partial ordering, marginal ordering, induced ordering and so on. A good critical review of the subject may be found in Barnett (1976). The fact that the different vectors in the model we are dealing with may have different dimensions lead us to consider the concept of induced or conditional ordering, i.e.

1. Order the random sample of the first components $Z_{1,j}$, $1\leq j\leq n$ (the largest components in this particular model).

2. Do not modify the random vectors $\underset{\sim}{Z}_j$, $1\leq j\leq n$, merely order them according to the ordering of the first components. We thus obtain what has been called concomitants of o.s. (David, 1973) or induced o.s. (Bhattacharya, 1974).

We shall use the terminology introduced by David (1973) in his study on normal bivariate samples. Observe that concomitants had been previously used by Watterson (1959) in the estimation of the unknown parameters in a binormal sample and by Conover (1965), when considering the ordering of $r$ random samples of size $n$, mutually independent, according to the ordering of the maximum values in each of the $r$ samples. Concomitants, either in binormal or in general models, have been studied in a great variety of ways, by David and Galambos (1974), O'Connell (1974), Barnett et al (1976), O'Connell and David (1976), Sen (1976), David et al (1977), Yang (1977), Gomes (1978,1981) and Yang (1981). It is still worth mentioning the review papers by David (1982) and Bhattacharya (1983). Development of the theory of concomitants may also be found in Galambos (1978, pp. 267-270) and in chapters 5 and 9 of David (1981).

Although some exact and asymptotic distributional theory is developed in § 3. for the multivariate sample of o.s. of largest values and their concomitants, our main purpose is the investigation in § 4.of its second order structure and the use of such a sample in statistical inference related to the multidimensional extremal model. In § 2. we shall make a brief review of the distributional theory of concomitants in a linear model.

## 2. A BRIEF REVIEW OF THE DISTRIBUTIONAL THEORY OF CONCOMITANTS IN A LINEAR MODEL

Let us now assume that we are dealing with random pairs $(X_j,Y_j)$, $1\le j\le n$, related by the usual model of linear regression.

$$Y_j = \mu_y + \rho\sigma_y \frac{X_j-\mu_X}{\sigma_X} + Z_j, \quad 1\le j\le n, |\rho|<1 \tag{2.1}$$

where $\rho=\mathrm{Corr}(X,Y)$, $\mu_X=E[X]$, $\mu_y=E[Y]$, $\sigma^2_X=\mathrm{Var}(X)$ and $\sigma^2_y=\mathrm{Var}(Y)$. The r.v.'s $Z_j$, $1\le j\le n$, independent of the $X_j$, $1\le j\le n$, are i.i.d. r.v.'s with zero mean and variance equal to $\sigma^2_y(1-\rho^2)$ - in the bivariate normal situation the $Z_j$, $1\le j\le n$, are obviously normal r.v.'s.

We then consider the associated bivariate sample $(X_{j:n},Y_{[j:n]})$, $1\le j\le n$, of the o.s. of the marginal sample $X_j$, $1\le j\le n$, and their concomitants. Notice that we are ordering the X's in descending order of magnitude. We obviously have

$$Y_{[j:n]} = \mu_y+\rho\, \sigma_y\cdot \frac{X_{j:n}-\mu_X}{\sigma_X} + Z_{[j]}, \quad 1\le j\le n \tag{2.2}$$

where $Z_{[j]}$, $1\le j\le n$, are i.i.d. r.v.'s, independent of

$X_{j:n}$, $1 \le j \le n$.

The exact distribution of the concomitants is thus easily derived from (2.2) - for the bivariate normal situation, results on this direction were obtained by O'Connell (1974).

We are now mainly interested in the asymptotic behaviour of $Y_{[j:n]}$, and let us put $\mu_X=\mu_y=0$, $\sigma_X=\sigma_Y=1$. Assume further that $F(x,y)$ the joint d.f. of $(X,Y)$ is absolutely continuous.

Then (David(1982)), if we are working with central concomitants, i.e., if $j/n \underset{n\to\infty}{\to} \lambda, 0<\lambda<1$, and if $f_X(F^{-1}(\lambda))>0$, where $f_X(x) = F'_X(x) = F'(x,+\infty)$, $X_{j:n} \underset{n\to\infty}{\overset{P}{\to}} F_X^{-1}(\lambda)$ and consequently $Y_{[j:n]} - \rho F_X^{-1}(\lambda) \underset{n\to\infty}{\overset{D}{\to}} Z$, independently of $j$.

In what regards extremal concomitants $Y_{[j:n]}$ - associated for instance to large o.s. -, i.e., such that $j/n \to 0$, as $n \to\infty$, we shall merely deal with the situation in which $F_X$ belong to the domain of attraction (for maximum values) of $G_k(x)$, $K \in \mathbb{R}$ - -which we shall denote as usual by $F \in \mathcal{D}(G_k)$.

1. Assume $k>0$, i.e., $F_X \in \mathcal{D}(\phi_{1/k})$ with attraction coefficients $\{a_n, b_n\}_{n\ge 1}$ $(a_n>0)$, $\phi_\alpha(x) = \exp(-x^{-\alpha})$, $x>0$, the Fréchet d.f. We may then choose $b_n=0$, $a_n = F_X^{-1}(1-1/n) \to\infty$, as $n\to\infty$. We have $Z/a_n \underset{n\to\infty}{\overset{P}{\to}} 0$ and consequently the asymptotic behaviour of $Y_{[j:n]}/a_n\rho$ is the same than the asymptotic behaviour of $X_{j:n}/a_n$.

2. If we assume $k<0$, i.e., $F_X \in \mathcal{D}(\psi_{1/|k|})$, $\psi_\alpha(x) = \exp(-(-x)^\alpha)$, $x<0$, the Weibull d.f., then $x_o^F = \sup\{x: F_X(x)<1\}<\infty$ and $X_{j:n} \underset{n\to\infty}{\overset{P}{\to}} x_o^F$. Consequently $Y_{[j:n]} - \rho x_o^F \underset{n\to\infty}{\overset{D}{\to}} Z$, once again independently of $j$.

3. Finally if we assume $k=0$, i.e., if $F_X \in \mathcal{D}(\Lambda)$, with

attraction coefficients $\{a_n, b_n\}_{n\geq 1}$ $(a_n>0)$, $\Lambda(x) = \exp(-\exp(-x))$, $x \in \mathbb{R}$, the Gumbel d.f., we distinguish the following situations:

a) If the attraction coefficients $a_n$ are such that $a_n \to a$, as $n \to \infty$ $0<a<\infty$, the convergence of $F^n(a_n x + b_n)$ to $\Lambda(x)$ is quick; then $X_{j:n} - b_n \overset{D}{\underset{n\to\infty}{\to}} U_j$, with known d.f. related to $\Lambda$, and consequently $Y_{[j:n]} - \rho b_n \overset{D}{\underset{n\to\infty}{\to}} \rho\, U_j * Z$.

b) If the attraction coefficients $a_n$ are such that either $a_n \underset{n\to\infty}{\to} \infty$ or $a_n \underset{n\to\infty}{\to} 0$, the convergence of $F^{n^n}(a_n x + b_n)$ towards $\Lambda(x)$ is rather slow. In case $a_n \underset{n\to\infty}{\to} \infty$, $Z/a_n \overset{p}{\underset{n\to\infty}{\to}} 0$ and the asymptotic behaviour of $\{Y_{[j:n]} - \rho b_n\}/\rho\, a_n$ coincides with the asymptotic behaviour of $(X_{j:n} - b_n)/a_n$. If $a_n \underset{n\to\infty}{\to} 0$, $X_{j:n} - b_n \overset{p}{\underset{n\to\infty}{\to}} 0$, and consequently $Y_{[j:n]} - \rho\, b_n \overset{D}{\underset{n\to\infty}{\to}} Z$.

Notice finally that there is a real difference between the behaviour of the o.s. and the concomitants both in what concerns the normalizing constants necessary for convergence towards a non-degenerate law and in what concerns the limit law itself.

## 3. A BRIEF ACCOUNT ON DISTRIBUTIONAL THEORY OF CONCOMITANTS IN A REDUCED EXTREMAL MODEL

Let us consider the model previously defined and let us denote by $\underset{\sim}{X}_j = (X_{1,j}, \ldots, X_{i_j,j})$, $1 \leq j \leq n$, $\underset{\sim}{X}_j = (\underset{\sim}{Z}_j - \lambda \underset{\sim}{1}_j)/\delta$, $1 \leq j \leq n$, $\underset{\sim}{1}_j$ a column vector of dimension $i_j$ with unit components, the <u>reduced multidimensional extremal model</u>. For the sake of simplicity let us put $i_j = 2$, $1 \leq j \leq n$, $X_j$ instead of $X_{1,j}$ and $Y_j$ instead of $X_{2,j}$, $1 \leq j \leq n$. We are thus dealing with pairs $(X_j, Y_j)$, $1 \leq j \leq n$, and the associated bivariate sample $(X_{j:n}, Y_{[j:n]})$, $1 \leq j \leq n$.

The distributional theory of the concomitants in this reduced model was developed by Gomes (1981). Since the pairs $(X_j, Y_j)$, $1 \leq j \leq n$, are independent, identically distributed (i.i.d.), we have

$$f_{Y_{[r:n]}|X_{r:n}}(y|x) = f_{Y|X}(y|x) = g_k(y)/G_k(x), \; y<x . \qquad (2.1)$$

With $f_{r:n}(x) = f_{X_{r:n}}(x) = r\binom{n}{r} G_k^{n-r}(x) \left[1-G_k(x)\right]^{r-1} g_k(x)$, we have

$$f_{Y_{[r:n]}}(y) = g_k(y) \int_y^{+\infty} \frac{f_{r:n}(x)}{G_k(x)} dx, \qquad (2.2)$$

and, with $f_{r,s:n}(x,y) = f_{(X_{r:n}, X_{s:n})}(x,y) = r(s-r)\binom{n}{r}\binom{n-r}{s-r}\left[1-G_k(x)\right]^{r-1}$ $\left[G_k(x)-G_k(y)\right]^{s-r-1} G_k^{n-s}(y)\, g_k(x)\, g_k(y)$ if $x>y$, $1 \leq r<s \leq n$, we also have, for $1 \leq r<s \leq n$,

$$f_{(X_{r:n}, Y_{[s:n]})}(x,y) = g_k(y) \int_y^x \frac{f_{r,s:n}(x,t)}{G_k(t)} dt, \; x>y \qquad (2.3)$$

$$f_{(X_{s:n}, Y_{[r:n]})}(x,y) = g_k(y) \int_{\max(x,y)}^{+\infty} \frac{f_{r,s:n}(t,x)}{G_k(t)} dt \qquad (2.4)$$

$$(x,y) \in \mathbb{R}^2$$

and

$$f_{(Y_{[r:n]}, Y_{[s:n]})}(x,y) = g_k(x) g_k(y) \iint_{\substack{u>v>y \\ u>\max(x,y)}} \frac{f_{r,s:n}(u,v)}{G_k(u) G_k(v)} du\, dv \qquad (2.5)$$

$$(x,y) \in \mathbb{R}^2.$$

The most interesting asymptotic property in this reduced extremal model is as follows: taking into account the joint d.f.

of $(X,Y)$, $f_{(X,Y)}(x,y)=g_k(x)\ g_k(y)/G_k(x)$ if $x>y$, we have $f_X(x)=g_k(x)$ and $f_Y(y)=g_k(y)\ [-\ln G_k(y)]$, and consequently $U_j=-\ln G_k(X_j)$, $1\leq j\leq n$, are i.i.d. standard exponential random variables (r.v.'s) and $V_j=-\ln G_k(Y_j)$, $1\leq j\leq n$, are i.i.d. gamma r.v.'s. with shape parameter equal to 2. We thus have a linear model,

$$V_j = U_j+W_j,\quad 1\leq j\leq n,$$

the $W_j$'s independent standard exponential r.v.'s, independent of $U_j$, $1\leq j\leq n$; hence

$$-\ln G_k(Y_{[j:n]}) = U_{n-j+1:n} + W_{[j]},\quad 1\leq j\leq n,$$

where $U_{n-j+1:n}$ is the j-th ascending o.s. associated to a standard exponential random sample of size $n$, and $W_{[j]}, 1\leq j\leq n$, are i.i.d. standard exponential r.v.'s, independent of $U_{n-j+1:n}$, $1\leq j\leq n$.

Since $U_{n-j+1:n}$ converges in probability, as well as almost surely as $n\to\infty$, to the left end point zero, we have that, for fixed j, $Y_{[j:n]}$ converges in distribution, as $n\to\infty$, to a non-degenerate r.v. with d.f. $G_k(x)$, the von Mises-Jenkinson standard d.f.

Observe that, for the Gumbel case ($k=0$), this result follows from Galambos (1978, Th. 5.5.1).

## 4. SECOND ORDER STRUCTURE; LINEAR ESTIMATION OF UNKNOWN PARAMETERS

We now investigate the second order structure of $\{\underset{\sim}{X}^0_n\}_{n\geq 1} = \{X_{j:n}, 1\leq j\leq n, Y_{[j:n]}, 1\leq j\leq n\}_{n\geq 1}$; we shall denote by $\underset{\sim}{\mu}_{k,n}$ and $\Sigma_{k,n}$, the column vector of the mean values and the covariance matrix respectively, of $\underset{\sim}{X}^0_n$, in the general von Mises-Jenkinson reduced extremal multidimensional model, with parameter $k$. We have

$$\underset{\sim}{\mu}_{k,n} = \begin{bmatrix} \underset{\sim}{\mu}^{(1)}_{k,n} \\ \\ \underset{\sim}{\mu}^{(2)}_{k,n} \end{bmatrix} \qquad \Sigma_{k,n} = \left[\begin{array}{c|c} \Sigma^{(1)}_{k,n} & \Sigma^{(c)}_{k,n} \\ \hline \Sigma^{(c)}_{k,n} & \Sigma^{(2)}_{k,n} \end{array}\right],$$

where $\underset{\sim}{\mu}^{(1)}_{k,n}$, $\Sigma^{(1)}_{k,n}$ are associated to the o.s. and $\underset{\sim}{\mu}^{(2)}_{k,n}$, $\Sigma^{(2)}_{k,n}$ associated to the concomitants.

The first algorithm presented in Gomes (1978) for the computation of $\underset{\sim}{\mu}_{0,n}$ and $\Sigma_{0,n}$ involves the use of direct formulas for the mean values, variances and covariances. However, even in the easiest situation, i.e. when computing the mean value of the r-th descending o.s., we have

$$E[X_{r:n}] = r \binom{n}{r} \sum_{m=0}^{r-1} \binom{r-1}{m} (-1)^{r-1-m} \theta_{1,k}(n-m), \quad 1 \leqq r \leqq n,$$

where

$$\theta_{1,k}(\alpha) = \int_R x\, G_k^{\alpha-1}(x)\, g_k(x)\, dx = \frac{\alpha^k \Gamma(1-k) - 1}{k\, \alpha} \quad \text{if} \quad k<1,$$

where $\Gamma(.)$ denotes, as usual, the complete gamma function; this sum contains terms large in magnitude and alternating in sign. This causes rounding errors and the accuracy of the results is poor. These rounding errors get worse when we compute $\underset{\sim}{\mu}^{(2)}_{0,n}$ and $\Sigma_{0,n}$ directly.

In the particular case $k=0$ (Gumbel model), such difficulties were slightly overcome in Gomes (1978,1981): a second algorithm, using recurrence relations, provided higher precision and the number of independent computations was considerably reduced. For a general parameter $k$, in von Mises-Jenkinson d.f., we thus do not intend to use the first algorithm, but the second one.

For the computation of $E[X_{r:n}]$, we have $E[X_{r+1:n}] = \{nE[X_{r:n-1}] - (n-r)E[X_{r:n}]\}/r$, $1 \leqq r \leqq n-1$, $n \geqq 2$, with the initial condition

$$E\left[X_{1:n}\right]=n\ \theta_{1,k}(n),\ k<1,\ n\geq 1.$$

Taking into account (2.2), this same recurrence relation remains valid for the computation of $E\left[Y_{[r:n]}\right]$, $1\leq r\leq n-1$, $n\geq 2$, with the initial conditions

$$E\left[Y_{[1:n]}\right]=\begin{cases}\left[\Gamma(2-k)-1\right]/k & \text{if } k<2,\ n=1\\ n\{\left[\Gamma(1-k)-1\right]/k-\theta_{1,k}(n)\}/(n-1) & \text{if } k<1,\ n>1.\end{cases}$$

The same recurrence relation holds true, as well, for the computation of the mean value of any power of $X_{r:n}$ or $Y_{[r:n]}$, $1\leq r\leq n-1$, $n\geq 2$, i.e. the computation of the diagonal elements of $\Sigma_{k,n}$, as well as for the computation of the diagonal elements of $\Sigma_{k,n}^{(c)}$. The initial values

$$E\left[X_{1:n}^2\right]=n\ \theta_{2,k}(n)\quad \text{if}\quad k<1/2$$

$$E\left[Y_{[1:n]}^2\right]=\begin{cases}\left[\Gamma(2-2k)-2\Gamma(2-k)+1\right]/k^2 & \text{if } k<1,\ n=1\\ n\{\left[\Gamma(1-2k)-2\Gamma(1-k)+1\right]/k^2-\theta_{2,k}(n)\}/(n-1) & \text{if}\end{cases}$$

$k<1/2$, $n>1$, where

$$\theta_{2,k}(\alpha)=\int_R x^2 G_k^{\alpha-1}(x)\ g_k(x)\ dx=\left[\alpha^{2k}\Gamma(1-2k)-2\alpha^k\Gamma(1-k)+1\right]/$$

$/(k^2\alpha)$ if $k<1/2$, and

$$E\left[X_{1:n}Y_{[1:n]}\right]=\begin{cases}\{\left[\Gamma(2-2k)-(2-k)\Gamma(2-k)\right]/(1-k)+1\}/k^2 & \text{if } k<1,\ n=1\\ n\ A_k(1,n-1) & \text{if } k<1,\ n>1,\end{cases}$$

are straightforward generalizations of the results in Gomes (1981). The function $A_k(\alpha,\beta)$, a generalization of the function $A(\alpha,\beta)=$ $=\lim_{k\to 0} A_k(\alpha,\beta)$, computed in Gomes (1981), is defined by

$$A_k(\alpha,\beta) = \iint_{x>y} xy\, G_k^{\beta-1}(x)\, G_k^{\alpha-1}(y) g_k(x) g_k(y)\, dxdy$$

$$=\{(\alpha\beta)^{k-1}\Gamma(2-2k)\, I_{\beta/(\alpha+\beta)}(1-k,1-k)-\Gamma(1-k)\left[(\beta-\alpha)(\alpha+\beta)^{k-1}\right.$$

$$\left.+\alpha^k\right]/(\alpha\beta)+1/(\alpha(\alpha+\beta))\}/k^2 \quad \text{if} \quad k<1,$$

where $I_x(a,b) = \left[\Gamma(a+b)/(\Gamma(a)\Gamma(b))\right] \int_0^x t^{a-1}(1-t)^{b-1}dt$ denotes, as usual, the incomplete beta function. We have $A_k(\alpha,\beta)+A_k(\beta,\alpha)= =\theta_{1,k}(\alpha)\theta_{1,k}(\beta)$ and consequently $A_k(\alpha,\alpha) =\theta_{1,k}^2(\alpha)/2$; the number of independent computations is then reduced.

For the computation of the remaining elements of $\Sigma_{k,n}^{(1)}$, it is advisable to use the recurrence relation $E\left[X_{r+1:n}\, X_{s+1:n}\right]= =\{nE\left[X_{r:n-1}\, X_{s:n-1}\right]-(s-r)E\left[X_{r:n}\, X_{s+1:n}\right]-(n-s)E\left[X_{r:n}\, X_{s:n}\right]\}/r$, $1\leq r<s\leq n-1$, $n\geq 2$, which, taking into account (2.3), (2.4} and (2.5) remains valid for the computation of the elements of $\Sigma_{k,n}^{(2)}$ and $\Sigma_{k,n}^{(c)}$ not yet computed. Here, the function $B(\alpha,\beta)$ in Gomes (1981) must be replaced by

$$B_k(\alpha,\beta) = \iint_{x>y} xy\left[-\ln G_k(x)\right] G_k^{\beta-1}(x)\, G_k^{\alpha-1}(y) g_k(x) g_k(y)\, dxdy$$

$$=\{(1-k)\beta A_k(\alpha,\beta)-\theta_{1,k}(\alpha)+\theta_{1,k}(\alpha+\beta)-\beta\left[(1-2k)\theta_{2,k}(\alpha+\beta)-\right.$$

$$\left.2\theta_{1,k}(\alpha+\beta)\right]/(\alpha+\beta)\}/\beta^2 \quad \text{if} \quad k<1/2,$$

and the initial conditions, although cumbersome, are an easy generalization of the formulas presented in Gomes (1981). Detailed results on the subject will be presented elsewhere, and such results will then be used in linear estimation of the three parameters $(\lambda,\delta,k)$.

Here, we turn back to the particular case $k=0$. The concomitants of o.s. may have a direct application in this location/

/scale dependent model. Lloyd's (1952) matricial approach to linear estimation through o.s. may be easily generalized to the multivariate samples we are dealing with. This is done taking into account the random vector $\underset{\sim}{Z}_n^0$ of the o.s. of largest values and concomitants associated to our sample $\underset{\sim}{Z}_j$, $1 \leq j \leq n$.

We thus get for B.L.U.E. of $\underset{\sim}{\theta}=\begin{pmatrix}\lambda\\ \delta\end{pmatrix}$, $\underset{\sim}{\theta}^* = (P_0' \Sigma_{0,n}^{-1} P_0)^{-1} P_0' \Sigma_0^{-1} \underset{\sim}{Z}_n^0$, where $P_0 = [\, \underset{\sim}{1} \; \underset{\sim}{\mu}_{0,n} \,]$, $\underset{\sim}{1}$ a column vector of size $2n$ and unit elements, $\underset{\sim}{\mu}_{0,n}$ and $\Sigma_{0,n}$ obtained previously.

The difficulties pointed out in Gomes (1981) with the second algorithm mentioned, where for sample sizes $n \geq 18$, in a CDC 7600, the matrix $\Sigma_{0,n}$ becomes "ill-conditioned", lead us to the use of numerical integration to obtain the initial values needed for the use of the recurrence relations.

The computation of the integrals which provide those initial mean values is thus performed numerically, after suitable variables transformations, by Gaussian quadrature techniques.

The partial results obtained so far indicate that this approach provides better and more reliable results than the ones obtained previously. Details regarding this approach will be presented elsewhere.

## REFERENCES


Barnett, V. (1976), J. Royal Statist. Soc. A 139, pp. 318-354.

Barnett, V., Green, P.J. and Robinson, A. (1976), Biometrika 63, pp. 323-328.

Bhattacharya, P.K. (1974), Ann. Statist. 2, pp. 1034-1039.

Bhattacharya, P.K. (1983), In Handbook of Statistics, vol. 4, P.R. Krishnaiah and P.K. Sen, eds., North-Holland, Amsterdam.

Conover, W.J. (1965), Ann. Math. Statist. 36, pp. 502-506.

David, H.A. (1973), Bull. Intern. Statist. Inst. 45, pp. 295-300.

David, H.A. (1981), Order Statistics. 2nd. ed., Wiley, New York.

David, H.A. (1982), In Some Recent Advances in Statistics, J. Tiago de Oliveira and B. Epstein, eds., Academic Press.

David, H.A. and Galambos, J. (1974), J. Appl. Probab. 11, pp. 762--770.

David, H.A., O'Connell, M.J. and Yang, S.S. (1977),Ann. Statist. 5, pp. 216-223.

Galambos, J. (1978), The Asymptotic Theory of Extreme Order Statistics. Wiley, New York.
Gomes, M.I. (1978),Some Probabilistic and Statistical Problems in Extreme Value Theory. Ph. D. Thesis, Univ. Sheffield.
Gomes, M.I. (1981),In Statistical Distributions in Scientific Work, vol. 6, pp. 389-410, C. Taillie et al, eds., D. Reidel, Dordrecht.
Lloyd, E.H. (1952),Biometrika 39, pp. 88-95.
O'Connell, M.J. (1974), Theory and Applications of Concomitants of Order Statistics. Ph.D. Thesis, Iowa.
O'Connell, M.J. and David, H.A. (1976),In Essays on Probability and Statistics, in Honor of J. Ogawa, pp. 451-466, S. Ikeda et al eds., Tokyo.
Sen, P.K. (1976),Ann. Probab. 4, pp. 474-479.
Watterson, G.A. (1959),Ann. Math. Statist. 30, pp. 814-824.
Weissman, I. (1978),J. Amer. Statist. Assoc. 73, pp. 812-815.
Yang, S.S. (1977),Ann. Statist. 5, pp. 996-1002.
Yang, S.S. (1981),J. Amer. Statist. Assoc. 76, pp. 658-662.

# A SHORT CUT ALGORITHM FOR OBTAINING COEFFICIENTS OF THE BLUE'S

Arne Fransén

National Defence Research Institute
Division 1, Section 123
P. O. Box 27322
S-102 54 Stockholm - Sweden

ABSTRACT. Theoretically there is no problem in obtaining coefficients of the simultaneous BLUE's of location and scale parameters for a known distribution, given the values of the means, variances and covariances of the order statistics. A matrix formula is readily available. The purpose of this paper is to show how this formula can be treated in practice for computer calculations.

KEY WORDS. Algorithm, Best linear unbiased estimate.

## 1. INTRODUCTION

Though the matrix formulae discussed in this paper are relevant for many other distributions I will consider chiefly the application for the Gumbel one (for maxima), i.e. that one proposed by Fisher and Tippett in 1928 which became known as the Extreme value distribution of Type I. When solving the Estimation problem for this distribution in the classical way using the maximum-likelihood method we are forced to use some iterative method, e.g. the one by Newton and Raphson. Of course,this is only a minor problem, but, nevertheless, we seek an alternative method. For simplicity we seek a linear method. Using order statistics and the theorem by Gauss and Markoff we are led to the simultaneously best linear (unbiased) estimates (BLUE's) of the location and scale paraters.

Given values of the means, variances and covariances of the order statistics we can produce the coefficients of these BLUE's simply by calculating a matrix expression giving the least-squares



*J. Tiago de Oliveira (ed.), Statistical Extremes and Applications, 365–371.*

solution. For the Gumbel distribution we can calculate the means and variances exactly and the covariances by additional use of a certain special function. Surely we will have precision problems for larger sample sizes n. In my report from 1971 I gave values up to $n = 19$ and in the one from 1974 up to $n = 31$. Using recurrence formulae in a most clever way we would have to start by roughly 75 digits to obtain 20 correct ones in the resulting values for $n = 100$. This suggests that we ought to have access to some high-precision software package for a conventional computer.

One reason why we could use the BLUE's instead of the maximum-likelihooh estimates is their comparatively good efficiency. (See my report from 1974.) Because of the difficulties in obtaining the covariances, approximate methods such as Blom's method from 1956 giving nearly BLUE's (NBLUE's) were frequently used. A table of comparison of several methods is given in my Swedish report from 1975. (See also that report for an extensive list of references.) The efficiency for the NBLUE's is quite good, too. However, to get this good efficiency we must of course have values of the coefficients not suffering from loss in precision.

In this paper I will mainly treat the uncensored case producing the coefficients of the BLUE's. At first we will assume that the covariances are explicitly given. In most practical cases, however, these values will be calculated from the mixed moments around zero. So we will alter our algorithm to cover that case.

## 2. PRELIMINARIES

Let X be a random variable and denote the corresponding reduced one by T for which the cdf F(t) is known. Denote the unknown location and scale parameters by $\lambda$ and $\delta$ respectively. An ordered random sample $x_1 \leq x_2 \leq \ldots \leq x_n$ is available. We seek the coefficients of the BLUE's $\hat{\lambda} = \underline{c}^T \underline{x}$ and $\hat{\delta} = \underline{d}^T \underline{x}$ having introduced the vectors $\underline{x} = (x_i)$, $\underline{c} = (c_i)$ and $\underline{d} = (d_i)$ gathering the latter two in the matrix $\underline{K} = (\underline{c}\ \underline{d})$. We introduce the vector of the means $\underline{m} = (E(T_{(i)}))$ and the matrix $\underline{A} = (E(T_{(i)}T_{(k)}))$ of the mixed moments around zero for the order statistics $T_{(i)}$ and further the vector $\underline{g} = (1)$, thus consisting purely of ones, denoting the matrix $\underline{H} = (\underline{g}\ \underline{m})$. Finally, denote the covariance matrix of the $T_{(i)}$'s by $\underline{M} = \underline{A} - \underline{mm}^T$.

We have chosen the sample size n and thus the order of the

matrices and vectors introduced is n. For simplicity, let us denote the elements of $\underline{A} = (a_{ik})$, $\underline{M} = (s_{ik})$ and $\underline{m} = (m_i)$. The matrix formula for the solution $\underline{K}$ given by Lloyd in 1952 and also in my report from 1971 is

$$\underline{K} = \underline{M}^{-1}\underline{H}(\underline{H}^T\underline{M}^{-1}\underline{H})^{-1}. \tag{1}$$

Introducing the covariance matrix of $\hat{\lambda}$ and $\hat{\delta}$ as $\delta^2\underline{E}$ we get

$$\underline{E} = \underline{K}^T\underline{M}\underline{K} = (\underline{H}^T\underline{M}^{-1}\underline{H})^{-1}. \tag{2}$$

Considering the matrix $\underline{E}^{-1}$ which is of order 2 we denote its elements by

$$P = \underline{g}^T\underline{M}^{-1}\underline{g},\ Q = \underline{g}^T\underline{M}^{-1}\underline{m} = \underline{m}^T\underline{M}^{-1}\underline{g} \text{ and } R = \underline{m}^T\underline{M}^{-1}\underline{m}\ . \tag{3}$$

We get its determinant

$$D = PR - Q^2, \tag{4}$$

and the elements of the inverse $\underline{E}$ denoted by

$$C_n = R/D,\ \ \tfrac{1}{2}B_n = -Q/D \text{ and } A_n = P/D, \tag{5}$$

which are useful when calculating the efficiency.

I refer to my above mentioned reports for further investigations and will in this paper concentrate on the matrix formula (1). Denoting further the vectors

$$\underline{x} = \underline{M}^{-1}\underline{g} \text{ and } \underline{y} = \underline{M}^{-1}\underline{m}, \tag{6}$$

we finally get

$$\underline{K} = (\underline{x}\ \underline{y})\underline{E} = (C_n\underline{x} + \tfrac{1}{2}B_n\underline{y}\quad \tfrac{1}{2}B_n\underline{x} + A_n\underline{y})\ . \tag{7}$$

Thus there remains simply the problem of inverting the covariance matrix. Of course, this is not done explicitly due to the loss in precision.

A symmetric and positive definite matrix such as $\underline{M}$ can always be factored as it stands by means of a numerically stable procedure, viz. the Cholesky factorization method. Let $\underline{R}$ denote an upper triangular matrix and we will get

$$\underline{M} = \underline{R}^T\underline{R} \text{ and } \underline{M}^{-1} = \underline{R}^{-1}\underline{R}^{-T}. \tag{8}$$

A small problem in this method is that we have to calculate n square roots. We can avoid that by using another decomposition of $\underline{M} = \underline{U}^T\underline{DU}$, where $\underline{U}$ is upper triangular and $\underline{D}$ diagonal. However, in that case we replace this difficulty by calculation of reciprocal values and furthermore we need extra storage. As we will use the Newton-Raphson method for solving both the square root and the reciprocity problem we find the former method more attractive.

We will seldom have trouble in solving (1). For moderately large n we can always use the double precision facilities usually available within the hardware of the computer. Occasionally we may need higher precision or the elements of the matrices involved have been calculated with only a few significant digits. As the algorithm presented below will take care of all cases, let us assume that we do have a high-precision software package available. Let us further assume that we have procedures only for performing addition, subtraction and multiplication of high-precision numbers.

## 3. PROCEDURES FOR THE RECIPROCAL AND THE SQUARE ROOT

There is only one reciprocal value we actually need. To get the elements of (5) we must have the value of 1/D. As we will see below we will only need the reciprocal value of the square root. Thus, we denote the functions

$$\mathrm{rec}(a) = 1/a \text{ and } \mathrm{rsq}(a) = 1/\mathrm{sqrt}(a), \tag{9}$$

and use a modified Newton-Raphson expression for their definition.

These two procedures will be very fast since we can easily get good starting values simply by using what we get with ordinary precision.

Assuming the starting values $x_0$ given in this way we use the recurrence formulae

$$x_{n+1} = x_n(2 - ax_n),\ \mathrm{rec}(a) = x_{n+1}, \tag{10}$$

$$x_{n+1} = x_n(3 - ax_n^2)/2,\ \mathrm{rsq}(a) = x_{n+1}. \tag{11}$$

## 4. ALGORITHM WITH KNOWN COVARIANCE MATRIX

Let us define the additional vectors

$$\underline{a} = \underline{R}^{-T}\underline{g} \text{ and } \underline{b} = \underline{R}^{-T}\underline{m}. \tag{12}$$

Using (8) in (6) we find that

$$\underline{x} = \underline{R}^{-1}\underline{a} \text{ and } \underline{y} = \underline{R}^{-1}\underline{b}, \tag{13}$$

and furthermore we now get from (3) that

$$P = \underline{a}^T\underline{a},\ Q = \underline{a}^T\underline{b} \text{ and } R = \underline{b}^T\underline{b}. \tag{14}$$

We get the vectors in (12) and (13) by solving the triangular linear systems

$$\underline{R}^T\underline{a} = \underline{g},\ \underline{R}^T\underline{b} = \underline{m},\ \underline{Rx} = \underline{a} \text{ and } \underline{Ry} = \underline{b}. \tag{15}$$

Finally, we calculate the values of (14), (4), (5) and (7).

We present this procedure as an algorithm. We denote the elements of the matrix $\underline{R} = (r_{ik})$ and introduce the auxiliary variables A, B and C. Note that we replace the diagonal elements of $\underline{R}$ by their reciprocals as well as D!

1. $r_{kk} = \mathrm{rsq}(s_{kk} - \sum_{j=1}^{k-1} r_{jk}^2)$, $r_{ki} = r_{kk}(s_{ki} - \sum_{j=1}^{k-1} r_{ji}r_{jk})$

   for i=k+1, ..., n and for k=1, 2, ..., n.

2. $a_i = r_{ii}(1 - \sum_{j=1}^{i-1} r_{ji}a_j)$, $b_i = r_{ii}(m_i - \sum_{j=1}^{i-1} r_{ji}b_j)$

   for i = 1, 2, ..., n.

3. $x_i = r_{ii}(a_i - \sum_{j=i+1}^{n} r_{ij}x_j)$, $y_i = r_{ii}(b_i - \sum_{j=i+1}^{n} r_{ij}y_j)$

   for i = n, n-1, ..., 1.

4. $P = \sum_{i=1}^{n} x_i$, $Q = \sum_{i=1}^{n} y_i$, $R = \sum_{i=1}^{n} b_i^2$.

5. $D = \mathrm{rec}(PR - Q^2)$.

6. $A = DR$, $B = -DQ$, $C = DP$.

7. $c_i = Ax_i + By_i$, $d_i = Bx_i + Cy_i$ for i=1, 2, ..., n.

8. $A_n = C$, $B_n = 2B$, $C_n = A$.

## 5. ALTERED ALGORITHM

We will now assume that the covariances have not been calculated, but that the matrix $\mathbf{A}$ is available. Surprisingly we will not need to make so many alterations of the algorithm in the preceding Section to get it working also in this case. Using the Sherman-Morrison formula for the inversion of $\mathbf{M}$ and defining now the vectors

$$\mathbf{x} = \mathbf{A}^{-1}\mathbf{g} \text{ and } \mathbf{y} = \mathbf{A}^{-1}\mathbf{m}, \tag{16}$$

we get that

$$\mathbf{M}^{-1} = \mathbf{A}^{-1} + \mathbf{y}\mathbf{y}^T/s \text{ with } s = 1-\mathbf{m}^T\mathbf{y}. \tag{17}$$

Letting $\mathbf{I}$ denote the unit matrix of order n we get that

$$\mathbf{M}^{-1}\mathbf{H} = (\mathbf{I} + \mathbf{y}\mathbf{m}^T/s)\ (\mathbf{x}\ \mathbf{y}), \tag{18}$$

and denoting further

$$p = \mathbf{g}^T\mathbf{x},\ q = \mathbf{g}^T\mathbf{y} \text{ and } r = \mathbf{m}^T\mathbf{y}, \tag{19}$$

we have thus that $s = 1-r$.

When calculating the matrix $\mathbf{E}^{-1}$ we will get the relations

$$P = p+q^2/s,\ Q = q/s \text{ and } R = r/s, \tag{20}$$

and denoting

$$d = pr - q^2, \tag{21}$$

we get that

$$D = d/s. \tag{22}$$

Putting

$$a = r/d,\ b = -q/d \text{ and } c = p/d, \tag{23}$$

we finally get that

$$\mathbf{c} = a\mathbf{x} + b\mathbf{y},\ \mathbf{d} = b\mathbf{x} + c\mathbf{y}, \tag{24}$$

and furthermore that

$$A_n = c-1,\ B_n = 2b \text{ and } C_n = a. \tag{25}$$

Thus, apart from new meanings of some variables it will suffice to alter the algorithm in the preceding Section only at 3 places to get it working in this case.

We assume that we can get the Cholesky factorization of

$\underline{A} = \underline{R}^T\underline{R}$. Then, in item number 1, we shall change $s_{kk}$ and $s_{ki}$ to $a_{kk}$ and $a_{ki}$ respectively. Finally, in item number 8, we shall change to $A_n = C - 1$.

## 6. CONCLUDING REMARKS

The algorithms presented in this paper could also be used in censored cases. Then we have only to exclude those rows and columns of the matrices involved where the censoring will take place. Thus we will work with a lower dimension than n.

This can be seen as a transformation of our matrices $\underline{H}$ and $\underline{M}$ and we may generally test other transformations. We have e.g. for Downton's method to work with a Vandermonde transformation matrix. However, we must be careful and analyse the resulting matrix to determine that it really is positive definite.

Denoting the vector $\underline{f}^T = (0\ 1)$ and the matrix $\underline{F} = \underline{f}\underline{f}^T$ in this paper we have proved the

THEOREM. With the notations of Section 2. we have that

$$\underline{K} = \underline{M}^{-1}\underline{H}(\underline{H}^T\underline{M}^{-1}\underline{H})^{-1} = \underline{A}^{-1}\underline{H}(\underline{H}^T\underline{A}^{-1}\underline{H})^{-1}, \tag{26}$$

$$\underline{E} = (\underline{H}^T\underline{M}^{-1}\underline{H})^{-1} = (\underline{H}^T\underline{A}^{-1}\underline{H})^{-1} - \underline{F}. \tag{27}$$

## REFERENCES

Fransén, A. (1971), "Estimation of Location and Scale Parameters from Ordered Samples, with Numerical Applications to the Gumbel and Weibull Distributions", FOA Reports 5, No 2.

Fransén, A. (1974), "Estimation of Location and Scale Parameters from Ordered Samples, with Numerical Application to the Gumbel and Weibull Distributions. Part II. Sample Sizes up to n = 31 inclusive -- Efficiency Investigation" + 2 Supplements, FOA Reports 8, No 2.

Fransén, A. (1975), "Estimation of Population Characteristics for the Gumbel Distribution"(in Swedish), FOA rapport C 10025-M4.

# STATISTICAL CHOICE OF UNIVARIATE EXTREME MODELS. PART II.

Arne Fransén

National Defence Research Institute
Division 1, Section 123
P. O. Box 27322
S-102 54 Stockholm - Sweden

and

J. Tiago de Oliveira

Faculty of Sciences
Department of Statistics, O. R. and Computing
58, Rua da Escola Politécnica
1294 Lisboa Codex - Portugal

ABSTRACT. Tiago de Oliveira presented the first paper in a series on statistical choice of univariate extreme models. This paper is a second one in that series. We present formulae for the general mean values and variances and some computer results. We investigated for how small values of the sample size the proposed test statistics may be useful by computer simulations applying the chi-square and Stephens'goodness of fit tests for normality. Some concrete data are further analysed in more detail. We conclude the paper with a general discussion.

KEY WORDS. Distribution of maxima, Weibull model, Gumbel model, Fréchet model, Asymptotic statistical decision, Simulation, Analysis of data.

## 1. INTRODUCTION

Tiago de Oliveira [12 ] presented the first paper in a series concerning statistical choice between univariate extreme models. The statistics $V_n(z)$ and $\hat{V}_n(x)$ useful for that purpose were introduced



*J. Tiago de Oliveira (ed.), Statistical Extremes and Applications, 373–394.*

among other points in that paper. Especially, the asymptotic results for large values of n presented in Appendix III in [12] seemed to confirm that conjecture.

This paper is a second one in which we will present some computer results and discuss the conjecture a little further. Thus, we will investigate the asymptotic normality of the two test statistics given above to see for how small values of n they might be useful. This is a rather difficult problem even when large computers are available. We have used a DEC-10 computer for the necessary simulations involving the chi-square test and Stephens'goodness of fit test. But, at first, we will take a closer look on how the general mean values and variances $\mu(k)$, $\sigma^2(k)$, $\hat{\mu}(k)$, and $\hat{\sigma}^2(k)$ may be accomplished.

Throughout this paper we will use the convention, when presenting numerical values, that they should be properly rounded so that the error will not exceed half a unit in the last digit shown. Thus, a bar over the last digit will indicate that this digit has been raised.

## 2. PREREQUISITES

We recall some formulae from the previous paper [12]. The von Mises-Jenkinson expression for the density, without location and dispersion parameters,

$$g(z|k)=\begin{cases}(1+kz)^{-1-1/k}\, e^{-(1+kz)^{-1/k}} & \text{for } (1+kz) > 0\\ 0 & \text{elsewhere}\end{cases} \tag{1}$$

contains the Weibull model for $k < 0$, the Fréchet model for $k > 0$ and the Gumbel model in the limiting case when $k \to 0$ from either side with, then,

$$g(z|0) = e^{-z-e^{-z}}. \tag{2}$$

The likelihood function leads to the locally most powerful test given by rejection regions, for the reduced models, based on the test statistic

$$V_n(z) = \sum_{i=1}^{n} v(z_i) \tag{3}$$

with

$$v(z) = z(-1+z(1-e^{-z})/2). \tag{4}$$

Note that

$$Z_n = V_n/(\sqrt{n}\ \sigma\ (0)) \tag{5}$$

is asymptotically standard normal, where $\sigma(0)$ denotes the standard deviation of $v(z)$ for $g(z|0)$ defined in Eqn. (2).

Denoting the three constants

$$p = \pi^2/6 = 1.64493\ 4066\underline{8}$$
$$q = \zeta(3) = 1.20205\ 6903\bar{2}$$
$$\gamma = 0.57721\ 56649$$

where $\zeta$ is Riemann's Zeta function and $\gamma$ is Euler's constant we may write the value of

$$\sigma^2(0) = p(1+3\gamma(-1+\gamma/2) + 27p/20) - 2q\ (1-\gamma)$$
$$+\gamma^2(1+\gamma(-1+\gamma/4)) \tag{6}$$
$$=2.42360\ 6055\bar{2}.$$

We denote further the constants

$$\beta = \Gamma''(2)/2 = \gamma(-1+\gamma/2) + p/2 = 0.41184\ 03304$$

$$\omega = 3(2\gamma - \Gamma'''(2) + (1-\gamma)\ \Gamma''(2))/\pi^2 = -1+\gamma + q/p$$
$$= 0.30797\ 86343.$$

After introducing the estimated location and scale parameters $\hat{\lambda}$ and $\hat{\delta}$ it was shown that, asymptotically, the test statistic (we let $x$ refer to observed values and $z$ to standardized ones)

$$\hat{V}_n(x) = \sum_{i=1}^{n} v(\hat{z}_i) \tag{7}$$

where

$$\hat{z}_i = (x_i - \hat{\lambda}(x))/\hat{\delta}(x) \tag{8}$$

was equivalent to $\sum_{i=1}^{n} \hat{v}(z_i)$ with

$$\hat{v}(z) = v(z) + (\beta+(1-\gamma)\omega)(e^{-z}-1) + \omega(1-z+ze^{-z}). \qquad (9)$$

This expression was denoted by $(v(z)+v_{\Delta}(z))$ in the previous paper.

Analogously

$$\hat{Z}_n = \hat{V}_n/(\sqrt{n}\ \hat{\sigma}(0)) \qquad (10)$$

where

$$\hat{\sigma}^2(0) = \sigma^2(0) - p\omega^2-\beta^2 = 2.09797\ 0220\bar{8} \qquad (11)$$

denotes the variance of $\hat{v}(z)$, is asymptotically standard normal.

Introducing the constant

$$B = \beta - \gamma\omega = 0.23407\ 0238\bar{3}$$

and the functions

$$a(z) = -z+(1+z)e^{-z} \qquad (12)$$

$$b(z) = -1+e^{-z} \qquad (13)$$

we can rewrite Eqn. (9) as

$$\hat{v}(z) = v(z) + \omega a(z) + B\ b(z). \qquad (14)$$

Using Eqns. (4) and (8) in Eqn. (7) together with the equations used for maximum likelihood estimation of the parameters for the Gumbel distribution we also obtained

$$v(\hat{z}) = ((x^2-nx^2\ e^{-x/\hat{\delta}}/\sum_{i=1}^{n} e^{-x_i/\hat{\delta}})/2-\hat{\delta}x)/\hat{\delta}^2. \qquad (15)$$

Eqn. (14) will be useful when we discuss the general mean values and variances whereas we will use Eqn. (15) in the simulations.

We will be interested in the functional

$$Iq = Iq(k) = \int_R q(z)\ g(z|k)\ dz \qquad (16)$$

for various functions q, particularly for $q=v$, $v^2$, $\hat{v}$, and $\hat{v}^2$. Changing the variable z to t by

$$t^{-k} = 1 + kz$$

and writing

$$h = -k$$

we obtain

$$Iq = \int_0^\infty q(-(t^h - 1)/h)\, e^{-t}\, dt. \qquad (17)$$

## 3. GENERAL MEAN VALUES AND VARIANCES

We will follow the outline in a private paper by Fransén from October 1979 omitting the details as far as possible. Necessary fundamental concepts of numerical analysis may be found in [1].

In functional notation we wish to calculate

$$\mu(k) = Iv \qquad (18)$$

$$\hat{\mu}(k) = I\hat{v} = \mu(k) + \omega\ Ia + B\ Ib \qquad (19)$$

$$\sigma^2(k) = Iv^2 - \mu^2(k) \qquad (20)$$

$$\hat{\sigma}^2\ (k) = I\hat{v}^2 - \hat{\mu}^2(k) = Iv^2 + 2\omega\ Iva + 2B\ Ivb + \omega^2\ Ia^2$$

$$+ 2\omega B\ Iab + B^2\ Ib^2 - \hat{\mu}^2\ (k). \qquad (21)$$

We see that it will suffice to consider

$$q(z) = z^r\ e^{-sz} \qquad (22)$$

in Eqn. (17) for various non-negative integers r and s. In fact, the values of r variates from 0 to 4 and those of s from 0 to 2. Denoting

$$I_r^s = Iq = \int_R z^r\ e^{-sz} g(z|k)\ dz \qquad (23)$$

according to Eqn. (22) we get our functionals

$$Iv = -I_1^0 + (I_2^0 - I_2^1)/2$$

$$Ia = -I_1^0 + I_0^1 + I_1^1$$

$$Ib = -I_0^0 + I_0^1$$

$$Iv^2 = I_2^0 - I_3^0 + I_4^0/4 + I_3^1 - I_4^1/2 + I_4^2/4$$

$$Iva = I_2^0 - I_3^0/2 - I_1^1 - I_2^1/2 + I_3^1 - (I_2^2 + I_3^2)/2$$

$$Ivb = I_1^0 - I_2^0/2 - I_1^1 + I_2^1 - I_2^2/2$$

$$Ia^2 = I_2^0 - 2(I_1^1 + I_2^1) + I_0^2 + 2\ I_1^2 + I_2^2$$

$$Iab = I_1^0 - I_0^1 - 2\ I_1^1 + I_0^2 + I_1^2$$

$$Ib^2 = I_0^0 - 2\ I_0^1 + I_0^2$$

Using the forward difference operator $\Delta$ identifying h as the displacement we finally get from Eqns. (17) and (23) the formal operator expression

$$I_r^s = (-1)^r\ (\Delta^r e^{s\Delta/h}/h^r)\ \Gamma(1) \tag{24}$$

thus operating on the Gamma function $\Gamma$. Developing in a Taylor series and differentiating we get

$$\Gamma^{(n)}(1 + h) = n!\ \sum_{j=0}^{\infty}\ \binom{n+j}{n}\ h^j\ b_{n+j+1} \tag{25}$$

of which accurate values of the coefficients

$$b_{j+1} = \Gamma^{(j)}\ (1)/j!,\ b_0 = 0 \tag{26}$$

are given in [4] and [5]. Using the differentiation operator D we have also the well-known formula

$$D^j\ \Gamma(n) = j\ D^{j-1}\ \Gamma(n-1) + (n-1)\ D^j\ \Gamma(n-1). \tag{27}$$

We may use Taylor series development in Eqn. (24) getting

$$I_r^s = (-1)^r\ (\sum_{j=0}^{\infty} s^j\ \Delta^{r+j}/(h^{r+j}\ j!))\ \Gamma(1). \tag{28}$$

From this equation we might step in two directions. Thus, either we may use divided differences on the function $\Gamma(1+x)$ defining the interpolation points $x_i = i\ h$ or we may introduce the operator D by the connection formula

$$\Delta = e^{hD} - 1 \tag{29}$$

and make use of the Stirling numbers of the second kind $\{{n \atop m}\}$ which may be defined by the formula

$$(e^x - 1)^m = m! \sum_{n=m}^{\infty} \{{n \atop m}\}\ x^n/n!\ . \tag{30}$$

In the sequel we will use both of these approaches obtaining a combined method which will work well for a good range values of h.

For $s = 0$ we use the first approach and we get from Eqn. (24)

$$I_r^0 = h^{-r} \sum_{i=0}^{r} (-1)^i \binom{r}{i} \Gamma(1 + i\,h). \tag{31}$$

Note that we get implicitly from Eqn. (31) the upper limiting conditions on the parameter k that $k < 1/2$ in Eqns. (18) and (19) and that $k < 1/4$ in Eqns. (20) and (21) because we must avoid the value of $\Gamma(0)$ .

With the usual notation for the divided differences and with $f(x) = \Gamma(1 + x)$ using the connection formula

$$f[x_i, x_{i+1}, \ldots, x_{i+j}] = (h^j\, j!)^{-1} \Delta^j f(x_i) \tag{32}$$

we finally get from Eqn. (28)

$$I_r^s = \sum_{j=0}^{\infty} s^j h^{-r} \sum_{i=0}^{r} (-1)^i \binom{r}{i} f[x_i, x_{i+1}, \ldots, x_{i+j}]. \tag{33}$$

By means of this formula we will get converging values in the interval $0.1 \leq h < 0.7$ whereas we get oscillation problems for smaller values of h when using the double precision of the computer.

For $h < 0.1$ and $s > 0$ we use the second approach. Defining tho polynomials (wo need them for integers m only)

$$P_n(m) = \sum_{k=0}^{n} d_k^n m^k \tag{34}$$

by means of the equation

$$\left\{{n+m \atop m}\right\} = \binom{n+m}{m} P_n(m) \tag{35}$$

we get the recurrence relation

$$(n + m + 1)\, P_{n+1}(m) - mP_{n+1}(m-1) = m\,(n + 1)\, P_n(m) \tag{36}$$

to calculate successive polynomials using the method of indeterminate coefficients and we also define

$$a_k^n = (n!)^{-1} \sum_{j=k}^{n} \left\{{j \atop k}\right\} d_k^n . \tag{37}$$

We list some of the first polynomials $P_n(m)$

$$P_0(m) = 1$$

$$P_1(m) = m/2$$

$$P_2(m) = (3m^2 + m)/12$$

$$P_3(m) = (m^3 + m^2)/8$$

$$P_4(m) = (15m^4 + 30m^3 + 5m^2 - 2m)/240$$

$$P_5(m) = (3m^5 + 10m^4 + 5m^3 - 2m^2)/96$$

$$P_6(m) = (63m^6 + 315m^5 + 315m^4 - 91m^3 - 42m^2 + 16m)/4032$$

$$P_7(m) = (9m^7 + 63m^6 + 105m^5 - 7m^4 - 42m^3 + 16m^2)/1152$$

Let us now write Eqn. (28) as

$$I_r^s = \sum_{n=0}^{\infty} c_n h^n \qquad (38)$$

where

$$c_n = A_n \, \Gamma(1) \qquad (39)$$

and we get the operators

$$A_n = (-1)^r \, r! \sum_{j=0}^{\infty} \binom{r+j}{r} \left\{ {n+r+j \atop r+j} \right\} \frac{s^j}{(n+r+j)!} D^{n+r+j}. \qquad (40)$$

Especially, for $n = 0$ we get, using Eqn. (26)

$$c_0 = (-1)^r \, r! \sum_{j=0}^{\infty} \binom{r+j}{r} s^j \, b_{r+j+1} = (-1)^r \, \Gamma^{(r)}(1 + s) \qquad (41)$$

with the aid of Eqn. (25).

After much algebra we finally get ($r$ is limited to 4 above)

$$c_n = (-1)^r \sum_{i=0}^{4} \binom{r}{i} i! \sum_{k=i}^{n} \binom{k}{i} a_k^n \, s^{k-i} \, D^{n+r+k-i} \, \Gamma(1+s). \qquad (42)$$

Finally, we use Eqn. (26) to get the coefficient for $s = 1$

$$c_n = (-1)^r \, m! \sum_{i=0}^{4} \binom{r}{i} \sum_{k=i}^{n} \binom{m+k-i}{k-i} a_k^n \, k! \cdot (b_{m+k-i} + b_{m+k-i+1}) \tag{43}$$

and for s = 2

$$c_n = (-1)^r \, m! \sum_{i=0}^{4} \binom{r}{i} \sum_{k=i}^{n} \binom{m+k-i}{k-i} a_k^n \, 2^{k-i} k! \cdot (b_{m+k-i-1} + 3b_{m+k-i} + 2b_{m+k-i+1}) \tag{44}$$

where we have abbreviated $m = n + r$.

In the special case for $h = 0$ we get $I_r^s = c_0$ and can use Eqn. (41) combined with Eqns. (27) and (26).

To sum up we use Eqn. (31) for $s = 0$. For $h < 0.1$ we use Eqn. (38) with the coefficients from Eqn. (43) for $s = 1$ and those from Eqn. (44) for $s = 2$. In the interval $0.1 \leq h < 0.7$ we use Eqn. (33) for $s = 1$ and $s = 2$. In the special case for $h = 0$ we use Eqn. (41). Moreover we may use Euler's transformation for negative values of h and repeated Aitken extrapolation for positive values of h to get more accurate values.

We present the results when using Eqn. (38) up to $n = 7$ in the following table, where within parentheses we have put the limiting values which will never be attained,

| k | $\mu(k)$ | $\hat{\mu}(k)$ | $\sigma^2(k)$ | $\hat{\sigma}^2(k)$ |
|---|---|---|---|---|
| -1 | $(-\infty)$ | $(-\infty)$ | $(+\infty)$ | $(+\infty)$ |
| -0.6 | -2.0468 | -2.0468 | 1747.7 | 2006.6 |
| -0.5 | -1.2447 | -1.1942 | 188.87 | 221.04 |
| -0.4 | -0.82299 | -0.75653 | 37.350 | 43.959 |
| -0.3 | -0.56552 | -0.50123 | 10.136 | 11.954 |
| -0.2 | -0.37800 | -0.32722 | 3.4133 | 4.0493 |
| -0.1 | -0.20571 | -0.17672 | 1.6730 | 1.8592 |
| -0.05 | -0.11053 | -0.09515 | 1.6879 | 1.6603 |
| -0.01 | -0.02376 | -0.02054 | 2.1964 | 1.9410 |
| -0.005 | -0.01200 | -0.01038 | 2.3038 | 2.0141 |
| -0.001 | -0.00242 | -0.00209 | 2.3986 | 2.0803 |
| 0 | 0 | 0 | 2.4236 | 2.0980 |
| 0.001 | 0.00243 | 0.00210 | 2.4491 | 2.1161 |
| 0.005 | 0.01224 | 0.01061 | 2.5568 | 2.1935 |

| $k$ | $\mu(k)$ | $\hat{\mu}(k)$ | $\sigma^2(k)$ | $\hat{\sigma}^2(k)$ |
|---|---|---|---|---|
| 0.01 | $0.0247\bar{5}$ | 0.02145 | $2.704\bar{7}$ | 2.3018 |
| 0.05 | 0.13565 | 0.11843 | $4.613\bar{3}$ | 3.8097 |
| 0.1 | $0.3106\bar{3}$ | $0.2742\bar{0}$ | $10.70\bar{7}$ | $9.067\bar{8}$ |
| 0.15 | $0.5471\bar{4}$ | 0.48928 | $30.53\bar{5}$ | $27.30\bar{5}$ |
| 0.2 | $0.8825\bar{5}$ | 0.80072 | $128.2\bar{3}$ | $121.5\bar{7}$ |
| 0.25 | $1.385\bar{1}$ | 1.2763 | $(+\infty)$ | $(+\infty)$ |
| 0.3 | 2.1933 | $2.054\bar{0}$ | | |
| 0.5 | $(+\infty)$ | $(+\infty)$ | | |

To get a better idea of the results obtained we also present two diagrams showing the general mean values and variances, respectively, as functions of the parameter $k$. Note that both curves intersect once, the former at the origin and the other one somewhere between $k = -0.1$ and $k = -0.05$.

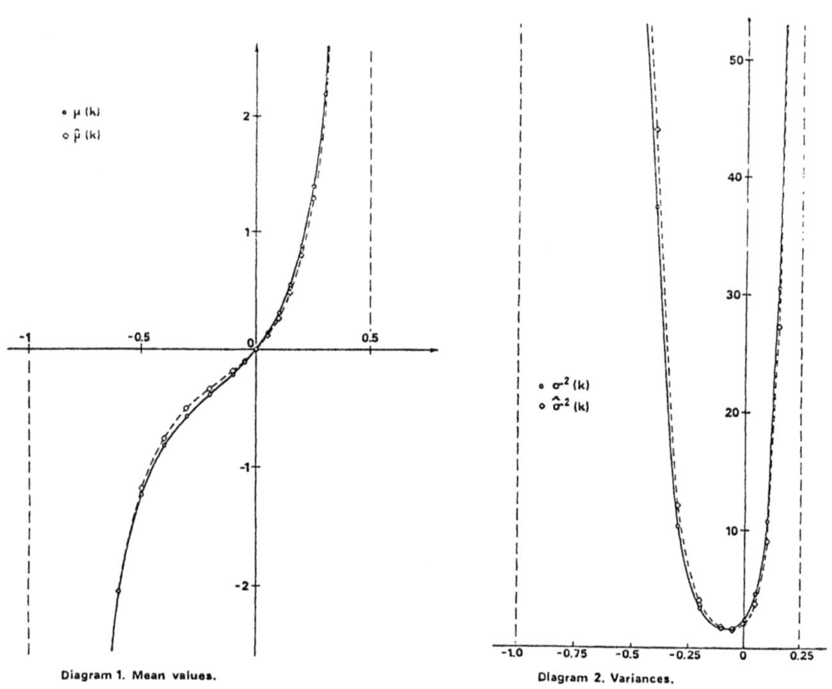

Diagram 1. Mean values.

Diagram 2. Variances.

## 4. DESCRIPTION OF THE SIMULATIONS

We were interested in the behavior of our test statistics presented in Eqns. (3) and (7) for various values of $n$. As they were suggested for asymptotic testing it was natural to try and get some idea of how large our sample should be, where the test statistics ought to be useful. Therefore we studied the expressions given in Eqns. (5) and (10) for some values of the sample size n testing for asymptotic normality.

For that purpose we have used the tables in [9]. Thus, for the chi-square test we use Table 3 on page 163 and for the Stephens' goodness of fit test Table 54 on page 359 in [9]. Performing the chi-square test we have divided our material into 9 subclasses. According to Fisher's theorem we then get 8 degrees of freedom in the case of $Z_n$ and 6 for $\hat{Z}_n$. Table 54 for Stephens'test gives the cases 1 and 2 used for $Z_n$ and $\hat{Z}_n$, respectively.

The principles for the simulations were the following

1. Fix N, the number of cases to run for the tests
2. For each of the N cases do the following
   a. Choose n, the value of the sample size
   b. Get n observations $z_i$ $(=x_i)$, i.e. generate n random standard Gumbel deviates
   c. Calculate $Z_n$ of Eqn. (5) using Eqns. (4) and (3)
   d. Estimate the location and scale parameters using Newton-Raphson's method on the maximum likelihood equations for the Gumbel distribution
   e. Calculate $\hat{Z}_n$ of Eqn. (10) using Eqns. (15) and (7)
   f. Save obtained values of $Z_n$ and $\hat{Z}_n$
   g. Put the values of $Z_n$ and $\hat{Z}_n$ in their appropriate subclasses
3. Use the values in the subclasses to perform the chi-square tests
4. Use the saved values to perform Stephens'tests

We have chosen the value of N = 103 to be sure of getting in the average at least 10 values in each subclass. Actually, we used two subdivisions for the chi-square tests, one with unequal (probability) subclasses called Case I and the other called Case II with equal (probability) subclasses. Thus, we put the abscissae in the points $\pm 0.14$, $\pm 0.4$, $\pm 0.7$, and $\pm 1.12$ in Case I. Apart from the reason given above this was the easiest way since the program for the normal distribution (See [3]) gave the corresponding probabilities and we could cut the mass in proper places.

However, we needed the abscissae in Case II with equal subclasses all of the mass 1/9. This was accomplished by means of inverse Bessel interpolation giving the values

$\pm$0.13971 02988 81862 04517
$\pm$0.43072 72992 95457 49009
$\pm$0.76470 96737 86387 04716
$\pm$1.22064 03488 47349 6985

Surely, both cases are interesting as they give us a possibility of eliminating wrong decisions when interpreting the results for certain values of the sample size n.

Applying Stephens test we must at first order the saved values of $Z_n$ and $\tilde{Z}_n$. We use the test statistic

$$U_N^2 = \sum_{i=1}^{N} (u_{(i)} - \frac{2i-1}{2N})^2 - N(\frac{1}{N}\sum_{i=1}^{N} u_{(i)} - \frac{1}{2})^2 + \frac{1}{12N}. \quad (45)$$

In Case I for $Z_n$ we use

$$u_{(i)} = N(t_{(i)})$$

with the notation N(x) for the standard normal distribution and where $t_{(i)}$ are the ordered values of $Z_n$. To obtain the right decision we then use the empirically modified statistic

$$(1.0 + 0.8/N)\ (U_N^2 - 0.1/N + 0.1/N^2)$$

according to the recommendations in [9]. In Case 2 for $\hat{Z}_n$ we use

$$u_{(i)} = N((t_{(i)} - \bar{t})/s),$$

where

$$\bar{t} = \frac{1}{N}\sum_{i=1}^{N} t_{(i)}$$

$$s^2 = \frac{1}{N-1}\left(\sum_{i=1}^{N} t^2_{(i)} - N\,\bar{t}^2\right)$$

and $t_{(i)}$ are the ordered values of $\hat{Z}_n$. Similarly to the preceding case we then use the empirically modified statistic

$$(1 + 0.5/N)\ U_N^2 .$$

Obtaining random standard Gumbel deviates is a rather difficult problem on most computers as, in particular, we mostly need very fast routines for it. Thus, this is the most expensive part in our simulation program as we use the routine in the inner-most loop there.

We use the ordinary inverse method. Let y denote a random uniform (0, 1) number. Then we get the sought deviate as

$$-\log\ (-\ \log\ y)$$

where log denotes the natural logarithm.

The first problem concerns the calculation of the logarithm. Most standard algorithms used for that will not give accurate precision so that we may use them to obtain our deviates without suffering from logarithmic errors. To fit our purpose we thus had to design a special routine for computing the logarithm by means of continued fractions. Due to the many necessary divisions this routine became rather slow, but it is hard to see a way avoiding them.

The second problem concerns the random number generator producing the values of y above. We have used Lehmer's multiplicative congruence generator (See [7])

$$u_{n+1} \equiv 5^{15} u_n \pmod{2^{35}} \tag{46}$$

with an odd integer value of $u_0$. After the division by the module we get pseudo-random uniform (0, 1) numbers. Some years ago there appeared a Swiss report in which some doubts about this generator were raised, but it is the one used by the Simula system and has been shown to pass most tests -- even the spectral test. But then it is important to use the stabilized generator. For that reason we have discarded the first 100 values. As the generator could be implemented in a low-level language this problem became of no importance.

Finally, when using Newton-Raphson's method we need an everywhere working starting value. We might then use the unbiased estimator obtained from the method of moments

$$\delta = \frac{\sqrt{6}}{\pi} s = 0.77969\ 68012\ s \tag{47}$$

thus using

$$\bar{x} = n^{-1} \sum_{i=1}^{n} x_i$$

$$s^2 = (n-1)^{-1} \sum_{i=1}^{n} (x_i - \bar{x})^2$$

but this starting value will occasionally lead to a wrong branch resulting in indefinite oscillation and thus no convergence. In our simulations, however, it will work well even when we use $\bar{s}$ instead of s in Eqn. (47), where

$$\bar{s}^2 = \frac{n-1}{n} s^2 .$$

To get on the right branch always resulting in convergence within 100 iterations we may use (See [2])

$$\delta = (\bar{x} + \mathrm{sqrt}(4\ \bar{s}^2 + \bar{x}^2))/2. \tag{48}$$

Although this starting value is far from the resulting value $\hat{\delta}$ it works excellently. In fact, for positive values of the location parameter it takes a value, roughly greater than $1.6032\bar{3}\ \hat{\delta}$. By deliberately introducing this bias, we are certain to get convergence to the right value $\hat{\delta}$.

To give an idea of how expensive the simulations may be we present the following table in which CPU time refers to the time spent for calculations only in the central processing unit

| Sample size n | CPU time using Eqn. (47) | CPU time using Eqn. (48) |
|---|---|---|
| 200 | 47.97 sec | 51.66 sec |
| 250 | 59.68 sec | 1 min 4.51 sec |
| 500 | 1 min 58.18 sec | 2 min 8.89 sec |
| 1000 | 3 min 56.28 sec | 4 min17.19 sec |

It might be possible to diminish the costs using some variance-reducing method, but it is very hard to see how this may be achieved in the actual simulation.

We have tried another way to reduce the execution time. Thus, at first, we produce a rather comprehensive list of random standard Gumbel deviates using the method described above. This list contained 20.000 values. We used consecutive values from that list with randomized starting points. In this way we could work with sample sizes less than n = 195.

We give the corresponding table of CPU time for this procedure

| Sample size n | CPU time using Eqn. (47) | CPU time using Eqn. (48) |
|---|---|---|
| 50 | 10.22 sec | 10.95 sec |
| 100 | 18.12 sec | 19.81 sec |
| 180 | 24.83 sec | 28.19 sec |
| 190 | 25.97 sec | 29.60 sec |

We see that we have almost halved the execution time. We could have produced an extensive list of random standard Gumbel deviates on a tape and used them in a similar way. This procedure will, however, involve many expensive transactions from the tape into the core which will be too much for our economic resources.

Though it is obvious we mention that of course we did get coinciding results independent of which one of our starting values for the Newton-Raphson iteration we have chosen.

The program for the simulations was designed in the most effective way. Thus, it was written in the FORTRAN language and was optimized by the compiler.

## 5. SIMULATION RESULTS

We present the results from the simulations below in two separate tables, one for $Z_n$ and the other for $\hat{Z}_n$. In both tables we shall give the values obtained for the chi-square tests under the headings Case I and Case II referring to the cases mentioned in the preceding section and the values obtained for the other goodness of fit test under the heading Stephens. Owing to what is said in [7] and other textbooks on statistical inference we should perform the test at least three times with different data. Although we have not managed to follow this rule due to the expensiveness, actually, we thus considered three different tests on the same data.

Using linear interpolation in the tables in [9] we give the value of p in our tables below. For the chi-square tests this corresponds to the p percent value and to the upper percentage point for the Stephens'test for which no values are given for $p > 15.0\%$. Thus p% is the significance level of rejecting a true normal (asymptotic) distribution. To put it in another way, the value of p% is the approximate probability that the statistic being used is larger than the obtained (computed) one.

| Simulation results for $Z_n$ | | | | Simulation results for $\hat{Z}_n$ | | | |
|---|---|---|---|---|---|---|---|
| n | Case I | Case II | Stephens | n | Case I | Case II | Stephens |
| 10 | < 0.01 | < 0.01 | < 1.0 | 10 | 0.04 | < 0.01 | < 1.0 |
| 10 | < 0.01 | < 0.01 | < 1.0 | 10 | <0.01 | < 0.01 | < 1.0 |
| 10 | 0.09 | 0.05 | < 1.0 | 10 | <0.01 | < 0.01 | >15.0 |
| 25 | 1.43 | 2.98 | < 1.0 | 25 | 2.45 | 3.81 | 1.87 |
| 50 | 23.08 | 5.25 | >15.0 | 50 | 34.81 | 44.92 | >15.0 |
| 50 | 65.37 | 72.84 | >15.0 | 50 | 22.64 | 71.82 | 8.53 |
| 75 | 43.86 | 42.90 | >15.0 | 75 | 10.98 | 10.67 | < 1.0 |
| 100 | 37.73 | 32.07 | >15.0 | 100 | 73.07 | 64.80 | >15.0 |
| 100 | 10.22 | 57.42 | >15.0 | 100 | 14.15 | 14.86 | 12.28 |
| 125 | 74.04 | 55.53 | >15.0 | 125 | 43.23 | 53.35 | >15.0 |
| 150 | 63.04 | 88.19 | >15.0 | 150 | 3.00 | 16.53 | >15.0 |
| 175 | 38.16 | 48.11 | >15.0 | 175 | 22.06 | 53.35 | >15.0 |
| 180 | 63.26 | 23.49 | >15.0 | 180 | 25.85 | 25.08 | >15.0 |
| 190 | 18.08 | 12.51 | >15.0 | 190 | 29.38 | 39.02 | >15.0 |
| 194 | 16.95 | 29.29 | >15.0 | 194 | 64.64 | 16.53 | >15.0 |
| 200 | 75.61 | 70.97 | >15.0 | 200 | 50.46 | 22.45 | >15.0 |
| 200 | 51.49 | 48.11 | >15.0 | 200 | 29.34 | 10.67 | >15.0 |
| 225 | 22.59 | 41.17 | >15.0 | 225 | 34.70 | 55.60 | >15.0 |
| 250 | 37.82 | 51.73 | >15.0 | 250 | 29.65 | 23.77 | >15.0 |
| 500 | 90.86 | 51.73 | >15.0 | 500 | 4.96 | 9.01 | >15.0 |
| 1000 | 89.77 | 83.44 | >15.0 | 1000 | 24.87 | 16.53 | >15.0 |

According to the recommendations in [7] of accepting our hypothesis of normality if two out of three values passed the tests we should thus get the results that already a sample size of n=50 could be useful for our purposes for both test statistics considered. We postpone further comments until the general discussion below.

## 6. FURTHER ANALYSIS OF SOME CONCRETE DATA

In this section we will analyse the concrete data considered in Appendix III of the preceding paper in more detail. We recall some more suggestions from [12].

Denoting $w_n$ = sqrt(2 log n) we considered the sequence

$$c_n = w_n - (\log \log n + \log(4\,\pi))/(2\,w_n) \qquad (49)$$

satisfying the consistency conditions of our tests and our decision rule was the following

if $\hat{Z}_n \leq -c_n$ then choose the Weibull model

if $|\hat{Z}_n| < c_n$ then choose the Gumbel model

if $\hat{Z}_n \geq c_n$ then choose the Fréchet model

The approximation of the error probability when choosing the Gumbel model was

$$\hat{E}_n = \hat{E}_n(k = 0) = 2(1-N(c_n)) \qquad (50)$$

with the further approximation

$$\hat{E}_n \simeq 2/n. \qquad (51)$$

The data were taken from the papers [6], [8], [10], and [11].

We have also used the data considered in [2] and through the kind assistance from the Central Bureau of Statistics at Örebro, Sweden we have obtained the following additional data

Oldest ages at death, Sweden

| Year | Men | Women | Year | Men | Women |
|---|---|---|---|---|---|
| 1959 | 104.23 | 104.77 | 1965 | 105.28 | 104.31 |
| 1960 | 103.59 | 106.13 | 1966 | 104.93 | 104.98 |
| 1961 | 103.74 | 107.10 | 1967 | 105.27 | 105.83 |
| 1962 | 103.00 | 104.56 | 1968 | 105.92 | 106.35 |
| 1963 | 104.25 | 110.07 | 1969 | 101.81 | 107.58 |
| 1964 | 104.12 | 106.15 | 1970 | 104.02 | 105.42 |

We have analysed the data of the following cases

a. Sample data from the Gumbel distribution [2, p. 48]
   Sample size n = 30
b. Yearly maximum precipitation during 24 hours measured in 0.1 mm in Belgium 1938-1972 [10, p. 6]
   Sample size n = 35
c. Floods of the Ocmulgee River in Georgia 1910-1949 [6, p. 797]
   Sample size n = 40
   1. Downstream at Macon
   2. Upstream at Hawkinsville
d. Floods of the North Saskatchewan River at Edmonton [8, pp. 426-427] or [11, p. 179]
   Sample size n = 47
   1. Ordinary raw data
   2. Natural logarithms of raw data
e. Oldest ages at death in Sweden 1905-1958 [6, p. 795]
   Sample size n = 54
   1. Men
   2. Women
f. Oldest ages at death in Sweden 1905-1970 [6, p. 795] plus the data given above
   Sample size n = 66
   1. Men
   2. Women

The results obtained will be presented in tabular form below. In the first table we give the values of Eqns. (49) - (51). As the last two equations refer to the level of significance we will give corresponding values in percent. We give also the ratio between the approximations in Eqns. (51) and (50).

| Case | n | $c_n$ | $\hat{E}_n$ | 2/n | $(2/n)/\hat{E}_n$ |
|---|---|---|---|---|---|
| a | 30 | $1.8882\bar{5}$ | 5.899 | $6.66\bar{7}$ | $1.1300\bar{9}$ |
| b | 35 | $1.9541\bar{7}$ | 5.068 | 5.714 | $1.1274\bar{9}$ |
| c | 40 | $2.0100\bar{1}$ | $4.44\bar{3}$ | 5.000 | $1.1253\bar{5}$ |
| d | 47 | 2.07598 | $3.79\bar{0}$ | $4.25\bar{5}$ | $1.1229\bar{1}$ |
| e | 54 | $2.1315\bar{7}$ | 3.304 | $3.70\bar{4}$ | 1.12091 |
| f | 66 | $2.2100\bar{7}$ | 2.710 | 3.030 | 1.11817 |

For comparison purpose we will use the sample sizes of our simulations, so give also the corresponding table

| $n$ | $c_n$ | $\hat{E}_n$ | $2/n$ | $(2/n)/\hat{E}_n$ |
|---|---|---|---|---|
| 10 | 1.36192 | 17.322 | 20.000 | 1.15459 |
| 25 | 1.80813 | 7.059 | 8.000 | 1.13337 |
| 50 | 2.10089 | 3.565 | 4.000 | 1.12200 |
| 75 | 2.25899 | 2.388 | 2.667 | 1.11652 |
| 100 | 2.36625 | 1.797 | 2.000 | 1.11302 |
| 125 | 2.44693 | 1.441 | 1.600 | 1.11051 |
| 150 | 2.51133 | 1.203 | 1.333 | 1.10856 |
| 175 | 2.56478 | 1.032 | 1.143 | 1.10699 |
| 180 | 2.57446 | 1.004 | 1.111 | 1.10671 |
| 190 | 2.59294 | 0.952 | 1.053 | 1.10618 |
| 194 | 2.60003 | 0.932 | 1.031 | 1.10598 |
| 200 | 2.61038 | 0.904 | 1.000 | 1.10568 |
| 225 | 2.65007 | 0.805 | 0.889 | 1.10456 |
| 250 | 2.68518 | 0.725 | 0.800 | 1.10358 |
| 500 | 2.90745 | 0.364 | 0.400 | 1.09774 |
| 1000 | 3.11647 | 0.183 | 0.200 | 1.09271 |

In the final table we present the estimated values $\hat{\lambda}$ and $\hat{\delta}$ of the location and scale parameters, respectively, the calculated mean value $\hat{V}_n/n$ obtained from Eqn. (7), the corresponding standard deviation $s(\hat{V}_n)$ based on the unbiased estimate of the variance, and the calculated value of $\hat{Z}_n$ in Eqn. (10). It should be noted that the starting value $\delta$ for the Newton-Raphson iteration given in Eqn. (47) will not work in the cases e and f. Wo romark that como of the values presented in Appendix III of the previous paper were incorrect.

| Case | $\hat{\lambda}$ | $\hat{\delta}$ | $\hat{V}_n/n$ | $s(\hat{V}_n)$ | $\hat{Z}_n$ |
|---|---|---|---|---|---|
| a | 10.130 | 17.508 | -0.04046 | 1.11626 | -0.15298 |
| b | 295.75 | 101.49 | 0.18348 | 0.92494 | 0.74941 |
| c1 | 26.378 | 17.042 | -0.03205 | 0.57692 | -0.13996 |
| c2 | 23.709 | 15.057 | -0.03277 | 0.71600 | -0.14310 |
| d1 | 38.151 | 17.740 | 0.80373 | 3.98761 | 3.80418 |
| d2 | 3.5489 | 0.3963 | -0.04167 | 0.89869 | -0.19725 |
| e1 | 102.47 | 1.4453 | -0.25274 | 0.68715 | -1.28226 |
| e2 | 103.81 | 1.3379 | -0.33398 | 0.93282 | -1.69440 |
| f1 | 102.64 | 1.4699 | -0.33185 | 0.77713 | -1.86127 |
| f2 | 104.04 | 1.4318 | -0.30347 | 1.11725 | -1.70210 |

Using our decision rule above comparing $\hat{Z}_n$ from this table to $c_n$ in corresponding cases in the table above we get the result that we shall choose the Gumbel model in all cases except dl where we should choose the Fréchet model. This result is in perfect agreement with those presented in the referenced papers. Note also, as could be expected, the increase in corresponding values of case f compared to e.

We have also used the list of standard Gumbel deviates from our simulations to produce the corresponding table

| $n$ | $\hat{\lambda}$ | $\hat{\delta}$ | $\hat{V}_n/n$ | $s(\hat{V}_n)$ | $\hat{Z}_n$ |
|---|---|---|---|---|---|
| 10 | -0.02895 | 1.14536 | -0.44365 | 0.45140 | -0.96858 |
| 25 | 0.17890 | 0.95363 | -0.04414 | 0.44156 | -0.15238 |
| 50 | 0.27046 | 1.03590 | 0.01154 | 2.32626 | 0.05634 |
| 75 | -0.01100 | 1.11001 | -0.11889 | 0.80962 | -0.71087 |
| 100 | -0.08584 | 0.90706 | -0.05372 | 1.00683 | -0.37091 |
| 125 | 0.01236 | 0.94559 | 0.10876 | 2.28671 | 0.83950 |
| 150 | 0.04125 | 0.97839 | 0.09614 | 2.03165 | 0.81290 |
| 175 | -0.02343 | 0.90042 | -0.03364 | 1.16960 | -0.30722 |
| 180 | -0.05909 | 0.97973 | 0.06863 | 1.54112 | 0.63570 |
| 190 | -0.02841 | 1.03201 | -0.10974 | 2.06560 | -1.04434 |
| 194 | 0.04748 | 1.06162 | -0.00998 | 1.26596 | -0.09599 |
| 200 | 0.06520 | 0.91521 | 0.08896 | 1.09688 | 0.86856 |
| 225 | -0.05108 | 1.02702 | 0.01375 | 1.07033 | 0.14244 |
| 250 | 0.06153 | 1.01731 | 0.01744 | 1.09845 | 0.19043 |
| 500 | 0.03078 | 1.04732 | -0.00286 | 2.10930 | -0.04420 |
| 1000 | 0.00917 | 1.04307 | 0.12704 | 1.84074 | 2.77363 |

From the results presented in this table and the corresponding one above we may conclude that our list really contains random standard Gumbel deviates and as our decision rule gives the result that we shall choose the Gumbel model in all cases we may rely on that rule as a useful one.

Although it appears as the values of $\hat{Z}_n$ for small sample sizes would have a tendency to have negative values we have found no indication of that when studying the individual observations in more detail. Further comments will be given in the next section.

## 7. GENERAL DISCUSSION

The methodology proposed by Tiago de Oliveira in [12] to obtain a trilemma decision for data relating to (assumed) i.i.d. maxima ($k < 0$, $k = 0$ or $k > 0$ in von Mises-Jenkinson's integrated formula, i.e. for choosing Weibull, Gumbel or Fréchet models) is, as is clear, a modification of the L.M.P.U. test of $k=0$ vs. $k \neq 0$ (or a joint case of the L.M.P. tests of $k = 0$ vs. $k<0$ and $k = 0$ vs. $k > 0$). As explained there, this is not a case of separation in Cox' sense -- see [12] for details -- and so was called statistical choice because the family of Gumbel distributions (with location and dispersion parameters) is, both, the limit of Weibull ($k < 0$) distributions and Fréchet ($k > 0$) distributions. The "contiguity" of the families, in contradistinction with the (metric) separation of the families used in Cox'formulation and the fact that we were dealing with a trilemma, and not a dilemma, as in the asymmetrical tests used by Cox, suggested that only for large values of the sample size n we could expect good efficiency for the L.M.P. statistic.

The exact simulations presented in section 5, with large accuracy, as well as the analysis of some concrete data in section 6 show that even for small samples, as small as n about 50, the statistical choice method has good efficiency, better than expected in the previous paper. The ratios $(2/n)/\hat{E}_n$ presented in the preceding section show that the error is smaller than expected.

As it is well known, if X is a Weibull or Fréchet random variable, the logarithm of some convenient linear transforms of X are Gumbel random variables. The analysis of d1 vs. d2 in section 6, using this fact, leads to the rejection of the Gumbel model and to the acceptance of the Fréchet model as was known and should be expected.

## REFERENCES

1. Dahlquist, G., Björck, A. and Anderson, N.(1974). "Numerical Methods", Prentice-Hall, Inc., Englewood Cliffs, New Jersey.
2. Fransén, A.(1975). "Estimation of Population Characteristics for the Gumbel Distribution" (in Swedish), FOA rapport C 10025-M4.
3. Fransén, A.(1976)."Calculation of the Probability Integral in Double Precision for DEC-10" (in Swedish), FOA rapport C 10047-M3(E5).
4. Fransén, A. and Wrigge, S.(1980). "High-Precision Values of the Gamma Function and of Some Related Coefficients", Math. Comp., v. 34, pp. 553-566.

5. Fransén, A. (1981). Addendum and Corrigendum to "High-Precision Values of the Gamma Function and of Some Related Coefficients", Math. Comp., v.37, pp. 233-235.
6. Gumbel, E.J. and Goldstein, N. (1964). "Analysis of Empirical Bivariate Extremal Distributions", Am. Stat. Assoc. Journal, September, pp. 794-816.
7. Knuth, D.E. (1969). "The Art of Computer Programming, Vol. 2: Seminumerical Algorithms", Addison-Wesley, Reading, Massachusetts.
8. van Montfort, M.A.J. (1970). "On Testing that the Distribution of Extremes is of Type I when Type II is the Alternative", Journ. of Hydrology 11, pp. 421-427.
9. Pearson, E.S. and Hartley, H.O. (ed.) (1972). "Biometrika Tables for Statisticians, Vol. 2", Cambridge at the University Press.
10. Sneyers, R. (1977). "L'intensité maximale des précipitations en Belgique", Institut Royal Météorologique de Belgique, Publications Série B, No 86, 15 pp.
11. (1967). "Statistical Methods in Hydrology, Discussion on Extreme Value Analysis", Proceedings of Hydrology Symposium No 5, Ottawa, Canada, pp. 171-181.
12. Tiago de Oliveira, J. (1981). "Statistical Choice of Univariate Extreme Models", Statistical Distributions in Scientific Work, Vol. 6, D. Reidel Publ. Co., pp. 367-387.

# DOUBLY EXPONENTIAL RANDOM NUMBER GENERATORS

Masaaki Sibuya

Department of Mathematics
Keio University,
3-14-1 Hiyoshi, Kohoku-ku, Yokohama, Japan 223

ABSTRACT. Six methods for generating doubly exponential random numbers are examined. They are a previously proposed one, four standard ones and a new asymmetrization method. Their features are compared, and guidelines for implementing a subroutine are given.

## 1. INTRODUCTION.

In a previous paper, the author proposed a method for generating doubly exponential (or Gumbel) random numbers (Sibuya, 1968). After that, efforts to generate random numbers following non-uniform distributions have continued, and some techniques have become standard in typical distributions. In this paper, these techniques are applied to the doubly exponential distribution, and the new generators are compared with the previous one.

The standard techniques here examined are von Neumann's rejection method, von Neumann and Forsythe's odd-even method, and Kinderman and Monahan's ratio method. These are described in a section of Knuth's book (Knuth, 1981). A new technique, the asymmetrization method, is also proposed. The author's previous method is based on the fact that if a random number of exponential random variables are observed, and if the random number follows a zero-truncated Poisson distribution, then the maximum of the observed values follows a truncated doubly exponential distribution.

In the last section, advantage and disadvantage of the methods are compared. It is recommended to divide the distribution range into intervals and to use different methods for the intervals. No definite conclusion is reached yet. Experiments are needed to decide how to divide into intervals and which method is used for each interval.



*J. Tiago de Oliveira (ed.), Statistical Extremes and Applications, 395–402.*

## 2. DOUBLY EXPONENTIAL RANDOM NUMBERS.

Suppose that a sequence of (0,1) uniform random numbers are given, and transform these to obtain doubly exponential random numbers. Assume the uniform random numbers to be ideal, then the problem is to get a random variable X with the probability distribution

$$F(x) = \exp(-\exp(-x)), \quad -\infty<x<\infty, \tag{2.1}$$

or the probability density function

$$f(x) = \exp(-x)F(x), \quad -\infty<x<\infty, \tag{2.2}$$

from the sequence of independent and identically distributed (0,1) uniform random variables $U_1, U_2, \ldots$

The random variable

$$Y = -\log(-\log(A+(B-A)U_1)), \tag{2.3}$$

where $0 \le A = F(a) < B = F(b) \le 1$, $(-\infty \le a < b \le \infty)$, follows a doubly exponential distribution possibly truncated to the interval (a,b); it has the distribution function $(F(x) - F(a))/(F(b) - F(a))$. So there is essentially no difficulty to generate X. However, using some techniques, we can avoid computation of the logarithmic functions. We cannot do it completely but can decrease the probability to compute the logarithmic or other transcendental functions. For the lower tail of f(x), no good technique is found and the transform (2.3) must be used with a small probability. It is assumed that a uniform random number is obtained at a cost quite less than logarithmic or exponential computation.

The main part of the previously proposed method generates a doubly exponential random variable X truncated on an interval $(a - \log M, b - \log M)$:

Algorithm A.

1. Generate a positive integer random variable L following the zero-truncated Poisson distribution with parameter $\lambda$. ($\Pr[L=k] = (\lambda^k/k!)/(e^\lambda-1)$; $k=1,2,\ldots$; $\lambda>0$.)
2. Generate a random sample of the random size L, $(Y_1,\ldots,Y_L)$, from the exponential distribution truncated to the interval (a,b). ($\Pr[Y_j \le y] = (e^{-y} - e^{-a})/(e^{-b} - e^{-a})$.)
3. Compute $X = \max(Y_1,\ldots,Y_L) - \log M$, where $M = \lambda/(e^{-a} - e^{-b})$.

The zero-truncated Poisson random variable for small value of $\lambda$ is generated by looking up a table of cumulative probabilities for a uniform random variable. The exponential random variable is generated by Marsaglia-Sibuya-Ahrens method or by von Neumann and Forsythe's odd-even method (Knuth, 1981).

## 3. REJECTION METHOD 1.

Let $\phi(x)$ denote the probability density function or standard normal distribution. The inequality

$$f(x)/F(0) \le \sqrt{\pi/2} \cdot 2\phi(x), \quad x \le 0, \tag{3.1}$$

holds since $\exp x \le 1+x+x^2/2$, $x\le 0$, and the negative part of $f(x)$ is generated by random rejection of the negative part of $\phi(x)$. The expected value of the necessary normal random variables, $\sqrt{\pi/2} \doteqdot 1.2533$, is not large. For the random rejection, however, a (0,1) uniform random variable is compared to the adoption ratio $r(x) = (e/\sqrt{2\pi})f(x)/\phi(x)$, the ratio of the left-hand side of (3.1) to the right-hand side, which decreases fast to zero when x decreases from zero. To avoid the computation of $r(x)$, approximation by broken lines below and above $r(x)$ may be used.

If f is restricted to an interval (a,b), $-\infty<a<b<\infty$, there exists a constant $C = C(a,b)$ such that

$$f(x) \le C\phi((x-b+e^b-1)/e^{b/2}), \quad a\le x\le b, \tag{3.2}$$

where the equality holds at $x=b$. Adoption ratio, the ratio of the left-hand side of (3.2) to the right-hand side, is minimum at $x=a$. The end points a and b can be positive, and the smaller the interval (a,b), the larger the adoption ratio.

Restricting to smaller intervals, however, we need truncated normal distributions on different intervals and the algorithm become more complex. It is recommended to use this method for an interval (a,0), where a is between -1.4 and -1.0.

## 4. REJECTION METHOD 2.

Since $f(x) \le e^{-x}$, the positive part of $f(x)$ is generated by rejecting the exponential random variable:

$$f(x)/(1-F(0)) = e^{-x}F(x)/(1-1/e) \le e^{-x}/(1-1/e), \quad 0<x<\infty, \tag{4.1}$$

where $1/(1-1/e) \doteqdot 1.5820$ is the expected value of necessary exponential random variables, and $F(x)$ equals the random rejection ratio having the minimum value $F(0) = 1/e \doteqdot 0.36788$. As the rejection method 1 of the previous section, the exponential function is calculated once more if $U\ge F(0)$.

To decrease the probability to calculate the exponential function, $f(x)$ must be restricted again to a smaller interval.

$$f(x)/(F(b)-F(a)) \le Ce^{-x}/(e^{-a}-e^{-b}), \quad a\le x\le b, \tag{4.2}$$

where $C = (e^{-a}-e^{-b})F(b)/(F(b)-F(a))$ and the equality holds at $x=b$. For a smaller interval, the rejection ratio $F(x)/F(b)$ is well approximated by straight lines below and above it. The inequality (4.2) holds even if a and b are negative, but then the method is worse than rejecting a uniform random variable. The method is effective for larger values of x.

The following is a method for generating the positive part of f(x).

Algorithm C.

1. Generate a random variable L with a probability distribution $Pr[L=\ell] = (F(\ell) - F(\ell-1))/(1-F(0))$, $\ell = 1,\ldots,k$, and $Pr[L=k+1] = (1 - F(k))/(1 - F(0))$.
2. Generate an exponential random variable Y truncated to the interval (0,1). Put $X = L + Y$.
3. Generate a (0,1) uniform random variable U.

4.1. If $F(L)U \le F(L-1)$, then adopt X.
4.2. If $F(L)U \le F(X)$, then adopt X.
4.3. $(F(L)U > F(X))$ If $L>k$, then increase L by 1, and go back to 2.

The value of k is 5 or 6 and table look-up is good for generating L. Steps 4.1 - 4.3 should be actually different for $L<k$ and for $L \ge k+1$ since we cannot prepare infinite number of $F(\ell)$'s. For $L<k$, a better 'squeeze method' can be used at Step 4.1.

## 5. ODD-EVEN METHOD.

The odd-even method by von Neumann and Forsythe is applied to the probability density on a finite interval of the form

$$C \exp(-h(x)), \quad a \le x \le b, \tag{5.1}$$

where $0 \le h(x) \le 1$ on the interval and C is a positive constant. Since $x + e^{-x}$ is a convex function with its minimum value at $x=0$, the method is applied to f(x) restricted, for example, to (a,b) such that $0<a<b$ and $a + e^{-a} + 1 = b + e^{-b}$;

$$f(x)/(F(b)-F(a)) = \exp(-x-e^{-x}+a+e^{-a})/f(a)(F(b)-F(a)), \quad a \le x \le b. \tag{5.2}$$

In the odd-even method, a random variable X, which is uniformly distributed on (a,b), is generated and $h(X) = X + \exp(X) - a - \exp(-a)$ is compared with a (0,1) uniform random variable U. The function h is approximated by straight lines below or above it.

To generate X on $(0,\infty)$, for example, define $b_n$ to be the positive solution of $x + e^{-x} = n$, $n = 1,2,\ldots$ The sequence $(b_n)$ satisfies $b_1=0$, $n-1 < b_n < n$, and $n - b_n$ decreases to zero. Put

$$h_n(x) = x + e^{-x} - n, \quad h_{n2}(x) = (x - b_n)/(b_{n+1} - b_n),$$

$$h_{n1}(x) = h_{n2}(x) + h_n(c_n), \text{ and } c_n = \log((b_{n+1} - b_n)/(b_{n+1} - b_n - 1)$$

where $c_n$ satisfies $h_n(c_n) = 1/(b_{n+1} - b_n)$. The latter two functions approximate $h_n$, and $h_{n1} \le h_n \le h_{n2}$, $n = 1,2,\ldots$

Algorithm D.

1. Generate a positive integer random variable L with probabilities $\Pr[L=\ell] = F(b_{\ell+1}) - F(b_\ell)$, $\ell = 1,\ldots,k$ and $\Pr[L=k+1] = 1 - F(b_{k+1})$.
2. Generate a uniform random variable $U_1$ and put $X=b_n+(b_{n+1}-b_n)U_1$.
3. Generate a uniform random variable $U_2$.

4.1. If $h_{n2}(X) < U_2$, then adopt X and return.
4.2. If $h_{n1}(X) > U_1$, then go to 5.
4.3. If $h(X) < U_2$, then adopt X and return.
5. Put $U_e = U_2$.
6. Generate a new uniform random number and put it to $U_o$.
7.1. If $U_e > U_o$, go to 8.
7.2. Otherwise, if $L \ge k+1$, then increase L by 1.
7.3. Go back to 2.
8. Generate a new uniform random number and put it to $U_e$.
8.1. If $U_o < U_e$, then adopt X and return.
8.2. Otherwise, go back to 6.

To generate X on $(-\infty,0)$, modify the above discussions remarking that $h(x)$ is decreasing. In Steps 2 - 4 of Algorithm D, the case $L \ge k+1$ must be treated differently since we cannot prepare many of $b_n$, $h_{n1}$, and $h_{n2}$.

## 6. RATIO OF UNIFORM RANDOM VARIABLES.

Kinderman and Monahan's method to obtain a random variable X with a probability density function $f(x)$, in general, is to generate a random point (S,T) uniformly distributed on a set W of the (u,v) plane defined by

$$W = \{(u,v);\ u^2 \le f(v/u)\} \tag{6.1}$$

and put $X = T/S$.

For f of (2.1), W has a shape of skewed egg and limited by a rectangle W';

$$W \subset W' = \{(u,v);\ 0 \le u \le e^{-1/2} \text{ and } v_1 \le v \le v_2\},$$

where $e^{-1/2} \doteqdot 0.6065$, $v_1 \doteqdot -0.4253$ and $v_2 \doteqdot 0.6929$. These values are $v_i = x_i\sqrt{f(x_i)}$ where $x_i$'s are solutions of the equation $x(1 - e^{-x}) = 2$. We generate a random point uniformly on W' and adopt it if it is within W.

The area of $W_+$, the upper part of W with positive v, is $(1 - e^{-1/2})/2$, and that of $W_-$, the lower part of W with negative v, is $e^{-1/2}/2$. The ratio of the area of W' to that of W is about 1.3564, which is the expected value of number of uniform random points (S',T') on W' until getting one within W.

The shape of W suggests sufficient conditions that $(u,v) \in W$. One is

$$0 \le v \le au(e^{-1/2} - u),$$

corresponding to $0 \le (T'/S') \le a(e^{-1/2} - S')$, where

$$a = \min_{0<x<\infty} x/(\sqrt{1/e} - \sqrt{f(x)}) \doteqdot 7.5298$$

which covers area $a/6e^{3/2} \doteqdot 0.2800$, 88.6% of $W_+$. Another is

$$0 \le -v \le bu(e^{-1/2} - u)^{1/2},$$

corresponding to $T' < 0$ and $(T'/S')^2 \le b^2(e^{-1/2} - S')$, where

$$b = \min_{-\infty<x<0} -x/(\sqrt{1/e} - \sqrt{f(x)})^{1/2} \doteqdot 2.3374.$$

Both together cover area about 0.4586 of that 0.5 of W.

A sufficient condition for $(u,v) \bar{\in} W$ is

$$x^2(3 - x)/12 \ge (1/u\sqrt{e}) - 1, \quad x = v/u,$$

which is based on the inequality $e^{-x} \ge 1 - x + x^2/2 - x^3/6$, $-\infty<x<\infty$. The condition covers corners of W' except for the upper left one. Other sufficient conditions for rejection can be constructed. The condition of (6.1) is checked only when both the sufficient conditions do not hold. For f of (2.1) the computation of f(v/u) may overflow for negative v and very small u.

## 7. ASYMMETRIZATION METHOD.

Let Z be a random variable with a probability density function h(z) which is symmetric about the origin; $h(-z) = h(z)$. Let s(z) be a function defined for z such that $h(z)>0$ and $0 \le s(z) \le 1$. A random variable X is defined by

$$X = \begin{cases} Z \text{ with the probability } s(Z), \\ -Z \text{ with the probability } 1 - s(Z). \end{cases} \quad (7.1)$$

Then X has the probability density function

$$f(x) = h(x) + h(x)t(x), \text{ where } t(x) = s(x) - s(-x).$$

Conversely, since t is an odd function

$$h(x) = f_e(x) = (f(x) + f(-x))/2,$$

the even part of f, and

$$t(x) = f_o(x)/f_e(x), \text{ where } f_o = (f(x) - f(-x))/2,$$

the odd part of f. The function s is not uniquely determined by t; an even function is added to s without changing t.

For f of (2.2)

$$h(x) = (\exp(-x - \exp(-x)) + \exp(x - \exp x))/2,$$
$$= \exp(-\cosh x)\cosh(x - \sinh x),$$

and

$$t(x) = \tanh(\sinh x - x).$$

The odd function t(x) is monotone increasing from -1 to 1, and a simple and effective choice of s is

$$s(x) = 1 + t(x) \text{ if } x<0, = 1 \text{ if } x\geq 0.$$

The function h is very close to $\exp(-1 - x^2/2)$ near the origin, and X is generated by random rejection of normal random variable. The adoption ratio r(x) is proportional to $f_e(x)\exp(-1 + x^2/2)$, which is an even function, decreases very slowly when x increases from zero, reaches the minimum value 0.9517... at $x \doteqdot 1.355$, then increases faster reaching 1 at $x \doteqdot 1.6746$.

The following algorithm generates X truncated to (-a,a) by random rejection of normal random variables and by random sign change. Here a=1.67 is used.

Algorithm F.

1. Generate a standard normal random variable X truncated to (-a,a).
2. Generate a uniform random variable $U_1$.

3.1. If $U_1 < .9517$, go to 4.

3.2. If $U_1 > h(x)/C\phi(x)$, go back to 1.
(C $\doteqdot$ 1.0135 is the expected number of necessary truncated normal random variables.)

4. If X>0, adopt it.
5. Generate a uniform random variable $U_2$.

6.1. If $U_2 \leq t(x)$, change the sign of X and adopt it.

6.2. Otherwise, adopt X.

In order to squeeze, good approximations of t on (-1.67,0) are

$$-(x^3/6)(1 - x^6/108) < t(x) < -0.1739x^3.$$

## 8. DESIGN OF A SUBROUTINE.

In previous sections, six methods for generating doubly exponential random variable X are suggested. We must examine them further to choose the best one.

Firstly, there is no one which generates X in the whole range $(-\infty,\infty)$ at a reasonable cost. The range must be divided at least into two parts, lower and upper ones. By dividing the range into smaller parts and preparing more constant data, the efficiency is improved in any of these methods. Marsaglia's rectangle-wedge-tail method can be used for a fast subroutine of this type. Our object is to construct fairly efficient subroutines of reasonable size.

The method using ratio of uniform random variables of Section 6 essentially divides the range into positive and negative parts, and is the simplest for programming. The method is criticized as unstable since the denominator random numbers are not ideal and the tail parts of the distribution are largely distorted. The trouble can be avoided, however, improving the uniform random number when the denominator is very small. Unfortunately this method is not fast enough. In an implementation the speed is less than two times of computing $-(\log(-\log U))$.

In the odd-even method of Section 5, the expected number of used uniform random variables is limited by $1+e$ for any interval, but the approximation of $h(x)$ is more difficult near the origin. While the maximum of random number of exponential random variables of Section 1 is effective near the median, and these two can be complementary.

The rejection method 1 of Section 3 is effective at the upper tail. Near the origin, $f(x)$ is close to normal density, and the rejection method 2 of Section 2 and the asymmetrization of Section 6 are preferable. A combination of these is another possibility. No method which fits to the lower tail was found.

To say more definitely, more calculations and experiments are needed. Some of them and details of data of this paper will be published in a technical report.

REFERENCES.

[1] Knuth, Donald E. (1981) *The Art of Computer Programming*, Vol. 2, Second ed., (Chapter 3, Random numbers, 3.4.1), Addison-Wesley.

[2] Sibuya, M. (1968) Generating doubly exponential random numbers, *Ann. Inst. Stat. Math.*, Supplement V, 1-7.

# Probability Problems in Seismic Risk and Load Combinations for Power Plants*

Laurence L. George

Lawrence Livermore National Laboratory
P. O. Box 808, L-140
Livermore, CA 94550 USA

ABSTRACT

This paper describes seismic risk, load combination, and probabilistic risk problems in power plant reliability, and it suggests applications of extreme value theory.

Seismic risk analysis computes the probability of power plant failure in an earthquake and the resulting risk. Components fail if their peak responses to an earthquake exceed their strengths. Dependent stochastic processes represent responses, and peak responses are maxima. A Boolean function of component failures and survivals represents plant failure.

Load combinations analysis computes the cdf of the peak of the superposition of stochastic processes that represent earthquake and operating loads. It also computes the probability of pipe fracture due to crack growth, a Markov process, caused by loads. Pipe fracture is an absorbing state.

Probabilistic risk analysis computes the cdf's of probabilities which represent uncertainty. These cdf's are induced by randomizing parameters of cdf's and by randomizing properties of stochastic processes such as initial crack size distributions, marginal cdf's, and failure criteria.

*This work was supported by the U. S. Nuclear Regulatory Commission under a memorandum of understanding with U. S. Department of Energy.



*J. Tiago de Oliveira (ed.), Statistical Extremes and Applications, 403–411.*

## 1. THE SEISMIC RISK PROBLEM

Seismic risk analysis computes the probability of earthquake and plant failure and the resulting risk. Since earthquakes and plant failures are rare, extreme value theory may help seismic risk computation.

A typical component reliability model is (1)

$$P[\text{failure}] = P[R \geq S] \text{ or } P[R/S \geq 1] \quad (1.1)$$

where R is the random response applied to the component and S is its random strength at failure. R and S may be dependent rv's. Traditionally, R and S are assumed to be independent lognormal random variables so probability computation is convenient. If the response is a random process, the model is

$$P[\text{failure}] = P[\sup(R(t) - S(t)) \geq 0;\ 0 \leq t \leq T] \quad (1.2)$$

where strength S(t) at time t may be a stochastically decreasing random process due to deterioration caused by the response process prior to time t. If the component can fail in several modes,

$$P[\text{failure}] = P[\ \cup\ (\sup(R_j(t)-S_j(t)) \geq 0;\ 0 \leq t \leq T] \quad (1.3)$$

where the union is over all modes j=1, 2,...,n. Sequential responses such as earthquake and then fire may be represented by

$$P[\text{Failure}] = P[R_{EQ} \geq S_{EQ} \cup R_F \geq S_F] \quad (1.4)$$

where $R_F$ and $S_F$ are dependent random variables or processes characterizing component fire response and strength. The modes are neither independent nor identically distributed.

A typical power plant system failure model is (2)

$$P[\text{system failure}] =$$

$$P[\ \phi(\underline{R} - \underline{S}) = 1] = P[\cup\ C_j] = P[\cup \cap\ B_{ij}] \quad (1.5)$$

where $\phi(.)$ is the system "structure function" (3), the $C_j$ are minimal cut sets (4) or perhaps disjoint intersections (5) and the $B_{ij}$ are "basic events" such as component survivals or failures.

A typical power plant failure model is a union of sequences of system survivals or failures represented by an "event tree", figure 1. Then

$$P[\text{Plant Failure}] = P[U \cap D_k] \qquad (1.6)$$

where the $D_k$ are system survivals or failures. Reference (6) describes a computer code for computing the probability of plant failure and doing "sensitivity" analyses.

Loss of load probability for a power generating system is the same as the probability of plant failure except electricity demands replace responses and generating or transmission capacities replace strengths (7).

There are several applications of statistical extremes. For a reliable component, the probability of failure, $P[R \geq S]$, is small. It is determined by the tails of cdf's. Which extreme value cdf's should we use to model the joint cdf's of R and S? Does it matter if we accurately model the whole cdf's? Is there a model of $P[R \geq S]$ assuming both R and S have extreme value cdf's? Similar questions apply to the model in equation (1.2). (Several approximations to this probability will be mentioned in the next section.) Is it adequate to consider only the extremes of R(t) if the R(t) and S(t) processes are dependent? What if there are many modes of failure in the model in equation (1.3)

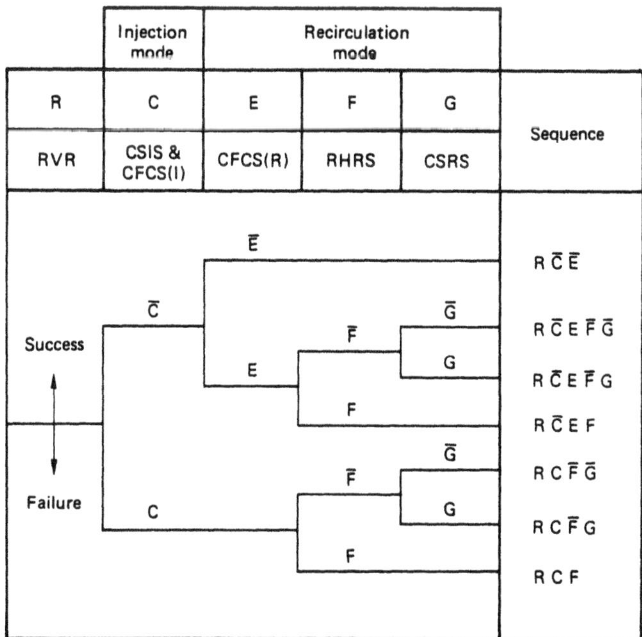


Figure 1. Reactor vessel-rupture event tree

that are approximately identically distributed rv's? What if there are many components in the model in equation (1.5) that have identically distributed rv's? Series or parallel system bounds won't suffice because the structure function $\phi(.)$ isn't "coherent".

Statistical problems arise because failures are scarce and testing is expensive. There is usually insufficient data to estimate component failure probabilities, the cdf of responses and the cdf of strengths. How do we estimate parameters of the random processes and the strength, resistance or capacity of components? It may be assumed that failures, if any are observed, are extremes. How should we estimate extreme value cdf's when data are scarce and censored?

Traditional fault and event tree methods find minimal cut sets and compute bounds on power plant reliability. Simulation or numerical integration methods give reliability bounds. These methods have limitations:

1. components have multiple states so representing them as good or bad is unrealistic,
2. reliability bounds are inaccurate if reliability is not high such as in an earthquake,
3. the conditional probabilities of dependent events are rarely represented accurately in computing probabilities of intersections, and fault tree analysts usually assume independent basic events,
4. fault and event trees give "snapshots" of plant conditions and aren't sufficient to represent dependent sequential failures,
5. competing modes of failures greatly enlarge fault trees,

One limitation which may be eliminated by extreme value theory is the computation of probability of plant failure due to competing modes (the multiple component analog of equation (1.3)). What is the reliability of a power plant subjected to loads in several modes? The loads are random processes, e.g. earthquake and fire, and, if peak responses to any load exceed component strengths, components fail.

Given the reliability model, how should we characterize uncertainty due to sampling error and modeling uncertainty. "Probabilistic" Risk Analysis randomizes parameters of input cdf's to represent uncertainty in plant failure probability. The probability bounds computed by traditional methods have cdf's induced by randomizing input parameters (8). Since the reliability model is complex it is difficult to compute the induced cdf of the bounds. Simulation of the effects of randomizing input parameters is unsatisfactory because the cdf's of uncertain parameters are arbitrary. Can we compute the maximum probability of plant failure given extremes of input parameters? Is there a "maximum entropy" or maximum uncertainty estimate of plant failure probability that requires no unwarranted assumptions regarding the limits or cdf's of unknown parameters?

## 2. THE SUPERPOSITION LOAD COMBINATION PROBLEM

Should piping design loads be combined by adding absolute values, by taking the square root of the sum of the squares (SRSS) or by similar formulas involving weights? Piping design loads include dead loads, operating loads, wind loads, snow loads, earthquake loads and tornado loads. Most loads are stochastic processes and no deterministic combination represents the superposition of stochastic processes, but piping designers insist on deterministic design codes. Probabilists try to accomodate with deterministic load combination formulas that insure a specified probability of "nonexceedance" for the superposition of a specified set of loads (9). They pick deterministic formulas that have probabilistic foundation. The sum of absolute values of upper bounds is an upper bound, but unfortunately design loads are "representative" values not upper bounds. The SRSS is the mean of the sum of independent Gaussian processes, but unfortunately many stochastic load processes are neither Gaussian nor independent.

Load process models often imbed stationary Poisson processes and some such load combination problems have exact solutions. A prototype load combination problem is to find the cdf of the peak load over a finite time of $X(t) + Y(t)$ where $X(t)$ is a Poisson Square Wave Process (PSW) and $Y(t)$ is a Poisson Shock Processes (PSP),

$$P[\sup(X(t) + Y(t)) \leq z;\ 0 \leq t \leq T]. \tag{2.1}$$

See figure 2. This problem has been solved exactly for independent $X(t)$ and $Y(t)$ by Jacobs and Gaver (10) and dependent $X(t)$ and $Y(t)$ (11). Jacobs and Gaver computed the limiting extreme value distribution but told me that convergence to it is slow.

In real power plants, there are many load processes, not all have imbedded Poisson processes, not all are stationary or independent, and the superposition is not summation. Can extreme value theory help find

$$P[\sup g(\underline{X}(t)) \leq z;\ 0 \leq t \leq T] \tag{2.2}$$

for $\underline{X}(t) = (X_1(t),...,X_k(t))$, large $z$ and simple scalar functions $g(.)$ such as SRSS, linear combinations or SRSS of linear combinations, for instance the resultant of three

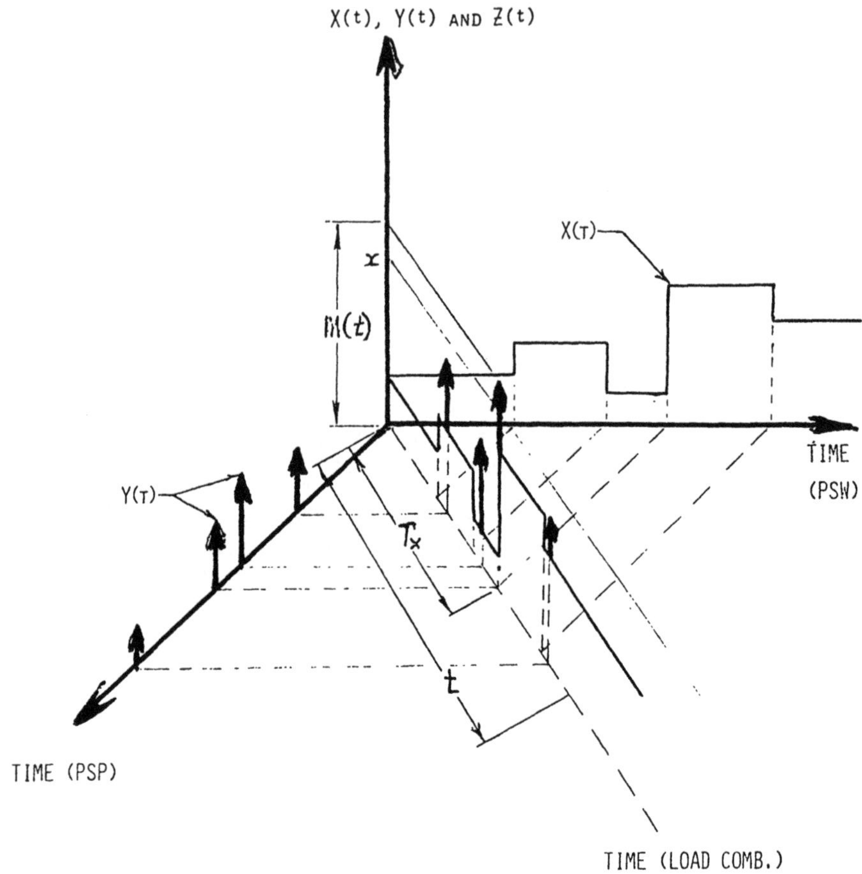


Figure 2. The Superposition of a PSW and A PSP

dimensional load processes? Is the typical 40 year plant life enough for extreme values approximations to be accurate? What if the durations of the shocks are short relative to the durations of the square waves? What if the square wave process alternates between big and little to represent the strengths of a system in operation and in maintenance? What if some load processes have no imbedded Poisson processes? Can the renewal theory of (10) and (11) still be used?

## 3. THE CRACK GROWTH AND PIPE FRACTURE PROBLEM

The probability of pipe fracture due to crack growth is

$$P[\text{pipe fracture}] = P[\sup A(t) \geq L;\ 0 \leq t \leq T] \quad (3.1)$$

where A(t) is the crack depth and L is the "ligature" of uncracked material that can withstand fracture. Sometimes A(t) is two dimensional to represent depth and length. Then

$$P[\text{Pipe fracture}] = P\ [\sup(g(A(t)) - g(\underline{L})) \geq 0;\ 0 \leq t \leq T]. \quad (3.2)$$

where g(.) may be a vector valued or scalar function such as area.

Fracture mechanics experts accept the crack growth model

$$dA(n)/dn = C(\Delta k(n))^M \quad (3.3)$$

where C and M are material property constants, $\Delta k(n)$ is the "stress intensity factor", and n is a discrete time variable counting stress cycles. The stress intensity factor depends on the random stress process, the ratio of peak positive and negative stresses, the crack size and pipe geometry,

$$\Delta k(n) = k(\sigma_n)/\sqrt{(1 - R^2)} \quad (3.4)$$

where $k(\sigma_n)$ is the stress intensity due to stress $\sigma_n$ at the crack tip on cycle n and R is the ratio of positive to negative stresses.

Probabilists recognize A(t) or A(n) as a Markov process with an absorbing state, but the process is so complicated so that pipe fracture probability, equations (3.1) or (3.2), is usually simulated (12, 13).

Probabilistic pipe fracture analysis asks the following questions. What is the induced cdf of pipe fractures probability assuming initial crack size cdf, crack probability, C, N, and L are random variables? What if there are several neighboring cracks growing simultaneously and perhaps merging? What if there are several locations in power plant piping where cracks may grow? What is the probability of pipe fracture due to crack growth anywhere? Can extreme value theory help answer these questions or provide more rational ways to quantify uncertainty to help in decision making?

REFERENCES


(1) K. C. Kapur and L. L. Lamberson, Reliability in Engineering Design, Wiley, New York, 1977.

(2) U. S. Nuclear Regulatory Commission, "An Assessment of Accident Risks in U. S. Commerical Nuclear Power Plants", WASH 1400 (NUREG-75/014), Washington, D. C., 1975.

(3) R. E. Barlow and F. Proschan, Statistical Theory of Reliability and Life Testing: Probability Models, Holt, Rinehart and Winston, New York, 1975.

(4) W. E. Vesely, F. F. Goldberg, N. H. Roberts and D. F. Haasl, Fault Tree Handbook, NUREG-0492 U. S. Nuclear Regulatory Commission, Washington, D. C. 1981.

(5) G. C. Corynen, "STOP: A Fast Procedure for the Exact Computation of the Performance of Complex Probabilistic Systems", Lawrence Livermore National Laboratory UCRL 53230, 1980.

(6) J. E. Wells, L. L. George, G. E. Cummings "Seismic Safety Margins Research Program - Phase I Final Report - Systems Analysis Project VII" UCRL-53021 Vol. 8. NUREG/CR2015 Vol. 8, Lawrence Livermore National Laboratory, 1982

(7) L. L. George and J. E. Wells "Loss of Load Probability for Systems of Dependent Transmission and Generating Components," Proceedings, Reliability Conference for th Electric Power Industry, Portland, OR, April 1981.

(8) A Dirk, G. P. Dahlgren, G. P. Steck, R. G. Easterling and R. L. Iman, "Uncertainty Propagations Through Computer Codes", Proceedings, Topical Meeting on Probabilistic Analysis of Nuclear Reactor Safety, 1978.

(9) B. Ellingwood, T. V. Galambos, J. G. MacGregor, and C. A. Cornell "Development of a Probability Based Load Criterion for American National Standard A58 Building Code Requirements for Minimum Design Loads in Buildings and Other Structures" U. S. National Bureau of Standards SP-577, Ap. 1980.

(10) P. Jacobs and D. Gaver, "On Combinations of Random Loads", SIAM J. of Applied Math. 40(3), 454-466 June 1981.

(11) L. L. George "Peak Combined Loads from a Bivariate Poisson Shock and Square Wave Process", UCRL 84150, ORSA/TIMS Meeting Colorado Springs, CO 1980.

(12) E. Lim and S. L. Basin, "Piping Reliability Analysis Including Seismic Events", Science Applications, Inc., Palo Alto, CA, June 1981

(13) L. L. George, E. Y. Lim, S. L. Basin and D. L. Iglehart, "A Fracture Mechanics Evaluation of Reactor Piping Reliability: Simulation of Pipe Fracture Probability", Proceedings of Structural Mechanics in Reactor Technology, Paris, Aug. 1981.

# THE DISTRIBUTION OF THE MAXIMAL TIME TILL DEPARTURE FROM A STATE IN A MARKOV CHAIN

Barry C. Arnold and Jose A. Villaseñor

University of California and Colegio de Postgraduados

Let $M_n$ be the maximal success run in the first n trials in a sequence of Bernoulli trials. Asymptotic bounds on the distribution of $M_n$ are given here. More generally, it is possible to get bounds on the asymptotic distribution of the maximal waiting time till departure from a given state in a finite Markov chain and from any state. The problem turns out to be more straightforward when dealing with the continuous time case. The limiting distributions are of double exponential type and the norming constants depend on the diagonal of the intensity matrix and the long run distribution.

## 1. INTRODUCTION

Consider a sequence of Bernoulli trials, each of which can result in one of two outcomes labelled success and failure. Let $M_n$ represent the length of the longest run of successes in the first n trials.

Although $M_n$ cannot be normalized to yield a non-degenerate limiting distribution, it is possible to provide bounds on its asymptotic distribution. The problem can be viewed as a particulary simple case of the more general problem which deals with the maximal time till departure from a given state of a finite Markov chain. Bounds on the asymptotic distribution of such random variables are presented in Section 3. In addition it is not difficult to consider the more general problem regarding the longest run of any kind (either of successes or of failures) or its



J. Tiago de Oliveira (ed.), Statistical Extremes and Applications, 413–426.

generalization: the maximal waiting time till departure from any state in a finite Markov chain. These results are also outlined in Section 3.

It turns out that if instead we consider a continuous time Markov chain, legitimate limiting distributions can be obtained for the maximal time till departure from a state or from any state. This material is presented in Section 2, before presenting the slightly more complicated results for the discrete time case. A key tool in both Sections 2 and 3, is a slight generalization of a theorem of Barndorff-Nielsen [2] which appears as Theorem 6.2.1 in Galambos [6, p. 282]. The limiting distributions, in the continuous time case, and the limiting bounds, in the discrete time case, are both of double exponential type. The appropriate normalizations depend only on the diagonal of the intensity (or transition) matrix and on the long run distribution of the chain. In Section 4, we sketch alternative derivations of some of the discrete time results using ideas related to recurrent events. These elementary techniques are quite tiresome to apply and Section 4 serves to highlight the power of the generalized Barndorff-Nielsen theorem.

In the final section we discuss the problem in a Semi-Markov process setting.

## 2. MAXIMAL TIME TILL DEPARTURE FROM A STATE IN A CONTINUOUS TIME MARKOV CHAIN

Consider an irreducible continuous time Markov chain, $X(t)$, with k states. Denote the corresponding intensity matrix by Q. The elements of Q satisfy

$$Q \underline{1} = \underline{0} .$$

Off diagonal elements are non-negative while diagonal elements are strictly negative. A typical sojourn in state i will thus have an exponential distribution with mean $-1/q_{ii}$ and, when the process leaves state i, the probability that it goes to state j ($\neq$i) is $q_{ij}/(-q_{ii})$. The long run distribution $\underline{\pi}$ is the unique solution to the equation $\underline{\pi}\, Q = \underline{0}$ satisfying $\sum_{i=1}^{k} \pi_i = 1$ .

During a time interval $(0, T]$, a particular state i will be visited a random number of times. Let $N_i(T)$ denote the number of visits to state i. Each visit will have an exponentially distributed duration with mean $-q_{ii}$. We denote the sequence of durations, sojourns or times till departure by $\{X_j^{(i)}, j=1,2,\ldots\}$. We focus attention on the asymptotic distribution of two random variables: $M_i(T)$ the maximal time till departure from state $i$ during the interval $(0, T]$ and $M(T)$, the maximal time till departure from any state during the interval $(0, T]$. Specifically we have

$$M_i(T) = \max_{j \leq N_i(T)} X_j^{(i)} \tag{2.1}$$

and

$$M(T) = \max_{i \leq k} M_i(T) . \tag{2.2}$$

We verify that, suitably normalized, each of these random variables has a double exponential asymptotic distribution. The appropriate normalization will be seen to depend only on $\{-q_{ii}, i = 1, 2, \ldots, k\}$ and $\underline{\pi}$.

The key observation is that visits to state i form a delayed recurrent event process (the distribution of the time till the first visit depends on the initial state of the process, but subsequent inter-visit times are i.i.d. random variables). It thus follows that $N_i(T)/T \xrightarrow{a.s.} \delta_i$. Considering that the proportion of time spent in state i converges to $\pi_i$ and that this time is representable as a sum of $N_i(T)$ i.i.d. exponential$(-q_{ii})$ random variables, we may conclude that $\delta_i = -q_{ii}\pi_i$.

Galambos [6, p.282] provides a useful theorem regarding the aymptotic distribution of the maximum of a random number of i.i.d. random variables. A slight extension of this theorem with analogous proof, may be stated as follows.

*Theorem 2.1.* Let $X_1, X_2, \ldots$ be i.i.d. random variables such that

$$a(n)(\max_{i \leq n} X_i) + b(n) \xrightarrow{d} \Lambda(x)$$

and let $\{N(T): T \in R^+\}$ be a stochastic process such that

$$N(T)/T \xrightarrow{P} \delta .$$

If we define

$$Z(T) = \max_{j \leq N(T)} X_j$$

it follows that

$$a(\delta T)\ Z(T) + b(\delta T) \xrightarrow{d} \Lambda(x) .$$

We may apply Theorem 2.1 directly to resolve the question of the asymptotic distribution of $M_i(T)$. Here $\delta = -q_{ii}\pi_i$ and the i.i.d. $X_j^{(i)}$'s are i.i.d. exponential random variables with mean $(-q_{ii})^{-1}$. It is well known that for such i.i.d. exponential random variables the appropriate normalization to obtain the asymptotic distribution of the maximum of $X_1^{(i)}, X_2^{(i)}, \ldots, X_n^{(i)}$ is

$$a(n) = -q_{ii}$$

$$b(n) = -\log n .$$

The resulting asymptotic distribution is of standard double exponential type.

We thus have as a corollary to Theorem 2.1.

*Theorem 2.2.* Let $M_i(T)$ denote the maximal time till departure from state i during the time interval $(0, T]$ in an irreducible continuous time Markov chain with intensity matrix Q and long run distribution $\underline{\pi}$. The asymptotic distribution of $M_i(T)$ is given by

$$\lim_{T\to\infty} P\left[(-q_{ii})M_i(T) - \log(-q_{ii}\pi_i) - \log T \leq z\right] = \exp\{-e^{-z}\},\ z\varepsilon R. \quad (2.3)$$

In order to resolve the question of the asymptotic distribution of M(T), the maximal time till departure from any state during the interval $(0, T]$, we need a slight extension of Theorem 2.1 which depends upon an elementary result regarding maxima of random variables with heterogenous exponential distributions. The results in question are:

*Lemma 2.3.* Let $\{X_j^{(i)}: j = 1,2,3, \ldots\}$, $i=1,2, \ldots, k$ be independent sequences of independent exponential random variables with $E(X_j^{(i)}) = 1/\lambda_i$ in which all the $\lambda_i$'s are distinct. Without loss of generality assume $\lambda_1 < \lambda_i$, $i\geq 2$. Define

$$M_n = \max_i \max_{j\leq\alpha_i n} X_j^{(i)}$$

where the $\alpha_i$'s are fixed but arbitrary positive numbers. It follows that

$$\lim_{n\to\infty} P(\lambda_1 M_n - \log\alpha_1 n \leq z) = \exp\{-e^{-z}\}, \quad z\varepsilon R .$$

This lemma states the intuitively obvious fact that the maximum is most likely to come from the sequence of exponential random variables with largest mean and the appropriate normalization is the normalization for observations drawn only from that sequence. As a consequence of this lemma and of the fact that according to Theorem 2.2, since $N_i(T)/T \xrightarrow{P} -q_{ii}\pi_i$, $\max_{j\leq N_i(T)} X_j^{(i)}$ has the same asymptotic distribution as $\max_{j\leq -q_{ii}\pi_i T} X_j^{(i)}$, we readily obtain the following:

*Theorem 2.4.* Let $M(T) = \max_{i<k} M_i(T)$ denote the maximal time till departure from any state during the interval $(0, T]$ in an irreducible continuous time Markov chain with intensity matrix Q and long run distribution $\underline{\pi}$ . Assume that, for some $k_1$, $-q_{11} = -q_{22} \cdots = -q_{k_1k_1} < -q_{jj}$, $j > k_1$. The asymptotic distribution of M(T) is given by

$$\lim_{T\to\infty} P\left[(-q_{11})M(T) - \log(-q_{11} \sum_{i=1}^{k_1} \pi_i) - \log T \leq z\right]$$

$$= \exp\{-e^{-z}\}, \quad z\varepsilon R . \tag{2.4}$$

*Example 2.1.* Two state process. Suppose that

$$Q = \begin{bmatrix} -\alpha & \alpha \\ \beta & -\beta \end{bmatrix}$$

where $\alpha, \beta>0$. One readily finds $\underline{\pi} = (\frac{\beta}{\alpha+\beta}, \frac{\alpha}{\alpha+\beta})$ and the correct normalizing constants for the longest time till departure from state 1 are $a(T) = \alpha$, $b(T) = -\log(\frac{\alpha\beta}{\alpha+\beta}) - \log T$. This process can be viewed as an alternating renewal process with exponentially distributed up and down times. If we consider the number of renewals $N^*(T)$ of the (uptime plus downtime) cycles then $N^*(T)/T \xrightarrow{P} \alpha^{-1} + \beta^{-1}$ and we are concerned with the maximum of $N^*(T)$ i.i.d. exponential($\alpha$) random variables. Galambos' Theorem 6.2.1 [6, p. 282] could be applied directly here. The correct normalization for the maximal time till departure from either state depends on whether $\alpha = \beta$ or not. If $\alpha < \beta$ then $a(T) = \alpha$ and $b(T) = -\log(\frac{\alpha\beta}{\alpha+\beta}) - \log T$. If $\alpha = \beta$, then $a(T) = \alpha$ and

$b(T) = -\log\alpha - \log T$.

*Example 2.2.* A simple circulant. Suppose Q is of the form

$$Q = \begin{bmatrix} -\alpha_1 & \alpha_1 & 0 & \cdot & \cdot & \cdot & 0 \\ 0 & -\alpha_2 & \alpha_2 & & & & \cdot \\ \cdot & & -\alpha_3 & \alpha_3 & & & \\ \cdot & & \cdot & \cdot & \cdot & \cdot & \cdot \\ 0 & & & & \cdot & & \cdot \\ \alpha_k & 0 & \cdot & \cdot & 0 & \cdot & -\alpha_k \end{bmatrix} .$$

It follows readily that $\pi_i = \alpha_i^{-1} / \sum_{j=1}^{k} \alpha_j^{-1}$ . The normalizing constants for the maximal time till departure from state 1 are thus $a(T) = \alpha_1$ and $b(T) = \log \sum_{i=1}^{k} \alpha_i^{-1} - \log T$ .

*Example 2.3.* Busy channels in a telephone exchange. Here the state space is $\{0,1,2, \ldots, k\}$ and

$$q_{00} = -\lambda , \qquad q_{01} = \lambda$$

$$q_{ii} = -\lambda - i\mu , \qquad q_{i,i+1} = \lambda, \; q_{i,i-1} = i\mu, \; i=1,2,\ldots,k-1$$

$$q_{kk} = -k\mu, \qquad q_{k,k-1} = k\mu .$$

The asymptotic distribution of the longest time till departure from any state depends on which of the $-q_{ii}$'s is smallest. The smallest $-q_{ii}$ will be either $-q_{00} = \lambda$ or $-q_{kk} = k\mu$ depending on whether $\lambda/\mu < k$ or not. ($\lambda/\mu$ would be the mean queue length in an unlimited queue).

## 3. MAXIMAL TIME TILL DEPARTURE FROM A STATE IN A DISCRETE TIME MARKOV CHAIN

In the case of a discrete time chain the sojourns in the states are geometrically distributed and the maximal sojourn in a particular state will correspond to a maximum of a random number of i.i.d. geometric random variables. If we could apply Theorem 2.1 to geometric random variables then our analysis of Section 2 would carry over with minor modifications. There is a fly in the ointment, however. It is not possible to normalize maxima of geometric random variables to obtain a non-degenerate limiting distribution. We can however get useful asymptotic bounds on probability statements regarding such maxima by considering certain exponential random variables intimately related to the geometric random variables. Specifically if $X \sim G(p)$ (i.e. $P(X=k) = pq^{k-1}$, $k=1,2,3, \ldots$), we may consider a random variable Y which has an exponential ($-\log q$) distribution truncated to the interval (0,1) and assumed independent of X. In such a setting it is readily verified that $W = X - 1 + Y \sim \exp(-\log q)$. Since by construction $0 < Y < 1$ we have $W < X < W + 1$ and consequently maxima of i.i.d. X's are bounded by maxima of i.i.d. W's (cf. Anderson [1]).

With this in mind we can quickly resolve the problem of obtaining asymptotic bounds on the distribution of the maximal time till departure from a given (and from any) state in a finite irreducible Markov chain. To avoid trivialities, we assume that it is possible to stay in at least one state, i.e. that $p_{ii} > 0$ for at least one i (note that this implies the chain is aperiodic). Thus suppose that our k state chain has transition matrix P and long run distribution $\underline{\pi}$. For a particular state i with $p_{ii} > 0$, let $M_i(n)$ be the maximal time till departure from state i during the interval of time $(0, n]$. Clearly each sojourn in state i has a geometric $(1-p_{ii})$ distribution and visits to state i form a delayed recurrent event process. If we let $N_i(n)$ be the number of visits to state i then $N_i(n)/n \xrightarrow{P} \delta_i$ and

$$M_i(n) = \max_{j \le N_i(n)} X_j^{(i)}$$

where $X_j^{(i)} \sim \text{geometric}(1-p_{ii})$. It follows that

$$\delta_i = \pi_i(1-p_{ii}) .$$

Now we may define $W_j^{(i)}$'s to be exponential($-\log p_{ii}$) such that $W_j^{(i)} < X_j^{(i)} < W_j^{(i)}+1$ and we have $\max\limits_{j \le N_i(n)} W_j^{(i)} \le M_i(n) \le \max\limits_{j \le N_i(n)} W_j^{(i)}+1$.

However $N_i(n)/n \xrightarrow{P} \pi_i(1-p_{ii})$, so that by Theorem 6.2.1 of Galambos [6], $\max\limits_{j \le N_i(n)} W_j^{(i)}$ has the same asymptotic distribution as

$\max\limits_{j \le \pi_i(1-p_{ii})n} W_j^{(i)}$. We may thus state:

*Theorem 3.1.* Let $M_i(n)$ denote the maximal time till departure from state i with $p_{ii} > 0$ in the time interval $(0, n]$ in an irreducible discrete time Markov chain with k states, transition matrix P and long run distribution $\underline{\pi}$. It follows that for every $z \in R$

$$\underline{\lim_{n\to\infty}} P[(-\log p_{ii})M_i(n) - \log\pi_i(1-p_{ii}) - \log n \le z]$$

$$\ge \exp\{-e^{-(z+\log p_{ii})}\} \tag{3.1}$$

and

$$\overline{\lim_{n\to\infty}} P[(-\log p_{ii})M_i(n) - \log\pi_i(1-p_{ii}) - \log n \le z]$$

$$\le \exp\{-e^{-z}\} . \tag{3.2}$$

Although we do not have an asymptotic distribution in this case, equations (3.1) and (3.2) can be used to construct asymptotically conservative tests based on the length of the longest observed time till departure from a given state.

If we consider the time till departure from any state, similar arguments regarding related exponential variables yield:

*Theorem 3.2.* Let $M_i(n)$ be as defined in Theorem 3.1 and define

$$M(n) = \max_{i\le k} M_i(n)$$

M(n) represents the maximal time till departure from any state in the time interval $(0, n]$. Assume that for some $k_1$

$$-\log p_{11} = -\log p_{22} = \dots = -\log p_{k_1k_1} < -\log p_{jj},\ j > k_1 .$$

Then for any $z \in R$

$$\underline{\lim_{n\to\infty}} P[(-\log p_{11})M(n) - \log(1-p_{11}) \sum_{i=1}^{k_1} \pi_i - \log n \le z]$$

$$\ge \exp\{-e^{-(z+\log p_{11})}\} \tag{3.3}$$

and

$$\overline{\lim_{n\to\infty}} P[(-\log p_{11})M(n) - \log(1-p_{11}) \sum_{i=1}^{k_1} \pi_i - \log n \le z]$$

$$\le \exp\{-e^{-z}\} . \tag{3.4}$$

*Example 3.1.* Two state chain. Here

$$P = \begin{bmatrix} \alpha & 1-\alpha \\ 1-\beta & \beta \end{bmatrix}$$

and $\underline{\pi} = (\frac{1-\beta}{2-\alpha-\beta}, \frac{1-\alpha}{2-\alpha-\beta})$. The correct normalizing constants to obtain the asymptotic bounds (3.1) and (3.2) for the distribution of the maximal time till departure from state 1 are $a(n) = -\log\alpha$ and $b(n) = -\log\frac{(1-\alpha)(1-\beta)}{2-\alpha-\beta} - \log n$. If we wish to consider $M(n)$ the maximal time till departure from any state, the correct normalizing constants are:

case (i) $\beta < \alpha$: $a(n) = -\log\alpha$, $b(n) = -\log\frac{(1-\alpha)(1-\beta)}{2-\alpha-\beta} - \log n$.

case (ii) $\alpha = \beta$: $a(n) = -\log\alpha$, $b(n) = -\log(1-\alpha) - \log n$.

The case of independent Bernoulli trials (coin tossing) corresponds to the choice $\alpha = 1-\beta$ (cf. [5]).

*Example 3.2.* Multinomial trials. Here

$$P = \begin{bmatrix} p_1 & p_2 & \cdots & p_k \\ p_1 & p_2 & \cdots & p_k \\ \vdots & & & \\ p_1 & p_2 & \cdots & p_k \end{bmatrix}$$

where $p_i > 0$, $\sum_{i=1}^{k} p_i = 1$.

Clearly $\underline{\pi} = (p_1, p_2, \ldots, p_k)$. The correct normalization for the longest time till departure from state 1 (i.e. the longest run of outcome 1) is $a(n) = -\log p_1$ and $b(n) = -\log p_1(1-p_1) - \log n$. If $k_1$ of the $p_i$'s are equal to $\tilde{p}$ and greater than the rest, then the correct normalization for the longest time till departure from any state (i.e. the longest run of any kind) is $a(n) = -\log\tilde{p}$ and $b(n) = -\log k_1\tilde{p}(1-\tilde{p}) - \log n$.

*Example 3.3.* A very simple circulant. Consider a k state chain with the following doubly stochastic transition matrix P (here of course $\underline{\pi} = (\frac{1}{k}, \frac{1}{k}, \ldots, \frac{1}{k})$).

$$P = \begin{bmatrix} \alpha & (1-\alpha) & & & \cdots & 0 \\ 0 & \alpha & (1-\alpha) & & & \cdot \\ \cdot & & \cdot & \cdot & & \cdot \\ \cdot & & & \cdot & \cdot & \cdot \\ \cdot & & & & \cdot & 0 \\ 0 & & & & \alpha & (1-\alpha) \\ (1-\alpha) & 0 & \cdot & \cdot & \cdot \; 0 & \alpha \end{bmatrix} .$$

The correct normalization for the maximal time till departure from state 1 is $a(n) = -\log\alpha$, $b(n) = -\log\frac{1-\alpha}{k} - \log n$. The normalization for the maximal time till departure from any state is $a(n) = -\log\alpha$, $b(n) = -\log(1-\alpha) - \log n$.

## 4. ELEMENTARY DERIVATIONS OF CERTAIN DISCRETE TIME RESULTS

Some, not all, of the results in Section 3 can be obtained by elementary techniques which do not rely on Theorem 6.2.1 in Galambos [6, p. 282]. The power of the cited theorem is much better appreciated after such struggles. A sampler of such computations follows.

### (i) Maximal Success Runs in Bernoulli Trials

Consider a sequence of Bernoulli trials with success probability p. As usual let $q = 1-p$. Let $M_n$ be the longest run of successes in the first n trials. Let $c_{m,n}$ denote the probability that there is not a success run of length m in the first n trials. By conditioning on the time of ocurrence of the first failure in the first m trials we find

$$\begin{aligned} c_{m,n} &= \sum_{j=1}^{m} q\, p^{j-1}\, c_{m,n-j}, \qquad n = m, m+1, \ldots \\ &= 1 \qquad\qquad\qquad\quad , \qquad n = 0, 1, 2, \ldots, m-1 . \end{aligned}$$

$\{c_{m,n}\}$ is thus a generalized m'th order Fibonacci sequence, cf. [4], [7]. If we define the generating function for $c_{m,n}$ to be

$$C_m(s) = \sum_{n=0}^{\infty} c_{m,n} s^n$$

it is readily verified that

$$C_m(s) = (1-s^m p^m)\,(1-s+q\,p^m s^{m+1})^{-1} . \tag{4.1}$$

It follows from (4.1), using Feller [3, p.275] that for large n (larger than m)

$$P(M_n \leq m) = c_{m+1,n} \doteq (1 + qp^{m+1})^{-n} . \tag{4.2}$$

Now from (4.2) we find

$$P\{(-\log p) M_n - \log qp - \log n < z\}$$
$$= P\{M_n \leq -(z + \log n + \log qp)/\log p\}$$
$$\doteq \{1 + qp^{1 - [(z + \log n + \log qp)/\log p]}\}^{-n} \tag{4.3}$$

where the square bracket in the exponent denotes "integer part". If we let $n \to \infty$ in (4.3) no limit exists but a $\underline{\lim}$ and $\overline{\lim}$ are obtainable confirming the result described at the end of Example 3.1. The key to the development is to obtain an expression like (4.2) for the distribution of the maximum. From it the correct normalizing constants for asymptotic bounds on the distribution of the maximum (as in Theorem 3.1 and Theorem 3.2) can be read off.

For example, consider $M^*_n$ the maximal run of either kind (success of failure). Let $c^*_{m,n}$ be the probability of no run of length m of either kind in the first n trials. Runs of length m of successes are recurrent events as are runs of length m of failures. Using material from Feller [3, p. 322] we conclude that the generating function of $c^*_{m,n}$ defined by

$$C^*_m(s) = \sum_{n=0}^{\infty} c^*_{m,n} s^n$$

is given by

$$C^*_m(s) = \frac{(1 - p^m s^m)(1 - q^m s^m)}{1 - s + qp^m s^{m+1} + pq^m s^{m+1} - p^m q^m s^{2m}} . \tag{4.4}$$

By identifying the smallest root of the denominator of (4.4), we conclude that for large n and large m

$$c^*_{m,n} \doteq (1 + qp^m + pq^m)^{-n} .$$

Thus if $p < q$ ,

$$P(M^*_n \leq m) \doteq (1 + pq^{m+1})^{-n} \tag{4.5}$$

while if $p = q$

$$P(M^*_n \leq m) \doteq (1 + (1/2)^{m+1})^{-n} \tag{4.6}$$

These results can be readily used to confirm the asymptotic bounds on the distribution of $M^*_n$ described at the end of Example 3.1.

(ii) Multinomial Trials

Consider a sequence of trials with k possible outcomes with associated probabilities $p_1, p_2, \ldots, p_k$. Define $M_n$ to be the longest run of outcomes of type 1 in the first n trials and let $M_n^*$ be the longest run of any type. Let $c_{m,n}$ denote the probability of no run of outcome type 1 of lenght m in the first n trials and $c^*_{m,n}$ be the probability of no run of length m (of any type) in the first n trials. Denote the corresponding generating functions by $C_m(s)$ and $C_m^*(s)$. Since we are dealing with recurrent events we readily verify that

$$C_m(s) = (1-p^m s^m)(1-s+(1-p_1)p_1^m s^{m+1})^{-1} \tag{4.7}$$

and

$$C_m^*(s) = \frac{\prod_{i=1}^{k} (1-p_i^m s^m)}{1 - s + \sum_{i=1}^{k} (1-p_i)p_i^m s^{m+1}} . \tag{4.8}$$

Again by identifying the smallest root in the denominator, we find that

$$P(M_n \le m) = c_{m+1,n} \doteq (1+(1-p_1)p_1^{m+1})^{-n} \tag{4.9}$$

and

$$P(M_n^* \le m) = c^*_{m+1,n} \doteq (1+ \sum_{i=1}^{k} (1-p_i)p_i^{m+1})^{-n} . \tag{4.10}$$

Now if $k_1$ of the $p_i$'s are equal to $\tilde{p}$ and greater than the rest then from (4.10) we find

$$P(M_n^* \le m) \doteq (1+ k_1(1-\tilde{p})\tilde{p}^{m+1})^{-n} . \tag{4.11}$$

Expressions (4.9) and (4.11) can be readily used to confirm the results described in Example 3.2.

(iii) Longest Stay in a Particular State in a Two State Chain

Here we consider $M_n$, the longest time till departure from state 1 in a two state chain with transition matrix

$$P = \begin{bmatrix} \alpha & 1-\alpha \\ 1-\beta & \beta \end{bmatrix} .$$

Let $c_{m,n}$ be the probability of no sojourn in state 1 of length m in the first n trials assuming an initial probability vector $(1-\beta, \beta)$. Let $C_m(s)$ be the corresponding generating funtion. By conditioning on the first visit to state 2 in the first m trials, we obtain

$$C_m(s) = \frac{1-(1-\alpha-\beta)s-(1-\beta)\alpha^{m-1}s^m}{1-(\alpha+\beta)s+(\alpha+\beta-1)s^2+(1-\beta)(1-\alpha)\alpha^{m-1}s^{m+1}} .$$

The smallest root of the denominator is $1+\frac{(1-\beta)(1-\alpha)\alpha^{m-1}}{2-\alpha-\beta}+o(\alpha^m)$ and consequently we find

$$P(M_n \leq m) = c_{m+1,n} \doteq \{1 + \frac{(1-\beta)(1-\alpha)}{2-\alpha-\beta}\alpha^m\}^{-n} . \tag{4.12}$$

This can be used to confirm the asymptotic bounds on the distribution of $M_n$ described in Example 3.1.

If we try to extend this analysis to cover the case of a k state chain, we have to solve k-1 equations in k-1 generating functions. Such a program is generally daunting though it can be carried through in some cases, for example:

(iv) A Simple 3x3 Circulant

Consider a three state Markov chain with transition matrix

$$P = \begin{bmatrix} \alpha & 1-\alpha & 0 \\ 0 & \alpha & 1-\alpha \\ 1-\alpha & 0 & \alpha \end{bmatrix} .$$

Let $c_{m,n}$ be the probability of no sojourn in state 1 of length m in the first n trials assuming an initial probability vector $(0, \alpha, 1-\alpha)$. Let $d_{m,n}$ be the analogous probability assuming an initial probability vector $(1-\alpha, 0, \alpha)$. Denote the corresponding generating functions by $C_m(s)$ and $D_m(s)$. By conditioning on the time of ocurrence of the first visit to state 2 or 3 in trials 1 to m we obtain two equations relating $C_m(s)$ and $D_m(s)$ with solution

$$C_m(s) = \frac{(1-\alpha s)^2+(1-\alpha)s(1-(2\alpha-1)s) - (1-\alpha)^2\alpha^{m+1}s^{m-1}}{(1-\alpha s)^3 - (1-\alpha)^3 s^3 + (1-\alpha)^3\alpha^{m-1}s^{m+2}} . \tag{4.13}$$

It is possible to verify that the smallest root of the denominator of (4.13) is given by $1 + \frac{1-\alpha}{3}\alpha^{m-1} + o(\alpha^m)$ and subsequently we find

$$P(M_n \leq m) = c_{m+1,n} \doteq \{1 + \frac{1-\alpha}{3}\alpha^m\}^{-n} . \tag{4.14}$$

This is consistent with the corresponding result included in Example 3.3.

## 5. EXTENSIONS

It is clear that the analysis in Section 2 does not depend crucially on the fact that the sojourn times in a state are exponentially distributed. With relatively minor modifications we can extend the analysis to a Semi-Markov setting in which sojourns in state $i$ are i.i.d. with common non-exponential distribution with mean $\mu_i$ (the simplest case of course being when all sojourns in all states are i.i.d. $F$ with mean $\mu$). Of course we will need to assume the existence of an asymptotic distribution for normalized maxima from the assumed distribution.

## REFERENCES

[1] Anderson, C.W. 1970, Extreme value theory for a class of discrete distributions with applications to some stochastic processes. J. Appl. Probability 7, pp. 99-113.

[2] Barndorff-Nielsen, O. 1964, On the limit distribution of the maximum of a random number of independent random variables. Acta Math. Acad. Sci. Hungar. 15, pp. 399-403.

[3] Feller, W. 1968, An introduction to probability theory and its application. Vol. I, 3rd. ed. John Wiley, New York.

[4] Fisher, P.S. and Kohlbecher, E.E. 1972, A generalized Fibonacci sequence. The Fibonacci Quarterly, 10, pp. 337-344.

[5] Földes, A. 1979, The limit distribution of the length of the longest head run. Periodica Mathematica Hungarica 10, pp. 301-310.

[6] Galambos, J. 1978, The asymptotic theory of extreme order statistics. John Wiley and Sons. New York.

[7] Szekely, G. and Tusnady, G. 1976-9, Generalized Fibonacci numbers and the number of "pure heads". Matematikai Lapok 27, pp. 147-151.

THE BOX-JENKINS MODEL AND THE PROGRESSIVE FATIGUE FAILURE OF LARGE PARALLEL ELEMENT STAY TENDONS

E. Castillo*, A. Fernández**, A. Ascorbe*, E. Mora*

* University of Santander, ** University of Oviedo
Spain

ABSTRACT

A model for analysing the progressive fatigue failure of a tendon made up of parallel elements is given. The logarithm of the number of cycles to failure of the sub-elements are assumed to follow an ARMA(p,q) Box-Jenkins model. Upper and lower bounds which converge to the exact solution, as the number of elements increase, are given. Finally, the asymptotic theory is applied to obtain an analytical expression for the cdf of the strength of the tendon.

## 1. INTRODUCTION

At present, large tendons made up of many parallel elements (wires or 7-core strands) showing lengths up to several hundred of meters represent a favourite solution as supporting elements in stayed long span bridges or similar structures. Due to the fact that the specialized literature,((1),(2)), presents the fatigue strength of the tendon as the governing criterion for the design of such structures, the knowledge of its statistical behaviour under fatigue becomes essential. The enormous costs involved in testing multiple tendons lead to the testing of specimens composed of a single element with short length. From these test results the total fatigue strength of the tendon must be estimated.
Some important works related to static and fatigue strength have been done in the past (references (3) to (12)), but most of them assume either independence among contiguous sub-elements or asymptotic cases.
In this paper, the cdf for the number of cycles to failure in the finite life region and the cdf for the stress range $\Delta\sigma$ in the endu-



J. Tiago de Oliveira (ed.), Statistical Extremes and Applications, 427–434.

rance zone of the Wöhler curve (S-N curve) of a multiple tendon will be derived from the cdf of the number of cycles to failure of one element of reference length $L_o$, assuming an ARMA(p,q) model of dependence.

## 2. MATHEMATICAL MODEL

The present model is based on the following assumptions:

a) The tendon is made up of m parallel elements (wires or strands) of identical length L.
b) It is considered that failure in the elements occurs due to fatigue, so that static failure is excluded.
c) An element with actual length L can be considered composed of n sub-elements with reference length $L_o$ :

$$L = n L_o \tag{1}$$

d) The failure of one element is associated with the first failure appearing in the n sub-elements.
e) The logarithm of the number of cycles to failure $r_1, r_2, \ldots, r_n$ for the different sub-elements of length $L_o$, when submitted to a given stress range $\Delta\sigma$, follows a stationary and invertible Box-Jenkins ARMA(p,q) model ((13),(14))

$$z_t = b_1 z_{t-1} + \ldots + b_p z_{t-p} + a_t - c_1 a_{t-1} - \ldots - c_q a_{t-q} \tag{2}$$

where

$$z_t = r_t - E(r_t) \tag{3}$$

and $a_i$ $(i \in \mathbb{Z})$ are iid rv with normal distribution $N(0, \sigma_a^2)$.
f) The S-N curve for a sub-element is assumed to be known. This curve is defined by the mean and standard deviation lines

$$\mu_{\log N} = g(\Delta\sigma) \tag{4}$$

$$\sigma_{\log N} = h(\Delta\sigma) \tag{5}$$

g) As a consequence of the successive failure of elements the stress range progressively increases, provided the applied total load P remains constant (load control). Since the S-N curve corresponds to one-step tests, i.e. tests with $\Delta\sigma$=const. , a so called cumulative damage hypothesis is needed. In the following two different cumulative damage hypotheses are studied. It is assumed that the previous damage history in the elements is transferred to the new stress range in such a way that the probability of failure or the Miner-number remain unaltered (15).
h) The number of cycles to failure for the different elements are assumed to be independent or quasi-independent rv.
i) The total strength of the tendon is considered to be reached only when the k-th element fails.

Let $N_{(1)}, N_{(2)}, \ldots, N_{(m)}$ be the order statistics , i.e. the number of cycles to failure of the m elements of the tendon ranged in an increasing order. Whilst the number of cycles increases and no failure in the elements occurs, the stress range in one element is equal to $\Delta\sigma_1$. As soon as the number of cycles reaches the value $N_{(1)}$, the first element fails and as a consequence a rearrangement in the stress range takes place in the rest of the elements, so that the new stress range in those becomes equal to $\Delta\sigma_1(m/m-1)$. According to the cumulative damage hypothesis the process goes on, as if the tendon were subjected to a stress range equal to $\Delta\sigma_1(m/m-1)$ from the beginning and for an equivalent number of cycles (see fig. 1). This process repeates until a stress range equal to $\Delta\sigma_1(m/m-k+1)$, corresponding to the k-th element failure, is attained.

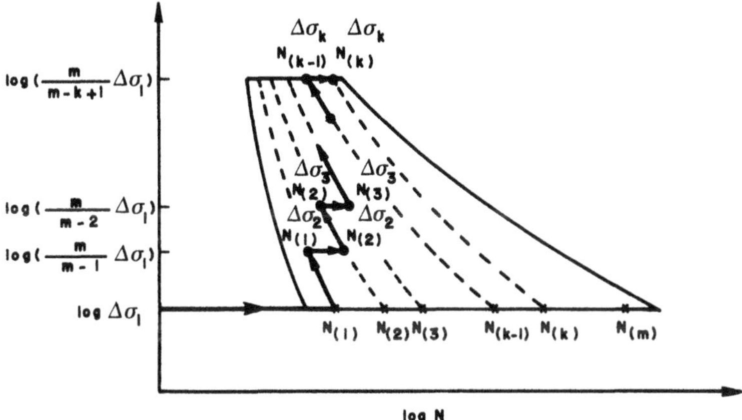


Fig. 1. Schematic variation of the stress range and the number of cycles till the admissible weakening of the tendon

The number of cycles associated with the admissible weakening of the tendon is given by (see fig. 1):

$$N^*_{f(k)} = N_{(1)} + \sum_{j=2}^{k} (N_{(j)}^{\Delta\sigma_j} - N_{(j-1)}^{\Delta\sigma_j}) \qquad (6)$$

which taken into account the two cumulative damage hypotheses becomes

$$N^*_{f(k)} = N_{(1)} + \sum_{j=2}^{k} A_j \, (N_{(j)}^{B_j} - N_{(j-1)}^{B_{j-1}}) \qquad (7)$$

where

$$\Delta\sigma_j = m \, \Delta\sigma_1 \, / \, (m-j+1) \qquad (8)$$

and $A_j$ and $B_j$ are given in table 1

| CUMULATIVE DAMAGE HYPOTHESIS | | |
|---|---|---|
| | ISOPROBABILITY | MINER'S NUMBER |
| $A_j$ | $\exp(g(\Delta\sigma_j) - B_j\, g(\Delta\sigma_1))$ | $\exp(g(\Delta\sigma_j) - g(\Delta\sigma_1))$ |
| $B_j$ | $h(\Delta\sigma_j)/h(\Delta\sigma_1)$ | 1 |

Table 1

Expression (7) gives the number of cycles to failure as a function of the number of cycles to failure of the component elements. In the following, they will be related to those for the sub-elements. The logarithm of the number of cycles to failure of the i-th element is given by:

$$\log N_i = \min (r_1^i, r_2^i, \ldots, r_n^i) \tag{9}$$

where $r_1^i, r_2^i, \ldots, r_n^i$ are the logarithms of the number of cycles to failure of the n sub-elements in the i-th element, which must fulfil equation (3), i.e. :

$$r_j^i = z_j^i + g(\Delta\sigma_1) \quad ; \quad i=1,2,\ldots,m \quad ; \quad j=1,2,\ldots,n \tag{10}$$

Hence, equation (9) can be written

$$\begin{aligned} \log N_i &= g(\Delta\sigma_1) + \min (z_1^i, z_2^i, \ldots, z_n^i) \\ &= g(\Delta\sigma_1) + \sigma_z \min (z_1'^i, z_2'^i, \ldots, z_n'^i) \\ &= g(\Delta\sigma_1) + h(\Delta\sigma_1)\, U_i \end{aligned} \tag{11}$$

where

$$z_j'^i = z_j^i / h(\Delta\sigma_1) \quad ; \quad i=1,2,\ldots,m \quad ; \quad j=1,2,\ldots,n \tag{12}$$

represent the values of the standardized time series, and $U_i$ can be deducted from the context.
From equation (11) the following holds:

$$\log N_i = g(\Delta\sigma_1) + h(\Delta\sigma_1)\, U_{(i)} \tag{13}$$

Equation (13) together with equation (7) allow the obtention of the pdf of $N^*_{f(k)}$ only by numerical or simulation techniques. Since these techniques can be very cumbersome , from a practical point of view it would be of interest to give upper and lower bounds for

$N^{*}_{f(k)}$ . These bounds are given by (see fig. 1):

$$N^{\Delta\sigma_k}_{(k)} \leqq N^{*}_{f(k)} \leqq N^{\Delta\sigma_1}_{(k)} \qquad (14)$$

Expression (14) allows the study of the rv $N^{*}_{f(k)}$ to be done by the, much easier to handle , analysis of the two order statistics $N^{\Delta\sigma}_{(k)}$ and $N^{\Delta\sigma_1}_{(k)}$

## 3. ASYMPTOTIC SOLUTIONS

Due to the presence of two parameters n and m, the following asymptotic cases can be considered:

### 3.1. Very long elements

As it is well known, if $n\to\infty$ the asymptotic theory ((16),(17)) shows that

$$F_{\frac{\log N_{\ell}^{\Delta\sigma_j} - g(\Delta\sigma_j)}{h(\Delta\sigma_j)}}(x) \cong 1 - \exp(-\exp(\frac{x-c_n}{d_n})) \qquad (15)$$

where F( ) is the cdf of the sub-index variable and

$$d_n - (2 \log n)^{-\frac{1}{2}} \qquad (16)$$

$$c_n = \frac{1}{d_n} + \frac{1}{2} d_n \; (\log(\log n) + \log 4\pi) \qquad (17)$$

According to Galambos (16) or David (18) and expression (15) the cdf of the upper and lower bounds, given in expression (14), can be obtained by making j=1 or k, respectively, in the following expression:

$$F_{\frac{\log N_{(k)}^{\Delta\sigma_j} - g(\Delta\sigma_j)}{h(\Delta\sigma_j)}}(x) =$$

$$= \sum_{i=k}^{m} \binom{m}{i} (1-\exp(-\exp(\frac{x-c_n}{d_n})))^{i} \; \exp(-(m-i)\exp(\frac{x-c_n}{d_n})) \qquad (18)$$

### 3.2. Very high number of elements

In this case the bounds converge to each other and any one of them can be used for a practical solution. Accordingly, (see Galambos (16)), the cdf of $(\log N^*_{f(k)} - g(\Delta\sigma_1))/h(\Delta\sigma_1)$ is given by:

$$G(x) = 1 - \exp(-\exp(\frac{x-c_{mn}}{d_{mn}})) \sum_{t=0}^{k-1} \frac{1}{t!} \exp(t(\frac{x-c_{mn}}{d_{mn}})) \qquad (19)$$

Expression (19) has been represented in fig. 2.

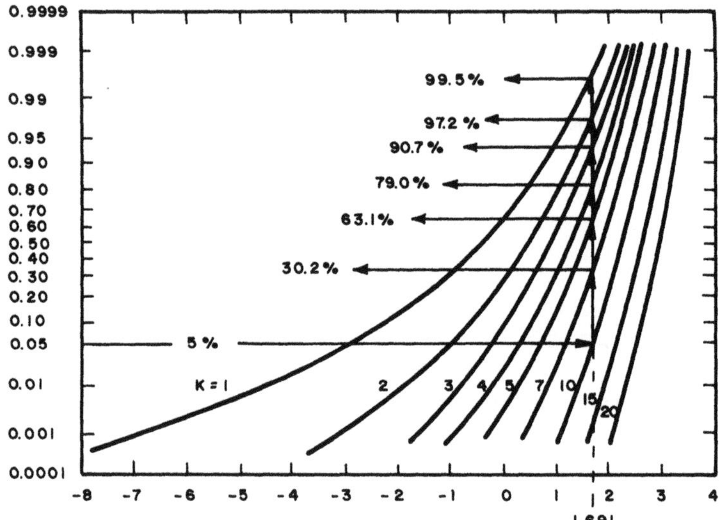


Fig. 2. Cdf of the number of cycles to failure for the k-th wire

The rate of convergence can be shown to be very fast using the convergence of the bounds. Faster normalising parameters $c_n$ and $d_n$ could be perhaps obtained as Gomes (19) and Veraverbeke (20) suggest

## 4. COMPATIBILITY CONDITIONS

The curves defined in equations (4) and (5) cannot be arbitrarily chosen if a given distribution is fixed for the stress range producing failure at N cycles. If this distribution is normal the following relation must be satisfied:

$$(\log N^*_k - g(\Delta\sigma))/h(\Delta\sigma) = (\Delta\sigma - \mu)/\sigma \qquad (20)$$

which explains the two scales in figure 2.

## 5. EXAMPLE OF APPLICATION

In order to illustrate the analytical results and their application to practical cases, let us consider an hypothetical tendon composed of 200 7mm-wires. If an admissible weakening of 5% (10 broken wires) and a probability of 0.05 is taken, a design value of 1.691 is obtained (see fig. 2). It is interesting to remark that high probabilities of failure up to the fith wire are associated with this design value (see fig. 2). This explains that tendon specimens when tested in the laboratories always show some broken wires even for low stress levels.

## 6. ACKNOWLEDGEMENTS

The authors wish to express their gratitude to the Fundación Leonardo Torres-Quevedo for financial support.

## 7. REFERENCES

(1) Hajdin N. (1976) "Vergleich zwischen den Paralleldrahtseilen und verschlossenen Seilen am Beispiel der Eisenbahnschrägseilbrücke über die Save in Belgrad". IVBH-Vorbericht zum 10. Kongress in Tokyo.

(2) Andrä W.,Saul R. (1979) "Die Festigkeit, insbesondere Dauerfestigkeit langer Paralleldrahtbündel". Bautechnik 4, 128-130.

(3) Daniels H.E. (1945) "The Statistical Theory of the Strength of Bundles of Threads". I. Proc. Roy. Soc. London A 183, 404-435.

(4) Harlow D.G., Phoenix S.L. (1978 a) "The Chain-of-Bundles Probability Model for the Strength of Fibrous Materials I: Analysis and Conjectures" . J. Composite Materials 12, 195-214.

(5) Harlow D.G., Phoenix S.L. (1978 b) "The Chain-of-Bundles Probability Model for the Strength of Fibrous Materials II: A Numerical Study of Convergence". J. Composite Materials 12, 314-334.

(6) Tierney L. (1982) "Asymptotic Bounds on the Time to Fatigue Failure of Bundles of Fibers under Local Load Sharing". Adv. Appl. Prob. 14, 95-121.

(7) Harlow D.G., Smith R.L., Taylor H.M. (1983) "Lower Tail Analysis of the Distribution of the Strength of Load-Sharing Systems". J. Appl. Prob. 20, 358-367.

(8) Smith R.L. (1983) "Limit Theorems and Approximations for the Reliability of Load-Sharing Systems". Adv. Appl. Prob. 15, 304-330.

(9) Castillo E., Fernández A.,Mora E.,Ascorbe A. (1983) "Influencia de la Longitud en la Resistencia a Fatiga de Tendones de Puentes Atirantados". Anales de Ing. Mecánica, Año 1.

(10) Castillo E.,Fernández A.,Ascorbe A., Mora E. (1983) "Aplicación de los Modelos de Series Temporales al Análisis Estadístico de la Resistencia de Tendones de Puentes Atirantados". Anales de Ing. Mecánica, Año 1.

(11) Mora E., Castillo E.,Ascorbe A. (1983) "Influence of the Stress Distribution of the Strength of Wires. Statistical Analysis". 44-th Session of ISI, Madrid.

(12) Castillo E., Ascorbe A.,Fernández A. (1983) "The Box-Jenkins Model and the Progressive Failure of a System of Parallel Wire Stay Tendon". 44-th Session of ISI, Madrid.

(13) Box G.E., Jenkins G.M. (1976) "Time Series Analysis.Forecasting and Control". Holden Day, S. Francisco.

(14) Tiago de Oliveira J. (1975-76) "Introduction to Stochastic Processes". Gaz. Mat. 155/156.

(15) Fernández A. (1982) "Statistical Interpretation of the Miner-Number using an Index of Probability of Total Damage". IABSE Colloquium Fatigue of Steel and Concrete Structures, Lausanne.

(16) Galambos J. (1978) "The Asymptotic Theory of Extreme Order Statistics". John Wiley & Sons, New York.

(17) Leadbetter M.R.,Lindgren G.,Rootzén H. (1983) "Extremes and Related Properties of Random Sequences". Springer Verlag, New York.

(18) David H.A. (1980) "Order Statistics". John Wiley & Sons.

(19) Gomes M.I. (1979) "Rates of Convergence in Extreme Value Theory". Rev. Univ. Santander 2, II.

(20) Veraverbeke N.,Teugels J.L. (1975) "The Exponential Range of Convergence of the Distribution of the Maximum of a Random Walk". J. Appl. Prob. 13.

# THE ASYMPTOTIC BEHAVIOUR OF THE MAXIMUM LIKELIHOOD ESTIMATES FOR UNIVARIATE EXTREMES

Jonathan Cohen

School of Statistics, University of Minnesota, 270 Vincent Hall, Minneapolis, MN 55455, U.S.A.

Let F be in the domain of attraction of the type I extreme value distribution. The behaviour of the maximum likelihood estimates when fitting the incorrect model $\Lambda(ax+b)$ to the distribution of the maximum $F^n(x)$ is investigated. The main result is that under widely applicable conditions, the estimated distribution function is consistent for an optimal type I distribution in the sense of [3] and that the parameter estimates of a and b are asymptotically normal. Some of the implications for statistical inference problems are discussed. Finally, similar results for the case where three-parameter extreme-value approximations are fitted are hypothesized. On this basis we give some general conclusions on the problem of choice of extrema model for $F^n(x)$.

In this paper we consider some statistical problems connected with the theory of extreme values. Suppose we have D observations of a maximum of n random variables. Thus we observe $Y_{nj} = \max \{X_{ij} : 1 \leq i \leq n\}$ for j = 1, 2, 3 ... D, where the $X_{ij}$'s are independent identically distributed variables with a common distribution function F. We assume that we know n is large but do not know F. The general procedure used is to fit an extreme value distribution to these observations by estimating the parameters, often by the method of maximum likelihood. In [1] and [3] we have effectively assumed that D is so large that the error in parameter estimation is negligible and that the only errors arise from the fact that the extreme value distributions are approximations (for large n) to the true distribution $F^n(x)$. In this paper we consider how the error depends upon both D and n. Note that all proofs appear in [2].



*J. Tiago de Oliveira (ed.), Statistical Extremes and Applications, 435–442.*

Assume that we fit the type I extreme value distribution $\Lambda((x-\beta)/\alpha)$ by the method of maximum likelihood. If the true distribution is of this form, then standard theory shows that the maximum likelihood estimates of $\alpha$ and $\beta$ are strongly consistent for $\alpha$ and $\beta$ and that they are asymptotically normally distributed with mean $(\alpha\ \beta)$ and variance - co-variance matrix

$$\frac{\alpha^2}{D} P^{-1} = \frac{\alpha^2}{D} \begin{pmatrix} \Gamma''(1) + 2\Gamma'(1) + 1 & -\Gamma'(1) - 1 \\ -\Gamma'(1)-1 & 1 \end{pmatrix}^{-1} \tag{1}$$

$$= \frac{6\alpha^2}{\pi^2 D} \begin{pmatrix} 1 & 1-\gamma \\ 1-\gamma & \frac{\pi^2}{6} + (1-\gamma)^2 \end{pmatrix}$$

where $\gamma$ is Euler's constant.

In this paper we consider what happens to these estimates when the true distribution is $F^n(x)$ (as in the first paragraph). In this case, since we are estimating based on an incorrect model it is not at all obvious what the precise behaviour of the estimates should be. However, one might expect that for fixed n, and D tending to infinity, the estimated distribution function would tend to a distribution close to $F^n(x)$. In fact we shall show that under certain conditions the estimated distribution function is consistent (as $D \to \infty$) for an optimal type I extreme value distribution in the sense of [3] and that the estimates are asymptotically normally distributed.

More precisely, the maximum likelihood estimation procedure is as follows: If we let

$$g(x,\alpha,\beta) = -\exp(-(x-\beta)/\alpha) - (x-\beta)/\alpha - \log \alpha \tag{2}$$

then $\alpha(\mathcal{Y})$, $\beta(\mathcal{Y})$, the estimates of $\alpha,\beta$, are the solutions of

$$\sum_{j=1}^{D} g(Y_{nj},\alpha(\mathcal{Y}),\beta(\mathcal{Y})) = \sup_{\substack{\alpha>0 \\ \beta}} \sum_{j=1}^{D} g(Y_{nj},\alpha,\beta), \tag{3}$$

assuming that the supremum can be attained. $\mathcal{Y}$ denotes the vector $(Y_{n1}, Y_{n2}, \ldots, Y_{nD})$.

It can be shown that (3) has a unique solution if and only if the observations $Y_{nj}$ are not all equal. Unless F is degenerate, with probability one we cannot have equality for all D; so the asymptotic properties (as $D \to \infty$) are unaffected. However, in order to have a well-defined procedure we shall define $\alpha(\mathcal{Y})$, $\beta(\mathcal{Y})$ to be the unique solution of (3) if it exists, and to have certain fixed values in the contrary case.

Assume that F is a twice differentiable function (for large x) in the domain of attraction of $\Lambda$ with $F(x) < 1$ for all x. Assume further that with

$$f(x) = \{\frac{d}{dx}(-\log-\log F(x))\}^{-1} = \frac{-F(x)\log F(x)}{F'(x)}, \tag{4}$$

$$b(x) = f'(x) \to 0 \text{ as } x \to \infty, \tag{5}$$

$b(x)$ has constant sign for large x, and (6)

$|b(x)|$ is regularly varying at infinity. (7)

Note that (4) and (5) together imply that F is in the domain of attraction of $\Lambda$. As proved in [3], (4)-(7) imply that with $s_n = b(b_n)$,

$$\sup_x |F^n(x) - \Lambda((x-b_n^*)/a_n^*)| = O(s_n) \tag{8}$$

if and only if

$$\frac{b_n^*-b_n}{a_n} = O(s_n) \; ; \; \frac{a_n^*-1}{a_n} = O(s_n), \tag{9}$$

where $a_n$ and $b_n$ satisfy

$$F(b_n) = \exp(-n^{-1}) \; ; \; a_n = f(b_n).$$

The convergence rate in (8) cannot be improved upon to $o(s_n)$ for any $a_n^*$ and $b_n^*$. In view of this it seems reasonable to suggest that $\alpha(y)$ and $\beta(y)$ should converge as $D \to \infty$ to some $a_n^*$ and $b_n^*$ satisfying (9) (since the estimated distribution function $\Lambda((x-\beta(y))/\alpha(y))$ would then be close, in a sense, to $F^n(x)$ for large D). A more precise formulation of this result together with a result on the asymptotic normality of the maximum likelihood estimates, under some additional conditions, is given in the following theorems proven in [2]. $\xrightarrow{a.s}$ denotes convergence almost surely and $\xrightarrow{D}$ denotes convergence in distribution.

<u>Theorem 1</u>. Assume F satisfies (4)-(7) and

$$F(y)\exp(-\lambda y) \to 0 \text{ as } y \to -\infty \text{ for some } \lambda>0, \tag{10}$$

Either (11)

(a) $f(x) \leq C_1 < \infty$ for large x

or, with $-1 < u < 0$ and $L > 0$,

(b) $b(x) \sim Lx^u$ as $x \to \infty$.

Define the constants P, Q, R (Q>0) by

$$P = \Gamma''(1) + \tfrac{1}{2}\Gamma'''(1) - \{\Gamma'(1)\}^2 - \tfrac{1}{2}\Gamma'(1)\,\Gamma''(1), \tag{12}$$

$$Q = \Gamma''(1) - \{\Gamma'(1)\}^2, \tag{13}$$

$$R = PQ^{-1}\,\Gamma'(2) - \tfrac{1}{2}\Gamma''(2). \tag{14}$$

If n is suitably large but fixed then the estimates $\alpha(\mathcal{Y})$, $\beta(\mathcal{Y})$ satisfy, as $D \to \infty$

$$\mathcal{T}_{\mathcal{D}} = (\alpha(\mathcal{Y})\ \beta(\mathcal{Y}))' \overset{a.s}{\to} (a_n^*\ b_n^*)' = \phi_0 \tag{15}$$

where, as $n \to \infty$ ,

$$\frac{a_n^*}{a_n} = 1 - PQ^{-1}s_n + o(s_n), \tag{16}$$

$$\frac{b_n^* - b_n}{a_n} = -Rs_n + o(s_n)\ . \tag{17}$$

Theorem 2. Let F satisfy the conditions in Theorem 1. Define $\mathcal{T}_{\mathcal{D}}$, $\phi_0$ as in Theorem 1. Use the notation $z_n = \mathcal{O}(s_n)$ to indicate that there is some finite constant $\delta$ (possibly zero) such that

$$z_n = \delta s_n + o(s_n)\ . \tag{18}$$

Then, as $D \to \infty$ for fixed n,

$$\sqrt{D}\,(\mathcal{T}_{\mathcal{D}} - \phi_0) \overset{D}{\to} N\left(\begin{pmatrix}0\\0\end{pmatrix},\ a_n^2\,\mathcal{P}_n^{-1}\right) \tag{19}$$

where, as $n \to \infty$ ,

$$\mathcal{P}_n = \mathcal{P} + \mathcal{O}\,(s_n)\ . \tag{20}$$

$\mathcal{P}$ is given by (1) and (20) is to be interpreted component-wise.

Note first of all that the conditions hold for many of the well known distributions in the domain of attraction of $\Lambda$. In particular they hold for the normal and all the gamma and Weibull distributions apart from the exponential distribution. Also note that these results may also be proven if we redefine f, b, $a_n$ and $b_n$ according to the definitions of [3] and replace the conditions (4)-(7) by the weaker conditions that $F(x) < 1$ for all x, F is in class N with $K > 1$ and $b_n/a_n = 0 \mid s_n \mid^{-\mu}$ for some $\mu$ in (0,K), in the notation of [3]. (Subtract $rs_n$ from the right hand side of (16)).

The next point is that although the error terms in (20) are given in the form $\mathcal{O}(s_n)$, it is easy in principle to calculate the exact formulae for these errors in the form (18), for which the constants $\delta$ will be given in terms of the gamma function and its derivatives evaluated at integer values. However the exact formulae are quite unwieldy and so they have not been given here.

The importance of Theorem 1 is that it shows that under fairly general conditions the method of maximum likelihood estimation is still a reasonable procedure even though the fitted model is known to be incorrect.

The effect of Theorem 2 is that under the given conditions the maximum likelihood estimates behave to first order in D and n as if the true distribution is $\Lambda((x-b_n^*)/a_n^*)$. Its theoretical and practical importance can be seen from the following considerations: Suppose we wish to estimate the a-Quantile (equivalently, to solve $F^n(z) = a$)). The natural estimate is given by the solution of

$$\Lambda\left(\frac{\hat{z} - \beta(\mathcal{Y})}{\alpha(\mathcal{Y})}\right) = a \tag{21}$$

so that

$$\hat{z} = (w\ 1)\ T_{\mathcal{D}} \tag{22}$$

where $w = -\log\text{-}\log a$. If the true distribution is $\Lambda((x-\beta)/\alpha)$ for some $\alpha>0$ and $\beta$, then confidence intervals for z will be based on the result that as $D \to \infty$,

$$\sqrt{D}\ (\hat{z} - z) \overset{D}{\to} N\ (0,\ \alpha^2(w\ 1)\ P^{-1}\ (w\ 1)') \tag{23}$$

which follows easily from the asymptotic normality of $T_{\mathcal{D}}$. The estimated 95% confidence interval for z will be

$$z \in \hat{z} + \frac{1.96}{\sqrt{D}} \{\alpha(\mathcal{Y})^2\ (w\ 1)\ P^{-1}\ (w\ 1)'\}^{\frac{1}{2}} \tag{24}$$

However, the true distribution of $Y_{nj}$ is $F^n(x)$, so that if the same estimate $\hat{z}$ is used, then we can apply Theorem 2 and prove that as $D \to \infty$,

$$\sqrt{D}\,(\hat{z}-z_o) \xrightarrow{D} N(0,\ a_n^{\,2}\ (w\ 1)\ P_n^{-1}\ (w\ 1)')\ , \tag{25}$$

where $P_n$ satisfies (20) and

$$z_o = z - a_n s_n\ \{R + PQ^{-1}\ w + \tfrac{1}{2}w^2 + o(1)\}. \tag{26}$$

Thus the estimated 95% confidence interval for z should be of the form

$$z \in \hat{z} + O(a_n s_n) \pm 1.96\ \{\alpha(Y)^2\ (w\ 1)\ P_n^{-1}\ (w\ 1'\}^{\frac{1}{2}}/\sqrt{D}. \tag{27}$$

For many well-known F, $a_n s_n$ decreases like a power of $(\log n)^{-1}$. Thus (24) and (27) will be significantly different even for moderately large n. If there is enough information on the tail behaviour of F so that reasonable estimates of $s_n$ may be made, but not enough information to allow the equation $F^n(z) = a$ to be solved directly, then (27) can be used to give more accurate confidence intervals than (24). Note that a possible method of estimation of $s_n$ is described in [6], whereas a reasonable estimate for $a_n$ is $\alpha(Y)$. Otherwise one should be reluctant to use (24) even if D is very large, unless n is also very large.

Similarly, theorems 1 and 2 can be used to compute the limiting distribution of other functionals of the true distribution which are estimated using the fitted distribution. (In particular one can consider the important case of estimating the true distribution $F^n(x)$ itself by $\Lambda((x-\hat{\beta})/\hat{\alpha})$ when x is fixed). In general the confidence intervals obtained by assuming the true distribution is of the form $\Lambda((x-\beta)/\alpha)$ will be inaccurate unless n is very large.

It would be wrong to finish this discussion without pointing out that there is much scope for further work on these problems. In this chapter we have only considered the case where estimation is based on the type I extreme value distribution. In view of the arguments in [1] and [3] it is clear that it is also important to consider the case where the type II or III extreme value distributions are used - in other words to consider a penultimate approximation rather than an ultimate approximation. In fact Jenkinson [4] advocates the use of maximum likelihood estimation based on a generalised extreme value distribution including all three types, with a distribution function

$$G(x) = \Lambda(y) \tag{28}$$

where

$$\left\{\begin{array}{ll} y = \frac{x-\beta}{\alpha} & , k = 0 \\ -\frac{1}{k}\log\{1 - k\frac{(x-\beta)}{\alpha}\} & , k \neq 0. \end{array}\right\} \tag{29}$$

Unfortunately it has not yet been possible to extend the present results to the three parameter case. Work on these problems is progressing but is not yet ready for publication. However, on the basis of these results and the results in [3] on the three-parameter 'penultimate' approximation it seems reasonable to suggest that the following heuristic arguments (whose form was based on a suggestion of R. Smith) might be made rigorous.

From the results in [3], under suitable conditions on F, $\sup_x |F^n(x) - G(x)|$ converges to zero at the fastest possible rate if $\alpha$, $\beta$, k are sequences satisfying

$$\left\{\begin{array}{l} \alpha(n) = a_n(1+o(s_n)) \\ \beta(n) = b_n+o(a_n s_n) \\ k(n) = -s_n+o(s_n). \end{array}\right\} \tag{30}$$

This suggests that the analogous version of Theorem 1 is that the maximum likelihood estimates $\hat{\alpha}(Y)$, $\hat{\beta}(Y)$ and $\hat{k}(Y)$ satisfy

$$(\hat{\alpha}\ \hat{\beta}\ \hat{k}) \overset{a.s.}{\rightarrow} (\alpha(n)\ \beta(n)\ k(n))$$

as $D \to \infty$, where $\alpha$, $\beta$ and k satisfy (30). In fact this result will not hold in general since we may deduce that if $k(n) > 0$ then almost surely $\alpha(n)/k(n) + \beta(n) + 1$ is an upper bound for all the observations. On the other hand if $k(n) < 0$ then almost surely $\alpha(n)/k(n) + \beta(n) - 1$ provides a lower bound. Nevertheless one might reasonably expect the result to hold if we allow n and D to vary together in a suitable fashion.

The analogous version of Theorem 2 is then that

$$\sqrt{D}\left\{\begin{pmatrix}\hat{\alpha}\\ \hat{\beta}\\ \hat{k}\end{pmatrix} - \begin{pmatrix}\alpha(n)\\ \beta(n)\\ k(n)\end{pmatrix}\right\} \overset{D}{\rightarrow} N\left\{\begin{pmatrix}0\\0\\0\end{pmatrix}, \ I_n^{-1}\right\} \tag{31}$$

as D and then n tends to infinity, where $I_n$ is the information matrix for the model (28)-(29) with $\alpha = a_n$, $\beta = b_n$ and $k = 0$. (See [5] and [7] for further details.)

If we consider the problem of estimating the a-quantile using the generalised extreme value distribution then the obvious estimate is $\hat{\hat{z}} = \hat{\beta} + \hat{\alpha}(1-\exp(-\hat{k}\ w))/\hat{k}$ where $w = -\log -\log a$. It then follows, using (30) and (31), that for large n and D,

$$a_n^{-1}\sqrt{D}\ (\hat{\hat{z}}-z) \underset{\sim}{\sim} N(0,L(w) = a_n^{-2}\ (w - \tfrac{1}{2}\ a_n w^2)\ I_n^{-1}\ (w - \tfrac{1}{2}\ a_n w^2)')$$

where L(w) is a quartic polynomial in w. Thus the mean square error of $\hat{\hat{z}}$ is approximately $a_n^2 L(w)/D$. On the other hand, if we let n and D tend to infinity in such a way that $s_n\sqrt{D} \to \mu$ then (25) and (26) suggest

$$\frac{\sqrt{D}}{a_n}\ (\hat{z}-z) = \frac{\sqrt{D}}{a_n}\ (\hat{z}-z_0) + \frac{\sqrt{D}}{a_n}\ (z_0-z) \underset{\sim}{\sim} N\ (-\mu p(w)\ ,\ m(w))$$

where $p(w) = R + PQ^{-1}w + \tfrac{1}{2}w^2$ and $m(w) = (w\ 1)P^{-1}(w\ 1)'$. Thus the mean square error of $\hat{z}$ is approximately $a_n^2\{\mu^2 p(w)^2 + m(w)\}/D$. Consequently, we would prefer to use $\hat{\hat{z}}$ rather than $\hat{z}$ if and only if $(L(w)-m(w))/p(w)^2 \geq s_n^2 D$.

The results in the last paragraph, assuming that they could be proven rigorously, suggest that even if F is in the domain of attraction of $\Lambda$, we should fit the 'penultimate' type II and III extreme value distributions and not the more natural type I distribution when $s_n\sqrt{D}$ is large. Since $s_n \to 0$ very slowly for many of the well known distributions in the domain of attraction of $\Lambda$, it follows that the type II and type III distributions should be used much more frequently than they have been in applications of extreme value theory.


[1] Cohen, J. P., 1982, Adv. Appl. Prob. 14, pp. 324-339.
[2] Cohen, J. P., 1982. *Ph.D. Thesis, University of London, September 1982.*
[3] Cohen, J. P., 1982, Adv. Appl. Prob. 14, pp. 833-854.
[4] Jenkinson, A. F., 1955, Q. J. Roy. Meteor. Soc. 87, pp. 158.
[5] Jenkinson, A. F., World Met. Office. Tech. Note 98, pp. 183-227.
[6] Pickands, J. III, 1983. *Proc. NATO ASI on Statistical Extremes and Applications*, this volume.
[7] Prescott, P. and Walden, A. T., 1980, Biometrika 67, 3, pp. 723-724.

# ON UPPER AND LOWER EXTREMES IN STATIONARY SEQUENCES

Richard A. Davis

Colorado State University

Unlike the classical result for iid sequences, the maximum and minimum of stationary mixing sequences can be asymptotically dependent. The class of non-degenerate limit distributions of the maximum and minimum from such processes turn out to be of the form $H(x, \infty) - H(x, -y)$ where $H(x, y)$ is a bivariate extreme value distribution. In some instances, the joint limiting distribution of any collection of upper and lower extremes may be determined as well. Moreover, using a special case of this result, it can be shown that the partial sums converge in distribution to a non-normal stable limit. Examples illustrating various aspects of the above results are also presented.

## 1. INTRODUCTION

The asymptotic behavior of extreme values from independent and identically distributed (iid) sequences of random variables is well known. For the past thirty years there has been an attempt to extend these classical results to the stationary process setting. In order to achieve such an extension, the processes are typically required to satisfy two types of dependence conditions. The first is a mixing condition requiring a certain class of events to become independent as their time separation increases. The second condition is more of a local condition restricting the dependence between any two of the random variables when both are large. One of the weakest and most workable forms of these two conditions are the hypotheses D and D' introduced by Leadbetter (1974). Under D and D', the maximum behaves asymptotically as through the underlying sequence is iid.



*J. Tiago de Oliveira (ed.), Statistical Extremes and Applications, 443–460.*

Although the mixing condition D is rather weak, one encounters many processes which do not meet the second dependence restriction D'. For example ARMA processes with non-normal noise rarely satisfy D'. However, the limiting distribution of say the maximum of these processes can still be ascertained in some instances (cf. Rootzen (1978), Chernick (1981), and Finster (1982)). Even for m-dependent sequences the asymptotic behavior of the extremes can be much different than in the iid case as evidenced by the examples in O'Brien (1974) and Mori (1976). In fact most of the anomalous behavior for extremes of dependent sequences can be detected in 1- and 2-dependent sequences.

The object of this paper is to discuss the joint limiting behavior of upper and lower extremes of stationary mixing sequences with particular emphasis on the maximum and minimum. Unlike the classical result for iid sequences, the maximum and minimum may be asymptotically dependent. In Section 2, we display the class of joint limiting distributions for the maximum and minimum and in the special case of m-dependent sequences, we give necessary and sufficient conditions for joint convergence. Also, a 1-dependent sequence is constructed in which both the maximum and minimum have non-degenerate limits yet there is no joint convergence.

Under mixing conditions similar to D, it is possible to prove the convergence of a certain sequence of point processes. This gives a more complete result which, in particular, permits us to find the limiting distribution of any collection of upper and lower extremes. We sketch out these results in Section 3 as well as give a sufficient condition for the asymptotic independence of the upper and lower extremes.

In certain cases, the point process result in Section 3 enables us to obtain a result about the limiting distribution of the partial sums. This may not be too surprising since if the common distribution function of an iid sequence belongs to a stable domain of attraction, then the asymptotic behavior of the partial sums can be described via the asymptotic properties of the extreme order statistics. Exploiting the ideas in LePage, Woodroofe, and Zinn (1981), we are then able to prove the analogous result for stationary sequences satisfying conditions similar to D and D'. This result is stated in Section 4 and demonstrated with two examples. The first process is an instantaneous function of a stationary Gaussin process with covariance function $r_n$ behaving like $r_n \log n \to 0$ as $n \to \infty$. The second example satisfies the conditions D and D' yet does not satisfy any of the other commonly used mixing conditions. These examples highlight the differences between the conditions which are needed for obtaining non-normal stable limits versus normal limits of the partial sums (cf. Ibragimov and Linnik, 1971).

## 2. CLASS OF LIMITING DISTRIBUTIONS FOR THE MAXIMUM AND MINIMUM

In this section we investigate the possible limit laws for the maximum and minimum of stationary mixing sequences. Let $\{X_n\}$ be a stationary sequence of random variables with common df F. Define $M_n = \max\{X_1,\ldots,X_n\}$ and $W_n = \min\{X_1,\ldots,X_n\}$ and let $u_n(x) = a_n x + b_n$, $v_n(y) = c_n y + d_n$ for some constants $a_n > 0, b_n, c_n > 0$ and $d_n$. These constants will eventually be used to normalize both $M_n$ and $W_n$. We shall say that the sequence $\{X_n\}$ satisfies condition C1 if for all x and y

$$P^k(M_{[n/k]} \leq u_n, W_{[n/k]} > v_n) - P(M_n \leq u_n, W_n > v_n) \to 0 \quad \text{(C1)}$$

as $n \to \infty$ for every integer k([s] - largest integer not greater than s).

Condition C1 is a type of mixing condition requiring that for every k, the k events

$$\{M_{[n/k]} \leq u_n, W_{[n/k]} > v_n\},$$

$$\{M_{[n/k], [2n/k]} \leq u_n, W_{[n/k], [2n/k]} > v_n\},$$

$$\ldots, \{M_{[(k-1)n/k], n} \leq u_n, W_{[(k-1)n/k], n} > v_n\}$$

are asymptotically independent where $M_{s,t} = \max_{s<j\leq t} \{X_j\}$, $W_{s,t} = \min_{s<j\leq t} \{X_j\}$. C1 is easily seen to be implied by $\overline{D}(v_n, u_n)$ in Davis (1979) and hence by strong mixing. In practice C1 may be harder to verify directly than $D(v_n, u_n)$ and is less appealing in that it already requires some knowledge of the maximum and minimum. However in what follows, C1 is the only mixing assumption necessary.

In order to specify the class of limit distributions for $M_n$ and $W_n$ above, we begin with an iid sequence $\{(Y_n^1, Y_n^2)\}$ of random 2-vectors. Let $(M_n^1, M_n^2)$ be the vector of component maxima $(\max_{1\leq j\leq n} \{Y_j^1\}, \max_{1\leq j\leq n} \{Y_j^2\})$ and assume there exist constants $a_n > 0$, $b_n$, $c_n > 0$, $d_n$ such that $P(a_n^{-1}(M_n^1-b_n) \leq x, c_n^{-1}(M_n^2-d_n) \leq y) \to H(x,y)$ where $H(x,y)$ is a non-degenerate distribution function. Such distribution functions H are called bivariate extreme value distributions (BEVD). A characterizing property for BEVD's is that there exist constants $A_k > 0$, $B_k$, $C_k > 0$, $D_k$ such that $H^k(A_k x + B_k, C_k y + D_k) = H(x,y)$ for all positive integers k. This is in complete analogy with univariate extreme value distributions. Although there are a variety of papers on multivariate extreme value distributions, the discussion contained in Chapter 5 of

Galambos (1978) will be sufficient for our purposes.

One direction of the following theorem is immediate from the characterizing property of BEVD's.

Theorem 2.1. Suppose the stationary sequence $\{X_n\}$ satisfies condition C1. Then the class of non-degenerate limiting distributions of $(a_n^{-1}(M_n-b_n), c_n^{-1}(W_n-d_n))$ is precisely $H(x,\infty) - H(x,-y)$ where H is a bivariate extreme value distribution.

Proof. First assume $P(M_n \leq u_n(x), W_n > v_n(-y)) \to H(x,y)$ where $u_n(x) = a_n x + b_n$, $v_n(y) = c_n y + d_n$, and $H(x,y)$ is a non-degenerate distribution. By the C1 assumption, $P(M_n \leq u_{nk}(x), W_n > v_{nk}(-y)) \to H^{1/k}(x,y)$ for $k = 1,2,\ldots$ . Employing the multivariate analogue of the convergence of types result, there exist constants $A_k > 0$, $B_k$, $C_k > 0$, $D_k$ such that $H(A_k x + B_k, C_k y + D_k) = H^{1/k}(x,y)$. Therefore, H is a bivariate extreme value distribution and

$$P(a_n^{-1}(M_n-b_n) \leq x, c_n^{-1}(W_n-d_n) \leq y)$$

$$= P(M_n \leq u_n(x)) - P(M_n \leq u_n(x), W_n > v_n(y))$$

converges to $H(x,\infty) - H(x,-y)$.

The proof is complete once we exhibit a stationary sequence satisfying C1 with $P(M_n \leq u_n(x), W_n > v_n(-y))$ converging to an arbitrary bivariate extreme value distribution H. A 2-dependent sequence having this property will be constructed later in this section (see Example 2.1). #

For m-dependent sequences it is possible to give necessary and sufficient conditions for the joint convergence of the maximum and minimum. For convenience, we shall assume the process also satisfies the m-dependent version of the local dependence condition $D'(u_n)$ (cf. Leadbetter, 1974). More specifically, assume

$$\lim_{n\to\infty} n \sum_{j=2}^{m+1} P(X_1 > u_n, X_j > u_n) = 0 \tag{2.1}$$

and

$$\lim_{n\to\infty} n \sum_{j=2}^{m+1} P(X_1 \leq v_n, X_j \leq v_n) = 0 \tag{2.2}$$

for all x and y where as before $u_n(x) = a_n x + b_n$, $v_n(y) = c_n y + d_n$. These two conditions guarantee that the limiting behavior of $a_n^{-1}(M_n-b_n)$ and $c_n^{-1}(W_n-d_n)$ is the same as the corresponding extremes from the associated independent sequence. In other words,

$$P(M_n \leq u_n) - F^n(u_n) \to 0 \tag{2.3}$$

and

$$P(W_n \le v_n) - (1-F(v_n))^n \to 0 \text{ as } n \to \infty. \tag{2.4}$$

Theorem 2.2. Suppose $\{X_n\}$ is a stationary m-dependent sequence satisfying (2.1) and (2.2). Further assume $P(a_n^{-1}(M_n-b_n) \le x) \to G_1(x)$ and $P(c_n^{-1}(W_n-d_n) \le y) \to 1 - G_2(-y)$, where $G_1$ and $G_2$ are extreme value distributions. Then

$$P(a_n^{-1}(M_n-b_n) \le x,\ c_n^{-1}(W_n-d_n) \le y) \to G(x,y), \tag{2.5}$$

G a non-degenerate df, if and only if

$$\lim_{n\to\infty} n\sum_{j=2}^{m+1} (P(X_1 > u_n, X_j \le v_n) + P(X_1 \le v_n, X_j > u_n)) = h(x,-y) \tag{2.6}$$

for all x and y, where $h(x,y) = \log(H(x,y)/(H(x,\infty)H(\infty,y)))$ for some bivariate extreme value distribution H. In particular, $G(x,y) = H(x,\infty) - H(x,-y)$. [The proof of this result may be found in Davis (1982).]

A similar result is also true if conditions (2.1) and (2.2) are relaxed. We state it only in the 1-dependent case.

Proposition 2.3. Let $\{X_n\}$ be a 1-dependent sequence and suppose $nP(X_1>a_nx+b_n) \to -\log G_1(x)$ and $nP(X_1\le c_ny+d_n) \to -\log G_2(-y)$ where $G_1$ and $G_2$ are extreme value distributions. Then

$$P(a_n^{-1}(M_n-b_n) \le x,\ c_n^{-1}(W_n-d_n) \le y) \to G(x,y)$$

G a non-degenerate df if and only if

$$P(X_1>x|X_2>x) \to 1-\alpha \text{ as } x \to x_o = \sup\{x: F(x) < 1\}, \tag{2.7}$$

$$P(X_1\le y|X_2\le y) \to 1-\beta \text{ as } y \to y_o = \inf\{x: F(x) > 0\}, \tag{2.8}$$

and

$$n\,(P(X_1>u_n, X_2\le v_n) + P(X_1\le v_n, X_2>u_n)) \to h(x,-y) \tag{2.9}$$

where $h(x,y) = \log(H(x,y)/(H(x,\infty)H(\infty,y)))$ for some BEVD H. In particular $G(x,y) = G_1^{\alpha}(x)(1-G_2^{\beta}(-y)e^{h(x,-y)})$.

Proof. By Proposition 2.2 in Davis (1982), (2.7) and (2.8) are equivalent to $n(P(X_1>u_n, X_2>u_n)) \to -(1-\alpha)\log G_1(x)$ and $nP(X_1\le v_n, X_2\le v_n) \to -(1-\beta)\log G_2(-y)$, respectively. The rest of the proof follows the same argument given in the preceding theorem.

Example 2.1. We start with an arbitrary bivariate extreme value df $H(x,y)$ and set $\tilde{H}(x,y) = H^2(x,y)$. Then $\tilde{H}(x,y)$ is also a BEVD and has the representation $\tilde{H}(x,y) = D_{\tilde{H}}(G_1(x), G_2(y))$ where $D_{\tilde{H}}(\cdot,\cdot)$ is the dependence function defined on p. 250 in Galambos (1978) and $G_1$ and $G_2$ are extreme value distributions. Let $F_1$ and $F_2$ be two df's which are symmetric about the origin and assume $F_i$ belongs to the domain of attraction of $G_i$, i=1,2. Then let $\{(Y_n^1, Y_n^2)\}$ be an iid sequence of 2-vectors with common df $D_{\tilde{H}}(F_1(x), F_2(y))$.

The sequence $\{X_n\}$ is then defined by $X_n = g(Y_n^1, -Y_{n+1}^2, -Y_{n+2}^2)$ where g is the function

$$g(u,v,w) = \begin{cases} u & \text{if } u > 0,\ v > 0 \\ v & \text{if } v < 0,\ w < 0 \\ 0 & \text{otherwise.} \end{cases}$$

It is clear that the sequence $\{X_n\}$ is stationary and 2-dependent. We shall show that

$$P(M_n \le u_n,\ W_n \le v_n) \to H(x,\infty) - H(x,-y) \tag{2.10}$$

where $u_n = a_n x + b_n$ and $v_n = c_n y + d_n$ are chosen so that $F_1^n(a_n x + b_n) \to G_1(x)$ and $F_2^n(c_n y - d_n) \to G_2(y)$.

For this sequence it is easy to verify that

(i) $P(X_1 > u_n) = P(Y_1^1 > u_n,\ Y_2^2 < 0) = -\log G_1(x)/(2n) + o(1/n)$.

(ii) $P(X_1 \le v_n) = P(Y_2^2 > -v_n,\ Y_3^2 < 0) = -\log G_2(-y)/(2n) + o(1/n)$.

(iii) $\lim_{n\to\infty} n \sum_{j=2}^{3} \{P(X_1 > u_n,\ X_j \le v_n) + P(X_1 \le v_n,\ X_j > u_n)\} = \tilde{h}(x,-y)/2$

where $\tilde{h}(x,-y) = \log(\tilde{H}(x,-y)/(\tilde{H}(x,\infty)\tilde{H}(\infty,-y)))$. Moreover, the conditions (2.1) and (2.2) are also easy to check. Now by applying the sufficiency direction of Theorem 2.2, whose proof does not rely on Theorem 2.1, the result (2.15) is established.

Example 2.2. Again start with an iid sequence of random 2-vectors having $F(x,y)$ as the common df. Assume $F(x,\infty) = F(\infty,x) = e^{-x^{-1}}$ for $x > 0$ and is zero for $x \le 0$. In addition to these marginal distributional assumptions on F, suppose $P(Y_1^1 > y \mid Y_1^2 > y)$ does not converge as $y \to \infty$. Since it may be shown that the convergence of $P(Y_1^1 > y \mid Y_1^2 > y)$ is equivalent to the convergence of $nP(Y_1 > nx, Y_1^2 > nx)$ as $n \to \infty$, for all $x > 0$, there exists an $\tilde{x} > 0$ such that $nP(Y_1^1 > n\tilde{x},\ Y_1^2 > n\tilde{x})$ does not converge. In this case, the sequence of numbers, $nP(Y_1^1 > n\tilde{x},\ Y_1^2 > n\tilde{x})$, is bounded so it necessarily has at least two distinct limit points.

To complete the construction of the $\{X_n\}$ sequence, let $\{J_n\}$ be an iid sequence of Bernoulli rv's, independent of the $\{(Y_n^1, Y_n^2)\}$ sequence with $P(J_n=1) = p$, $P(J_n=0) = q = 1 - p$, $0 < p < 1$. Now define

$$X_n = \begin{cases} Y_n^1 & \text{if } J_n = 1 \\ -Y_{n+1}^2 & \text{if } J_n = 0. \end{cases}$$

It is clear that $\{X_n\}$ is a stationary 1-dependent sequence. In addition, the following properties hold for this process.

(i) $nP(X_1>nx) = nP(Y_1^1>nx) \to px^{-1}$ for all $x > 0$.
(ii) $nP(X_1<ny) = nqP(Y_2^2>-ny) \to q(-y)^{-1}$ for all $y < 0$.
(iii) $nP(X_1>nx, X_2>nx) \to 0$ for all $x > 0$.
(iv) $nP(X_1\le ny, X_2\le ny) \to 0$ for all $y < 0$.
(v) $nP(X_1>nx, X_2\le ny) \to 0$ for all $x > 0$ and $y < 0$.
(vi) $nP(X_1<ny, X_2>nx) = npqP(Y_2^2>-ny, Y_2^1>nx)$ does not converge for all $x > 0$ and $y < 0$ (e.g. $x = \tilde{x}$, and $y = -\tilde{x}$). From equations (2.3) and (2.4), we have $P(M_n\le nx) \to e^{-px^{-1}}$, $x > 0$ and $P(W_n\le ny) \to 1 - e^{qy^{-1}}$ for $y < 0$. However by Theorem 2.2, $(n^{-1}M_n, n^{-1}W_n)$ does not have a non-degenerate limit distribution. From the proof of the theorem, it follows that $P(M_n\le n\tilde{x}, W_n\le n(-\tilde{x}))$ has two distinct limits. So although the maxima and minima individually have proper limit distributions, the dependence in the sequence $\{X_n\}$ is enough to destroy any joint convergence.

Frequently, it is desirable to know when the maximum and minimum are asymptotically independent. If the process satisfies the hypotheses of Theorem 2.2, then the maximum and minimum are asymptotically independent if and only if $h(x,y) = 0$ for all $x$ and $y$. The following theorem provides a sufficient condition for asymptotic independence for general mixing sequences.

<u>Theorem 2.4.</u> Assume that $P(M_n\le u(x)) \to G(x)$ and $P(W_n\le v_n(y)) \to H(y)$ where $G$ and $H$ are non-degenerate distribution functions. Also, in addition to C1, suppose $\{X_n\}$ satisfies the condition

$$\limsup_{n\to\infty} n \sum_{j=2}^{[n/k]} \{P(X_1>u_n, X_j \le v_n) + P(X_1\le v_n, X_j \le u_n)\}=o(1)$$

as $k \to \infty$ for all $x$ and $y$. Then

$$P(M_n\le u_n, W_n\le v_n) \to G(x)H(y).$$

The proof of this theorem can be found in [8]. Notice that if $X_i$ and $X_j$ are associated for each $i\neq j$ and $1 - F(u_n) = 0(1/n)$

and $F(v_n) = O(1/n)$ then condition C2 is satisfied because

$$n \sum_{j=2}^{[n/k]} \{P(X_1>u_n, X_j\le v_n) + P(X_1\le v_n, X_j>u_n)\}$$

$$\le n \sum_{j=2}^{[n/k]} \{P(X_1>u_n)P(X_j\le v_n) + P(X_1\le v_n)P(X_j>u_n)\}$$

$$\le \frac{n^2}{k} \cdot O(1/n^2).$$

For example, the AR(1) process $X_n = \phi X_{n-1} + Z_n$ where $\{Z_n\}$ is an iid sequence and $0 < \phi < 1$ can be written as $X_n = \sum_{j=0}^{\infty} \phi^j Z_{n-j}$. In this case $X_i$ and $X_j$ are associated (see Barlow and Proschan, 1975) and hence the maximum and minimum will be asyptotically independent This may not be the case, however, if $-1 < \phi < 0$ (see [6]).

3. SOME POINT PROCESS RESULTS

Recalling the definition of BEVD's (see the discussion preceding Theorem 2.1) we have for H a BEVD,

$$P(a_n^{-1}(M_n^1-b_n) \le x,\ c_n^{-1}(M_n^2-d_n) \le y) \to H(x,y)$$

where $(M_n^1, M_n^2)$ are the component maxima from some iid sequence of random 2-vectors $\{(Y_n^1, Y_n^2)\}$. As a consequence of Theorem 5.3.1 in [12], we have

$$\left.\begin{aligned} &nP(Y_1^1>a_n x+b_n) \to -\log G_1(x) \\ &nP(Y_1^2>c_n y+d_n) \to -\log G_2(y) \end{aligned}\right\} \tag{3.1}$$

$$nP(Y_1^1>a_n x+b_n,\ Y_1^2>c_n y+d_n) \to h(x,y), \tag{3.2}$$

and

$$H(x,y) = G_1(x)G_2(y)e^{h(x,y)}, \tag{3.3}$$

for all x and y with $H(x,y) > 0$, where $G_1$ and $G_2$ are extreme value distributions. Letting $G_i^* = \inf\{x: G_i(x) > 0\}$ denote the endpoint of $G_i$, $i = 1,2$, the exponent measure $\nu$ corresponding to H (cf. de Haan and Resnick, 1977) may be chosen to have support on $[G_1^*,\infty) \times [G_2^*,\infty)$ and has the properties:

$$\left.\begin{aligned} &\nu((x,\infty) \times (y,\infty)) = h(x,y) \text{ for all } x > G_1^*,\ y > G_2^*, \\ &\nu(\overline{\mathbb{R}} \times (y,\infty)) = -\log G_2(y) \\ &\nu((x,\infty) \times \overline{\mathbb{R}}) = -\log G_1(x) \text{ for all y and x}, \end{aligned}\right\} \tag{3.4}$$

where $\bar{\mathbb{R}} = [-\infty,\infty)$.

From (3.2) and (3.4), it follows that

$$nP\left(\left\{\frac{Y_1^1-b_n}{a_n}\varepsilon R_1\right\}\cup\left\{\frac{Y_2^1-d_n}{c_n}\varepsilon R_2\right\}\right) \to \nu(R_1 \times \bar{\mathbb{R}} \cup \bar{\mathbb{R}} \times R_2),$$

where for each $i = 1,2$, $R_i$ is a finite union of half-open intervals ($R_i = \sum_{j=1}^{k_i} (\alpha_j^i,\beta_j^i]$) whose closure is contained in $(G_i^*,\infty)$. For more details, see [12, 14].

Now define the sequence of point processes

$$\tilde{I}_n(B) = \#\{1\le j\le n:\ (a_n^{-1}(Y_j^1-b_n),c_n^{-1}(Y_j^2-d_n))\ \varepsilon\ B\}$$

where $B$ is a Borel subset of $\bar{\mathbb{R}} \times \bar{\mathbb{R}}$. If $\tilde{I}$ is a Poisson process on $\bar{\mathbb{R}} \times \bar{\mathbb{R}}$ with intensity $\nu$, then by virtue of equations (3.1) - (3.4), it may be shown that $\tilde{I}_n \xrightarrow{d} \tilde{I}$ (see [14, 23]). Because of the resemblance between the conditions of Theorem 2.2 and (3.1)-(3.3), it seems plausible that a more complete result may be obtained for at least m-dependent sequences.

Let $\{X_n\}$ be a stationary sequence of random variables and for the current discussion assume $\{X_n\}$ satisfies the hypotheses of Theorem 2.2 (i.e. $\{X_n\}$ is m-dependent satisfying (2.1) and (2.2)). For our situation, the pertinent sequence of point processes will be defined on $\{1,2\} \times \mathbb{R}$ by

$$I_n(1 \times B) = \#\{1\le j\le n:\ a_n^{-1}(X_j-b_n)\varepsilon B\} \tag{3.6}$$

and

$$I_n(2 \times B) = \#\{1\le j\le n:\ -c_n^{-1}(X_j-d_n)\varepsilon B\} \tag{3.7}$$

where $B$ is a subset of $\mathbb{R}$. The limit point process $I$ is defined by $I(1 \times B) = \tilde{I}(B \times \bar{\mathbb{R}})$ and $I(2 \times B) = \tilde{I}(\bar{\mathbb{R}} \times B)$ where $\tilde{I}$ is the Poisson process defined above (i.e. with intensity $\nu$ corresponding to the BEVD $H$ in Theorem 2.2). In other words $I$ is just the respective projections of $\tilde{I}$ onto the two axes. In order to establish $I_n \xrightarrow{d} I$, it suffices to show (see Kallenberg, 1976)

$$E\ I_n(i\times(a,b]) \to \begin{cases} \nu((a,b]\times\bar{\mathbb{R}}) = E\ I(1\times(a,b]) & i = 1 \\ \nu(\bar{\mathbb{R}} \times(a,b]) = E\ I(2\times(a,b]) & i = 2, \end{cases} \tag{K1}$$

and

$$P(I_n(1\times R_1) = 0,\ I_n(2\times R_2) = 0) \to$$

$$P(I(1\times R_1) = 0,\ I(2\times R_2) = 0) = e^{-\nu(R_1\times\bar{\mathbb{R}}\cup\bar{\mathbb{R}}\times R_2)}, \tag{K2}$$

where $R_i$ is either the empty set or a finite union of half-open intervals with closure contained in $(G_i^*,\infty)$, i=1,2. The proof of the following theorem relies on this result and may be found in Davis (1983).

Theorem 3.1. Under the assumptions of Theorem 2.2 and (2.5) or, equivalently (2.6), $I_n \overset{d}{\to} I$.

Remark 1. As a consequence of this result, the joint limiting distribution for any collection of upper and lower extreme values can be evaluated. For example, if we choose $B_1 = (x,\infty)$ and $B_2 = (-y,\infty)$, then $\{I_n(1\times B_1) \le k-1\} = \{M_n^k \le a_n x + b_n\}$ and $\{I_n(2\times B_2) > \ell - 1\} = \{W_n^\ell \le c_n y + d_n\}$ where $M_n^k$ and $W_n^\ell$ denote respectively the $k^{th}$ largest and $\ell^{th}$ smallest among $X_1,\ldots X_n$. The limit distribution of $(a_n^{-1}(M_n^k - b_n),\ c_n^{-1}(W_n^\ell - d_n))$ is then given by $P(I(1\times B_1) \le k-1,\ I(2\times B_2) > \ell - 1)$ which is equal to

$$P(\tilde{I}((x,\infty)\times\bar{\mathbb{R}}) \le k-1,\ \tilde{I}(\bar{\mathbb{R}}\times(-y,\infty)) > \ell - 1)$$

$$= \sum_{j=0}^{k-1}\{P((\tilde{I}((x,\infty)\times y,\infty)) = j)P(\tilde{I}((-\infty,x]\times(-y,\infty)) > \ell - 1 - j)$$

$$\times P(\tilde{I}((x,\infty)\times(-\infty,y]) \le k-1-j)\}.$$

Each term in the sum may now be routinely calculated.

Remark 2. The 2-vector of point processes $(I_n(1\times\cdot), I_n(2\times\cdot))$ converges in distribution to $(I(1\times\cdot), I(2\times\cdot)) = (\tilde{I}(\cdot\times\bar{\mathbb{R}}), \tilde{I}(\bar{\mathbb{R}}\times\cdot))$. It is easy to see that the components of the limit vector, $I(1\times\cdot)$ and $I(2\times\cdot)$ and hence the upper and lower extremes are independent if and only if $\nu((x,\infty)\times(y,\infty)) = 0$ or, equivalently, if $h(x,y) = 0$ for all $x > G_1^*$ and $y > G_2^*$. Also note that if this is the case, then the limit process I is a Poisson process with intensity measure

$$\lambda(i\times B) = \nu(B\times\bar{\mathbb{R}}), \quad i = 1,$$

$$= \nu(\bar{\mathbb{R}}\times B), \quad i = 2, \text{ for all Borel sets } B.$$

Remark 3. The assumptions (2.1) and (2.2) of Theorem 2.2 are required in order to guarantee that $I_n(1\times\cdot) \overset{d}{\to} I(1\times\cdot)$ and $I_n(2\times\cdot) \overset{d}{\to} I(2\times\cdot)$. Without such a condition, $I_n(1\times\cdot)$ and $I_n(2\times\cdot)$ may not converge in distribution to a point process. This is best illustrated by the 1-dependent example in Mori (1976) where the maximum has a non-degenerate limit while the largest and second largest does not have a joint limiting distribution, and hence precludes the convergence in distribution of $I_n(1\times\cdot)$.

Remark 4. This theorem is also valid if an additional time dimension is included in the definition of the point processes. That is redefine $I_n$ so that $I_n(1\times B) = \#\{j: (j/n, a_n^{-1}(X_j-b_n)\varepsilon B\}$ and $I_n(2\times B) = \#\{j: (j/n, -c_n^{-1}(X_j-d_n))\varepsilon B\}$ where B is a Borel subset of $[0,1]\times\mathbb{R}$. If $\tilde{I}$ is a Poisson process on $[0,1]\times\bar{\mathbb{R}}\times\bar{\mathbb{R}}$ with intensity $dt\times\nu(dx_1,dx_2)$, then $I_n \xrightarrow{d} I$, where I is the process defined by $I(1\times B) = \tilde{I}(B\times\mathbb{R})$ and $I(2\times B) = \tilde{I}(B^*)$ (here $B^* = \{(t,x,y): (t,y,x)\varepsilon B\times\mathbb{R}\}$) for all Borel subsets B of $[0,1]\times\mathbb{R}$.

It is also possible to obtain similar results under much weaker mixing conditions than m-dependence. We start with a stationary sequence $\{X_n\}$ and the sequence of point processes $\{I_n\}$ defined in (3.6) and (3.7). The mixing condition that will be used is defined as follows. Set $u_n = a_nx+b_n$, $v_n = c_ny+d_n$ and assume

$$\begin{aligned} n(1-F(u_n)) &\to -\log G_1(x) \\ nF(v_n) &\to -\log G_2(-y) \end{aligned} \qquad (3.8)$$

where $G_1$ and $G_2$ are extreme value distributions. Now let $B_1$ and $B_2$ be a finite union of disjoint intervals (possibly infinite) of the form (a,b]. Put

$$Y_{jn}^1 = I_{\{a_n^{-1}(X_j-b_n)\varepsilon B_1\}}, \quad Y_{jn}^2 = I_{\{-c_n^{-1}(X_j-d_n)\varepsilon B_2\}}.$$

Then condition D is said to hold if for any choice of integers $1\le i_1<\dots<i_p<j_1<\dots<j_q\le n$, $j_1-i_p>\ell$,

$$\begin{aligned} &|E(Y_{i_1n}^1\cdots Y_{i_pn}^1Y_{i_1n}^2\cdots Y_{i_pn}^2Y_{j_1n}^1\cdots Y_{j_qn}^1Y_{j_1n}^2\cdots Y_{j_qn}^2) \\ &- E(Y_{i_1n}^1\cdots Y_{i_pn}^1Y_{i_1n}^2\cdots Y_{i_pn}^2)E(Y_{j_1n}^1\cdots Y_{j_qn}^1Y_{j_1n}^2\cdots Y_{j_qn}^2)| \le \alpha_{n,\ell}, \end{aligned}$$

where $\alpha_{n,\ell}$ satisfies the usual properties as set forth by Leadbetter (1974). Namely, $\alpha_{n,\ell}$ is nonincreasing in $\ell$ and $\alpha_{n,\ell_n}\to 0$ as $n\to\infty$ for some sequence $\ell_n\to\infty$ and $\ell_n = o(n)$. Also, $\alpha_{n,\ell_n}$ may depend on the choice of $B_1$ and $B_2$ above.

Lemma Under condition D and (3.8)

$$E\left(\prod_{j=1}^{n} Y_{jn}^{1} Y_{jn}^{2}\right) - E^{k}\left(\prod_{j=1}^{[n/k]} Y_{jn}^{1} Y_{jn}^{2}\right) \to 0 \quad \text{as } n \to \infty$$

for every positive integer k.

The proof of this lemma is omitted for it follows routinely the same argument given in the proof of [18, Lemma 2.3].

We now give a slightly more general result than Theorem 4.2 in [8].

Theorem 3.2. Assume $\{X_n\}$ satisfies condition D and C2 of Theorem 2.4 for all $x > G_1^*$ and $y > G_2^*$ and suppose the two sequences of point processes $\{I_n(1\times\cdot)\}, \{I_n(2\times\cdot)\}$ have the property that

$$I_n(1\times\cdot) \xrightarrow{d} N_1,$$

and

$$I_n(2\times\cdot) \xrightarrow{d} N_2$$

where $N_1$ and $N_2$ are simple point processes on $\mathbb{R}$. Then $(I_n(1\times\cdot), I_n(2\times\cdot)) \xrightarrow{d} (N_1,N_2)$ where $N_1$ and $N_2$ are independent. More generally, $I_n \xrightarrow{d} N$ where N is a point process on $\{1,2\} \times \mathbb{R}$ defined by $N(1\times\cdot) \stackrel{d}{=} N_1(\cdot)$ and $N(2\times\cdot) = N_2(\cdot)$.

Proof. We shall show $I_n \xrightarrow{d} N$. Observe that $\{I_n\}$ is tight and N is simple. By Theorem 4.7 in [16], it remains to show

$$P(I_n(1\times R_1) = 0,\ I_n(2\times R_2) = 0) \to P(N(1\times R_1) = 0)P(N(2\times R_2) = 0)$$

where $R_i$ is either the empty set or a finite union of half-open intervals with closure contained in $(G_i^*,\infty)$, i = 1,2.

From the lemma, it can be shown that

$$P^k(I_{[n/k]}(i\times R_i) = 0) - P(I_n(i\times R_i) = 0) \to 0, \quad i=1,2 \tag{3.9}$$

and

$$P^k(I_{[n/k]}(1\times R_1) = 0,\ I_{[n/k]}(2\times R_2) = 0) - P(I_n(1\times R_1) = 0,\ I_n(2\times R_2) = 0) \to 0 \tag{3.10}$$

for every positive integer k. Moreover.

$$\begin{aligned} &P(I_{[n/k]}(1\times R_1) = 0,\ I_{[n/k]}(2\times R_2) = 0) \\ &= P(I_{[n/k]}(1\times R_1) = 0) - P(I_{[n/k]}(2\times R_2) > 0) \\ &+ P(I_{[n/k]}(1\times R_1) > 0,\ I_{[n/k]}(2\times R_2) > 0). \end{aligned}$$

Using (3.9) and (3.10) and setting $\alpha = -\log P(I(1\times R_1) = 0)$ and $\beta = -\log P(I(2\times R_2) = 0)$, the $\liminf_{n\to\infty}$ of the right hand side is bounded below by

$$e^{-\alpha/k} - (1-e^{-\beta/k}). \qquad (3.11)$$

On the other hand, the $\limsup_{n\to\infty}$ of the same side is bounded above by

$$e^{-\alpha/k} - (1-e^{-\beta/k}) + \limsup_{n\to\infty} P(I_{[n/k]}(1\times R_1) > 0, \; I_{[n/k]}(2\times R_2) > 0). \qquad (3.12)$$

But,

$$P(I_{[n/k]}(1\times R_1) > 0, \; I_{[n/k]}(2\times R_2) > 0)$$

$$= P\Big(\bigcup_{i=1}^{[n/k]} \{a_n^{-1}(X_i-b_n)\varepsilon R_1\} \bigcup_{j=1}^{[n/k]} \{-c_n^{-1}(X_i-d_n)\varepsilon R_2\}\Big)$$

$$\leq [n/k] \sum_{j=2}^{[n/k]} P(X_1 > u_n(x), X_j \leq v_n(-y)) + P(X_1 \leq v_n(-y), X_j > u_n(x))$$

where $x = \inf\{w: w\varepsilon R_1\}$ and $y = \inf\{w: w\varepsilon R_2\}$. By the C2 assumption the $\limsup_{n\to\infty}$ of this bound is $o(1/k)$. Hence, using (3.11) - (3.12), we have

$$(e^{-\alpha/k}-(1-e^{-\beta/k}))^k \leq \liminf_{n\to\infty} P(I_n(1\times R_1) = 0, \; I_n(2\times R_2) = 0)$$

$$\leq \limsup_{n\to\infty} P(I_n(1\times R_1) = 0, \; I_n(2\times R_2) = 0)$$

$$\leq (e^{-\alpha/k}-(1-e^{-\beta/k}) + o(1/k))^k.$$

Upon letting $k\to\infty$ the two sides of the inequality approach $e^{-\alpha}e^{-\beta} = P(I(1\times R_1) = 0)P(I(2\times R_2) = 0)$ as desired. This concludes the proof. #

Remark 5. A sufficient condition for the convergence of $I_n(1\times\cdot)$ and $I_n(2\times\cdot)$ (see Adler, 1978) is

$$\limsup_{n\to\infty} n \sum_{j=2}^{[n/k]} (P(X_1>u_n, X_j>u_n) + P(X_1\leq v_n, X_j\leq v_n)) = o(1) \text{ as } k \to \infty.$$

If this is the case, then N is a Poisson process with intensity defined by

$$\lambda(i \times B) = \begin{cases} \nu(B \times \bar{\mathbb{R}}) & i = 1 \\ \nu(\bar{\mathbb{R}} \times B) & i = 2. \end{cases}$$

Remark 6. Mori (1977) has determined the class of limit point processes of $I_n(1\times\cdot)$ when $\{X_n\}$ is strongly mixing. In particular the limit point process belongs to a special class of infinitely divisible point processes. The same result is expected under condition D.

Remark 7. As indicated in Remark 4, the point process $I_n$ could be modified to include an additional time dimension. However, condition D would have to be strengthened somewhat (as in [1]) in order to get the more general convergence result.

## 4. APPLICATION TO PARTIAL SUMS

We now apply the results of the preceding section to partial sums. The setup is as follows.

Let $\{X_n\}$ be a stationary sequence of random variables with common df $F(x)$ and let $1-G(x)$ be the distribution function of $|X_1|$. We shall assume F belongs to the stable domain of attraction with index $\alpha$, $0 < \alpha < 2$. This translates into the following assumptions on F and G:

$$G(x) = P(|X_1| > x) = x^{-\alpha}L(x) \tag{4.1}$$

where $L(x)$ is a slowly varying function at $\infty$, and

$$\frac{1-F(y)}{G(y)} \to p \text{ and } \frac{F(-y)}{G(y)} \to q \text{ as } y \to \infty, \tag{4.2}$$

where $0 \le p \le 1$ and $q = 1 - p$. Then the eventual normalizing constants, $a_n > 0$ and $b_n$, for the partial sums $S_n = X_1+\ldots+X_n$ will be determined by

$$nG(a_n x) \to x^{-\alpha} \text{ as } n \to \infty,\ x > 0,$$

and

$$b_n = \int_{-a_n}^{a_n} x \, dF(x).$$

Let $\{\hat{X}_n\}$ be the associated independent sequence of $\{X_n\}$ (i.e. $\{\hat{X}_n\}$ is an iid sequence and $\hat{X}_1 \overset{d}{=} X_1$). Then the above assumptions on F imply that $a_n^{-1}\hat{S}_n \to S$ if $0 < \alpha < 1$, and $a_n^{-1}(\hat{S}_n - b_n) \to S^*$ for $1 \le \alpha < 2$ where $\hat{S}_n = \hat{X}_1+\ldots+\hat{X}_n$ and S and S* have stable distributions. Under the following mixing and dependence conditions, the analogous result can be established for the $S_n$'s.

We shall assume that the $\{X_n\}$ sequence satisfies the mixing condition D of the preceding section and the local dependence condition D'. The condition D' is said to hold if for all $x > 0$

$$\limsup_{n\to\infty} S_{k,n}(x) = o(1) \text{ as } k \to \infty \text{ where}$$

$$S_{k,n}(x)=n \sum_{j=2}^{[n/k]} \{P(X_1>a_n x,\ X_j>a_n x)+P(X_1>a_n x,\ X_j\le -a_n x) +$$

$$P(X_1\le -a_n x,\ X_j>a_n x)+P(X_1\le -a_n x,\ X_j\le -a_n x)\}.$$

This D' hypothesis combines the two conditions, C2 and the one contained in Remark 5, and thus guarantees that the upper and lower extremes have the same limiting behavior as the corresponding statistics from the $\{X_n\}$ sequence. Together with the techniques in LePage, Woodroofe and Zinn (1981), this is the basic fact that is needed in proving the convergence of the partial sums to a stable limit.

Theorem 4.1. If $0 < \alpha < 1$, then $a_n^{-1} S_n \overset{d}{\to} S$.

For the case when $1 \le \alpha < 2$, a further dependence assumption is needed. Condition D" is said to hold if

$$\lim_{\varepsilon\to 0} \limsup_{n\to\infty} \frac{n}{a_n^2} \sum_{j=2}^{n} \max(0, \mathrm{Cov}(X_1 I_{(-\varepsilon a_n, \varepsilon a_n)}(X_1), X_j I_{(-\varepsilon a_n, \varepsilon a_n)}(X_j)))$$

$$= 0. \tag{4.3}$$

Each term in the sum is bounded by $a_n^{-2} n \,\mathrm{Var}(X_1 I_{(-\varepsilon a_n, \varepsilon a_n)}(X_1))$ which in turn is bounded by $\frac{n}{a_n^2} \int_{-\varepsilon a_n}^{\varepsilon a_n} x^2\, dF(x)$. Using properties of regularly varying distributions (de Haan (1970)),

$$\lim_{\varepsilon\to 0} \limsup_{n\to\infty} \frac{n}{a_n^2} \int_{-\varepsilon a_n}^{\varepsilon a_n} x^2\, dF(x) = 0 \text{ so that each of the sum-}$$

mands in (4.3) has the desired property and, thus, an m-dependent sequence satisfies (4.3).

We then have

Theorem 4.2. For $1 \le \alpha < 2$, suppose the stationary sequence $\{X_n\}$ also satisfies D". Then $a_n^{-1}(S_n - b_n n) \xrightarrow{d} S^*$.

The proofs of the preceding theorems can be found in [10].

We now give two examples illustrating these results.

Example 4.1. Let $\{Y_n\}$ be a stationary Gaussian sequence with zero mean, unit variance, and covariance function $r_n = EY_1Y_{n+1}$. Assume $r_n \log n \to 0$ as $n \to \infty$. Now define $X_n = H(Y_n)$ where H is chosen so that the distribution function of $X_1$ belongs to a stable domain of attraction with index $\alpha$, $0 < \alpha < 1$. Note that the dependence structure of the $\{X_n\}$ sequence is completely determined by the covariance function $r_n$. Using methods similar to those in Berman (1964, 1971) and Leadbetter, Lindgren and Rootzen (1979), it can be shown that conditions D and D' are satisfied by the $\{X_n\}$ sequence and thus $a_n^{-1} \sum_{j=1}^{n} X_j \xrightarrow{d} S$.

It is worth remarking that if the instantaneous function of $Y_n$ was chosen such that $\mathrm{Var}(H(Y_n)) < \infty$, then the normalized partial sums of $X_n = H(Y_n)$ may not be asymptotically normal even under the assumption $r_n \log n \to 0$. The special case where $H(Y_n) = Y_n^2 - 1$ and $r_n \approx n^{-\gamma}$, $0 < \gamma < \frac{1}{2}$, was dealt with by Rosenblatt (1979). This is contrasted with the above example where H is chosen to make the df of $H(Y_1)$ belong to a stable domain of attraction.

Example 4.2. We begin with an iid sequence of 2-vectors $\{(Z_n^1, Z_n^2): n \ge 1\}$ with common distribution defined as follows. The density $f(x,y)$ of this distribution is supported on the unit square and is given by

$$f(x,y) = \begin{cases} 0 \text{ if } (x,y) \in (3/8,1/2)\times(1/8,2/8) \cup (1/2,5/8)\times(0,1/8) \\ 2 \text{ if } (x,y) \in (3/8,1/2)\times(0,1/8) \cup (1/2,5/8)\times(1/8,2/8) \\ 1 \text{ elsewhere on the unit square.} \end{cases}$$

In other words, $f(x,y)$ is an altered uniform density on $(0,1) \times (0,1)$ with twice as much mass on the squares $(3/8,1/2)\times(0,1/8)$ and $(1/2,5/8)\times(1/8,2/8)$ and no mass on the adjacent squares $(3/8,1/2)\times(1/8, 2/8)$ and $(1/2,5/8)\times(0,1/8)$. It is easy to check that both marginal densities are uniform on $(0,1)$ and $P(a < Z_1^1 \le b, a < Z_1^2 \le b) = (b-a)^2$ for $b > 5/8$ and $a < 3/8$. Define the sequence $(Y_1, Y_2, Y_3, \ldots)$ to be equal to $(Z_1^2, Z_3^1, Z_3^2, Z_5^1, Z_5^2, \ldots)$ and $(Z_2^1, Z_2^2, Z_4^1, Z_4^2, \ldots)$ each with probability 1/2. It is easy to check that $\{X_n\}$ is stationary and $Y_j$ is independent of $(Y_{j+2}, Y_{j+3}, \ldots)$ for all j. Moreover if A is a Borel subset of $(0,1)$ such that

$$A \cap (3/8,\ 5/8) = \phi \text{ or } (3/8,\ 5/8) \subset A, \tag{4.4}$$

then $P(Y_{i_1} \varepsilon A, \ldots, Y_{i_s} \varepsilon A) = P^s(A)$ for all choices of integers $1 \le i_1 < \ldots < i_s$. Now define $X_n = F^{-1}(Y_n)$ where F is a df satisfying properties (4.1) and (4.2). If B is a finite union of disjoint intervals, then the event $\{a_n^{-1} X_j \varepsilon B\} = \{Y_j \varepsilon A_n\}$ where $A_n = \{x: a_n^{-1} F^{-1}(x) \varepsilon B\}$. For large n, $A_n$ eventually satisfies (4.4) so that $P(X_{i_1} \varepsilon B, \ldots X_{i_s} \varepsilon B) = P^s(X_1 \varepsilon B)$. It follows that conditions D and D' are fulfilled for this sequence and since $X_1$ and $X_j$ are independent for $j>2$, condition D" is also satisfied by the remark preceding Theorem 4.2. Thus

$$a_n^{-1}\left(\sum_{j=1}^{n} X_j - n\, b_n\right) \xrightarrow{d} S^*.$$

One other interesting feature of this example is that the process $\{X_n\}$ is not mixing. If T denotes the shift operator, then a sequence of random variables is said to be mixing if for any two events A and B, $P(A \cap T^{-j}B) - P(A)P(B) \to 0$ as $j \to \infty$. "Mixing" is far weaker than most of the mixing conditions (e.g. strong mixing, maximal correlation mixing, uniform mixing) which are commonly used in proving central limit theorems.

The above results and examples suggest that the mixing assumptions required for proving stable limits of partial sums need not be as stringent as those for establishing central limit theorems. However, a price is paid in the form of a local dependence hypothesis (condition D') which occasionally is not even satisfied by 1-dependent sequences.

ACKNOWLEDGMENT

This research was supported in part by the National Science Foundation Grant MCS 8202335.

REFERENCES

[1] Adler, Robert J. (1978). Weak convergence results for extremal processes generated by dependent random variables. Ann. Prob., 6, 660-667.

[2] Barlow, R. E. and Proschan, F. (1975). Statistical Theory of Reliability and Life Testing. (Holt, Rinehart and Winston, New York).

[3] Berman, S. M. (1964). Limit theorems for the maximum term in stationary sequences. Ann. Math. Statist., 35, 502-516.

[4] Berman, S. M. (1971). Asymptotic independence of the numbers of high and low level crossings of stationary Gaussian processes. Ann. Math. Statist., 42, 927-947.

[5] Chernick, M. R. (1981). A limit theorem for the maximum of autoregressive processes with uniform marginal distributions. Ann. Prob., 9, 145-149.

[6] Chernick, M. R. and Davis, R. A. (1982). Extremes in autoregressive processes with uniform marginal distributions. Statistics and Probability Letters, 1, 85-88.

[7] Davis, R. A. (1979). Maxima and minima of stationary sequences. Ann. Prob., 7, 453-460.

[8] Davis, R. A. (1982). Limit laws for the maximum and minimum of stationary sequences. Z. Wahr. verw. Geb., 61, 31-42.

[9] Davis, R. A. (1983). Limit laws for upper and lower extremes from stationary mixing sequences. J. Multivariate Analysis, 13, pp. 273-286.

[10] Davis, R. A. (1983). Stable limits for partial sums of dependent random variables. Ann. Prob., 11, pp. 262-269.

[11] Finster, M. (1982). The maximum term and first passage times for autoregressions. Ann. Prob., 10, 737-744.

[12] Galambos, J. (1978). The Asymptotic Theory of Extreme Order Statistics. John Wiley, New York.

[13] Haan, L. de (1970). On Regular Variation and Its Applications to the Weak Convergence of Sample Extremes. Mathematical Centre Tract 32, Mathematics Centre, Amsterdam.

[14] Haan, L. de and Resnick, S. I. (1977). Limit theory for multivariate sample extremes. Z. Wahr. Verw. Geb., 40, 317-37.

[15] Ibragimov, I. A. and Linnik, Y. V. (1971). Independent and Stationary Sequences of Random Variables. Wolters-Noordhoff, Groningen.

[16] Kallenberg, O. (1976). Random Measures. Akademie-Verlag, Berlin.

[17] Leadbetter, M. R. (1974). On extreme values in stationary sequences. Z. Wahr. verw. Geb., 28, 289-303.

[18] Leadbetter, M. R., Lindgren, G. and Rootzen, H. (1979). Extremal and related properties of stationary processes. Part I. Statistical Research Report No. 1979-2, University of Umea, Sweden.

[19] LePage, R.,Woodroofe, M. and Zinn, J. (1981). Convergence to a stable distribution via order statistics. Ann. Prob., 9, 624-632.

[20] Mori, T. (1976). Limit laws for maxima and second maxima from strong-mixing processes. Ann. Prob., 4, 122-126.

[21] Mori, T. (1977). Limit distributions of two-dimensional point processes generated by strong mixing sequences. Yokohama Math. J., 25, 155-168.

[22] O'Brien, G. L. (1974). The maximum term of uniformly mixing stationary processes. Z. Wahr. verw. Geb., 30, 57-63.

[23] Resnick, S. I. (1983). Point processes, regular variation, and weak convergence. (To appear Adv. in Applied Prob.)

[24] Rosenblatt, M. (1979). Some limit theorems for partial sums of quadratic forms in stationary Gaussian variables. Z. Wahr. verw. Geb., 49, 125-132.

# MODELLING EXCESSES OVER HIGH THRESHOLDS, WITH AN APPLICATION

Anthony C. Davison

Department of Mathematics, Imperial College, London SW7

In many areas of application the extremes of some process may be modelled by considering only its exceedances of a high threshold level. The natural parametric family for such excesses for continuous parent random variables, the generalized Pareto distribution, is closely related to the classical extreme-value distributions. Here its basic properties are discussed, with some ideas for graphical exploration of data. Maximum likelihood estimation of parameters in the presence of covariates is considered, and techniques for checking fit based on residuals and a score test developed.

An application in modelling high exposures to radionuclides due to releases from notional sites in W. Europe is described.

## 1. INTRODUCTION

This paper presents techniques for the statistical modelling of sample excesses over thresholds and is closely related to the work of Smith [28]. The problem of statistical inference for such excesses may arise in any area of science where the analysis of extremes of sequences of observations is important, for example in hydrology, meteorology, oceanography, and air pollution. The sequences of observations may be independent or may exhibit trend, seasonality, and long- or short-term dependence, all of which will probably complicate analysis. The sequences may be related, for example several sequences of water levels at different points along a river, or air pollution levels at a number of receptors at different locations relative to a common point source of contaminant, may be available. It may then be required to link



J. Tiago de Oliveira (ed.), Statistical Extremes and Applications, 461–482.

the sequences using covariates, which might be hydrological variables in the first case, and meteorological and physical ones in the second. From now on we confine attention to upper thresholds since excesses below lower thresholds may be converted into upper ones simply by negating the data.

A number of approaches to modelling upper extremes of such sequences may be possible, depending on the structure and complexity of the data. If the sequences are fairly long, the classical method treating annual maxima of consecutive periods of equal length, for example years, months or days, of the series as independently and identically distributed in one of the extreme value distributions is often used, following the work of Gumbel [16]. The method is common and often sucessful in environmental applications. As at present used, it has some drawbacks:

(i) parameter estimation in the presence of covariates, formal tests of goodness of fit (as opposed to informal graphical checks), and studies of influence have yet to be developed for the method, except in special cases, see Stephens [29];

(ii) more seriously, its use of data is rather uneconomical and inferences based on short sequences are likely to be unreliable. This last difficulty is common to all methods of analysis for sample extremes; the point here is that if, say, k years data are available, then inference based on the upper k order statistics ought to be at least as good as inference based on the annual maxima. This is because order statistics are at least as great as the annual maxima and so might be expected to be more informative about the upper tail of the distribution.

Methods of analysis based on fixed numbers of sample order statistics have been mooted by Weiss [30], Hill [18], Weissman [31] and Hall [17], but estimation must then be based on possibly complicated densities of dependent observations. The "Peaks Over Threshold" method developed in the Flood Studies Report [23], based on a fixed number of order statistics per year, is related but somewhat ad hoc. Here a third possibility is investigated: choosing a high threshold level and modelling the process of exceedances of it, conditionally upon the level chosen. Thus the sequence is reduced to a random number N of exceedances, their values $Y_i$ and the epochs $T_i$ at which they occur. The natural parametric family of distributions for such exceedances is the generalized Pareto distribution (GPD),

$$F(y) = \begin{cases} 1 - (1-ky/\sigma)^{1/k} & (k \neq 0) \\ 1 - \exp(-y/\sigma) & (k=0), \end{cases}$$

$$(-\infty < k < \infty, \sigma > 0; 1-ky/\sigma > 0),$$

found by Pickands [25], which is closely related to the generalized extreme value distribution (GEV)

$$G(y) = \begin{cases} \exp\{-(1-k(y-\alpha)/\sigma)^{1/k} & (k\neq 0) \\ \exp\{-\exp(-(y-\alpha)/\sigma) & (k=0), \end{cases}$$

$$(-\infty < k,\ \alpha < \infty,\ \sigma > 0;\ 1-k(y-\alpha)/\sigma > 0)\ ,$$

for which Prescott and Walden [26] studied maximum likelihood estimation. Smith [28] enlarges on their relationship, discusses maximum likelihood estimation of the parameters of the GPD in detail, gives a treatment of possible procedures for modelling the point process $T_i$, and describes an application of the method. His paper should be read in conjunction with this one. Here, following Smith, we note only that if a homogeneous Poisson process with independent excesses identically distributed as generalized Pareto with parameter k is observed for a fixed time-period, then the largest excess during the period has exactly the GEV with the same value of k. This result establishes a strong connection between the present topic and the classical statistical analysis of extremes.

Here we study statistical analysis for sets of independent observations from the GPD in the presence of covariates, ignoring possible dependence of the observations. This restriction is not so great as it appears, since Leadbetter, Lindgren, and Rootzen [20] show that under some general conditions the excesses of stationary sequences over high thresholds are asymptotically independent, provided they are not too close together.

Estimation and the testing of goodness of fit in the presence of covariates is discussed. Thus the class of models proposed is flexible in the approach it allows to complex data; because the GPD has a simple hazard function it may have applications in the analysis of medical and industrial time-failure data. The model has possible uses in medicine where it may be intended to model the exceedances of some variable, for example a patient's blood pressure, over a threshold level. There are also potential applications in the analysis of ground metal surfaces, since engineers whose aim is to produce as smooth a surface as possible may focus on exceedances of some tolerance level as a measure of roughness. More generally any area of science where attention concentrates on some threshold and the process of the excesses over it is a field of application for the model.

Section 2 of the paper gives simple properties of the GPD with some ideas for graphical exploration of data, while maximum likelihood estimation of the GPD for independent observations with covariates is studied in section 3. Methods for assessing goodness of fit and analysis of residuals are presented in section 4, and section 5 illustrates the techniques with an application in modelling extreme values of exposures in the environment to radionuclides due to hypothetical discrete releases from a site in Northern Europe. Finally, the results are discussed in section 6.

## 2. PROPERTIES OF THE GPD.

The generalized Pareto distribution was first proposed by Pareto [24] for $k<0$, and consequently is known as a Pareto type II as well as Pearson type VI distribution. Davis and Feldstein [15] call the Pareto type III a generalized Pareto distribution, but Pickand's use of the term generalized Pareto predates this and so we follow Pickands. For $k<0$ the distribution has a heavy Pareto-type upper tail, and so may give relatively large values with appreciable probabilities. The case $k=0$ is the exponential distribution for which many statistical techniques are available, including tests of fit based on the empirical distribution and procedures for detecting outliers. When $k>0$ the distribution has an upper endpoint at $\sigma/k$. For $k = \frac{1}{2}$ and 1 the distribution is triangular and uniform respectively. The rth moment of the distribution about the origin exists when $-k < 1/r$, and then is

$$E[X^r] = \sigma^r(-k)^{r+1}\Gamma(1+r)\Gamma(-1/k-r)/\Gamma(1-1/k).$$

The variance of the distribution exists only for $k>-\frac{1}{2}$, the kurtosis for $k>-\frac{1}{4}$ with implications for estimation by the method of moments. The hazard function of the GPD has the simple form $(\sigma-ky)^{-1}$. The distribution was derived by Maguire, Pearson and Wynn [21] for $k<0$ as the compound of exponential variables with gamma-distributed random hazard.

Suppose that Y has generalized Pareto distribution with parameters k and $\sigma$. Then it is easily verified that if $s < \sup\{x: F(x)<1\}$, then

$$E[Y-s \mid Y>s] = (\sigma-sk)/(1+k)$$

provided the expectation exists. This suggests a graphical procedure to determine $\sigma$ and k in a simple random sample of size n as follows: for a succession of increasing thresholds s form the mean excess y(s) of the sample above s and plot it against s. This graph of $\Sigma(Y_i-s)/|A_s|$, where the sum is over $i\epsilon A_s=\{j:y_j>s\}$, versus s should be an approximate straight line of slope $-k/(1+k)$ and intercept $\sigma/(1+k)$. This may be useful when determining the appropriate threshold since the plot may be expected to be non-linear for low values of s, but to become increasingly straight as s increases and the excesses approach the GPD. This suggests choosing a threshold above any non-linear lower portion of the graph.

It may also be useful for obtaining initial estimates of the parameters for numerical routines. For $k>-\frac{1}{2}$ a weighted least squares procedure with correlated observations can be based on this graph, but is only likely to be useful in simple cases, when less cumbersome techniques could more easily be used.

The expected value of the $r^{th}$ order statistic in a simple random sample of size n exists provided that $n+1-r>-k$, and is then

$$E[Y_{r,n}] = \sigma[\prod_{1}^{r}(n+1-i)/(n+1-i+k)-1].$$

Provided the expectations exist, the difference between two successive order statistics is

$$E[Y_{r+1,n}-Y_{r,n}] = (n-r+k)^{-1}\sigma\prod_{1}^{r}(n+1-i)/(n+1-i+k).$$

The corresponding quantity for the exponential distribution is $\sigma(n-r)^{-1}$, motivating the idea of plotting ordered sample values against exponential order statistics. Such a graph should be concave for $k>0$, a straight line for $k=0$, and convex for $k<0$, providing another means of assessing the weight of the sample upper tail.

Estimation procedures based on sample order statistics can be developed which sometimes have high efficiency in other applications, and may be useful for the GPD when k is large and positive. In most enviromental applications k will lie in the range $(-\frac{1}{2},\frac{1}{2})$, in which case maximum likelihood estimation will be regular and asymptotically efficient. In addition maximum likelihood estimators are asymptotically sufficient statistics for the parameters, so will be the only type of estimators considered here. Bayesian techniques could be developed for the distribution.

## 3. MAXIMUM LIKELIHOOD ESTIMATION WITH COVARIATES.

A major aspect of statistics is the explanation of systematic variation in response variables in terms of variation in explanatory variables: thus here there will often be the need to assess the extent to which the distributions of exceedances for different but related series depend on external factors whose distributions are fixed or at least irrelevant to the distribution of the excesses. The most successful and widely used technique for this is the general linear model, which we now explore for the GPD, considering only estimation by maximum likelihood.

One important issue to be faced immediately is this: if the random variable $Y_i$ has shape parameter $k_i$, the expectation of the $r^{th}$ partial derivatives of the likelihood of $Y_i$ with respect to

$k_i$ are not finite if $k_i \geq 1/r$. Thus if $k_i \geq \frac{1}{2}$ the Fisher information for the observation is infinite, the usual results of asymptotic maximum likelihood theory do not hold, and in particular the variance of the estimators is not order 1/n in probability. Smith [ 28 ] discusses this in more detail, giving rates of convergence of the maximum likelihood estimators for different values of $k_i$.

A consequence of this difficulty is that when $0 \leq k_i \leq \frac{1}{2}$, and especially when $1/3 \leq k_i$, confidence intervals for parameters and computational procedures based on their estimates which depend on quadratic approximations to the loglikelihood may be poor, and it may be desirable to find likelihood-based confidence regions rather than ones using the asymptotic normal distribution of the maximum likelihood estimates. Plotting the loglikelihood is a useful guide as to the need for this.

Details of computing the estimates in the presence of covariates will not be discussed here. Nevertheless note that if each $k_i \leq 0$, methods for unconstrained optimization of the likelihood may be used, but that if some or all of the $k_i$ are positive, maximization is subject to the n constraints $1 > k_i y_i/\sigma_i$, and a suitable algorithm may be needed.

Suppose, then, that the n variables $Y_i$ are independently distributed according to the GPD with parameters $k_i$ and $\sigma_i$, and that $k_i \leq \frac{1}{2}$ for all i. Let $k_i = \gamma_0 + \gamma^t z_i$ and $\sigma_i = \exp(\beta_0 + \beta^t x_i)$, where $z_i$ and $x_i$ are vectors of q and p covariates respectively. Then $\vartheta = (\gamma, \beta, \gamma_0, \beta_0)$ is a p+q+2 vector of parameters to be estimated from the data. An important special case is when the $z_i$ are all absent or zero and the observations have a common shape parameter k. Without loss of generality say that $\Sigma_i x_{iu} = \Sigma z_{it} = 0$ (u=1,...,p, t=1,...,q).

The loglikelihood of the data is then

$$L(\vartheta) = -\sum_{i=1}^{n} [\beta_0 + \beta^t x_i + \{1-(\gamma_0+\gamma^t z_i)^{-1}\} \ln\{1-(\gamma_0+\gamma^t z_i)\exp(-\beta_0-\beta^t x_i) y_i\}]$$

which must be maximized to find estimates $\hat{\vartheta}$ of $\vartheta$. The elements of the Fisher information matrix $i_\vartheta$ are:

$$E[-\partial^2 L/\partial\gamma_s \partial\gamma_t] = 2\Sigma z_{is} z_{it} a_i b_i, \quad E[-\partial^2 L/\partial\gamma_s \partial\beta_u] = -\Sigma z_{is} x_{iu} a_i b_i,$$

$$E[-\partial^2 L/\partial\gamma_s \partial\gamma_0] = 2\Sigma z_{is} a_i b_i, \quad E[-\partial^2 L/\partial\gamma_s \partial\beta_0] = -\Sigma z_{is} a_i b_i,$$

$$E[-\partial^2 L/\partial\beta_u \partial\beta_v] = \Sigma x_{iu} x_{iv} b_i, \quad E[-\partial^2 L/\partial\beta_u \partial\gamma_0] = -\Sigma x_{iu} a_i b_i,$$

$$E[-\partial^2 L/\partial\beta_u \partial\beta_0] = \Sigma x_{ul} b_i, \quad E[-\partial^2 L/\partial\gamma_0^2] = 2\Sigma a_i b_i,$$

$$E[-\partial^2 L/\partial\gamma_0 \partial\beta_0] = -\ a_i b_i, \quad E[-\partial^2 L/\partial\beta_0^2] = \Sigma b_i,$$

$(1 \leq s,t \leq q;\ 1 \leq u,v \leq p),$

where $a_i = (1-k_i)^{-1}$ and $b_i = (1-2k_i)^{-1}$. In the case when $k_i = k$ for all i, $i_\vartheta$ becomes the matrix

$$(1-2k)^{-1} \begin{vmatrix} X^tX & 0 & 0 \\ 0 & 2n(1-k)^{-1} & -n(1-k)^{-1} \\ 0 & -n(1-k)^{-1} & n \end{vmatrix}$$

of side p+2, which is easily inverted provided $X^tX$ is full rank.

The form of $i_\vartheta$ has implications for robust and optimal design in those situations where the data are the result of a designed experiment or trial, not usually the case with enviromental data.

The usual theory of hypothesis testing for the effect of particular covariates based on the differences of maximized loglikelihoods may be used, though care should be taken if any $k_i$ is positive. One hypothesis of interest when the data are divided into homogeneous subsamples is that the shape parameter k is constant throughout the data or some collection of subsamples, the alternative being that it differs in each. More generally it may be required to test the hypothesis of constant k against the alternative that some explanatory factors influence the values of the $k_i$.

Although the difference of the maximized loglikelihoods under the models can be found, it will usually be more economical to construct a score test for the null against the alternative hypothesis.

Such a test is based on partitioning the vector $\vartheta$ into parameters $\xi$ of interest and nuisance parameters $\zeta$, and testing the possibility that $\zeta=\zeta_0$, some particular null value. If $\vartheta_0 = (\hat{\xi},\zeta_0)$ is the maximum likelihood estimate of $\vartheta$ on the hyperplane $\zeta=\zeta_0$, then the statistic $W_u = U(\vartheta_0)^t i(\vartheta_0)^{-1} U(\vartheta_0)$ is asymptotically $\chi^2_{\dim\zeta}$ when the true value of $\zeta$ is $\zeta_0$. Here $U(\vartheta_0)$ and $i(\vartheta_0)$ are the score statistic and Fisher information evaluated at $\vartheta_0$. The test has power properties similar to the likelihood ratio test but the likelihood need only be maximized on the null hypothesis. See Cox and Hinkley [11, Ch.9] for more details.

For testing $\gamma=0$, $i(\vartheta_0)$ is the p+q+2 matrix

$$(1-2k)^{-1} \begin{vmatrix} 2(1-k)^{-1}Z^tZ & -(1-k)^{-1}Z^tX & 0 \\ -(1-k)^{-1}X^tZ & X^tX & 0 \\ 0 & 0 & M \end{vmatrix}$$

where M is (1-2k) times the asymptotic variance-covariance matrix for $\hat{\gamma}_0, \hat{\beta}_0$. The upper left qxq element of $i^{-1}(\vartheta_0)$ is $Q=\frac{1}{2}(1-k)(Z^tZ)^{-1}Z^t[I+\frac{1}{2}(1-k)^{-1}XE^{-1}X^t]Z^t(Z^tZ)^{-1}$, where E is the matrix $X^t\{I-\frac{1}{2}(1-k)^{-1}Z(Z^tZ)^{-1}Z^t\}X$. The $s^{th}$ $(1 \leqslant s \leqslant q)$ element of $U(\vartheta_0)$ is

$$k^{-2}\Sigma_i z_{is}\{k(k-1)y_i(\sigma_i-ky_i)^{-1}-\log(1-ky_i/\sigma_i)\} ,$$

giving $W_u=U_q^t(\vartheta_0)QU_q(\vartheta_0)$, where $U_q$ is the vector of the q uppermost elements of the score U, to be compared with $\chi^2_q$.

Finally we deal with the computation of $\hat{\vartheta}$ in the very special case of a simple random sample. By reparameterizing in terms of $\eta=k/\sigma$ and k, the maximization can be reduced to a one-dimensional search, since the unique value of k which maximizes $L(\eta,k)$ given an estimate $\hat{\eta}$ of $\eta$ is $k(\hat{\eta})=n^{-1}\Sigma\log(1-\hat{\eta}y_i)$. It follows that the best value of L lies on the locus $(\eta,k(\eta))$ within the plane, for some $\eta<(\max_i\{y_i\})^{-1}$. I am grateful to Richard Smith for pointing this out to me.

The maximum likelihood estimate of $\eta$ may easily be found numerically or graphically and the best $\vartheta$ calculated. In cases where it is required to fit the GEV to maxima of sequences over some period but where suitable data is available, it may be valuable to find an initial estimate of k by estimating it from the sample exceedances of some high threshold.

A fairly thorough account of maximum likelihood estimation for the GPD and some related issues has been given, but some topics have not been tackled:

(i) consideration of the computational details of multiparameter estimation when some or all of the $k_i$ are positive;

(ii) asymptotic properties of the estimators and the shape of the loglikelihood when some or all of the $k_i$ are close to but less than $\frac{1}{2}$;

(iii) small sample properties of the estimators in the regular case, a study of which could be based on saddlepoint approximations [Barndorff-Nielsen, 6] or simulation.

## 4. GOODNESS OF FIT

Techniques commonly used for checking goodness of fit in the linear model include:

(i) the inspection of residuals;

(ii) embedding the fitted model in a plausible more

comprehensive alternative and testing fit against the larger model.

We consider some possible modifications of these ideas suitable in the present context.

### 4.1 Residuals

The first possibility is the construction of quantities analagous to the usual residuals of the linear model, random quantities whose distributions are known approximately when the fitted model is correct, which may be inspected, usually graphically, to detect various forms of departure. Cox and Snell [13] gave a general definition of residuals for a broad class of models, and suggested a method of approximating their means, variances, and covariances. For a model where the observations $Y_i$ have some expression

$$Y_i = g_i(\vartheta, \varepsilon_i) \qquad (i=1,\dots,n)$$

in terms of a vector of unknown parameters and the independent identically distributed variables $\varepsilon_i$, and the equations

$$Y_i = g_i(\hat{\vartheta}, R_i)$$

have a unique solution for $R_i$ in terms of $Y_i$ and the maximum likelihood estimate $\hat{\vartheta}$ of $\vartheta$, they define the crude or unadjusted residuals $R_i$ as that solution and go on to find means, variances and covariances of the $R_i$ to order $n^{-1}$, which can then be used to modify the $R_i$ to have the same mean and variances to that order. Their results suggest that even in fairly small samples these adjustments make little difference to the $R_i$ for plotting purposes.

With this in mind define

$$R_i = -\hat{k}_i^{-1} \log(1 - \hat{k}_i y_i / \hat{\sigma}_i) \qquad (i=1,\dots,n).$$

which are crude residuals in the sense above. These should be distributed approximately as independent unit exponential variables when the model is correct, and so may be checked for outliers, dependence, and distributional form in the usual way using graphical techniques for the exponential distribution. The values of test statistics should be interpreted with great care, and can be misleading in many cases, as Durbin's contribution to the discussion following Cox and Snell [13], and their 1971 paper [14] make clear. The levels of significance of test statistics may be seriously underestimated if they are formed from unadjusted residuals, so their systematic use is not recommended.

In some circumstances it may be useful to use the $U_i = 1 - \exp(-R_i)$, which should be approximately uniform on (0,1), rather than the $R_i$. Two cases when this may be informative are:

(i) when the data is divided into subsamples, in which case plotting the $U_i$ against subsample number or some meaningful physical quantity may reveal distortions in the fitting of the GPD to specific subsamples, or discrepancies between them;

(ii) when plotting the data to investigate the possibility of serial dependence or clustering into groups (possibly occurring together in time) of the observations.

If in these situations the plots are made on the original scale, the skewness of the $R_i$ may lead to difficulties of interpretation which plotting the $U_i$ instead alleviates.

Clearly there is much scope for further work, and in particular for obtaining the adjustments to residuals for the linear model outlined in section 3 and for developing plots and test statistics specifically for checking the critical assumption of independence of the observations.

## 4.2 A Score Test

Possible more formal methods of checking fit of a model to data are to test the fit of the model against alternatives:

(i) which make the same distributional assumptions but include more explanatory variables or combinations of them in the systematic part of the model;

(ii) representing plausible but separate families of hypotheses [Cox, 9,10];

(iii) which embed the random component of the model in some more general distribution;

(iv) in which the systematic part of the model applies to some transformation of the data [Box and Cox, 7].

Here the possibilities are logically distinct; we shall pursue (iii) but tests for alternatives of type (i) can be made simultaneously by generalizing the procedure we propose.

The question of transformation is an important one from the point of view of improving the GPD approximation to a set of sample excesses over some threshold, which need not be good for low thresholds, although it may be expected to improve as the threshold increases. As this happens, however, less of the sample will be available and the uncertainty of eventual conclusions will increase, so it may be wise to transform the data and then to apply techniques based on thresholds, thus retaining a reasonable proportion of the data for analysis whilst improving the approximation. Ideally such a transformation should be based on knowledge of the physical processes underlying the data, but in situations where this cannot be, the possibility of more or less empirical transformations should be considered.

Here a score test for the possibility that not $Y_i/\sigma_i$, but $\varphi^{-1}\{\exp(\varphi(Y_i/\sigma_i)^{\lambda})-1\}$ has the generalized Pareto distribution with parameters k and 1, for some $\lambda>0$ and $-\infty<\varphi<\infty$, is considered. For $\varphi=0$ and $\lambda=1$ this reduces to the null hypothesis. Atkinson [4,5] gives similar tests in case (iv) for the Box-Cox family of transformations and for the folded-power family for proportions [Mosteller and Tukey, 22], and Cook and Weisberg [8] give a score test for homogeneity of variance in the normal-theory linear model. The test is locally uniformly most powerful for the alternatives considered, but is specific to them, so should be used together with residual plots to reveal other possible discrepancies of the data. Significance of the test means that a transformation of the form $\varphi^{-1}\{\exp(\varphi(y_i/\sigma_i)^{\lambda})-1\}$ for some $\sigma_i$, $\varphi$ and $\lambda$ should improve the fit of the GPD to the data.

After some tedious calculations, it emerges that provided $k\neq 0$ and $|k|<\frac{1}{2}$, the inverse information matrix for the parameters $(\beta,\gamma_0,\beta_0,\lambda,\varphi)$ when $\lambda=1$ and $\varphi=0$ is

$$\begin{vmatrix} (X^tX)^{-1}(1-2k) & 0 \\ 0 & P \end{vmatrix}$$

where the 2x2 corner submatrix of P corresponding to $(\lambda,\varphi)$ is

$$2\,\Delta^{-1}\begin{vmatrix} 2k^{-2}(1+2k)(1+k)^2(b-k^2-1) & (k+1)(1-4k^2)(1-2k+2c)k^{-1} \\ (k+1)(1-4k^2)(1-2k+2c)k^{-1} & 2(1-2k)(4k^2+2k+1) \end{vmatrix}$$

with

$$\Delta=(16bk^2+8bk+4b-32ck^3-24k^3+16k^2c^2-20k^2+8kc+16ck^2-4k-4c^2-4c-5)$$

and $b=\pi^2/6+\psi'(-1/k)$, $c=\ln(-k)+\psi(-1/k)-\psi(2)$ when $k<0$ and $b=\pi^2/6+\psi'(1+1/k)$, $c=\ln(k)+\psi(1+1/k)-\psi(2)$ when $k>0$, with $\psi(.)$ and $\psi'(.)$ respectively the first and second derivatives of the log-gamma function.

The corresponding elements of the score vector U are

$$U_\lambda=\Sigma_i(y_i/\sigma_i)(1-\tfrac{1}{2}(1-k)(y_i/\sigma_i)(1-ky_i/\sigma_i)^{-1}) \quad \text{and}$$

$$U_\varphi=\Sigma_i[1+\ln(y_i/\sigma_i)-(1-k)(y_i/\sigma_i)\ln(y_i/\sigma_i)(1-k_iy_i/\sigma_i)^{-1}].$$

When in addition it is required to test $k=0$, the parameters k and

$\varphi$ give the same type of first-order departure from the model, so that their components of the score vector are equal and $i_\vartheta$ is singular. One of them must be dropped, in which case we have the p+3-sided matrix

$$i_\vartheta^{-1} = \begin{vmatrix} (X^tX)^{-1} & 0 \\ 0 & P^* \end{vmatrix}$$

where $P^*$ is the matrix

$$\Delta^{-1} \begin{vmatrix} \pi^2/3-(\gamma-2)^2 & -\gamma & \pi^2/6-(\gamma-1)(\gamma-2) \\ -\gamma & 1 & -1 \\ \pi^2/6-(\gamma-1)(\gamma-2) & -1 & \Delta+1 \end{vmatrix}$$

with $\Delta=\pi^2/6-(\gamma-1)^2-1$ and $\gamma$ is Euler's constant, for the parameter vector $\vartheta=(\beta,\beta_0,\lambda,\varphi)$. The components of the score statistic are $U_\varphi = \Sigma_i[\frac{1}{2}(y_i/\sigma_i)^2-y_i/\sigma_i]$, $U_\lambda = \Sigma_i[1+(1-y_i/\sigma_i)\ln(y_i/\sigma_i)]$, so that the test for the hypothesis $\varphi=0,\lambda=1$ is based on the statistic

$$W_u=n^{-1}\Delta^{-1}[U_\varphi^2(\Delta+1)-2U_\varphi U_\lambda+U_\lambda^2]$$

which has asymptotic $\chi_2^2$ distribution when it is true.

There are analytical difficulties with this test which arise when $\lambda$ is included since then maximum likelihood theory is not regular if $k\leqslant-\frac{1}{2}$. Since in most enviromental applications $|k|\leqslant\frac{1}{2}$ [Jenkinson, 19], these difficulties are unlikely much to restrict the use of the test.

Finally, note that contributions to the test made by individual observations can be plotted in order to identify particular points or groups thereof influencing the statistic unduly; see Atkinson [5] or Cook and Weisburg [8] for discussion of this idea. This may help to distinguish aberrant values if a single datum or group of data contributes overmuch to the significance of the test, or conversely may show that evidence of failure to fit is spread throughout the entire data.

## 5. AN APPLICATION

The assessment of the potential consequences to the public of actual or hypothetical releases of material from nuclear installations in Europe is an important issue. Depending on the nature of the release and the location of the source, the potential consequences may spread across national frontiers. Accordingly article 37 of the Euratom treaty requires member states of the European Community to submit to the European Commission information on any plan for the disposal of radioactive waste, in order that the possible consequences for other member states may be determined. One aspect of such determinations is often the estimation of the long-range atmospheric transport of radionuclides; the modelling of such transport and its implications have accordingly been studied at Imperial College with the computer model MESOS, under a series of CEC contracts.

In the trajectory model MESOS the paths of a sequence of discrete puffs each beginning at a point source with a unit quantity of some radionuclide are estimated using adjusted geostrophic winds. These winds are deduced from consecutive 3-hourly pressure fields over Western Europe reconstructed from real weather data over a period of one year. In the present case the hypothetical puffs are each deemed to contain one Curie of the longlived inert nuclide $K_r^{85}$ (half-life ten years), and to be released at 3-hourly intervals from a hypothetical source in Western Europe through 360 days of the year 1976. Figure 1 shows the area covered by the calculations.

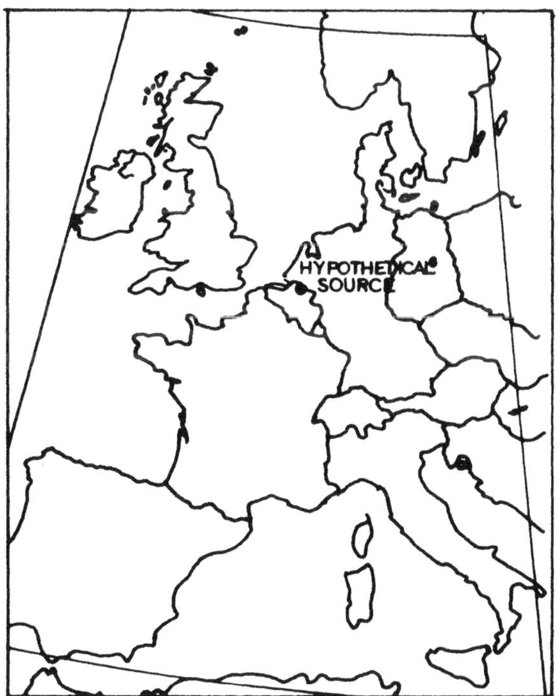


Figure 1: 1976 database area showing hypothetical source location.

The puffs then develop according to the real-time weather conditions along their trajectories at the time of passage: within the limits of computing feasibility and the physical and meteorological assumptions underlying the model they are moved by the current wind, develop laterally and vertically, are diluted by expansion of the puff, and depleted by radioactive decay and by dry and wet deposition to the ground. The depletion aspect is not relevant to $K_r^{85}$ as it is inert, and so does not deposit, and becuase almost all puffs leave the map area within 4 days, a negligible time by comparison with its radioactive half-life.

These trajectories are then used to build time-series of hypothetical exposures at a large number of points throughout the map area, so building a comprehensive database of "receptor histories". From these histories, the probability of exposure and the distribution of exposure levels when each receptor is contaiminated may be estimated. Characteristics of hypothetical releases of various nuclides with a wide range of decay and deposition rates over several different release durations from these sources in each of two years may be investigated using the MESOS database.

The model is dealt with in more detail by ApSimon et al. [2,3]; see ApSimon et al. [1] for a more detailed description of the exposure database.

One aspect of the general problem is the statistical study of episodes of particularly high contamination which happen infrequently due to particular combinations of meteorological factors but pose the maximum potential risk to the public, and it is to this that we now turn. We consider data for 16 receptors for which 3-hourly exposure data is available and attempt to find a model for the occurrence of contamination levels over some high threshold value. A summary of receptor positions and the numbers of values available is given in table 1. Figure 2 shows the time-series at receptor 4, 800 km North of the source through 1976. The marginal distribution of exposures is rather skew, and the exposures themselves occur in episodes due to short-term persistence of the weather conditions leading to contamination at the receptors.

Clearly exposures happening in the same episode at a receptor are not likely to be fully independent, nor are exposures occurring at about the same time at receptors fairly close together. For the purpose of this illustration, however, exposures at different receptors will be treated as independent.

| Receptor | Position (km) | Total no. of positive exposures | Threshold (Cism$^{-3}$10$^{-4}$) | No. of obs. over threshold | No. of clusters over threshold |
|---|---|---|---|---|---|
| 1 | 100 N | 303 | 316. | 5 | 4 |
| 2 | 300 N | 281 | 66. | 21 | 4 |
| 3 | 500 N | 270 | 33. | 19 | 7 |
| 4 | 800 N | 187 | 17. | 17 | 7 |
| 5 | 100 E | 422 | 316. | 8 | 5 |
| 6 | 300 E | 309 | 66. | 10 | 5 |
| 7 | 500 E | 268 | 33. | 20 | 11 |
| 8 | 800 E | 193 | 17. | 11 | 6 |
| 9 | 100 S | 349 | 316. | 5 | 4 |
| 10 | 300 S | 239 | 66. | 7 | 3 |
| 11 | 500 S | 232 | 33. | 14 | 8 |
| 12 | 700 S | 248 | 21. | 9 | 6 |
| 13 | 100 W | 356 | 316. | 9 | 6 |
| 14 | 300 W | 252 | 66. | 13 | 7 |
| 15 | 500 W | 216 | 33. | 15 | 9 |
| 16 | 800 W | 117 | 17. | 4 | 4 |

Table 1: Summary of receptors and high exposures for hypothetical puffs initially containing 1 Ci of $K_r^{85}$ released every 3 hours through 1976.

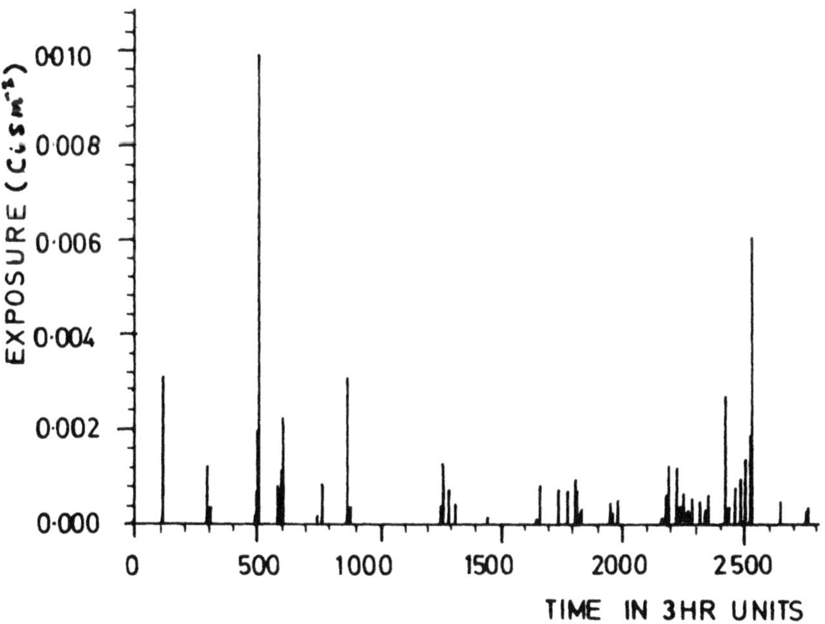


Figure 2: Exposure time-series 800 km North of the source through 1976.

One simple way of dealing with the episodic nature of the single-receptor process is to model the occurrence of cluster maxima and to treat different clusters as independent. Here we say that a cluster begins with a positive exposure and ends with an uncontaminated period of 12 hours or more, and retain only the cluster maxima for analysis. This leads to a total of 96 observations at 16 different receptors ; the threshold may conveniently be taken to be power-law function of distance from the source. For convenience the data are multiplied by $10^4$ throughout.

Inspection of the data shows no evidence that the distribution of the excesses depends on their direction from the source, so a model treating the excesses as having common shape parameter k but $\ln\sigma_i = \beta_0 + \beta_1$ x distance was fitted. The resulting parameter estimates, their standard errors, and their correlation matrix are displayed in table 2.

| Covariate | Parameter | Estimate | Standard error | | Correlation matrix | | |
|---|---|---|---|---|---|---|---|
| | | | | | $\beta_1$ | k | $\beta_0$ |
| scale | $\beta_0$ | 3.508 | .168 | | | | |
| distance | $\beta_1$ | -0.459 | .057 | $\beta_1$ | 1 | 0 | 0 |
| shape | k | -0.363 | .139 | k | | 1 | .607 |
| | | | | $\beta_0$ | | | 1 |

Table 2: Parameter estimates, their estimated standard errors, and correlations for the fitted model.

The value -.363 of k shows that the Pareto power-law tail is appropriate, obvious from figure 2, and confirmed by the likelihood ratio statistic for the hypothesis that k=0, highly significant at 10.08 as chi-squared on one degree of freedom. Figure 3 shows the likelihood maximised for different fixed values of k in the range-1.0 to 0. It is not quadratic, falling sharply as k increases. Confidence intervals for k based on the observed and expected information, and on Wilks' statistic are given. That based on the asymptotic chi-squared distribution with one degree of freedom of Wilks' statistic, $2[\ln(L_{max})-\ln(L_{max}(k))]$, where $L_{max}$ is the overall maximum and $L_{max}(k)$ the maximum for fixed k, it the same length as that based on the observed information but is not symmetric. The interval based on the asymptotic Normal distribution of the estimator using the expected or Fisher information as covariance matrix is smallest.

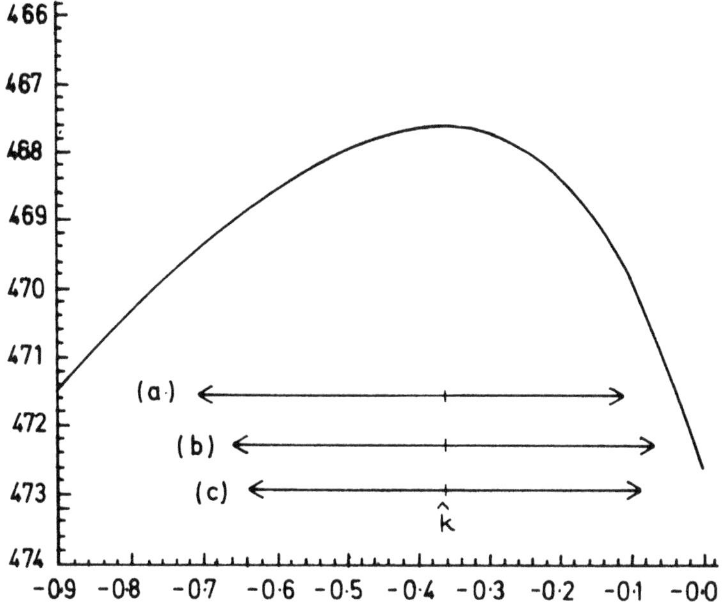


Figure 3: Maximized loglikelihood $L_{max}$ as a function of k, with 95% confidence intervals based on:
(a) Wilks' (likelihood ratio) statistic;
(b) the observed information statistic;
(c) the expected (Fisher) information.

This fact is interesting in the light of the close connexion between the GPD and GEV and some recent work by Prescott and Walden [27], which shows that confidence intervals for the shape parameter k of the GEV based on the observed information tend to be smaller than those for the expected information for $k>-.3$, but that the difference decreases as k decreases. In the present example all the intervals are rather large, and the asymmetry of the loglikelihood about its maximum renders the interval based on Wilks' statistic preferable to the others.

The ordered crude residuals $R_{(i)}$ plotted against exponential order statistics in figure 4 show no strong departure from the distributional assumptions made, although examination of the supposedly uniform $U_i$ reveals a slight preponderance of both large and small residuals at the 100 km receptors. The score statistic for dependence of k on inverse distance from the source has value 3.13, significant at between the nominal 10% and 5% levels of $\chi^2_1$.

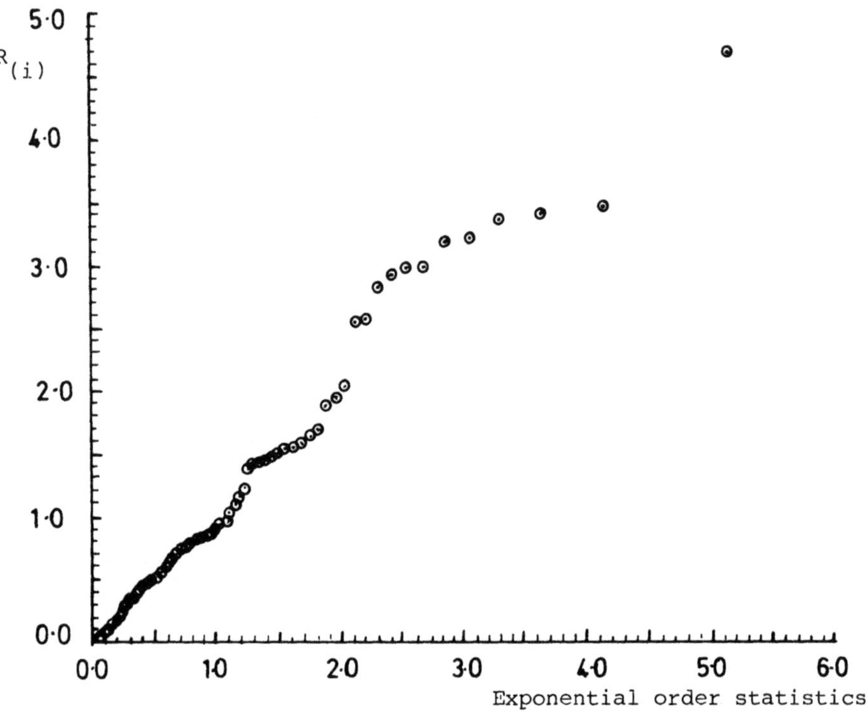


Figure 4: Ordered crude residuals $R_{(i)}$ plotted against exponential order statistics.

Thus there seems to be some evidence in favour of allowing k to increase a little with distance and so making the distribution of the excesses relatively less heavy-tailed further from the source.

The score goodness-of-fit statistic (nominally $\chi^2_2$) for the fitted model is 0.78, showing no evidence of poor fit of the model.

Overall, with mild misgivings about the possible change of k with distance, the fitted model seems to be adequate to the modelling of the magnitude of the cluster maxima. However, assumptions of indpendence between receptors have been made which seem fairly reasonable on meteorological grounds, but which merit closer examination for some specific uses of the model, such as the study of the joint distribution of excesses at nearby receptors.

## 6. DISCUSSION

One topic important in any practical use of the general family of models discussed in this paper is the choice of a suitable threshold. Here physical considerations specific to the subject matter are pertinent and so it is impossible to give more than rather general guidelines. However it is clear that from a statistical viewpoint enough sample information about the extremes to answer questions central to the analysis should be retained but the asymptotic nature of the generalized Pareto approximation to the excesses borne in mind. Combination of information from different but related samples using covariates will often be valuable if feasible since higher thresholds than justifiable in single smaller samples may be taken. A threshold level with a direct physical interpretation should be chosen where possible, provided of course that a useful model results. As usual in the study of sample extremes, it will be necessary to avoid affecting eventual inference based on the model by including data which do not reflect tail behaviour: hence the need for care in the choice of threshold. An idea of possible statistical importance would be some criterion for choice of threshold based on balancing the opposing requirements of maximizing sample information on the extremes whilst ensuring the adequacy of the asymptotic approximation on which the model rests.

Clustering of extreme values is a phenomenon observed in many situations and so the evolution of ideas for reflecting this characteristic is of direct interest. In some applications it may be sufficient to treat only the cluster maxima, but in others it may be important to predict the length of the episode and the sizes of its component excesses. There is scope here for existing probabilistic models for clustering point processes [see eg. Cox and Isham, 12, Ch.3], but suitable models are more likely to be suggested in specific applications by a detailed understanding of the process being observed, than by any purely mathematical considerations.

A number of remarks have been made in earlier sections about extensions to this work. Some other possibilities are:

(i) the study of the influence individual observations have on parameter estimates;

(ii) the development of exploratory techniques for complex data;

(iii) extensions to the study of excesses for bi- and multivariate observations;

(iv) extensions to censored data, both for medical applications and those in which the measuring instrument has an upper bound dictated by its construction.

ACKNOWLEDGEMENTS

I am grateful to Drs. H.M. ApSimon and R.L. Smith, and to Professor D.R. Cox, for many helpful remarks and suggestions. George Fraser of the Health and Safety Directorate of the CEC made useful comments about section 5. The work was supported by EEC contract number 1205-82-9 L/V.

REFERENCES

1. ApSimon, H.M., Davison, A.C. and Goddard, A.J.H. 1982. The probability distribution of individual exposure due to hypothetical accidental releases of various radionuclides to the atmosphere. Proc. 3rd Int. Symp. Soc. Rad. Protect., Inverness., Vol 2, 731-737.

2. ApSimon, H.M., Goddard, A.J.H. and Wrigley, J. 1983a. Lagrangian atmospheric dispersion of radioisotopes, part I: the MESOS model. Accepted for Atmospheric **Environment.**

3. ApSimon, H.M., Goddard, A.J.H., Wrigley, J. and Crompton, S. 1983b. Lagrangian atmospheric dispersion of radioisotopes, part II: applications of the MESOS model. Accepted for Atmospheric Environment.

4. Atkinson, A.C. 1973. Testing transformations to normality. J.R. Statist, Soc., B, 35, 473-79.

5. Atkinson, A.C. 1982. Regression diagnostics, transformations and constructed variables (with discussion). J.R. Statist. Soc., B, 44, pp. 1-36.

6. Barndorff-Nielsen, O. 1983. On a formula for the distribution of the maximum likelihood estimator. Biometrika, 70, pp. 343-365.

7. Box, G.E.P. and Cox, D.R. 1964. An analysis of transformations (with discussion). J.R. Statist. Soc., B, 26, pp. 211-246.

8. Cook, R.D. and Weisburg, S. 1983. Diagnostics for heteroscedasticity in regression. Biometrika, 70, pp. 1-10.

9. Cox, D.R. 1961. Tests of separate families of hypotheses. Proc. 4th Berkeley Symp., 1, pp. 105-123.

10. Cox, D.R. 1962. Further results on tests of separate families of hypotheses. J.R. Statist. Soc., B, 24, pp. 406-424.

11. Cox, D.R. and Hinkley, D.V. 1974. Theoretical Statistics. Chapman and Hall, London.

12. Cox, D.R. and Isham, V. 1980. Point Processes. Chapman and Hall, London.

13. Cox, D.R. and Snell, E.J. 1968. A general definition of residuals (with discussion). J.R. Statist. Soc., B, 30, 248-275.

14. Cox, D.R. and Snell, E.J. 1971. On test statistics calculated from residuals. Biometrika, 58, pp. 589-594.

15. Davis, H.T. and Feldstein, M.L. 1979. The generalized Pareto law as a model for progressively censored survival data. Biometrika, 66, pp. 299-306.

16. Gumbel, E.J. 1958. Statistics of Extremes. Columbia Univ. Press, New York.

17. Hall, P. 1982. On estimating the endpoint of a distribution. Ann. Statist. 10, pp. 556-568.

18. Hill, B.M. 1975. A simple general approach to inference about the tail of a distribution. Ann. Statist. 3, pp. 1163-1174.

19. Jenkinson, A.F. 1969. Statistics of Extremes. In: World Met. Office Technical Note 98, Chapter 5, pp. 183-227.

20. Leadbetter, M.R., Lindgren, G., and Rootzén, H. 1983. Extremes and Related Properties of Random Sequences and Processes. Springer-Verlag, New York.

21. Maguire, B.A., Pearson, E.S. and Wynn, A.H.A. 1952. The time intervals between industrial accidents. Biometrika, 39, 168-180.

22. Mosteller, F. and Tukey, J.W. 1977. Data Analysis and Linear Regression. Addison-Wesley, Reading, Massachusetts.

23. NERC 1975. Flood Studies Report, Vol 1. National Enviromental Research Council, London.

24. Pareto, V. 1897. Cours d'Economie Politique. Rouge et Cie, Lausanne and Paris.

25. Pickands, J. 1975. Statistical inference using extreme order statistics. Ann. Statist. 3, pp. 119-131.

26. Prescott, P. and Walden, A.T. 1980. Maximum likelihood estimation of the parameters of the generalized extreme-value distribution. Biometrika, 67, pp. 723-724.

27. Prescott, P. and Walden, A.T. 1983. Maximum likelihood estimation of the parameters of the three-parameter generalized extreme-value distribution from censored samples. J. Statist. Comput. Simul., 16, pp. 241-250.

28. Smith, R.L. 1984. Threshold methods for sample extremes. NATO-ASI, Statistical extremes and applications, Vimeiro, Portugal, August, 1983, this volume.

29. Stephens, M.A. 1977. Goodness of fit for the extreme value distribution. Biometrika, 64, pp. 583-588.

30. Weiss, L. 1971. Asymptotic inference about a density function at the end of its range. Nav. Res. Log. Quart. 18, pp. 111-114.

31. Weissman, I. 1978. Estimation of parameters and large quantiles, based on the k largest observations. J. Amer. Statist. Assoc. 73, pp. 812-815.

# STATIONARY MIN-STABLE STOCHASTIC PROCESSES

L. F. M. de Haan
Erasmus University
and
J. Pickands III
Erasmus University and
University of Pennsylvania

Summary:

We consider the class of stationary stochastic processes whose margins are jointly min-stable. We show how the scalar elements can be generated by a single realization of a standard homogeneous Poisson process on the upper half-strip $[0,1] \times R_+$ and a group of $L_1$ - isometries. We include a Dobrushin-like result for the realizations in continuous time.

## 1. Introduction.

Multivariate extreme value distributions and their domains of attraction have been studied extensively. See de Haan & Resnick (1977). For a general source on extreme value theory and applications see the book by Galambos (1978). We are not concerned here with domains of attraction. The univariate limiting extreme value types can be transformed into one another by means of simple functional transformations ($\log x$, $1/x$, $x^\alpha$ etc.). The same is true for multivariate extreme value distributions. That is a distribution is determined by its margins and independently, by its dependence function. The choice of marginal type is one of convenience. We use the negative exponential family. That is we use X which is such that $- \log p\{X > x\} \equiv x/EX$. Notice that these are limiting distributions of <u>smallest</u> values.

We say that a random vector $Z^\sim$ with elements $Z_k$, is "min-



*J. Tiago de Oliveira (ed.), Statistical Extremes and Applications, 483–489.*

stable" if and only if it is a limiting extreme value distribution with negative exponential margins. In fact $\tilde{Z}$ is min-stable if and only if $\min_n (Z_n/a_n)$ has a negative exponential distribution for any nonrandom vector $\tilde{a}$, with elements $a_n \varepsilon [0, \infty)$ at least one of which is positive. Notice that if $a_n \equiv 0$, then $Z_n/a_n = \infty$ and the term plays no role in the minimization.

In Section 2, we present a representation for any finite or infinite dimensional min-stable random vector $\tilde{Z}$. It depends upon a standard homogeneous Poisson process on the strip $[0,1] \times R_+$ and a set of non-random functions $f_n: [0,1] \rightarrow R_+$ which correspond to the components $Z_n$ of $\tilde{Z}$. We consider the nonuniqueness of $\{f_n\}$ and introduce the group of "pistons": a class of function

transformations $\Gamma$ which are such that $\int_0^1 \Gamma(f)(t)dt = \int f(t)dt$ for all non-negative f. We discuss this group in Section 3. In Section 4 we consider the implications of strict stationarity. We consider continuous time stationary processes in Section 5. Finally we have a general discussion in Section 6. The proofs and detailed derivations can be found in de Haan (1983) and de Haan and Pickands (1983).

## 2. Representation.

Let $\{S_\ell, U_\ell\} \varepsilon [0,1] \times R_+$ be the points of a standard homogeneous Poisson process. Let the random vector $\tilde{Z}$, with elements $Z_n$, be min-stable in our sense. Then there exists a non-unique sequence of functions $f_n: [0,1] \rightarrow R_+$ such that $\{Z_n\}$ and $\bigwedge_{\ell=1}^{\infty}(U_\ell/f_n(S_\ell))$ have the same joint distribution. Since the distributions are the same we can without loss of generality let

$$Z_n \equiv \bigwedge_{\ell=1}^{\infty}(U_\ell/f_n(S_\ell)) \tag{2.1}$$

for all n. For $z_n \varepsilon (0,\infty)$, the event

$$\{Z_n > z_n\} \equiv \bigcap_{\ell=1}^{\infty} \{U_\ell/f_n(S_\ell) > z_n\} \equiv \bigcap_{\ell=1}^{\infty} \{U_\ell > z_n f_n(S_\ell)\}$$

$$\equiv \mathrm{Emp}\ \{s,u \mid u \varepsilon [0,z_n f_n(s))\}$$

where Emp A means that $A \subset [0,1] \times R_+$ contains no points of the homogeneous Poisson process on $[0,1] \times R_+$. But the number of points of the process in A has the Poisson distribution with mean (parameter) $\lambda_2(A)$ where $\lambda_2$ is 2-dimensional Lebesgue measure. It follows that

$$-\log p\{Z_n > z_n\} \equiv \lambda_2 \{s,u | u \varepsilon [0,z_n f_n(s))\} \tag{2.2}$$

$$= \int_0^1 z_n f_n(s)ds = z_n \int_0^1 f_n(s)ds.$$

Notice that $Z_n$, then, has a negative exponential distribution with mean $EZ_n \equiv 1/\int_0^1 f_n(s)ds$.

Let the event $\{Z^\sim > z^\sim\} \equiv \bigcap_{n=1}^{\infty} \{Z_n > z_n\} \equiv \text{Emp} \bigcup_{n=1}^{\infty} A_n$ where, for each n, $A_n \equiv \{s, u | u \varepsilon [0, z_n f_n(s)]\}$. But

$$\bigcup_{n=1}^{\infty} A_n \equiv \{s, u | u \varepsilon [0, \bigvee_n z_n f_n(s)]\}$$

and

$$\lambda_2(\bigcup_{n=1}^{\infty} A_n) \equiv \int_0^1 [V_n z_n f_n(s)]ds$$

and so

$$(2.3) \qquad -\log p\{Z^\sim > z^\sim\} \equiv \int_0^1 [V_n z_n f_n(s)]ds .$$

Thus min-stable joint distributions are determined by the values of all integrals of the form (2.3) above. For every min-stable (joint) distribution there exists a representation of the form (2.1) but the sequence $\{f_n\}$ is not uniquely determined. Another sequence $\{g_n\}$ yields the same distribution if and only if integrals of the form (2.3), above are all unchanged if $f_n$ are replaced by $g_n$. We say, then, that $\{f_n\}$ and $\{g_n\}$ are equivalent and we write $\{f_n\} \sim \{g_n\}$.

Clearly, a sufficient condition for equivalence is that

$$g_n \equiv \Gamma(f_n)$$

where $\Gamma$ is a function mapping such that if $f: [0,1] \to R_+$, then $\Gamma(f): [0,1] \to R_+$ and

$$(2.4) \qquad \int_0^1 \Gamma(f)(s)ds = \int_0^1 f(s)ds .$$

If such a mapping can be written in the following way we call it a "piston". We let

$$(2.5) \qquad \Gamma(f)(t) = r(t)f(H(t))$$

where $r(t)$ and $H(t)$ are measurable and $H(t)$ is a $1 \leftrightarrow 1$ mapping of $[0,1]$ onto itself. By (2.4) $r(t)$ must exceed 0 except on a set of measure 0.

## 3. Pistons.

As examples we could let

$$H(t) \equiv 1-t,$$

$$\equiv t+\theta \bmod 1, \ \theta \in (-\infty, \infty)$$

or

$$\equiv t^a, \ a \in (0,\infty) .$$

By (2.4), for the first 2 examples $r(t) \equiv 1$. For the third $r(t) \equiv a\, t^{a-1}$. We can always write

$$H(t) \equiv P(R(t))$$

where

$$R(t) \equiv \int_0^t r(s)ds$$

and P is a 1-1 mapping of [0,1] into itself which is not only measurable but measure preserving.

The pistons form a noncommutative group. If

$$\Gamma_{H,r}: f(t) \to r(t)f(H(t)),$$

then

$$\Gamma_{H,r} \equiv \Gamma_M \Gamma_P$$

where

$$\Gamma_P: f(t) \to f(P(t))$$

and

$$\Gamma_M: f(t) \to r(t)f(R(t)).$$

The former is of "permutation" type and the latter is of "monotone" type.

## 4. Stationarity.

If $\Gamma$ is a piston, then

$$\{f_n\} \sim \{\Gamma f_n\}.$$

Is the converse true? If

$$\{f_n\} \sim \{g_n\},$$

does it follow that there exists a piston $\Gamma$, such that

$$g_n \equiv \Gamma(f_n)$$

for all n? The answer is yes, if $\{f_n\}$ and $\{g_n\}$ are both "proper", that is if they both satisfy a technical condition which includes the following:

$$\lambda_1(t \in (0,1) \mid \bigvee_{n=1}^{\infty} f_n(t) > 0) = 1,$$

where $\lambda_1(.)$ is 1-dimensional Lebesgue measure. A proper representation always exists. That is we can always write (2.1) with proper sequence $\{f_n\}$.

Suppose $\tilde{Z}$ has elements $Z_n$, $n = 0, \pm 1, \pm 2, \ldots$ . Then there exists a representation with proper sequence $\{f_n, n = 0, \pm 1, \pm 2, \ldots\}$. Suppose we assume strict stationarity. Then, by definition,

$$\{f_n\} \sim \{f_{n+1}\}$$

and so there exists a piston $\Gamma$ such that

$$f_{n+1} \doteq \Gamma(f_n)$$

for all integers n. It follows that

$$f_n \equiv \Gamma^n(f_0) \tag{4.1}$$

for all n. Thus, if $\tilde{Z}$ is strictly stationary, then we have the representation (2.1) with (4.1), where $\Gamma$ is a piston, that is a transformation of the form (2.5) which satisfies (2.4).

## 5. Continuous time.

Suppose $\tilde{Z}$ has elements Z(t) where t spans the reals. Then an appropriate representation would be the following: Let

$$Z(t) \equiv \bigwedge_{\ell=1}^{\infty} U_\ell / f_t(S_\ell)$$

where

$$f_t(s) \equiv \Gamma^t(f_0)(s).$$

A proper representation of the form above exists provided that we assume continuity in probability.

We have a Dobrushin-type result. The sample functions are of either of two types:

1.) With probability 1, the sample functions are bounded away from 0 on every finite interval.

2.) With probability 1, the sample functions are arbitrarily close to 0 on every finite interval.

The realizations are of the former or latter kind according as

$$\lim_{\lambda\to 0} \int_0^1 [\bigvee_{t\in[0,\lambda]} f_t(s)]\, ds < \text{ or } = \infty.$$

It was not necessary to use a local 0-1 law to prove this.

We consider a class of processes in continuous time which are analogues of moving average processes. In effect these are "moving minimum" processes. Let $H(t)$ be a differentiable increasing measurable $1 \leftrightarrow 1$ map of $R \equiv (-\infty, \infty)$ onto $[0,1]$. Now $H(t)$ is a cumulative distribution function. Let $\phi: R \to R_+$ with $\int_{-\infty}^{\infty} \phi(s)ds \in (0,\infty)$. Recall that $\{S_\ell, U_\ell\}$ are the points of a homogeneous Poisson process on the strip $[0,1] \times R_+$. Let

$$\bar{S}_\ell \equiv H^{-1}(S_\ell)$$

and

$$\bar{U}_\ell \equiv U_\ell/(H^{-1})'(S_\ell).$$

Now $\{\bar{S}_\ell, \bar{U}_\ell\}$ are the points of a homogeneous Poisson process with unit intensity on $R \times R_+$. Inverting

$$S_\ell \cdot H(\bar{S}_\ell)$$

and

$$U_\ell \equiv U_\ell(H^{-1})'(H(\bar{S}_\ell)) \equiv U_\ell/H'(\bar{S}_\ell).$$

Let $Z$ have elements $Z(t)$, where

$$Z(t) \equiv \bigwedge_\ell \bar{U}_\ell/\phi(t+\bar{S}_\ell)$$

$$= \bigwedge_\ell U_\ell/(H^{-1})'(S_\ell)\phi(t + H^{-1}(S_\ell))$$

$$= \bigwedge_\ell U_\ell/f_t(S_\ell) .$$

Here

$$f_t(s) \equiv (H^{-1})'(s)\phi(t+H^{-1}(s)) \equiv \Gamma^t(f_0)(s),$$

where

$$f_0(s) \equiv (H^{-1})'(s)\phi(H^{-1}(s))$$

and

$$\Gamma^t \equiv \Delta^{-1} L^t \Delta.$$

Here $\Delta$ maps functions from $[0,1] \to R_+$ into functions from $R \to R_+$ and $L^t$ is the "shift" group: $L^t(\phi)(s) \equiv \phi(t+s)$ . Clearly

$$\Delta^{-1}(f)(s) \equiv (H^{-1})'(s)\ f\ (H^{-1}(s)).$$

For such processes the sample functions are of the first or second kind above according as

$$\lim_{\lambda \to 0} \int_{-\infty}^{\infty} \left[ \bigvee_{t \in [0,\lambda]} \phi(t+s)\right] ds < \text{ or } = \infty .$$

6. Discussion.

There is considerable scope for further investigation of stationary min-stable processes. If in (2.1) we consider not only the minimum but all of the order statistics, we have, for each t, the entire realization of a homogeneous Poisson process with intensity $\int_0^1 f_t(s)ds$ ( $= \int_0^1 f_0(s)ds$ in the stationary case). Although the Dobrushin-type result does not require a local 0-1 law, it would be interesting to investigate this. Ergodic properties of our processes are undoubtedly strongly related to the piston power groups $\Gamma^t$. Estimation and forecasting will be more feasible in the Poisson process context than in the minimum one.

REFERENCES

Galambos, J. (1978) The Asymptotic Theory of Extreme Order Statistics. Wiley, New York.

de Haan, L. and Resnick, S. J. (1977) Limit Theory for Multivariate Sample Extremes. Zeitschrift fur Wahrscheinlichkeitstheorie und verwandte Gebiete. Vol. 40, 317-337.

de Haan, L. (1983) A Spectral Representatin for Max-Stable Processes. Ann. Probability (to appear).

de Haan, L. and Pickands III, J. (1983) Stationary Min-Stable Stochastic Processes. Technical Report, Erasmus University Rotterdam.

STRONG APPROXIMATIONS OF RECORDS AND RECORD TIMES

Paul Deheuvels

Université Paris VI

ABSTRACT. This note reviews some recent results obtained for the strong approximation of records and record times with emphasis on the case of k-th records.

1. RECORD TIMES AND RECORD VALUES. Let $X_1, X_2, \ldots$ be a sequence of random variables with a continuous distribution function $F(x)=P(X_1 \leq x)$. Let $X_{1,n}<X_{2,n}<\ldots<X_{n,n}$ denote the order statsitic of $X_1,\ldots,X_n$. For a fixed $k \geq 1$, let

$$n_1^{(k)}=k \ , \ n_m^{(k)}=\mathrm{Inf}\{n>n_{m-1}^{(k)};X_{n-k+1,n}>X_{n_{m-1}^{(k)}-k+1,n_{k-1}^{(k)}}\}, \ m \geq 2,$$

and let

$$R_m^{(k)}=X_{n_m^{(k)}-k+1,n_m^{(k)}}$$

The sequence $\{n_m^{(k)}, m \geq 1\}$ is called the sequence of <u>k-th (upper) record times</u>, while $\{R_m^{(k)}, m \geq 1\}$ is called the sequence of <u>k-th (upper) records</u>.

A similar definition could be made for k-th lower records and record times, replacing above $X_{n-k+1,n}$ by $X_{k,n}$ and $>$ by $<$. There is clearly no need to discuss separately both cases.



*J. Tiago de Oliveira (ed.), Statistical Extremes and Applications, 491–496.*

It will be convenient to reduce the study of k-th records to the case where the underlying sequence is exponentially distributed. This can be done sithout loss of generality by a change of scale since $\{\xi_n=-\mathrm{Log}(1-F(X_n)),n\geq 1\}$ defines an i.i.d. exponential sequence. We shall put accordingly

$\xi_n=-\mathrm{Log}(1-F(X_n))$, $\xi_{\ell,n}=-\mathrm{Log}(1-F(X_{\ell,n}))$, $1\leq\ell\leq n$,

$z_m^{(k)}=-\mathrm{Log}(1-F(R_m^{(k)}))$, $m\geq 1$, noting that

$0 < \xi_{1,n} < \ldots < \xi_{n,n}$, and that

$\{z_m^{(k)},m\geq 1\}$ is the k-th (upper) record sequence of $\{\xi_n,n\geq 1\}$, with record times $\{n_m^{(k)},m\geq 1\}$.

With the preceding notations, we have the following results.

Theorem 1. 1°) For any $k\geq 1$, $(0<)\ z_1^{(k)}<z_2^{(k)}<\ldots$ are the times of arrivals of a homogeneous Poisson process $\{N_k(t),t>0\}$ such that $N_k(t)=m$ iff $z_m^{(k)}<t<z_{m+1}^{(k)}$, $m\geq 1$, and $E(N_k(t))=kt$.

2°) The Poisson processes $N_1(.)$, $N_2(.)-N_1(.),\ldots,N_k(.)-N_{k-1}(.)$ are independent.

Proof. See Deheuvels, 1983b, Theorem 1.

Theorem 2. For any $k\geq 1$, the fact that, for some $m\geq 1$, $R_m^{(k)}$ and $R_{m+1}^{(k)}-R_m^{(k)}$ are independent characterizes F (assumed to be continuous) as an exponential distribution function.

Proof. The case k=1 is due to Tata, 1969. Resnick, 1973, has proved that $\{N_1(t),t>0\}$ is a standard Poisson process. Dziubdziela and Kopocinski, 1976, obtained the distribution of $R_m^{(k)}$. For $k\geq 2$, the proof is due to Deheuvels, 1984.

Theorem 3. Let $\xi_1,\xi_2,\ldots$ be defined on a probability space $(\Omega,A,P)$. Then, on a possibly enlarged probability space, it is possible to define for each $k\geq 1$ an i.i.d. sequence $\{\omega_m^{(k)},m\geq 1\}$ of exponentially distributed random variables, independent of $z_m^{(k)},m\geq 1$, and such that, for any $m\geq 1$,

$$n_m^{(k)}=k+\sum_{i=1}^{m-1}\{[\frac{\omega_i^{(k)}}{-\mathrm{Log}(1-\exp(-z_i^{(k)}))}]+1\},$$

Proof. See Deheuvels, 1981, 1983. Note here that [u] stands for the integer part of u, and $\sum_1^0(.)=0$.

2. RECORD PROCESSES. Let $a=\inf\{t:F(t)>0\}$ and $b=\sup\{t;F(t)<1\}$. Put for $a<t<b$ and $k\geq 1$:

$$\tau^{(k)}(t)=\inf\{n\geq k;X_{n-k+1}\geq t\}.$$

The process $\{\tau^{(k)}(t),a<t<b\}$ is called the empirical record process of $X_1,X_2,\ldots$ . It has been proved in Deheuvels, 1973, 1974 that this process possesses independent increments. The corresponding finite dimensional distributions are also given in the same references.

We shall be concerned here with the asymptotic convergence of $\tau^{(k)}(t),a<t<b$ toward a limiting process as $t\uparrow b$. For that, it can be seen from Theorems 1,2, and 3, that

$$\tau^{(k)}(t)=k+\sum_{i=1}^{N_k(-\mathrm{Log}(1-F(t))}\{[\frac{\omega_i^{(k)}}{-\mathrm{Log}(1-\exp(-z_i^{(k)}))}]+1\},$$

which leads to the introduction of the processes

$$Y_k^e(t) = k + \sum_{i=1}^{N_k(\mathrm{Log}\ t)}\{[\frac{\omega_i^{(k)}}{-\mathrm{Log}(1-\exp(-z_i^{(k)}))}]+ 1\},$$

and

$$Y_k(t) = \sum_{i=-\infty}^{N_k(\mathrm{Log} t)}\omega_i^{(k)}\exp(z_i^{(k)}),\ t\geq 0.$$

This process will be called the <u>standard k-th record process</u>. Its properties are given in the following theorem.

Theorem 4. 1°) $\{Y_k(t),t\geq 0\}$ is a nondecreasing step process with independent increments, and such that $Y_k(0)=0$.

2°) For any $0\leq s<t$, the characteristic function of $Y_k(t)-Y_k(s)$ is $((1-ius)/(1-iut))^k$. The distribution of $Y_k(t)-Y_k(s)$ is that

of $\sum_{i=1}^{k}\theta_i$, where $\theta_1,\ldots,\theta_k$ are independent and such that

$$P(\theta_i=0)=s/t\ ,\ P(\theta_i>u)=(1-\frac{s}{t})e^{-u/t}\ ,\ u>0,\ 1\leq i\leq k.$$

3°) $E(Y_k(t))=kt$, and, for any $\lambda>0$, the process $\{\lambda^{-1}Y_k(\lambda t),t\geq 0\}$ has a distribution independent of $\lambda$.

Proof. See Deheuvels, 1981, 1983.

Note that the inverse of $Y_k(t)$, defined as

$$Y_k^{-1}(s)=\inf\{t;Y_k(t)>s\},$$

is an extremal process, and can be interpreted as the limiting process as $n\to\infty$ of $\{nU_{k,[nt]},t\geq k/n\}$, where $U_{1,n}<\ldots<U_{n,n}$ stands for the k-th lower order statistic of an i.i.d. sequence of uniformly distributed random variables on (0,1).

The fact that, for k=1, $Y_1(t)$ has independent increments has been proved by Resnick, 1974.

An interesting consequence of Theorem 4 is that the empirical record process can be approximated by $Y_k(t)$ with the following strong bounds:

Theorem 5. With the notations above, we have, almost surely as $t\to\infty$

$$|Y_k^e(t)-Y_k(t)| = O\{t\mathrm{LogLog}\ t\}^{1/2}.$$

Proof. See Deheuvels, 1982, 1983, where further applications of strong approximations for record and extremal processes are discussed.

We shall limit ourselves to the following corollary.

Corollary 1. Let $s=s_n$ and $t=t_n$, n=1,2,... be such that $0<F(s)<F(t)<1$, and that $F(s)\to 1$ as $n\to\infty$. If $\lim_{n\to\infty}\frac{1-F(t)}{1-F(s)}=\alpha\in(0,1)$, then, the distribution of $(1-F(t))(\tau_k(s)-\tau_k(t))$ has a limit as $n\to\infty$, given by the distribution of $Y_k(1)-Y_k(\alpha)$, with characteristic function $((1-iu\alpha)/(1-iu))^k$.

An easy computation gives this distribution in closed form.

3. UPPER AND LOWER STRONG BOUNDS FOR RECORD TIMES. In Deheuvels, 1982, 1983, the following theorems were proved.

Theorem 6. Without loss of generality, for any $k \geq 1$, it is possible to define a standard Wiener process $\{W_k(t), t \geq 0\}$ such that

$$n_m^{(k)} = \exp\left(\frac{m}{k} + \frac{W(m)}{k} + O(\text{Log } m)\right), \text{ a.s. as } m \to \infty.$$

Theorem 7. For any $k \geq 1$, $p \geq 5$,

$$P(\text{Log } n_m^{(k)} - \frac{m}{k} \geq \frac{1}{k}(2m(\text{Log}_2 m + \frac{3}{2}\text{Log}_3 m + \text{Log}_4 m + \ldots + (1+\varepsilon)\text{Log}_p m))^{\frac{1}{2}} \text{ i.o.})$$

$$= P(\text{Log } n_m^{(k)} - \frac{m}{k} \leq \frac{-1}{k}(2m(\text{Log}_2 m + \frac{3}{2}\text{Log}_3 m + \text{Log}_4 m + \ldots + (1+\varepsilon)\text{Log}_p m))^{\frac{1}{2}} \text{ i.o.})$$

= 0 or 1, according as $\varepsilon > 0$ or $\varepsilon \leq 0$. Here, $\text{Log}_j$ is the $j^{th}$ iterated logarithm.

Proof. See Deheuvels, 1982, 1983. This precises a result of Rényi, 1962, who proved Theorem 7 for p=2 and k=1. In general the same result is true with $\Delta_m^{(k)} = n_m^{(k)} - n_{m-1}^{(k)}$ instead of $n_m^{(k)}$.

4. APPLICATIONS. There has been many attempts to use records for the detection of trends of breaks of homogeneity in time series (see Chandler, 1952, Foster and Stuart, 1954), and the discussion in Galambos, 1978). The main draw-back in this approach lies in the fact that the records become scarce as the length of the series increases (for i.i.d. sequences).

Rényi was among the first to note in 1962 that (see Theorem 7) $n_m^{(1)}$ behaves roughly as $e^m$ as m increases.

For sequences of "reasonable" size (i.e. 50-500), this argument breaks up if one uses the k-th records instead of the first (k=1). In this case, $n_m^{(k)}$ behaves as $e^{m/k}$, and one could choose k=k(n) increasing, in order to derive distribution-free tests.

This has yet to be done, since the structure of k-th records has been understood fully only since the last years.

REFERENCES.

Deheuvels, P. (1973) Sur les sommes de minima de variables aléatoires indépendantes de même loi, C.R. Acad. Sci. Paris,

276, Ser.A, 309-312

Deheuvels, P. (1974) Majoration et minoration presque sûre optimale des éléments de la statistique ordonnée d'un échantillon croissant de variables aléatoires indépendantes, Rendi Conti Acad. Nazionale dei Lincei 8, 56, 707-719

Deheuvels, P. (1981) The strong approximation of extremal processes, Z. Wahrscheinlichkeit. verw. Gebiete 58, 1-6

Deheuvels, P. (1982) Strong approximation in extreme value theory and applications, Coll. Math. J. Bolyai 36, to appear

Deheuvels, P. (1983) The complete characterization of the upper and lower classes of the record and inter-record times of an i.i.d. sequence, Z.W.G. 62, 1-6

Deheuvels, P. (1983) The strong approximation of extremal processes (II) Z. Wahrscheinlichkeit. verw. Gebiete 62, 7-15

Dziubdziela, W. and Kopociński, 1976, Limiting properties of the k-th record values, Zastosowania Mat. 15, 187-190

Galambos, J. (1978) The Asymptotic Theory of Extreme Order Statistics, Wiley, New York

Rényi, A. (1962) On outstanding values of a sequence of observations, in: Selected Papers of A. Rényi, Vol. 3, 50-65, Akademiai Kiádo, Budapest

Resnick, S.I. (1973) Limit laws for record values, Stoch. Processes Appl. 1, 67-82

Resnick, S.I. (1974) Inverses of extremal processes, Adv. Appl. Probability 6, 392-406

Tata, M.N. (1969) On outstanding values in a sequence of random variables, Z.W.G. 12, 9-20

Deheuvels, P. (1984) The characterization of distributions by order statistics and record values, a unified approach, to appear in the J. Applied Probability

Chandler, K.N. (1952) The distribution and frequency of record values, J. Royal Stat. Soc. B, 14, 220-228

Other references are to be found in Galambos, 1978.

## HIGH PERCENTILES ATMOSPHERIC $SO_2$-CONCENTRATIONS IN BELGIUM

Goossens Chr., Berger A., Mélice J.L.

U.C.L., Institut d'Astronomie et de Géophysique
Georges Lemaitre, B-1348 Louvain-la-Neuve, Belgium

## ABSTRACT

From data of 210 stations covering whole Belgium, it is demonstrated that, at 99% of the stations, the distribution of large concentrations of atmospheric pollutants is well fitted by a 2-parameter exponential law. Analysis of the whole data set of atmospheric concentration shows also that a 2-parameter Gamma distribution provides a better representation of this whole ensemble than the usual log-normal.

## INTRODUCTION

In a previous paper, Berger et al. (1) showed that the distribution of large values of atmospheric concentration in pollutants (exceedances) is well fitted by a 2-parameter exponential law, a distribution directly related to the theory of extreme values. It was also demonstrated that the $SO_2$ -concentrations exceeding relatively low percentiles, continue to be well fitted by the same distribution. This has lead to the use of the 2-parameter Gamma distribution (instead of the usual log-normal distribution (2)) for the whole set of $SO_2$ -data, on the basis that this Gamma distribution is a generalization of the 2-parameter exponential law. The data analysed were those available at 12 stations of the Gent region (Belgium) between 1 January 1968 and 30 March 1969.

The purpose of this paper is to extent this analysis to the other Belgian stations. The data used are the 24h-averaged



*J. Tiago de Oliveira (ed.), Statistical Extremes and Applications, 497–502.*

$SO_2$ concentrations recorded since the creation of the Belgian network (1 January 1968) to 30 April 1980 at 210 stations (Fig. 1). All data are measured with the same type of recorder and are expressed in $\mu g\ m^{-3}$ (Belgian Sulphur Dioxide-Smoke Network, IHE, 1968-1980 (3)). However, it must be outlined that the length of individual series varies from one station to the other due, for example, to late equipment.

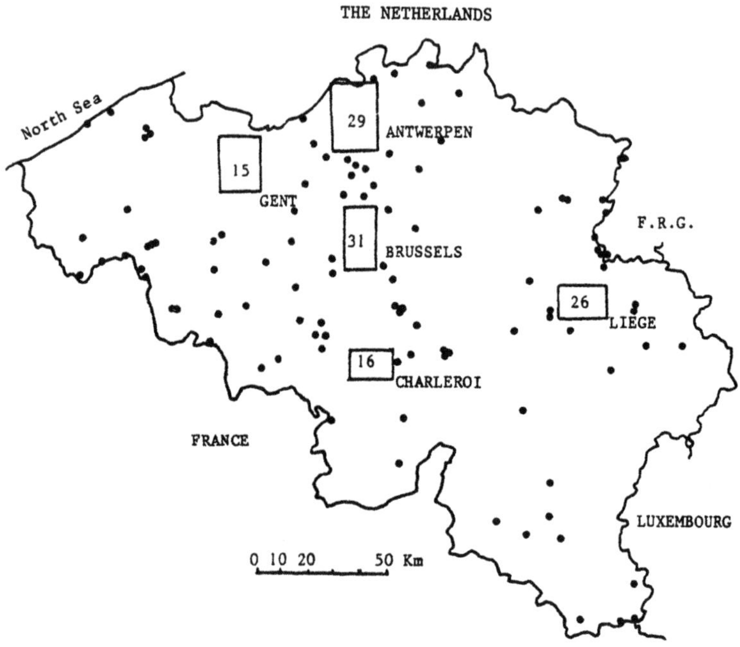


Fig. 1. $SO_2$-monitoring network for all Belgian stations. The squares represent urban regions with a very dense network which number of stations is given.

## DISTRIBUTION OF LARGE VALUES (EXCEEDANCES)

Let us define the subsample of large values as the series of values ( among the whole set of N observations) which exceed a certain level, the 95-th percentile for example.

Berger et al. (1) have extended the empirical relationship between the distributions of yearly extreme values and of annual exceedances (the n top-values when arranged in a decreasing order) first investigated by (4), to the series of large values . This gives the following relationship between the extreme values series and the large values series:

$$P_E = 1-e^{-P_L k}$$

where $P_E$ is the probability that an annual extreme value of magnitude x will be exceeded and $\exp(-P_L k)$ is the probability that an event of magnitude x becomes an extreme of the m events in one year, $P_L$ being the probability of a variate in the large value subsample being eqal or greater than x and k the average number of exceedances per year (mathematical details can be found in Berger et al. (1)).

The values exceeding the 95-th percentile for each of the 210 stations were fitted by the first 2 parameter exponential distribution (ED1), the initial distribution belonging obviously to the ''exponential'' type and $P_L k = (P_L)(0.05*365)$ being << 365. This kind of distribution gives a remarkable fit for all stations as it may be seen in Fig. 2. Examples for some individual stations are given in Fig. 3 a to d where the parameters are listed for each stations. The goodness of the fit is verified by a Kolmogorov-Smirnov test or by a test based on the median and the greatest values (1;5).

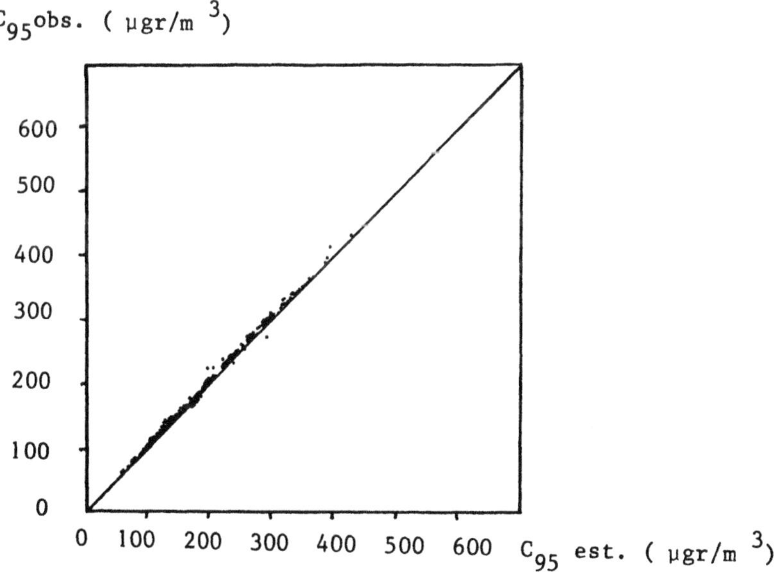


Fig. 2. Comparison of the observed 95th percentiles with the estimated ones by the first 2-parameter exponential distribution for all the 210 stations. The diagonal represents the perfect agreement.

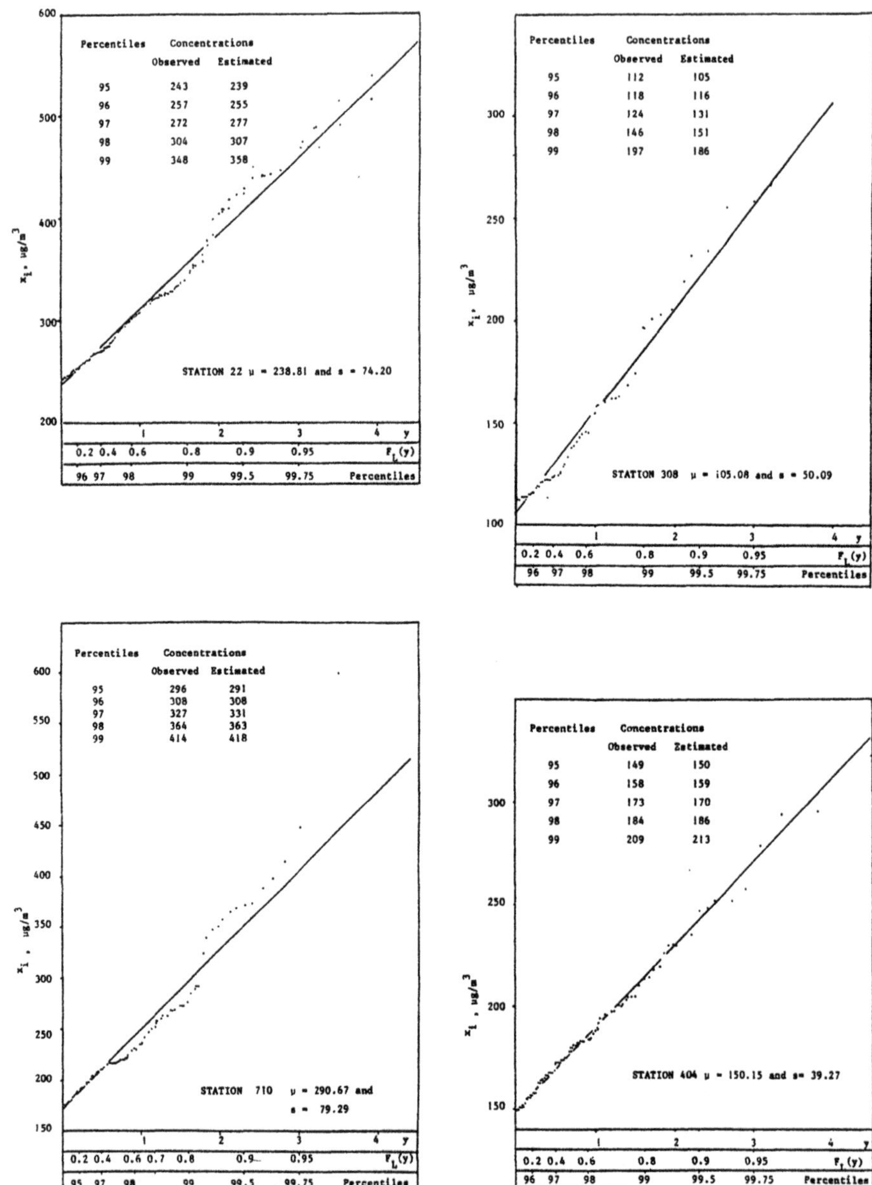


Fig. 3 a to d . Fitting the 2-parameter exponential distribution (ED1) to 24h-averaged $SO_2$-concentrations selected over the 95th percentile level. $F_L(y)=1-\exp(-y)$ with $y=(x-\mu)/s$ is the cumulative frequency function, $\mu$ and s paramters are estimated by the weighted least squares method.

THE TWO PARAMETER GAMMA DISTRIBUTIONS FOR THE WHOLE SAMPLE OF CONCENTRATION

Berger et al. (1) showed that all concentrations selected even over the 20-th percentile are also well fitted by ED1. This lead then to the assumption that the entire sample of atmospheric $SO_2$ -concentrations could be fitted by a 2-parameter Gamma distribution, a generalization of this exponential distribution (1). This distribution provides a good fit for most of the stations (Fig.4) and is accepted by a test based on the median and the largest values at the 0.2 significance level for 71% of

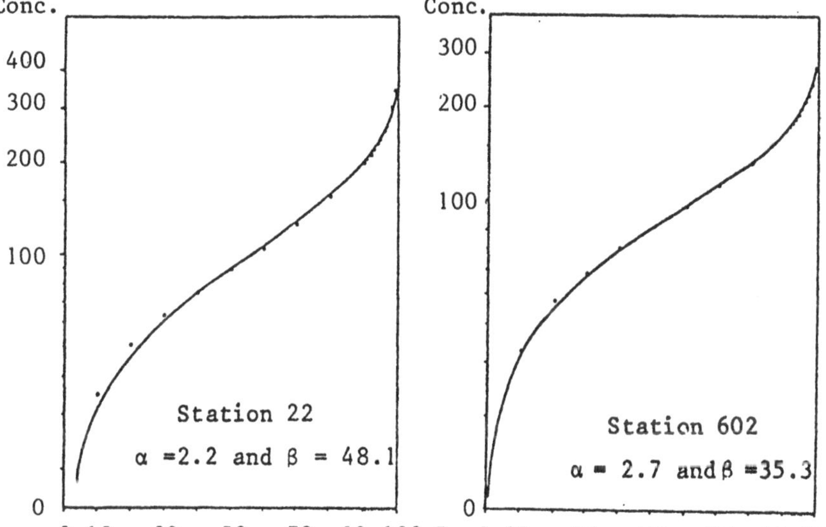


Fig.4 a to b . Fitting the 2-parameter Gamma distribution to all 24h-averaged $SO_2$ -concentrations. The shape parameter α and the scale parameter β were estimated by the maximum likelihood functions.

the stations. Moreover, it has been show that the 2-parameter Gamma distribution gives better results than the log-normal distribution even for high concentration level as the 95th-percentile, the log-normal distribution overestimating the extreme concentration.

CONCLUSIONS

In summary,the probability distribution of extreme values being a Gumbel distribution, the series of large atmospheric pollution of exceedances ( large value selected over the 95-th percentile of the sample size N) is well fitted by the first 2-parameter exponential distribution .

For the frequency distribution of the whole sample of atmospheric concentration in pollutants, the significant results obtained argue for the use of the 2-parameter Gamma distribution.

The distribution of extremes, as the annual maxima of $SO_2$ -concentration or of other meteorological variables, needs long series of observations (usually more than 30 years) to be investigated. The extension of the theory of extremes to the theory of large values, the series of values (among N observations) which exceed a certain level, allows in turn to determine the distribution law of the maxima using relatively short series. Sneyers and Vandiepenbeeck (6) have shown that, as a consequence of this principle, it is possible to give for the annual maximum of meteorological variates, a good estimation of fractiles with large return period, i.e. the secular maximum using short series of observations.

ACKNOWLEDGEMENTS

Thanks are due to the ''Institut de Recherches Appliquées à l'Industrie et à l'Agriculture'' for financial help to one of the authors (Chr.G.) and to the ''Institut d'Hygiène et d'Epidémiologie, Bruxelles'' for providing all the data.

REFERENCES

1. Berger, A., Mélice, J.L., Demuth, Cl. : 1982, Atmospheric Environment 16, 2863-2877.
2. Larsen, R.I. : 1969, J. Air Pollut. Control Ass. 19, 24-30
3. IHE : 1968-1980, Réseau soufre-fumée. Bulletin Mensuel, Institut d'Hygiène et d'Epidémologie, Bruxelles.
4. Langbein, W.B. : 1949, Trans. Am. Geophys., Un. 30, 879-881.
5. Sneyers, R. : 1975, Note technique n° 143, Secrétariat de l'Organisation Météorologique Mondiale, Genève, Suisse, OMM n° 145
6. Sneyers, R. and Vandiepenbeeck M. : 1982, Sur la détermination de la répartition des valeurs maximales à partir de celle des grandes valeurs, 17e Congrès de la météorologie, Berchtesgaden, 21-25, Septembre 1982.

# EXTREME RESPONSE OF THE LINEAR OSCILLATOR WITH MODULATED RANDOM EXCITATION

A.M. Hasofer and P. Petocz

University of New South Wales, Australia

ABSTRACT

Corotis and Marshall (1977) have studied the extreme response of a single degree of freedom linear oscillator to Gaussian white noise modulated by a deterministic function in the form of a sum of exponential terms. This form of excitation provides a broad range of descriptive capabilities for different physical situations -- in particular for earthquake excitation. In this paper the question of uniqueness of definition of the envelope of such a process is raised, following Hasofer (1979) and a solution presented. It is then argued that the use of stationary formulae for estimating the upcrossing rate of such a response at high levels might seriously underestimate the actual rate. Exact formulae are then presented for the upcrossing rate of the response and its envelope. It is shown that Vanmarcke's (1975) concept of "qualified upcrossings" can be extended to the case under consideration, resulting in a considerable improvement in the estimation of the reliability function of the response. Finally numerical results are presented to illustrate the main conclusions.

## 1. THE MODEL

The model studied in this paper can be described by the following equation

$$X''(t) + 2\,\zeta\,\omega_0\,X'(t) + \omega_0^2\,X(t) = n(t)\,W(t)\,. \qquad (1)$$

In most of the work the initial conditions will be taken to be $X(0) = X'(0) = 0$ .



J. Tiago de Oliveira (ed.), Statistical Extremes and Applications, 503–512.

The stochastic process $X(t)$ is thus the response of an initially quiescent linear oscillator. The parameter $\omega_0$ represents the undamped natural frequency of the oscillator, and $\zeta$ is the fraction of critical damping. Light damping is assumed, i.e. $\zeta << 1$, and $\omega_1$ is written for $\omega_0\sqrt{1-\zeta^2}$, the damped natural frequency.

The excitation of the linear oscillator is made up of two parts:

(i) A random excitation $W(t)$, which consists of a real, stationary Gaussian process with zero mean and covariance $r_W(t)$. An important special case will be that of white noise excitation for which $r_W(t) = \delta(t)$, the Dirac delta function. The corresponding (one-sided) spectral density function is $dG_W(\omega) = d\omega/\pi$.

(ii) A deterministic modulating function $n(t)$ which can be written as a sum of exponentials

$$n(t) = \sum_{j=1}^{m} a_j \exp(b_j t) . \tag{2}$$

The response $X(t)$ can be expressed in the form

$$X(t) = \int_0^t W(\tau)\, n(\tau)\, h(t-\tau)d\tau, \; t \geq 0 , \tag{3}$$

where the impulse response function is given by

$$h(t) = \frac{1}{\omega_1} \exp(-\zeta\omega_0 t) \sin \omega_1 t . \tag{4}$$

The process $X(t)$ is thus a real, non-stationary Gaussian process, with zero mean. If $W(t)$ is white noise, then the covariance function of $X(t)$, $R_X(t;s)$ is given by

$$R_X(t,s) = \int_0^{t\wedge s} n^2(\tau)\, h(t-\tau)\, h(s-\tau)\, d\tau , \tag{5}$$

where $t\wedge s = \min(t,s)$.

The quantity of interest is the first passage time at some high level $u$ by the response $X(t)$.

## 2. THE UPCROSSING RATE METHOD

One method for calculating the first passage of a stochastic process, which dates back to the work of S.O. Rice [8] consists in calculating the upcrossing rate $q_X(t;u)$ of the process at level $u$. One then relies on the fact that, provided the covariance function at two points of the process decays rapidly enough when the distance of the two points increases, the upcrossings tend to occur independently, and tend to a Poisson process when $u \to \infty$. If $T_X(u)$ is the first passage time, its distribution is then given approximately by the formula

$$L_X(t;u) = P\{T_X(u) > t\} = \exp[-\int_0^t q_X(\tau;u)d\tau] . \tag{6}$$

However, formula (6) gives poor results for a stationary process having a narrow band spectrum, because then the upcrossings tend to occur in clusters. A better approximation is obtained by using the envelope of the process, in essence "demodulating" it.

It has been pointed out by Vanmarcke [9] that the use of the envelope for approximating the distribution of the first passage time may be misleading because some upcrossings of the envelope do not contain any upcrossing of the process, as the envelope peaks tends to be narrow at high levels. To compensate for this tendency, Vanmarcke has introduced the concept of "qualified" envelope crossings, reducing the envelope upcrossing rate in an appropriate ratio to compensate for the "empty" envelope upcrossings.

Vanmarcke's method will be illustrated in this application.

## 3. OSCILLATORY PROCESSES

The definition of the envelope of a Stochastic Process has traditionally been restricted to Stationary Processes having a spectral representation of the form

$$\xi(t) = \int_{-\infty}^{\infty} e^{i\omega t} dZ(\omega) , \tag{7}$$

with $Z(\omega)$ a process with orthogonal increments with almost surely no jumps at the origin. A "quadrature" process is defined by

$$\hat{\xi}(t) = 2 \, \mathrm{Im}\{\int_{0+}^{\infty} e^{i\omega t} \, dZ(\omega)\} , \tag{8}$$

and the envelope $S(t)$ is defined by

$$S(t) = [\xi^2(t) + \hat{\xi}^2(t)]^{\frac{1}{2}} . \tag{9}$$

This procedure obviously does not apply to general non-stationary processes, since they do not have a spectral representation. However, the concept of envelope has been extended to a subclass of non-stationary processes introduced by Priestley [7]. This subclass can be defined in many ways, but the most convenient for the purposes of this paper is

$$X(t) = \int_{-\infty}^{+\infty} h(t,\tau)\ \xi(\tau)d\tau\ , \tag{10}$$

where $\xi(t)$ is a stationary process, and $h(t,\tau)$ is the impulse response function of a linear time-varying filter. It is easily seen that formula (3) for the response of the linear oscillator is exactly of that form.

The definition of the envelope of an oscillatory process was originally given by Arens [1], and subsequently by Yang [10]. The quadrature process for $X(t)$ is defined by

$$\hat{X}(t) = \int_{-\infty}^{+\infty} h(t,\tau)\ \hat{\xi}(\tau)d\tau \tag{11}$$

where $\hat{\xi}(t)$ is the quadrature process of the stationary process $\xi(t)$. Then the envelope $S(t)$ is defined by

$$S(t) = [X^2(t) + \hat{X}^2(t)]^{\frac{1}{2}}\ . \tag{12}$$

It was not noted by Arens and Yang that, unlike the stationary case, the definition of the quadrature process depends on the representation (10), of which there is usually an infinity. For details see Hasofer [4].

However in the case of the oscillator response considered here, it is possible to distinguish one "natural" envelope.

The case of the oscillator response to an unmodulated stationary excitation was considered in Hasofer [4], where the concept of a "transient process" is introduced. This is a process which tends to a stationary process asymptotically as $t$ tends to infinity. It is shown that for such processes, there is a unique transient envelope which will tend to the envelope of the asymptotic stationary process, and this is taken to be the "natural" envelope.

Specifically let $t_0$ be the point of time at which the process $X(t_0;t)$ is initiated, and let it have the representation

$$X(t_0;t) = \int_{-\infty}^{+\infty} h(t_0;t,\tau)\xi(\tau)d\tau\ . \tag{13}$$

Suppose that as $t_0 \to -\infty$, the kernel $h(t_0;t,\tau)$ tends to an invertible kernel of the form $h(t-\tau)$, representing a time-invariant filter. Then a "natural" envelope can be defined for $X(t_0;t)$.

The above procedure applies to the linear oscillator with modulated excitation as well. If the initial conditions are taken to be $X(t_0) = X'(t_0) = 0$, then the solution can be written as

$$X(t_0;t) = \int_{t_0}^{t} W(\tau)\, n(\tau - t_0)\, h(t-\tau)d\tau \ , \ t \geq t_0 \ . \tag{14}$$

If, in addition, we let the $a_j$ and the $b_j$ be functions of $t_0$ in such a way that $n(t)$ tends to the Heaviside unit function $H(t)$ as $t_0 \to -\infty$, then $X(t_0;t)$ turns out to be a transient process, and a unique "natural" envelope can be chosen. For example one can let $b_j \to 0$ for all $j$, $a_1 \to 1$ and $a_j \to 0$ for all $j > 1$.

## 4. THE EFFECT OF NON-STATIONARITY ON THE UPCROSSING RATE

The first-passage time of the oscillator response to unmodulated and modulated excitation has been studied by Corotis, Vanmarcke and Cornell [3] and by Corotis and Marshall [2]. They used the concept of oscillatory process, but used upcrossing rate formulae which are correct only in the stationary case.

In this section, we compare the stationary and non-stationary upcrossing rate formulae and show that the stationarity formulae consistently underrate the correct upcrossing rate.

The upcrossing rate of a non-stationary Gaussian process $X(t)$ with zero mean at level $u$, denoted by $q_X(t;u)$, is a function of the following parameters:

$$\sigma_0(t) = E\, X^2(t) = R(t,t)\ ,$$

$$\tau_0(t) = E\, X(t)\, X'(t) = \partial R(t,s)/\partial s|_{s=t},$$

$$\sigma_2(t) = E\, X'^2(t) = \partial^2 R(t,s)/\partial s\, \partial t|_{s=t}\ ,$$

where $R(t,s)$ is the covariance function $E\, X(t)\, X(s)$.

Let

$$\gamma(t) = [1 - (\tau_0^2/\sigma_0\sigma_2)]^{\frac{1}{2}} \tag{15}$$

and

$$\eta(t) = \tau_0 u / \left( \gamma \sqrt{\sigma_2 \sigma_0^2} \right) \tag{16}$$

Then it follows from Leadbetter [6] that

$$q_X(t;u) = \frac{\gamma}{\sqrt{2\pi}} \sqrt{\frac{\sigma_2}{\sigma_0}} \exp[- \frac{u^2}{2\sigma_0}] [\phi(\eta) + \eta\Phi(\eta)] \tag{17}$$

where $\phi(.)$ and $\Phi(.)$ are the standard normal density and distribution functions respectively.

When the process $X(t)$ is stationary, $\tau_0 \equiv 0$, $\gamma = 1$ and $\eta = 0$. The formula becomes

$$q_X(t;u) = \frac{1}{2\pi} \sqrt{\frac{\sigma_2}{\sigma_0}} \exp[-\frac{u^2}{2\sigma_0}] . \tag{18}$$

As $u \to \infty$, formula (17) becomes asymptotically

$$q_X(t;u) \sim \frac{1}{\sqrt{2\pi}} \frac{\tau_0}{\sigma_0} \frac{u}{\sqrt{\sigma_0}} \exp[- \frac{u^2}{2\sigma_2} ] . \tag{19}$$

Thus the non-stationary formula contains terms with a higher power of $u$ than the corresponding stationary formula, and at high levels these terms dominate the expressions for the upcrossing rates.

The same considerations apply to the envelope. Here we need a fourth parameter: $\tau_1(t) = E\, X(t)\, \hat{X}'(t)$ .

Let

$$\gamma^*(t) = [1 - ((\tau_0^2 + \tau_1^2)/\sigma_0\sigma_2)]^{\frac{1}{2}} \tag{20}$$

and

$$\eta^*(t) = \tau_0 u / (\gamma^* \sqrt{\sigma_2 \sigma_0^2} ) . \tag{21}$$

Then the envelope upcrossing rate, $q_s(t;u)$, is given by

$$q_s(t;u) = \gamma^* \sqrt{\frac{\sigma_2}{\sigma_0}} \cdot \frac{u}{\sqrt{\sigma_0}} \exp[- \frac{u^2}{2\sigma_0}][\phi(\eta^*) + \eta^*\Phi(\eta^*)] . \tag{22}$$

For a proof of these formulae, see Hasofer and Petocz [5]. For a stationary process $\tau_0 \equiv 0$, and $\eta^* = 0$ .

When X(t) is stationary

$$q_s(t;u) = \sqrt{\frac{\sigma_0\sigma_2 - \tau_1^2}{2\pi\ \sigma_0^2}}\ \frac{u}{\sqrt{\sigma_0}} \exp[-\frac{u^2}{2\sigma_2}] \ . \tag{23}$$

However, as $u \to \infty$ , (22) becomes

$$q_s(t;u) \sim \frac{\tau_0}{\sigma_0}\frac{u}{\sigma_0} \exp[-\frac{u^2}{2\sigma_0}] . \tag{24}$$

with a $u^2$ instead of a $u$ .

Numerical calculations with the above formulae reveal that the stationarity assumption is always <u>unconservative</u>, and may therefore lead to serious underestimation of failure risks.

One such calculation is illustrated in Table 1.

<u>Table 1</u>. Ratio of Process Upcrossing Rates for non-Stationary and Stationary Formula

| $u/\sqrt{\sigma_0}$ \ $\gamma$ | 0.6 | 0.8 | 0.99 |
|---|---|---|---|
| 1 | 2.07 | 1.77 | 1.18 |
| 2 | 4.01 | 3.07 | 1.39 |
| 3 | 6.02 | 4.52 | 1.62 |
| 4 | 8.02 | 6.02 | 1.86 |

## 5. EXACT FORMULAE FOR UPCROSSING RATES.

The expressions for $\sigma_0(t)$ , $\tau_0(t)$ and $\sigma_2(t)$ are given, when $W(t)$ is white noise, by

$$\sigma_0(t) = \int_0^t n^2(t-\tau)h^2(\tau)d\tau \ , \tag{25}$$

$$\tau_0(t) = \int_0^t n^2(t-\tau)h'(\tau)d\tau , \tag{26}$$

$$\sigma_2(t) = \int_0^t n^2(t-\tau)h'^2(\tau)d\tau \ , \tag{27}$$

where $n(t)$ and $h(t)$ are given by (2) and (4) respectively. Explicit formulae in terms of the parameters of $n(t)$ and $h(t)$ are easy but tedious to calculate.

The formula for $\tau_1(t)$ is more difficult to obtain. It

turns out to be:

$$\tau_1(t) = \frac{1}{\pi}\int_0^{\infty} d\omega \int_0^t \int_0^t d\tau \; d\sigma \; n(\tau)\; n(\sigma)\; h(t-\tau)h'(t-\sigma)\sin\omega(\sigma-\tau) \; . \quad (28)$$

The upcrossing rates for the process and its envelope can then be calculated, using formulae (17) and (22) respectively.

## 6. QUALIFIED UPCROSSINGS

As pointed out above (Section 2) it is necessary to compensate for envelope upcrossings which do not contain process upcrossings. This is done as follows:

Let $p(t;u)$ be the probability that an excursion of the envelope above level $u$ starting at time $t$ contains at least one upcrossing of the modulus of the process.

Since the process considered is narrow-band, the mean distance between two successive crossings of zero of the process is given by

$$\beta = \beta(t) = 1/2q_X(t;0) \; . \quad (29)$$

This is approximately the same as the distance between two successive peaks of the modulus of the process. On the other hand, the expected length of an excursion $T_S$ of the envelope above $u$, starting at time $t$ is approximately

$$\alpha = \alpha(t) = \frac{P\{S(t) > u\}}{\text{upcrossing rate}} = q_S^{-1}(t;u)\exp[-u^2/2\sigma_0] \; . \quad (30)$$

Given that $T_S = x$, the probability of no upcrossing will be given by $1 - (x/\beta)$ if $x \leq \beta$ and $0$ if $x > \beta$.

Assuming $T_S$ to have a negative exponential distribution with mean $\alpha$, it follows that

$$p(t;u) = 1 - \int_0^{\beta}(1 - \frac{x}{\beta})\,\frac{1}{\alpha}\,e^{-x/\alpha}dx$$

$$= \frac{\alpha}{\beta}\,(1 - e^{-\beta/\alpha}) \; . \quad (31)$$

## 7. SOME NUMERICAL RESULTS

For numerical calculations, two modulating functions $n_1(t)$ and $n_2(t)$ were used. One was a "fast rise" function

$$n_1(t) = 2.32(e^{-0.009t} - e^{-1.49t}) ,$$

while the modulating function $n_2(t)$ rose less fast.

Its formula is

$$n_2(t) = 12.80(e^{-0.14t} - e^{-0.19t}) .$$

The amplitudes of the modulating functions were chosen in such a way as to deliver the same amount of energy over a 20 second interval as a Heaviside function modulation. The parameter $\omega_0$ was taken to be $0.4\pi$, and the parameter $\zeta$ was taken to be 0.05.

The first passage distribution $L_X(t;u)$ defined by equation (6) was calculated for various values of $u$ and $t$.

The corresponding first passage distributions for the envelope, $L_S(t;u)$, and for the qualified upcrossings of the envelope, $\tilde{L}_S(t,u)$, were also calculated.

These distributions are also called the "reliability functions" of the response process and its envelope.

Table 2 shows a comparison of the values of $\tilde{L}_S(t;u)$ obtained by the present method, compared with the values given by Corotis and Marshall [2] .

Table 2. Comparison of Values of the Reliability Function $\tilde{L}_S(t;u)$ of Qualified Envelope Upcrossings

CM = Corotis and Marshall - values read from their graphs (those in brackets by extrapolation).

P = Present results as calculated.

LEVEL $u = \kappa\sqrt{\sigma_0^*(20)}$ , where $\sigma_0^*$ is from Heaviside modulation.

| $n_1(t)$ |
|---|
| $\omega_0=0.4$ |
| $\zeta=0.05$ |

| | level $\kappa=1.2$ | | level $\kappa=2.0$ | | level $\kappa=3.0$ | |
|---|---|---|---|---|---|---|
| t | CM | P | CM | P | CM | P |
| 5 | 0.50 | 0.65 | 0.97 | 0.86 | 0.98 | 0.98 |
| 10 | 0.17 | 0.33 | 0.80 | 0.60 | 0.94 | 0.91 |
| 15 | (0.06) | 0.11 | 0.64 | 0.45 | (0.89) | 0.87 |
| 20 | (0.02) | 0.09 | 0.52 | 0.39 | (0.88) | 0.86 |

| $n_2(t)$ $\omega_0=0.4$ $\zeta=0.05$ | t | level $\kappa = 1.2$ | | level $\kappa = 2.0$ | | level $\kappa = 3.0$ | |
|---|---|---|---|---|---|---|---|
| | | CM | P | CM | P | CM | P |
| | 5 | 0.89 | 0.89 | 1.00 | 0.99 | 1.00 | 1.00 |
| | 10 | 0.35 | 0.47 | 0.94 | 0.79 | 0.98 | 0.98 |
| | 15 | 0.15 | 0.25 | 0.80 | 0.58 | 0.94 | 0.93 |
| | 20 | (0.08) | 0.14 | 0.70 | 0.47 | 0.93 | 0.90 |

REFERENCES


[1] Arens R. [1957]. "Complex processes for envelopes of normal noise." IRE Trans. Inf. Th. 3, pp. 204-207.

[2] Corotis R.B., and Marshall T.A. [1977]. "Oscillator response to modulated random excitation." J. Eng. Mech. Div. ASCE EM4, pp. 501-513.

[3] Corotis R.B., Vanmarcke E.H. and Cornell C.A. [1972]. "First passage of non-stationary random processes." J. Eng. Mech. Div. ASCE, pp. 401-414.

[4] Hasofer A.M. [1979]. "A uniqueness problem for the envelope of an oscillatory process." J. Appl. Prob. 16, pp. 822-829.

[5] Hasofer A.M. and Petocz P. [1978]. The envelope of an oscillatory process and its upcrossings." Adv. Appl. Prob. 10, 4, pp. 711-716.

[6] Leadbetter, M.R. [1966]. "On crossings of levels and curves by a wide class of stochastic processes." Ann. Math. Stat. 37, pp. 260-267.

[7] Priestley M.B. [1965]. "Evolutionary Spectra and non-stationary processes." J. Royal Stat. Soc. B, 27, pp. 204-229.

[8] Rice S.O. [1945]. "Mathematical Analysis of Random Noise." Bell System Technical J. 24, pp. 46-156.

[9] Vanmarcke E.H. [1975]. "On the distribution of first passage time for normal stationary random processes." J. App. Mech. 42, pp. 215-220.

[10] Yang J.N. [1972]. "Non-stationary Envelope Process and First Excursion Probability." J. Struct. Mech. 1 (2), pp. 231-248.

# FROST DATA: A CASE STUDY ON EXTREME VALUES OF NON-STATIONARY SEQUENCES

Jürg Hüsler

Department of math. Statistics, University of Berne

The non-stationary sequence of daily minimum temperature is studied to estimate the last frost day, i.e. the last day in spring time with a minimum temperature below a critical temperature. Instead of the usual order statistics to estimate a certain fractile of the last frost distribution, a model of extreme values is used to give a more reliable estimator, which is compared with some other simple estimators, including also an estimator of a computer simulation.

## 1. INTRODUCTION

In meteorology and in particular in agriculture the last frost nights of the spring period are of great importance. They vary naturally from place to place. Since for some places only a few years of observations are available, it is a statistical problem to give a sufficiently reliable estimator for a certain fractile. For, if we consider the usual order statistic of the last frost observations for a 0.9-fractile, we observe that only a one-sided confidence interval may be found for sample sizes $n \leq 35$ ($\alpha = 5\%$). This would give us only a anti-conservative estimator of the fractile; but more reasonable would be a conservative estimator. At the common sites of observations not only the information of the last frost days is given, but a much larger information, e.g. the minimum-temperature of each day (observed early in the morning). With this larger information a more reliable estimator may be given for the 0.9-fractile. This estimator is found by applying a theorem of extreme values of non-stationary sequences, given in Hüsler (1983 a). Based on the data of a special site in Switzerland, Breitenhof ($n = 18$), we discuss the reliability of the estimator, by



*J. Tiago de Oliveira (ed.), Statistical Extremes and Applications, 513–520.*

assuming in the following some additional assumptions on the sequence of minimum temperature. Finally we add a comparison of the frost fractile estimator with the fractile found by a computer simulation of sequences of minimum temperatures.

## 2. THE PROBABILISTIC MODEL

By analyzing the data of the minimum temperatures $X_i$ of the i-th day per year, we found it reasonable to assume the following model. Let $\mu_i$ be the expectation of the minimum temperature at the i-th day and $\sigma_i$ its standard deviation. Then $X_i = \mu_i + \sigma_i \cdot Y_i$ where $\{Y_i\}$ is a standardized random sequence ($EY_i^2 = 1$, $EY_i = 0$). In our case study we used the empirical values $\overline{x}_i$ and $s_i$. The trend $\overline{x}_i$ is plotted in Fig. 1.

Fig. 1: Mean trend of minimum temperatures $\overline{x}_i$ of 18 years, site Breitenhof

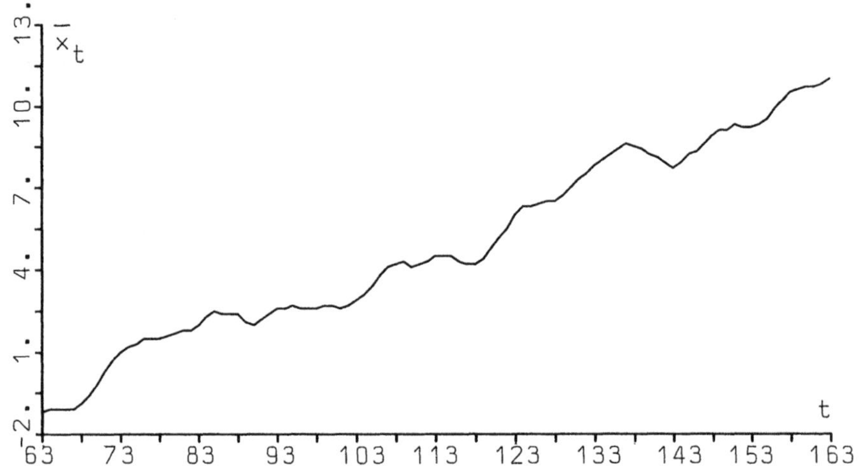


The statistical analysis of the residuals showed that we may assume $\{Y_i\}$ as a Gaussian sequence. The details are given in Hüsler (1983 b). In the following we will assume for some calculations that $\{Y_i\}$ is a Gauss-AR(1)-sequence, since the empirical correlation $r_k \approx (0.73)^k$ for small k; the remaining $r_k$'s vary around 0. The AR(p)-models had been also used in Ginsburg (1970) for other sequences of temperatures.

Let T be the critical temperature (e.g. $0^\circ, -1^\circ, -2^\circ, -3^\circ, -4^\circ$C) and $\tau$ be the last frost day

$$\tau = \max_{i \leq t_0}\{i: X_i < T\}, \text{ if such an i exists, 0, else,}$$

where $t_0$ is a summerday after which no temperature below T occurs (in our case $t_0 = 163$). Then the fractile is defined by $t_p = \inf\{t: P\{\tau \leq t\} \geq p\}$. Thus we have to evaluate the distribution $P_t$ of $\tau$.

$$P_t = P\{\tau \leq t\} = P\{X_i \geq T,\ i \in I_t\} = P\{Y_i \geq (T-\mu_i)/\sigma_i,\ i \in I_t\}$$
$$= P\{Y_i' \leq (\mu_i - T_i)/\sigma_i,\ i \in I_t\}$$

where $Y_i' = -Y_i$ and $I_t = \{t+1, t+2, \ldots, t_0\}$. The theorem of Hüsler (1983 a) states that $P_t$ may be approximated under certain conditions by the product probabilities

$$\tilde{P}_t = \prod_{i \in I_t} P\{Y_i' \leq u_i\} = \prod_{i \in I_k} \Phi(u_i)$$

where $u_i = (\mu_i - T)/\sigma_i$. Since we assume $\{Y_i'\}$ as a Gauss sequence, the conditions

$$r_n \log n \to 0 \quad \text{as } n \to \infty \tag{1}$$

and $u_i \to \infty$ should be satisfied. We know that AR(p)-sequences satisfy the condition (1). The goodness of the approximation depends thus mainly on the given empirical values. Thus we find by this model a new estimator $\tilde{t}_p$ for $t_p$:

$$\tilde{t}_p = \min\{t: \tilde{P}_t \geq p\} \tag{2}$$

Fig. 2: Distribution $\tilde{P}_t$ of $\tau$ for T = $0^o, -1^o, -2^o, -3^o, -4^o$C.

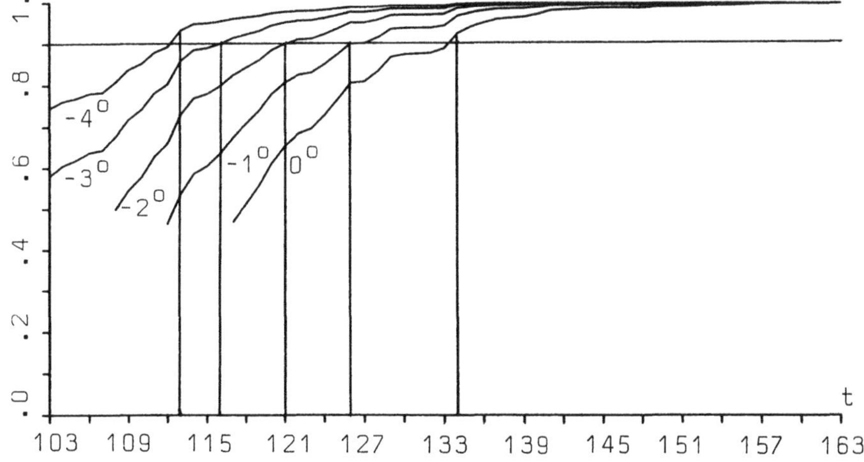

From the calculations of $\tilde{P}_t$, given in Fig. 2, we find $\tilde{t}_{0.9}$. A comparison with the order statistics $t^*_{0.9}$ for $t_{0.9}$, given in Tab. 2, shows that $\tilde{t}_{0.9}$ are all larger than $t^*_{0.9}$. In the following we study the reliability of the estimator $\tilde{t}_{0.9}$.

## 3. THE RELIABILITY OF THE ESTIMATOR

### 3.1. Bonferroni-Inequalities

We begin with a procedure which may be used for any random sequence $\{Y_i\}$ as long as the one and two-dimensional distributions are known or may be reasonably approximated. The Bonferroni inequalities give

$$P_t \leqslant 1 - \sum_{i \in I_t} P\{Y_i' > u_i\} + \sum_{i<j \in I_t}\sum P\{Y_i' > u_i, Y_j' > u_j\} = \hat{P}_t$$

$$P_t \geqslant 1 - \sum_{i \in I_t} P\{Y_i' > u_i\} = \check{P}_t$$

These distributions $\hat{P}_t$ and $\check{P}_t$ define each an estimator $\hat{t}_p$ resp. $\check{t}_p$ of $t_p$ as in (2). We remark that $\hat{P}_t$ may take values larger than 1 or no value of $\hat{P}_t$ is smaller than p, (leaving $\hat{t}_p$ undefined) depending on the values of $u_i$ and the distributions of $Y_i'$. Similar, $\check{P}_t$ may take values smaller than 0, nevertheless $\check{t}_p$ is well defined.

Fig. 3: Bonferroni-estimation $\hat{P}_t$ and $\check{P}_t$ of the distribution of $\tau$ for $T = -1^{o}C$.

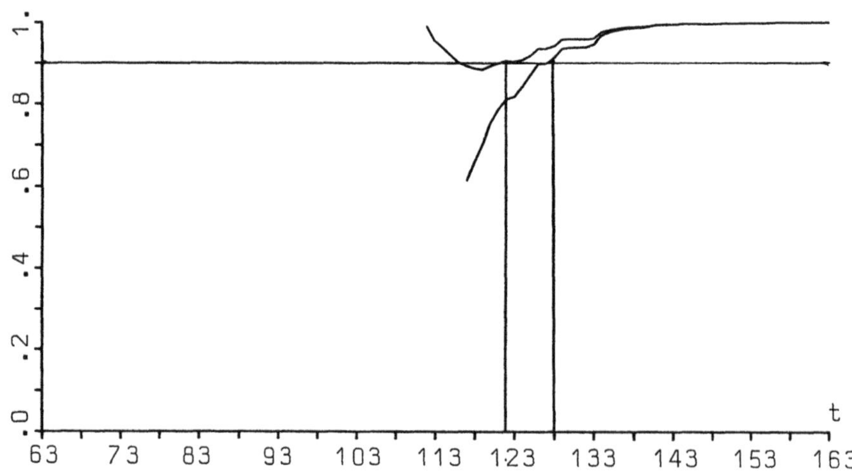

In our case we used for the Gauss-sequence $Y_i'$ the one and two-dimensional normal distributions (using a series expansion for the two-dimensional one with $r_k = \rho^k$, $\rho = 0.73$; see e.g. Abramowitz + Stegun (1972)). Fig. 3 gives for $T = -1^oC$ the values $\hat{t}_{0.9} = 122$ and $\check{t}_{0.9} = 128$. (For other T values see Tab. 2 and figures in Hüsler (1983 b)).

Thus by the Bonferroni-inequality we know that $\check{t}_p$ is a conservative estimator of $t_p$. In some cases the anti-conservative estimator $\hat{t}_p$ may be improved by using the following inequalities

$$P_t \leqslant P\{Y_i' \leqslant u_i, i \in I\} \leqslant 1 - \sum_{i\in I} P\{Y_i' > u_i\} + \sum_{i<j\in I}\sum P\{Y_i' > u_i, Y_j' > u_j\} =: \hat{P}_t' \quad (3)$$

where $I \subset I_t$ is chosen such that $\hat{P}_t'$ is minimal. Since $\tau$ is a discrete random variable we did not find in our case an improved estimation $\hat{t}_p$ by this inequality.

### 3.2. Inequalities of Gauss-sequences

Since we assume $\{Y_i\}$ as a Gauss-sequence, we may use particular inequalities of such sequences. In the following we use Slepian's inequality (1962). Since $r_k \approx (0.73)^k \geqslant 0$, the first application of Slepian's inequality gives $P_t \geqslant \Pi_{i \in I_t} \Phi(u_i) = \tilde{P}_t$. This implies that $\tilde{t}_p$ is a conservative estimator of $t_p (\tilde{t}_p \leqslant t_p)$, which we found already in our calculations of the example.

Let $\{Z_i\}$ be an equally correlated Gauss sequence, i.e. $Z_i = \sqrt{1-\rho}\, U_i + \sqrt{\rho}\, V$ where $U_i$, $V$ are iid. unit normal random variables and $\rho > 0$. Since $r_k = \rho^k \leqslant \rho$, a second application of Slepian's Lemma gives

$$P_t \leqslant P\{Z_i \leqslant u_i,\ i \in I_t\} = \int_{-\infty}^{\infty} \prod_{i \in I_t} \Phi\left(\frac{u_i - \sqrt{\rho}v}{\sqrt{1-\rho}}\right) \varphi(v)dv = P_t(Z)$$

with $\varphi(v) = \exp(-v^2/2)/\sqrt{2\pi}$. The integral may be numerically evaluated, given in Fig. 4. $P_t(Z)$ defines $t_p(Z) = \min\{t\colon P_t(Z) \geqslant p\}$, another anti-conservative estimator of $t_p$ with values: $t_{0.9}(Z)$: $126(0^o)$, $120(-1^o)$, $114(-2^o)$, $111(-3^o)$ and $104(-4^oC)$.

The differences between $\tilde{t}_p$ and $t_p(Z)$ are mainly due to the high value $\rho = 0.73$ (in case $\rho \approx 0.3$, the differences would be of about one day). An improvement for $P_t(Z)$ may be found by using again the inequality (3):

$$P_t \leqslant P\{Y_i \leqslant u_i,\ i \in I\} \leqslant P\{Z_i \leqslant u_i,\ i \in I\}$$

where I is of the form $I = \{t+1, t+k+1, t+2k+1, \ldots, t+k_0 \cdot k+1\} \subset I_k$, for any k, and $\{Z_i\}$ with correlation $s_\ell = \rho' = \rho^k$ for all $\ell \neq 0$.

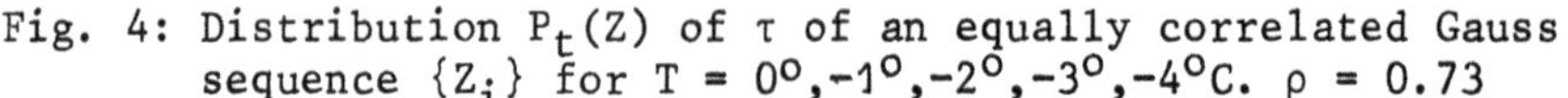

Fig. 4: Distribution $P_t(Z)$ of $\tau$ of an equally correlated Gauss sequence $\{Z_i\}$ for $T = 0^o, -1^o, -2^o, -3^o, -4^o C$. $\rho = 0.73$

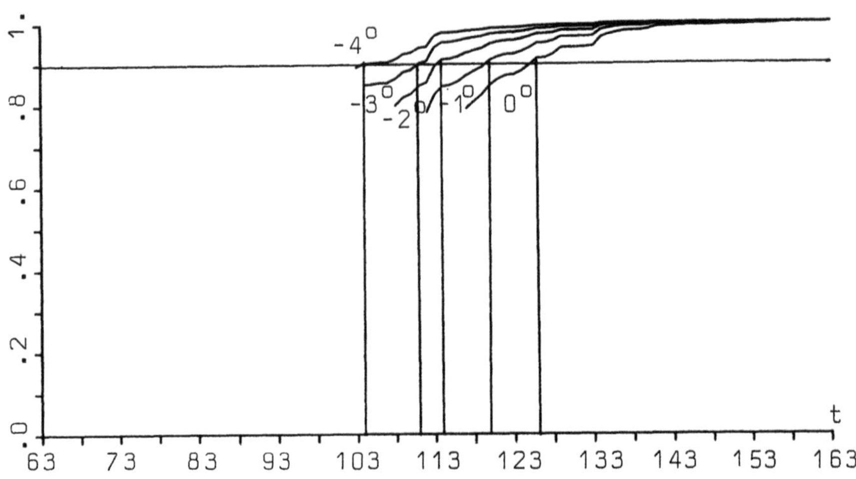


4. SIMULATION OF AR(1)-SEQUENCES

In general it is difficult to evaluate $P_t$ for any random sequences $Y_i$. This is true also for Gauss-AR(1)-sequences. But a simulation study is much easier and quite accurate with large simulation sizes. The simulated distribution functions $P_t(S)$ of $\tau$ are shown in Fig. 5,

Fig. 5: Simulated distribution $P_t(S)$ of $\tau$ of a Gauss-AR(1)-sequence for $T = 0^o, -1^o, -2^o, -3^o, -4^o C$, $\rho = 0.73$.

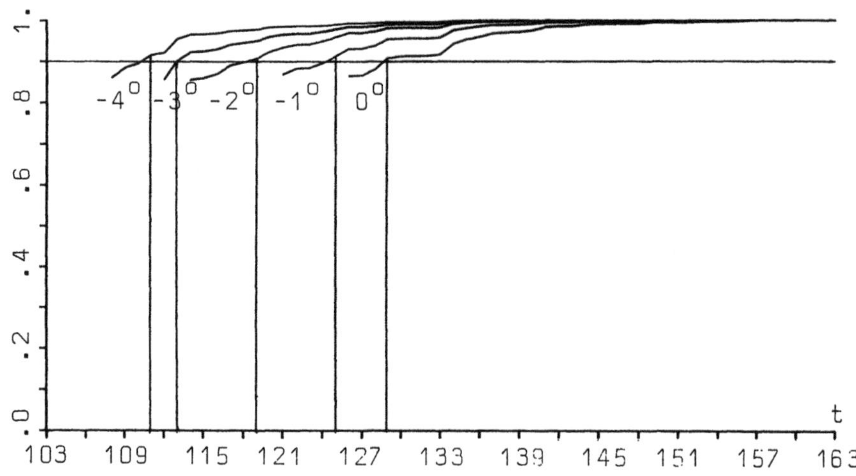

where the simulation size is 5'000 and $\rho = 0.73$. The values $t_p$ depend naturally on these assumptions. But the comparison with simulations with size 100'000 or with $\rho = 0.7$ shows that $t_{0.9}(\rho)$ varies by ± 1 day.

The comparison of the different fractile estimators is given in Tab. 1. It shows that the values are better than $t^*_{0.9}$ and are sufficiently reliable for the use in agriculture. A quite accurate estimator is found by taking the mean of $\tilde{t}_{0.9}$ and $t_{0.9}(Z)$.

Table 1: The comparison of estimators of $t_{0.9}$

| T | $t^*_{0.9}$ | $\tilde{t}_{0.9}$ | $\hat{t}_{0.9}$ | $\check{t}_{0.9}$ | $t_{0.9}(Z)$ | $t_{0.9}(S)$ | $\bar{t}_{0.9}$ |
|---|---|---|---|---|---|---|---|
| $-0^o$ | 125 | 134 | 128 | 134 | 126 | 129 | 130 |
| $-1^o$ | 125 | 126 | 122 | 128 | 120 | 125 | 123 |
| $-2^o$ | 121 | 121 | 114 | 122 | 114 | 119 | 117.5 |
| $-3^o$ | 105 | 116 | 113 | 117 | 111 | 113 | 113.5 |
| $-4^o$ | 102 | 113 | 102 | 113 | 104 | 111 | 108.5 |

$t^*_{0.9}$ : Estimator w.r.t. the order statistics

$\tilde{t}_{0.9}$ : " " the product probabilities

$\hat{t}_{0.9}$ : " " the Bonferroni-inequalities: upper bound

$\check{t}_{0.9}$ : " " " " " : lower bound

$t_{0.9}(Z)$ : " " the equally correlated Gauss sequence $\{Z_i\}$

$t_{0.9}(S)$ : 0.9-fractile of the simulated distribution w.r.t. the Gauss-AR(1)-sequence

$\bar{t}_{0.9}$ : Mean of $\tilde{t}_{0.9}$ and $t_{0.9}(Z)$.

The estimator $\tilde{t}_p$ with respect to the product probabilities does not consider any correlation of the random sequence. The estimator $t_p(Z)$ uses a too strong correlation structure, thus being anti-conservative. A simple approximative procedure, which uses an intermediate correlation structure, is the following: By denoting $A_i = \{X_i > T\} = \{Y_i' \leq u_i\}$ we approximate

$$\tilde{P}_t = P(A_{t+1} \cap A_{t+2} \cap \ldots \cap A_{t_0})$$

$$= P(A_{t+1}) \cdot \prod_{i=t+2}^{t_0} P(A_i | A_{i-1} \cap \ldots \cap A_{t+1}) \approx P(A_{t+1}) \cdot \prod_{i=t+2}^{t_0} P(A_i | A_{i-1})$$

The conditional probabilities are for a Gauss-AR(1)-sequence equal to

$$P(A_i|A_{i-1}) = \int_{-\infty}^{u_{i-1}} \Phi\left(\frac{u_i - \rho v}{\sqrt{1-\rho^2}}\right) \varphi(v)dv/\Phi(u_{i-1})$$

The numerical evaluations of $P_t(M)$ for various T showed that $P_t(M) - P_t(S)$ is in the range of the simulation error, which implied that $|t_{0.9}(S) - t_{0.9}(M)| \leq 1$, where $t_p(M)$ is obviously the estimator based on $P_t(M)$.

We remark also that if the seasonal trend is also smoothed, the numerical values of the estimators in the smoothed case differ from the unsmoothed estimators by 1 day, showing a robust feature of the proposed procedures (see the details in Hüsler (1983 b)). Finally we remark that our estimators had been used with equally satisfactory results by R. Volz and P. Villiger (1982) for other sites, for data with a larger number of observed years, too.

References:

Abramowitz, M. and Stegun, I. A. 1965, Handbook of mathematical functions. Dover, New York.

Ginsburg, T. 1970, Die statistische Auswertung von langjährigen Temperaturreihen. Veröff. der Schweiz. Met. Zentralanst., Nr. 19, Zürich.

Hüsler, J. 1983 a, Asymptotic approximation of crossing probabilities of random sequences. Zeitschrift f. Wahrscheinlichkeitstheorie verw. Geb. 63, pp. 257 - 270

Hüsler, J. 1983 b, Anwendung der Extremwerttheorie von nicht-stationären Zufallsfolgen bei Frostdaten. Techn. Bericht 10, Inst. für math. Statistik, Universität Bern

Slepian, D. 1962, The one-sided barrier problem for Gaussian noise. Bell System Techn. J. 41, pp. 463 - 501.

Volz, R. and Filliger, P. 1982, Ein Wahrschinlichkeitsmodell zur Bestimmung des letzten Spätfrosttermins. Geophysik, Beiheft zum Jahrbuch der SNG, pp. 21 - 25.

# ESTIMATION OF THE SCALE AND LOCATION PARAMETERS OF THE EXTREME VALUE (GUMBEL) DISTRIBUTION FOR LARGE CENSORED SAMPLES

Peter Kubat

Graduate School of Management
University of Rochester
Rochester, New York 14627 U.S.A.

ABSTRACT. Simple estimators of the scale and location parameters in large right censored samples are considered. These estimators are based on two contiguous blocks of order statistics. For the Extreme-Value (Gumbel) distribution it is shown how to select the "best blocks" in order to obtain the most efficient estimators within the class studied.

## 1. INTRODUCTION

In many statistical problems commonly arising in engineering, medical and biological applications the estimates of quantities of interest are based on samples which are only partially observed. Examples of such problems are life tests: at the time $t_0 = 0$ n items are placed on test and test is discontinued as soon as one has observed the first M (M < n) failures. Thus, only the first M order observations (statistics) $x_{(1)} \leq x_{(2)} \leq \cdots \leq x_{(M)}$ have been completely observed and there are still n - M items for which the lifte times have to be observed. Since the right tail of the observations is missing we say that the sample is censored from the right. For a more detailed discussion concerning the incomplete data see, Epstein [4].

Except for the case in which the underlying distribution is exponential the maximum likelihood (m.l.) estimates of unknown parameters in censored samples will require some numerical solution. To overcome this complication various estimators based on selected order statistics were proposed by numerous authors. A bibliography



*J. Tiago de Oliveira (ed.), Statistical Extremes and Applications, 521–534.*

of estimation theory using order statistics may be found in David [3] and Sarhan and Greenberg [8]; the applications for commonly used distributions are detailed by Johnson and Kotz [6].

In this paper simple estimators for the scale ($\delta$) and the location ($\lambda$) parameters based on large right censored samples are presented. More specifically, for a location-scale family of continuous distributions $F((x-\lambda)/\delta)$ we will find estimators of $\lambda$ and $\delta$ which will be simple linear combinations of the means of the blocks of ordered observations. This estimation procedure is then applied to estimate the unknown parameters of the Extreme-Value (Gumbel) distribution.

Estimation procedures based on blocks of order statistics for complete samples were investigated by Kubat [7]. The asymptotic distribution theory for the means of the order statistics blocks can be derived from the results of Chernoff, et al. [1] and can be found conveniently summarized in [7].

## 2. CHOICE OF ESTIMATORS

Let $X_{(1)} \leq X_{(2)} \leq \cdots \leq X_{(M)}$, $M = [nq]$, $0 < q < 1$ be the M smallest order statistics in a sample of very large size n ($n \to \infty$) taken from a distribution having c.d.f $F((x-\lambda)/\delta)$. Here $[x]$ denotes the largest integer not exceeding x. The censoring level q is assumed to be known.

In order to estimate the unknown scale ($\delta$) and location ($\lambda$) parameters the following procedure is proposed:

The censored sample is divided into two parts. The first part will include the N smallest order statistics $X_{(1)} \leq \cdots \leq X_{(N)}$, $N = [np]$, $0 < p < q$, (left tail) and the second part, the $M - N$ remaining order statistics, $X_{(n+1)} \leq \cdots \leq X_{(M)}$. (Right end of the censored sample, or alternatively, middle part of uncensored sample.)

We now compute the means for each part, namely

$$X_N^* = \frac{1}{N} \sum_{i=1}^{N} X_{(i)} \tag{2.1}$$

and

$$\tilde{X}_{N,M} = \frac{1}{M-N} \sum_{i=N+1}^{M} X_{(i)}. \tag{2.2}$$

Assume that the c.d.f. $F(\cdot)$ has twice differentiable density $f(\cdot)$ a.s. and the integrals

$$\int_{-\infty}^{z_p} x^2 dF(x) \quad \text{and} \quad \int_{z_p}^{z_q} x^2 dF(x)$$

are finite. Then, for $n \to \infty$, we have that $X_N^*$ and $\tilde{X}_{N,M}$ are jointly normal (Kubat [7]), i.e.

$$\begin{pmatrix} X_N^* \\ \tilde{X}_{N,M} \end{pmatrix} \sim N\left( \begin{bmatrix} \lambda + \delta\mu_1(p) \\ \lambda + \delta\mu_3(p,q) \end{bmatrix}, \frac{\delta^2}{n} \begin{bmatrix} \sigma_1^2(p)/p & \sigma_{13}(p,q) \\ \sigma_{13}(p,q) & \sigma_3^2(p,q)/(q-p) \end{bmatrix} \right),$$

where

$$\mu_1(p) = \frac{1}{p} \int_{-\infty}^{z_p} x df(x), \tag{2.3}$$

$$\mu_3(p,q) = \frac{1}{q-p} \int_{z_p}^{z_q} x dF(x), \tag{2.4}$$

$$\sigma_1^2(p) = \frac{1}{p} \int_{-\infty}^{z_p} x^2 dF(x) - \mu_1^2(p) + [z_p - \mu_1(p)]^2 (1-p), \tag{2.5}$$

$$\sigma_3^2(p,q) = \frac{1}{q-p} \int_{z_p}^{z_q} x^2 dF(x) - \mu_3^2(p,q) +$$

$$+ \frac{1}{q-p} \{p(1-p)[\mu_3(p,q)-z_p]^2 + q(1-q)[z_q-\mu_3(p,q)]^2$$

$$+ 2p(1-q)[\mu_3(p,q)-z_p][z_q-\mu_3(p,q)]\} \tag{2.6}$$

and

$$\sigma_{13}(p,q) = (z_p-\mu_1(p))[\mu_3(p,q) + \frac{1-q}{q-p} z_q - \frac{1-p}{q-p} z_p]. \tag{2.7}$$

Here $z_p = F^{-1}(p)$ and $z_q = F^{-1}(q)$ are the corresponding p and q quantiles of the distribution F(x). Since $X_N^* \xrightarrow{P} \lambda + \delta\mu_1(p)$ and $\tilde{X}_{N,M} \xrightarrow{P} \lambda + \delta\mu_3(p,q)$ as $n \to \infty$, we get asymptotically unbiased estimators of the parameters $\lambda$ and $\delta$:

$$\lambda_n^* = \frac{\mu_3(p,q)X_N^* - \mu_1(p)\tilde{X}_{N,M}}{\mu_3(p,q) - \mu_1(p)}, \tag{2.8}$$

$$\delta_n^* = \frac{\tilde{X}_{N,M} - X_N^*}{\mu_3(p,q) - \mu_1(p)}. \tag{2.9}$$

For brevity let $\mu_1 = \mu_1(p)$, $\mu_3 = \mu_3(p,q)$, $v_1 = \delta_1^2(p)/p$, $v_3 = \sigma_3^2(p,q)/(q-p)$ and $c = \sigma_{13}(p,q)$. Then the variances and the covariance of the estimators $\lambda_n^*$ and $\delta_n^*$ are:

$$\mathrm{Var}(\lambda_n^*) \cong \frac{\delta^2}{n} \cdot \frac{\mu_3^2 v_1 + \mu_1^2 v_3 - 2\mu_1\mu_3 c}{(\mu_3 - \mu_1)^2}, \tag{2.10}$$

$$\mathrm{Var}(\delta_n^*) \cong \frac{\delta^2}{n} \cdot \frac{v_1 + v_3 - 2c}{(\mu_3 - \mu_1)^2}, \tag{2.11}$$

and

$$\mathrm{Cov}(\lambda_n^*,\delta_n^*) \cong \frac{\delta^2}{n} \cdot \frac{(\mu_1+\mu_3)c - \mu_1 v_3 - \mu_3 v_1}{(\mu_3 - \mu_1)^2}. \tag{2.12}$$

Following Cramer [2, p. 493] we define the joint asymptotic relative efficiency (A.R.E.) of the estimators $\lambda_n^*$ and $\delta_n^*$ as

$$e_c(p,q) = \lim_{n\to\infty} [n^2 \Delta_c(q)G]^{-1}, \tag{2.13}$$

where

$$G = \mathrm{Var}(\lambda_n^*)\ \mathrm{Var}(\delta_n^*) - \mathrm{Cov}^2(\lambda_n^*,\delta_n^*) \tag{2.14}$$

and $\Delta_c(q)$ measures the amount of information for $\lambda$ and $\delta$ contained in the right censored sample (for details see Halperin [5]). Note that the quantity $1/n^2\Delta_c(q)$ is in fact the generalized variance of the m.l. estimators of the scale and location parameters in the right censored samples. It is shown in Appendix A that for a scale- location distribution $F((x-\lambda)/\delta)$ the information $\Delta_c(q) = \delta^{-4}\Delta(q)$. When (2.10-2.12) are substituted into (2.13), after some algebra we obtain that

$$e_c(p,q) = \frac{[\mu_3(p,q) - \mu_1(p)]^2}{\Delta(q)[\sigma_1^2(p)\sigma_3^2(p,q)/(pq-p^2) - \sigma_{13}^2(p,q)]}. \tag{2.15}$$

For given censoring level q one can look for a value of p, say p*, which maximizes (2.15).

## 3. ESTIMATION OF $\lambda$ AND $\delta$ FOR THE EXTREME VALUE (GUMBEL) DISTRIBUTION

The Gumbel distribution has c.d.f.

$$F(x;\lambda,\delta) \equiv F(\tfrac{x-\lambda}{\delta}) = \exp\{-\exp\{-\tfrac{x-\lambda}{\delta}\}\}, \quad -\infty < x < \infty. \tag{3.1}$$

This distribution is regular and satisfies all the assumptions required for a successful application of the estimation procedure introduced in the previous section. The coefficients $\mu_1(p)$ and $\mu_3(p,q)$ occurring in (2.8) and (2.9) are:

$$\mu_1(p) = \frac{1}{p}\int_{-\infty}^{-\ln(-\ln p)} x\exp\{-x\}\exp\{-\exp\{-x\}\}dx \tag{3.2}$$

and

$$\mu_3(p,q) = [q\mu_1(q) - p\mu_1(p)]/(q-p). \tag{3.3}$$

The integral in (3.2) requires a numerical integration. The values of $\mu_1(p)$ were tabulated for p = 0.02 (0.02) 0.98 and can be found in [7]. The values of $\Delta(q)$ for q = 0.05 (0.05) 0.95 are tabulated in Appendix A. The joint A.R.E. $e_c(p,q)$ is tabulated in Table 1.

Using Table 1, it is easy to find the optimal partition of a censored sample. For instance, when q = 0.5 (i.e., 50% of the largest observations are censored), the maximum efficiency is attained at p = 0.2, where $e_c(p,q)$ = 74%. Hence, $X_N^*$ is then based

| q \ p | 0.05 | 0.10 | 0.15 | 0.20 | 0.25 | 0.30 | 0.35 | 0.40 | 0.45 | 0.50 | 0.55 | 0.60 | 0.65 | 0.70 | 0.75 | 0.80 | 0.85 | 0.90 |
|---|---|---|---|---|---|---|---|---|---|---|---|---|---|---|---|---|---|---|
| 0.10 | 72 | ** | ** | ** | ** | ** | ** | ** | ** | ** | ** | ** | ** | ** | ** | ** | ** | ** |
| 0.15 | 67 | 76 | ** | ** | ** | ** | ** | ** | ** | ** | ** | ** | ** | ** | ** | ** | ** | ** |
| 0.20 | 64 | 73 | 76 | ** | ** | ** | ** | ** | ** | ** | ** | ** | ** | ** | ** | ** | ** | ** |
| 0.25 | 63 | 71 | 74 | 74 | ** | ** | ** | ** | ** | ** | ** | ** | ** | ** | ** | ** | ** | ** |
| 0.30 | 62 | 70 | 73 | 75 | 77 | ** | ** | ** | ** | ** | ** | ** | ** | ** | ** | ** | ** | ** |
| 0.35 | 62 | 70 | 73 | 74 | 74 | 73 | ** | ** | ** | ** | ** | ** | ** | ** | ** | ** | ** | ** |
| 0.40 | 61 | 70 | 73 | 73 | 73 | 72 | 70 | ** | ** | ** | ** | ** | ** | ** | ** | ** | ** | ** |
| 0.45 | 61 | 70 | 73 | 74 | 73 | 72 | 70 | 68 | ** | ** | ** | ** | ** | ** | ** | ** | ** | ** |
| 0.50 | 61 | 70 | 73 | 74 | 73 | 72 | 71 | 69 | 67 | ** | ** | ** | ** | ** | ** | ** | ** | ** |
| 0.55 | 62 | 71 | 74 | 75 | 74 | 73 | 71 | 69 | 66 | 63 | ** | ** | ** | ** | ** | ** | ** | ** |
| 0.60 | 62 | 72 | 75 | 76 | 75 | 74 | 71 | 69 | 66 | 63 | 60 | ** | ** | ** | ** | ** | ** | ** |
| 0.65 | 62 | 72 | 76 | 77 | 76 | 74 | 72 | 70 | 67 | 64 | 61 | 57 | ** | ** | ** | ** | ** | ** |
| 0.70 | 63 | 73 | 77 | 78 | 77 | 76 | 73 | 71 | 68 | 65 | 61 | 57 | 53 | ** | ** | ** | ** | ** |
| 0.75 | 64 | 74 | 78 | 79 | 78 | 77 | 75 | 72 | 69 | 65 | 62 | 58 | 54 | 50 | ** | ** | ** | ** |
| 0.80 | 64 | 76 | 80 | 81 | 80 | 78 | 76 | 73 | 70 | 66 | 63 | 59 | 55 | 50 | 46 | ** | ** | ** |
| 0.85 | 65 | 77 | 81 | 82 | 82 | 80 | 78 | 75 | 71 | 68 | 64 | 60 | 55 | 51 | 46 | 41 | ** | ** |
| 0.90 | 66 | 78 | 83 | 84 | 83 | 82 | 79 | 76 | 73 | 69 | 65 | 61 | 56 | 51 | 46 | 41 | 35 | ** |
| 0.95 | 67 | 80 | 85 | 86 | 85 | 84 | 81 | 78 | 74 | 71 | 66 | 62 | 57 | 52 | 46 | 41 | 34 | 27 |

TABLE 1: Joint asymptotic relative efficiency $e_c(p,q)$, of the estimators $\lambda_n^*$ and $\delta_n^*$ in right censored samples (Gumbel distribution).

on the smallest 20% of the order statistics (the smallest 40% of the censored data) and $\tilde{X}_{N,M}$ uses 30% of the observations (the largest 60% of the censored sample). In this case $\mu_1(0.2) = -0.901$ and $\mu_3(0.2,0.5) = [(0.5)(-0.391) - (0.2)(-0.901)]/(0.5-0.2) = -0.051$. The estimators (2.8) and (2.9) will then become

$$\lambda_n^* = -0.06\, X_N^* + 1.06\, \tilde{X}_{N,M}$$

and

$$\delta_n^* = 1.18(\tilde{X}_{n,M} - X_N^*).$$

## 4. ONE-PARAMETER CASE

Let $X_{(1)} \leq X_{(2)} \leq \cdots \leq X_{(M)}$, $M = [nq]$ be the M smallest order statistics drawn from a distribution having c.d.f $F(x/\delta)$.

We can adapt the method described in Section 2, to obtain an estimator of the scale parameter $\delta$ having the form

$$\tilde{\delta}_n = a_1 X_N^* + a_2 \tilde{X}_{N,M}. \tag{4.1}$$

Here the coefficients $a_1$ and $a_2$ will be chosen in such a way that $\tilde{\delta}_n$ will be asymptotically unbiased and will have the smallest variance among all unbiased estimators of the above form.

We may also consider another estimator of $\delta$ having the form

$$\hat{\delta}_n = c_1 X_M^* + c_2 X_{(M)}. \tag{4.2}$$

This estimator is an extreme case of (4.1). Note that $X_M^*$ and $X_{(M)}$ are jointly asymptotically normally distributed:

$$N\left(\delta\begin{bmatrix}\mu_1(q)\\ x_q\end{bmatrix}, \frac{\delta^2}{n}\begin{bmatrix}\sigma_1^2(q)/q & \sigma_{12}(q)\\ \sigma_{12}(q) & q(1-q)/f^2(z_q)\end{bmatrix}\right),$$

where $\mu(\cdot)$ and $\sigma_1^2(\cdot)$ are given by (2.3) and (2.5) respectively, $f(x) = F'(x)$ is the density and $z_q = F^{-1}(q)$. The covariance term $\sigma_{12}(q)$ is given by

$$\sigma_{12}(q) = \left(\int_o^q \frac{w}{f(z_w)}\, dw\right)(1-q)/(qf(z_q)) \qquad (4.3)$$

The above assertion follows from Corollary 4 in Chernoff, et al. [1].

In order for $\tilde{\delta}_n$ to be an asymptotically unbiased estimator of $\delta$ we must have

$$a_1\mu_1(p) + a_2\mu_3(p,q) = 1. \qquad (4.4)$$

As in Section 2, for brevity we let $\mu_1 = \mu_1(p)$, $\mu_3 = \mu_3(p,q)$, $v_1 = \sigma_1^2(p)/p$, $v_3 = \sigma_3^2(p,q)/(q-p)$ and $c = \sigma_{13}(p,q)$. Then the variance of $\tilde{\delta}_n$ for large n is

$$\mathrm{Var}(\tilde{\delta}_n) \cong \frac{\delta^2}{n}\,[a_1^2 v_1 + a_2^2 v_3 + 2a_1a_2c]. \qquad (4.5)$$

The optimal weights $a_1$ and $a_2$ are then obtained by minimizing (4.5) subject to (4.4) and are:

$$a_1 = (1 - a_3\mu_3)/\mu_1, \qquad (4.6)$$

$$a_2 = (\mu_3v_1-\mu_1c)/(\mu_3^2v_1+\mu_1^2v_3-2\mu_1\mu_2c). \qquad (4.7)$$

The weights $c_1$ and $c_2$ in (4.2) are calculated in the same way with $\mu_1 = \mu_1(q)$, $\mu_3 = z_q$, $v_1 = \sigma_1^2(q)/q$, $v_3 = q(1-q)/f^2(z_q)$ and $c = \sigma_{12}(q)$.

## 5. ESTIMATION OF $\delta$ FOR THE ONE-PARAMETER GUMBEL DISTRIBUTION

Consider a Gumbel distribution with the c.d.f.

$$F(x;\delta) \equiv F(x/\delta) = \exp\{-\exp\{-x/\delta\}\}, \quad -\infty < x < \infty. \qquad (5.1)$$

For q = 0.1 (0.1) 0.9 the estimators $\tilde{\delta}_n$ and $\hat{\delta}_n$ were found, and their A.R.E.'s relative to the m.l. estimate of $\delta$ in censored samples were computed. The results are shown in Table 2. The estimator $\tilde{\delta}_n$ is based on the best partition of the censored sample, p*.

Comparing the efficiencies it can be seen that $\tilde{\delta}_n$ is superior to $\hat{\delta}_n$ for all q.

| q | p* | $\tilde{\delta}_n$ | A.R.E (%) | $\hat{\delta}_n$ | A.R.E. (%) |
|---|---|---|---|---|---|
| 0.1 | 0.05 | $-0.58\ X^*_N - 0.23\ \tilde{X}_{N,M}$ | 97 | $-0.95\ X^*_M + 0.12\ X_{(M)}$ | 95 |
| 0.2 | 0.05 | $-0.62\ X^*_N - 0.21\ \tilde{X}_{N,M}$ | 96 | $-1.41\ X^*_M + 0.57\ X_{(M)}$ | 87 |
| 0.3 | 0.10 | $-0.94\ X^*_N + 0.17\ \tilde{X}_{N,M}$ | 90 | $-1.64\ X^*_M + 0.88\ X_{(M)}$ | 79 |
| 0.4 | 0.15 | $-1.10\ X^*_N + 0.44\ \tilde{X}_{N,M}$ | 85 | $-1.67\ X^*_M + 1.03\ X_{(M)}$ | 71 |
| 0.5 | 0.15 | $-1.04\ X^*_N + 0.49\ \tilde{X}_{N,M}$ | 82 | $-1.57\ X^*_M + 1.06\ X_{(M)}$ | 64 |
| 0.6 | 0.20 | $-1.05\ X^*_N + 0.60\ \tilde{X}_{N,M}$ | 80 | $-1.38\ X^*_M + 1.00\ X_{(M)}$ | 57 |
| 0.7 | 0.20 | $-0.95\ X^*_N + 0.59\ \tilde{X}_{N,M}$ | 80 | $-1.14\ X^*_M + 0.88\ X_{(M)}$ | 50 |
| 0.8 | 0.20 | $-0.86\ X^*_N + 0.56\ \tilde{X}_{N,M}$ | 82 | $-0.83\ X^*_M + 0.71\ X_{(M)}$ | 42 |
| 0.9 | 0.20 | $-0.76\ X^*_N + 0.52\ \tilde{X}_{N,M}$ | 85 | $-0.38\ X^*_M + 0.49\ X_{(M)}$ | 32 |

TABLE 2: Estimators $\tilde{\delta}_n$ and $\hat{\delta}_n$ and their efficiencies for the One-parameter Gumbel Distribution and the right censored sample.

## 6. CONCLUDING REMARKS

In this paper only a simple estimation procedures for right censored samples based on two contiguous groups of order statistics were considered. There are estimators, similar in nature, that may be of interest to consider as well. For instance, the estimators may be based on:

(a) More than two blocks (Kubat [7] considered these estimators for complete samples);

(b) Overlapping blocks;

(c) One (two) block(s) and one (two) suitably selected order statistic(s).

These estimators may be more efficient (and far more complicated) then estimators considered here.

For a sample censored from the left, the estimators of $\lambda$ and $\delta$ can be also easily obtained since a left censored sample becomes a right censored if all the observations are multiplied by (-1).

A question arising in practice concerns the determination of a reasonable sample size n for which one can successfully apply an asymptotic method. For the Gumbel distribution random samples were generated for various values of n. It was found that if n is about 60, then $X^*_{(N)}$, $N = [0.2n]$ is very close to being normally distributed. Thus we recommend that sample block means should contain at least 15 observations.

## APPENDIX A: COMPUTATION OF INFORMATION CONTENT IN THE RIGHT CENSORED SAMPLES FOR A REGULAR SCALE-LOCATION DISTRIBUTION.

For large right censored samples $X_{(1)} \leq X_{(2)} \leq \dots \leq X_{(M)}$, $M = [nq]$, $0 < q < 1$, Halperin [5] found the amount of information about unknown parameters contained in these observations. In particular, for a regular scale-location distribution $F((x-\lambda)/\delta)$ the amount of information about the parameters $\lambda$ and $\delta$ contained in a right censored sample is

$$\Delta_c(q) = I_{\lambda\lambda}(q)\ I_{\delta\delta}(q) - [I_{\lambda\delta}(q)]^2, \tag{A.1}$$

where

$$I_{\lambda\lambda}(q) = \frac{1}{\delta^2}\left\{\int_{-\infty}^{z_q}\left[\frac{f'(u)}{f(u)}\right]^2 f(u)du + \frac{1}{1-q}\left[\int_{-\infty}^{z_q} f'(u)du\right]^2\right\} \tag{A.2}$$

$$I_{\delta\delta}(q) = \frac{1}{\delta^2}\left\{\int_{-\infty}^{z_q}\left[1 + \frac{f'(u)}{f(u)}\right]^2 f(u)du + \right.$$

$$\left. + \frac{1}{1-q}\left[\int_{-\infty}^{z_q}(f'(u) + uf'(u))du\right]^2\right\} \tag{A.3}$$

and

$$I_{\lambda\delta}(q) = \frac{1}{\delta^2}\left\{\int_{-\infty}^{z_q}\left[1 + \frac{f'(u)}{f(u)}\right]f'(u)du + \right.$$

$$\left. + \frac{1}{1-q}\left[\int_{-\infty}^{z_q}(f'(u)u + f(u))du\right]\cdot\left[\int_{-\infty}^{z_q}f'(u)du\right]\right\}. \tag{A.4}$$

Here we denote by $f(u)$ the density of the distribution $F(u)$, i.e., $f(u) = F'(u)$ and $z_q$ is the solution of the equation

$$q = \int_{-\infty}^{z_q} f(u)du.$$

Gumbel Distribution

For the extreme value (Gumbel) distribution, with c.d.f.

$$F(x) = \exp\{-\exp\{(x-\lambda)/\delta\}\}, \qquad -\infty < x < \infty$$

we get

$$H_1(q) \equiv \delta^2\, I_{\lambda\lambda}(q) = (\log q)^2 q/(1-q) + q \tag{A.5}$$

| q | $H_1(q)$ | $H_2(q)$ | $H_3(q)$ | $\Delta(q)$ | Eff. |
|---|---|---|---|---|---|
| 0.05 | 0.522 | 0.849 | -0.636 | 0.039 | 0.024 |
| 0.10 | 0.689 | 0.882 | -0.707 | 0.108 | 0.066 |
| 0.15 | 0.785 | 0.883 | -0.709 | 0.191 | 0.116 |
| 0.20 | 0.848 | 0.891 | -0.688 | 0.281 | 0.171 |
| 0.25 | 0.891 | 0.911 | -0.660 | 0.376 | 0.229 |
| 0.30 | 0.921 | 0.943 | -0.628 | 0.473 | 0.288 |
| 0.35 | 0.943 | 0.985 | -0.598 | 0.572 | 0.348 |
| 0.40 | 0.960 | 1.036 | -0.569 | 0.670 | 0.408 |
| 0.45 | 0.972 | 1.094 | -0.543 | 0.768 | 0.467 |
| 0.50 | 0.980 | 1.157 | -0.519 | 0.865 | 0.526 |
| 0.55 | 0.987 | 1.224 | -0.499 | 0.959 | 0.583 |
| 0.60 | 0.991 | 1.293 | -0.481 | 1.051 | 0.639 |
| 0.65 | 0.995 | 1.364 | -0.466 | 1.140 | 0.693 |
| 0.70 | 0.997 | 1.435 | -0.453 | 1.225 | 0.745 |
| 0.75 | 0.998 | 1.507 | -0.443 | 1.308 | 0.795 |
| 0.80 | 0.999 | 1.576 | -0.435 | 1.385 | 0.842 |
| 0.85 | 1.000 | 1.644 | -0.430 | 1.459 | 0.887 |
| 0.90 | 1.000 | 1.708 | -0.426 | 1.527 | 0.928 |
| 0.95 | 1.000 | 1.769 | -0.424 | 1.589 | 0.966 |

TABLE A1: Amount of information in right censored samples for the extreme value (Gumbel) distribution.

$$H_2(q) \equiv \delta^2 I_{\delta\delta}(q) = \int_{-\ell nq}^{\infty} [1 + (1-t)\ell nt]^2 e^{-t} dt +$$

$$+ \frac{1}{1-q} [q + \int_{-\ell nq}^{\infty} (1-t)\ell nt\ e^{-t} dt]^2 \qquad (A.6)$$

and

$$H_3(q) \equiv \delta^2 I_{\lambda\delta}(q) = -q\ell nq - \int_{-\ell nq}^{\infty} (1-t)^2 \ell nt\ e^{-t} dt -$$

$$- \frac{q\ell nq}{1-q} [q + \int_{-\ell nq}^{\infty} (1-t)\ell nt\ e^{-t} dt] \qquad (A.7)$$

The integrals in (A.6) and (A.7) require numerical integration. For q = 0.05 (0.05) 0.95 the values of $H_1(q)$, $H_2(q)$, $H_3(q)$ and $\Delta(q) \equiv H_1(q)\, H_2(q) - H_3^2(q)$ are tabulated in Table A1. The last column in Table A1, labeled "Eff" is defined as the ratio of the asymptotic generalized variances of the m.l. estimators of $\lambda$ and $\delta$ for censored and complete samples, i.e., Eff $=$ $\Delta(q)/\Delta(1)$, where $\Delta(1) = \pi^2/6$.

REFERENCES

[1] Chernoff, H., Gastwirth, J. L. and Johns, M. V. Jr. (1967), "Asymptotic Distribution of Linear Combinations of Order Statistics with Application to Estimation", *Ann. Math. Statist.*, 38, pp. 52-72.

[2] Cramer, H. (1946), *Mathematical Methods of Statistics*, Princeton University Press.

[3] David, H. A. (1981), *Order Statistics*, Second ed., J. Wiley & Sons, New York.

[4] Epstein, B. (1982), "The Statistical Analysis of Incomplete Life Length Data". In J. Tiago de Oliveira and B. Epstein (ed.): *Some Recent Advances in Statistics*, Academic Press, London.

[5] Halperin, M. (1952), "Maximum Likelihood Estimation in Truncated Samples", *Ann. Math. Statist.*, 23, pp. 226-238.

[6] Johnson, L. N. and Kotz, S. (1970), *Distributions in Statistics: Continuous Univariate Distribution - 1 & 2*, Houghton Mifflin Comp. Boston.

[7] Kubat, P. (1982), "Simple Large Sample Estimators of Scale and Location Parameters Based on Blocks of Order Statistics", *Trab. Estad. Y. Inv. Oper.*, 33, pp. 86-118.

[8] Sarhan, A. E. and Greenberg, B. G. (1962), *Contributions to Order Statistics*, J. Wiley & Sons, New York.

# ASYMPTOTIC BEHAVIOUR OF THE EXTREME ORDER STATISTICS IN THE NON IDENTICALLY DISTRIBUTED CASE

D. Mejzler

The Hebrew University, Jerusalem

## 1. INTRODUCTION

Let $X_1,\cdots,X_n$ be independent and identically distributed random variables and let $X_{1n} \leqslant \cdots \leqslant X_{nn}$ be the corresponding order statistics. For every $k$ $(1 \leqslant k \leqslant n)$ let $F_{kn}$ denote the distribution function (df) of $X_{kn}$.

Much work has been done on the asymptotic behaviour of the normalized statistics $X_{kn}$ as $n \to \infty$. The study of the maximal term $X_{nn}$ was started by M. Frechet [3] and definite results were obtained by B.V. Gnedenko [4]. This class of the well known Gnedenko's extreme value limit laws will be denoted by $A$.

N.V. Smirnov [13] considered the term $X_{kn}$, where $k = k(n)$ is a function of $n$. He investigated the possible proper limits of a sequence of the form $\{F_{kn}(a_n x + b_n)\}$, under the condition that

$$k/n \to \lambda \tag{1.1}$$

and $a_n > 0$, $b_n$ are numberical sequences. A sequence of terms $X_{kn}$ is called a sequence of *central* terms, if $0 < \lambda < 1$, and *extreme* terms if $\lambda = 0$ or $\lambda = 1$, respectively. With an additional restriction on the sequence $\{k(n)\}$ Smirnov found the class of all possible limits for the central terms, but his investigation of the extreme terms was restricted to the cases, when $k = \text{Const}(\lambda = 0)$, or, what is essentially the same, when $p = n-k = \text{Const}(\lambda = 1)$. This class of df's is denoted here by $A_p$.

The study of the limits for the extreme terms in the remaining case, when



*J. Tiago de Oliveira (ed.), Statistical Extremes and Applications, 535–547.*

$$k \to \infty \text{ and } p = n-k \to \infty \tag{1.2}$$

was initiated by D.M. Chibisov [2], under rather strong restrictions on the sequence $\{k(n)\}$. Chibisov's class was obtained in [11] under less restrictive conditions. But the best results in this direction were given by A.A. Balkema and L. de Haan [1]. They developed a general theory of limit laws for order statistics in case (1.2) and they got the classes of Smirnov and Chibisov under the more general condition concerning the $\{k(n)\}$:

$$k(n+1) - k(n) = o(\min\{\sqrt{k(n)}, \sqrt{n-k(n)}\}).$$

Moreover, they proved that if apart of (1.1) no other restriction is imposed on $k(n)$, then for a given $\lambda \in [0,1]$ any df may be considered as a limit law for some $X_{kn}$.

Now, let us remove the restriction that the initial random variables $X_1,\cdots,X_n$ are identically distributed. Let $F_i$ be the df of $X_i$. For every df $H$ we introduce the notations

$$\underline{H} = \inf\{x:H(x) > 0\}, \ \overline{H} = \sup\{x:H(x) < 1\}. \tag{1.3}$$

A non trivial extension of the class $A(A_p)$ is the class $G(G_p)$ of all proper limits $H$ of a sequence $F_{kn}(a_n x+b_n)$ with $k=n$ ($p=n-k=$ Const) under the additional requirement that for every $x > \underline{H}$

$$\min_{1\leqslant i\leqslant n} F_i(a_n x+b_n) \to 1.$$

The classes $G$ and $G_p$ were studied in [6] - [10] and [12]. It seems to be interesting to consider a similar extension concerning the extreme terms in case (1.2).

We consider here the particular case, when $a_n \equiv 1$. Keeping the previous notations we denote by $S^*$ the class of all proper df's $H$, which have the following property:

There exist a sequence of independent random variables $X_n$ with corresponding df's $F_n$, a numberical sequence $b_n$ and positive integers $k(n)$ such that we have (1.2), where

$$p/n \to 0 \tag{1.4}$$

$$\sqrt{p(n+1)} - \sqrt{p(n)} \to 0 \tag{1.5}$$

$$F_{kn}(x+b_n) \to H(x) \tag{1.6}$$

and for every $x > \underline{H}$

$$\min_{1\leqslant i\leqslant n} F_i(x+b_n) \to 1. \tag{1.7}$$

Let us point out that here and in the sequel convergence of monotone functions means weak convergence, i.e. convergence at each point of continuity of the limit function.

We prove the following

Theorem 1.1. Let $H$ be a proper df of the form

$$H = N(u), \tag{1.8}$$

where $N$ is the standard normal df and $u$ is a concave function in $(\underline{H},\infty)$. Then
(A) $H$ belongs to $S^*$.
(B) The sequences $\{b_n, p(n)\}$, which appear in the definition of the class $S^*$, can be chosen to satisfy the following relations: If for some integer-valued function $m = m(n)$ we have $b_{n+m}-b_n \to \beta$, where $0 < \beta < \infty$, then also $p(n+m)/p(n) \to 1$.
(C) If $H$ belongs to $S^*$ and it can be obtained by sequences $\{b_n, p(n)\}$, which are related as in (B), then $H$ must be of the form (1.8).

## 2. AN AUXILIARY THEOREM

Let $X_1,\cdots,X_n$ be independent random variables and let $F_i$ be the df of $X_i$. Then

$$F_{kn} = \left\{1+\sum_{s=1}^{n-k}\sum{}^* \prod_{i\in(n,s)} (1/F_i-1)\right\}\prod_{i=1}^{n} F_i, \tag{2.1}$$

where $(n,s)$ is any subset of the set $(1,\cdots,n)$, consisting of $s$ indices and the summation $\sum^*$ is over all such subsets. This expression is quite complicated and we prove here a proposition, which may simplify the investigation of $X_{kn}$. For simplicity we will write $\sum$ and $\prod$ instead $\sum_{i=1}^{n}$ and $\prod_{i=1}^{n}$. Put

$$u_n = \sqrt{n}(\sum F_i-k)/[k(n-k)]^{\frac{1}{2}}. \tag{2.2}$$

Theorem 2.1. (cf. [13], Theorem 4). Let $H$ be a proper df (i.e. $\underline{H} < \overline{H}$). Let (1.2) hold and assume, for every $x \in (\underline{H},\overline{H})$,

$$F_i(a_nx+b_n) \to \lambda = \text{Const} \tag{2.3}$$

uniformly in $i$ $(1\leqslant i\leqslant n)$. Then relation (1.6) holds if and only if

$$u_n(a_nx+b_n) \to u(x) \tag{2.4}$$

and the non decreasing function $u$ is uniquely determined from $H$ by the equation

$$H = N(u), \tag{2.5}$$

where $N$ is the standard normal df. (By the premises of the theorem relation (1.1) follows from (1.6) or from (2.4) and (2.5)).

Proof. We shall first prove that if (1.6) or if (2.4) and (2.5) hold, then for every $x \in (\underline{H},\overline{H})$

$$V_n(x) = \sum F_i(a_nx+b_n)[1-F_i(a_nx+b_n)] \to \infty. \tag{2.6}$$

If $0 < \lambda < 1$, then (2.6) follows immediately from (2.3) since

$$V_n(x)/[n\lambda(1-\lambda)] \to 1. \tag{2.7}$$

Let $\lambda = 1$, then (2.3) takes the form (1.7). We shall show that

$$\prod F_i(a_nx+b_n) \to 0 \tag{2.8}$$

for $x < \overline{H}$. Obviously, for $x < \overline{H}$

$$\limsup \prod F_i(a_nx+b_n) < 1.$$

Let $\xi \in (\underline{H},\overline{H})$ be any continuity point of $H$ and assume that for some subsequence $n'$

$$\prod_{i=1}^{n'} F_i(a_{n'}\xi+b_{n'}) \to e^{-a},$$

where $a > 0$. Then it can be shown ([12], Lemma 4.1) that by (2.3), for any positive integer $s$

$$\sum{}^{*}\prod_{i\in(n',s)}[1/F_i(a_{n'}\xi+b_{n'})-1] \to a^s/s!$$

Therefore, by (2.1), for any natural $m$

$$F_{n'-m,n'}(a_{n'}\xi+b_{n'}) \to e^{-a}\sum_{s=0}^{m} a^s/s!$$

Taking $m$ large enough we shall have

$$\lim F_{n'-m,n'}(a_{n'}\xi+b_{n'}) > H(\xi), \tag{2.9}$$

since $H(\xi) < 1$. On the other hand, by condition (1.2) we have $n-k>m$ from some $n$ on. Hence, again by (2.1), from some $n$ on

$$F_{n-m,n}(a_nx+b_n) \leqslant F_{kn}(a_nx+b_n)$$

and we get

$$\limsup F_{n-m,n}(a_n\xi+b_n) \leqslant H(\xi),$$

which contradicts (2.9). This proves (2.8) and we conclude ([12], Lemma 4.2) that for $x < \overline{H}$

$$\sum[1-F_i(a_nx+b_n)] \to \infty. \tag{2.10}$$

Hence, by (2.3) we get (2.6), since

$$1 \geqslant V_n(x)/\sum[1-F_i(a_nx+b_n)] \geqslant \min_{1\leqslant i\leqslant n} F_i(a_nx+b_n) \to 1 \qquad (2.11)$$

for every $x \in (\underline{H},\overline{H})$.

Conversely, let us assume now that (2.4) and (2.5) hold. The expression (2.2) can be rewritten as

$$u_n = \sqrt{n/k}\left\{\sum[F_i-1]/(n-k)^{\frac{1}{2}}+(n-k)^{\frac{1}{2}}\right\}.$$

Hence, by (1.2) we get (2.10) and, consequently, (2.6).

In a similar way we deal with the case $\lambda = 0$. We consider the function $G_{kn} = 1-F_{kn}$ and prove that for $x > \underline{H}$ we have $\prod[1-F_i(a_nx+b_n)] \to 0$.

Now, for any $x \in (\underline{H},\overline{H})$ consider a system of random variables $\xi_{in} = \xi_{in}(x)$, $(1\leqslant i\leqslant n,\ n = 1,2,\cdots)$ defined by

$$P\{\xi_{in}=1\} = F_i(a_nx+b_n),\ P\{\xi_{in}=0\} = 1-F_i(a_nx+b_n).$$

Denote $s_n = \sum\xi_{in}$. Then in view of (2.1)

$$F_{kn}(a_nx+b_n) = P\{s_n \geqslant k\}. \qquad (2.12)$$

We have $\mathrm{Var}(s_n) = V_n(x)$, where $V_n(x)$ is given by (2.6). Thus the normalized system

$$\overline{\xi}_{in} = [\xi_{in}-F_i(a_nx+b_n)]/\sqrt{\mathrm{Var}(s_n)}$$

satisfies the Lindeberg condition and by (2.12) we get

$$F_{kn}(a_nx+b_n)-N(h_n(a_nx+b_n)) \to 0,$$

where $h_n = [\sum F_i-k]/[\sum F_i(1-F_i)]^{\frac{1}{2}}$. Hence, relation (1.6) holds if and only if

$$h_n(a_nx+b_n) \to u(x),$$

where $u$ satisfies equation (2.5).

If $0 < \lambda < 1$, then in view of (2.7) $h_n$ can be replaced by

$$u_n = [\sum F_i-k]/[n\lambda(1-\lambda)]^{\frac{1}{2}}. \qquad (2.13.1)$$

If $\lambda = 1$, then by (2.11), putting $p(n)$ instead of $(n-k)$, we can replace $h_n$ by

$$u_n = \sum[F_i-1]/\sqrt{p(n)}+\sqrt{p(n)}. \qquad (2.13.2)$$

Finally, if $\lambda = 0$, then in a similar way we get that $h_n$ can be replaced by

$$u_n = \sum F_i/\sqrt{k}+\sqrt{k}. \tag{2.13.3}$$

It is easy to verify that all the particular expressions (2.13.1) - (2.13.3) are included in the general form (2.2).

## 3. REMARKS

3.1. We will consider in the sequel only the case $\lambda = 1$ and we will use the form (2.13.2) of $u_n$. By equation (2.6) the limit function $u$ in (2.4) is a non-decreasing function, which assumes finite or infinite values in $(-\infty,\infty)$ and satisfies the conditions $u(-\infty) = -\infty, u(+\infty) = +\infty$. Such a function will be called an *s*-function (sf). For every sf $u$ let us denote

$$\underline{u} = \inf\{x:u(x)>-\infty\}, \ \overline{u} = \sup\{x:u(x)<\infty\}.$$

An sf will be called proper if $\underline{u} < \overline{u}$. It follows from (2.5) that $H$ is a proper df if and only if $u$ is a proper sf, since $\underline{u} = \underline{H}$, $\overline{u} = \overline{H}$ ($\underline{H},\overline{H}$ were defined by (1.3)).

3.2. The notion of type of sf's can be introduced in the same way, as it was done in the class of df's. The theorems of Khintchine and Gnedenko [5, §10, Theorems 1-2] about sequences of df's, which converge to a proper type, can be extended to sequences of non decreasing functions (not necessarily sf's), which converge to a proper sf [11, Theorems 2.1-2.2]. It follows from (2.13.2) that

$$u_{n+1}(x) = A_n u_n(x) + B_n(x),$$

where, by (1.5), $A_n \to 1$ and $B_n(x) \to 0$. Therefore, if relations (1.2), (1.5) and (2.4) hold, then

$$u_{n+1}(a_n x+b_n) \to u(x).$$

Hence, by the above mentioned theorems we conclude that if, in addition, $u(x)$ is a proper sf, then necessarily

$$a_{n+1}/a_n \to 1, (b_{n+1}-b_n)/a_n \to 0.$$

3.3. In contrast to the df $F_{kn}$ the expression (2.13.2) makes sense for $p(n)$, which are not necessarily integers. The limits of the sequence $\{u_n(a_n x+b_n)\}$ do not change if $p(n)$ is replaced by any function $g(n)$ such that

$$\sqrt{p(n)} - \sqrt{g(n)} \to 0.$$

Obviously, each such $g(n)$ will satisfy, together with $p(n)$, the conditions (1.4) and (1.5).

3.4. Let $c$ be any constant. Put

$$\tilde{p}(n) = p(n) + c\sqrt{p(n)},\ \tilde{u}_n = \sum[F_i - 1]/\sqrt{\tilde{p}(n)} + \sqrt{\tilde{p}(n)}.$$

If for some $\{a_n, b_n\}$ we have (2.4), then

$$\tilde{u}_n(a_n x + b_n) \to u(x) + 2c.$$

3.5. Let $\{F_n\}$ be any sequence of df's, $\{a_n, b_n\}$-numerical sequences. If for some $x$

$$a_{n+1}x + b_{n+1} \geqslant a_n x + b_n,\ F_n(a_n x + b_n) \to 1$$

and for each $i$ constant $F_i(a_n x + b_n) \to 1$, then

$$\min_{1 \leqslant i \leqslant n} F_i(a_n x + b_n) \to 1.$$

## 4. PROOF OF THEOREM 1.1.

(A) Let $u$ be a proper sf, which is concave in $(\underline{u}, \infty)$. By Theorem 2.1 it is enough to show that there exists a triple $\{F_n, p(n), b_n\}$ such that we have (1.2), (1.4), (1.5), (1.7) and

$$u_n(x + b_n) \to u(x), \tag{4.1}$$

where $u_n$ is given by (2.13.2). Since $u$ is concave, it increases continuously in $(\underline{u}, \infty)$ and has there non-increasing left- and right-hand derivatives. These properties of $u$ enable us to construct effectively the desired triple $\{F_n, p(n), b_n\}$. By (2.13.2) and (1.7) a finite number of terms of this triple may be chosen arbitrarily. Therefore, some relations, which hold from some $n$ on will be treated as valid for every $n$.

We shall consider four cases depending on the behaviour of the sf $u$ on the right of $\underline{u}$. For the sake of brevity we mark now those properties and designations, which will hold in each of the considered cases. The numerical sequences will always satisfy

$$p(n+1)/p(n) < p(n)/n \to 0 \text{ and } 0 < [p(n+1) - p(n)] \to 0 \tag{4.2}$$

$$b_n \to \infty \text{ and } 0 < (b_{n+1} - b_n) \to 0. \tag{4.3}$$

We introduce auxiliary numerical sequences defined by

$$c_n = -n/\sqrt{p(n)} + \sqrt{p(n)},\ u(d_n) = \sqrt{p(n)}. \tag{4.4}$$

By (4.2) - (4.3) we have

$$c_{n+1} < c_n \to -\infty, d_{n+1} > d_n \to +\infty \tag{4.5}$$

The df's $F_n$ are defined by

$$F_n = 1-p(n)+p(n-1) + \sqrt{p(n)}u_n - \sqrt{p(n-1)}u_{n-1} \quad (p(0) = 0), \tag{4.6}$$

where the $u_n$ are non-decreasing functions continuous for $x \neq 0$, which will be defined later according to the considered case. However, we will always have

$$u_n(x) = c_n \text{ if } x < 0; u_n(x) = \sqrt{p(n)} \text{ if } x > b_n + d_n. \tag{4.7}$$

(In the sequel we will give the determination of $u_n$ for $0 < x < b_n + d_n$ only). By (4.4) - (4.7) we conclude that

$$F_n(x) = 0 \text{ if } x < 0; \ F_n(x) = 1 \text{ if } x > b_n + d_n$$

and $\sum_{i=1}^{n} F_i = n-p(n) + \sqrt{p(n)}u_n$, i.e. $u_n$ has the form (2.13.2). Moreover, it will be shown that in each of the cases considered below

$$F_n(0+) \to 1. \tag{4.8}$$

Hence we will conclude the validity of (1.7) and that $F_n(0+) \geqslant 0$ for every $n$.

I. *Let* $\underline{u} = -\infty$. We define

$$u(-b_n) = -lg\ n \tag{4.9}$$

$$p(n) = b_n \tag{4.10}$$

$$u_n(x) = u(x-b_n), \quad 0 < x < b_n + d_n. \tag{4.11}$$

Since $u$ is concave, there exists a constant $c > 0$ such that

$$b_n < c\ lg\ n, \ b_{n+1} - b_n < c\ lg\frac{n+1}{n} < \frac{c}{n},$$

and the sequences $\{p(n), b_n\}$ satisfy the conditions (4.2) - (4.3).

In order to show that $F_n$ is a df, we examine its behaviour in the intervals $(0, b_{n-1}+d_{n-1})$ and $(b_{n-1}+d_{n-1}, b_n+d_n)$. The expression of $F_n$ for $0 < x < b_{n-1} + d_{n-1}$ can be presented in the form

$$\begin{aligned} F_n(x) = {} & 1-p(n)+p(n-1)+(\sqrt{p(n)} - \sqrt{(p(n-1)})u(x-b_n) + \\ & +\sqrt{p(n)}[u(x-b_n) - u(x-b_{n-1})]. \end{aligned} \tag{4.12}$$

Since $u$ is concave, then in view of (4.3) the difference

$u(x-b_n) - u(x-b_{n-1})$ does not decrease and by (4.2) we get that $F_n$ does not decrease in the considered interval. The monotonicity of $F_n$ in the next interval is obvious. The functions $u_n$ and, consequently, $F_n$ are continuous for $x \neq 0$. By (4.6), (4.9) - (4.11) and simple calculations we get that $F_n(0+)$ has the form

$$F_n(0+) = 1 - \varepsilon_n \tag{4.13}$$

where $0 < \varepsilon_n \to 0$. Thus we have (4.8) and $F_n$ is a df. Now, for any $x$ we have from some $n$ on $-b_n < x < d_n$. Hence by (4.11), from some $n$ on

$$u_n(x + b_n) = u(x). \tag{4.14}$$

Thus condition (4.1) is satisfied too.

II. *Let* $\underline{u} > -\infty$. In this case we define

$$p(n) = \sqrt{(n)},\ b_n = \sqrt[4]{p(n)} = \sqrt[8]{n}. \tag{4.15}$$

We may assume that $\underline{u} = 0$. Moreover, if in addition, $u(0+) > -\infty$, then by Remark 3.4, we can assume that $u(0+) = 0$.

II 1). *Let* $\underline{u} = 0$, $u(0+) = 0$, $u'(0) < \infty$.

($u'$ and $F'$ denote right-hand derivatives.)

We define the function $u_n$ by

$$u_n(x) = \begin{cases} b_n(x-b_n) & \text{if } 0 < x < b_n \\ u(x-b_n) & \text{if } b_n < x < b_n + d_n. \end{cases} \tag{4.16}$$

By the definition of the sequences $\{b_n\}$ and $\{d_n\}$ the "critical points" of $F_n$ are ordered as follows:

$$0 < b_{n-1} < b_n < b_{n-1}+d_{n-1} < b_n + d_n.$$

The expression of $F_n$ for $b_{n-1} < x < b_n$ is

$$F_n(x) = 1-2p(n)+p(n-1) + \sqrt{p(n)}x - \sqrt{p(n-1)}u(x-b_{n-1}).$$

Hence $F_n'(x) = \sqrt{p(n)}b_n - \sqrt{p(n-1)}u'(x-b_{n-1})$ and we conclude that $F_n$ increases in this interval, since $u'(x-b_{n-1}) \leq u'(0+)$ for $x > b_{n-1}$ and $u'(0+) < \infty$. In the interval $(b_{n-1}+d_{n-1}, b_n+d_n)$ $F_n$ has the form (4.12). The monotonicity of $F_n$ in the other intervals is obvious.

We have again (4.13), which implies (4.8). By (4.16) if

$x > 0$, then from some $n$ on we get (4.14), while if $x < 0$, then $u_n(x+b_n) \to -\infty$. Thus (4.1) holds.

II 2). *Let* $\underline{u} = 0, u(0+) = 0, u'(0+) = \infty$.

Let $\phi(x)$ be a continuous and non-increasing function in $(0,\infty)$, satisfying the inequality $\phi(x) \geqslant u'(x)$. Then the function

$$f(x) = \phi(x)/u(x) \tag{4.17}$$

is positive, decreases in $(0,\infty)$ and $f(0+) = \infty$. We define a numerical sequence $\{\beta_n\}$ by

$$f(\beta_n) = b_n = \sqrt[8]{n}. \tag{4.18}$$

Obviously, $\beta_{n-1} > \beta_n \to 0$ and we can assume that $u(\beta_n) < 1$. Let

$$\Delta_n = f(\beta_n) - f(\beta_{n-1}). \tag{4.19}$$

By (4.18) we have the estimate

$$1/8n^{7/8} < \Delta_n < 1/8(n-1)^{7/8}. \tag{4.20}$$

On the other hand, by (4.17) - (4.19) we can rewrite $\Delta_n$ in the form

$$\Delta_n = \frac{\phi(\beta_n)}{u(\beta_n)} \frac{u(\beta_{n-1}) - u(\beta_n)}{u(\beta_{n-1})} + \frac{\phi(\beta_n) - \phi(\beta_{n-1})}{u(\beta_{n-1})},$$

where each of the above summands is non-negative. Hence, by (4.20) we get

$$0 < u(\beta_{n-1}) - u(\beta_n) < 1/8(n-1) \tag{4.21}$$

$$0 < \phi(\beta_n) - \phi(\beta_{n-1}) < 1/8(n-1)^{7/8}. \tag{4.22}$$

We have $\beta_{n-1} - \beta_n < u(\beta_{n-1}) - u(\beta_n)$, since in the present case $u'(0+) = \infty$. Therefore by (4.19) - (4.21)

$$\beta_{n-1} - \beta_n < b_n - b_{n-1}. \tag{4.23}$$

Now we define the functions $u_n(x)$ by

$$u_n(x) = \begin{cases} u(\beta_n) + \phi(\beta_n)(x - \beta_n - b_n) & \text{if } 0 < x < b_n + \beta_n \\ \\ u(x - b_n) & \text{if } b_n + \beta_n < x < b_n + d_n. \end{cases} \tag{4.24}$$

In view of (4.23) we have

$$0 < b_{n-1} + \beta_{n-1} < b_n + \beta_n < b_{n-1} + d_{n-1} < b_n + d_n. \tag{4.25}$$

The monotonicity of $F_n$ is proved as in the previous cases and it remains to show the validity of (4.8). The expression of $F_n(0+)$ is now somewhat more complicated and it can be written in the form

$$F_n(0+) = 1-p(n) + p(n-1) + \ell_1(n) + \ell_2(n) + \ell_3(n) \tag{4.26}$$

where

$$\ell_1(n) = \sqrt{p(n)}u(\beta_n) - \sqrt{p(n-1)}u(\beta_{n-1})$$

$$\ell_2(n) = \sqrt{p(n-1)}\phi(\beta_{n-1})b_{n-1} - \sqrt{p(n)}\phi(\beta_n)b_n$$

$$\ell_3 = \sqrt{p(n-1)}\phi(\beta_{n-1})\beta_{n-1} - \sqrt{p(n)}\phi(\beta_n)\beta_n.$$

It follows from (4.17) - (4.18) that $\phi(\beta_n) < \sqrt[8]{n}$. Hence, using the estimate (4.22) we conclude that $\ell_i(n) \to 0$ $(i = 1,2,3)$. If $x > 0$ then we get (4.14) from (4.24). For $x < 0$ we get $u_n(x+b_n) \to -\infty$, since $\phi(\beta_n) \to \infty$, $u(\beta_n) \to 0$.

II 3). *Let* $\underline{u} = 0$, $u(0+) = -\infty$, $u'(0+) = +\infty$

By Remark 3.4 we can assume that $u(1) = 0$ and, therefore, $u(x) < 0$ for $0 < x < 1$. We take a function $\phi(x)$ as in the previous case and we use the function

$$f(x) = -u(x)\phi(x), \tag{4.27}$$

which is positive and decreases in (0,1) and $f(0+) = \infty$. We define a sequence $\{\beta_n\}$ by the equation (4.18), where now $f(x)$ is the function (4.27). The function $u_n(x)$ is defined again by (4.24). In the present case we can assume that $-u(\beta_n) < \sqrt[8]{n}$ and $\phi(\beta_n) < \sqrt[8]{n}$. The arguments are similar to those of the previous case. We have again the estimate (4.20) for the difference $\Delta_n$, which can be now presented in the form

$$\Delta_n = \phi(\beta_n)[u(\beta_{n-1}) - u(\beta_n)] - u(\beta_{n-1})[\phi(\beta_n) - \phi(\beta_{n-1})].$$

Thus we get (4.21) - (4.23) and (4.25) - (4.26).

(B) The proof of this part of our theorem follows immediately from (4.10) and (4.15).

(C) Let $H$ be a proper df of the class $S^*$. Then by definition we have (1.2), (1.4) - (1.7). Hence, by Theorem 2.1, $H$ must be of the form (2.5), where $u(x)$ is a proper sf given by (4.1). Thus it remains to show that $u$ is concave in $(\underline{u},\infty)$.

It is easy to see that under our conditions

$$b_n \to \infty. \tag{4.28}$$

Indeed, let us assume that for some subsequence $n'$ we have $b_{n'} \to b$. By (1.7) $b > -\infty$. If $|b| < \infty$, then by the theorems, mentioned in Remark 3.2, we have also $u_{n'}(x+b) \to u(x)$ and for every $i$ and $x > \underline{u}$ we have $F_i(x+b) = 1$. Hence we conclude that $u_{n'}(x+b) = \sqrt{p(n')}$ and, therefore, $u(x) = \infty$ for $x > \underline{u}$. This is impossible, since $u$ is a proper sf.

Now, by (1.5) and Remark 3.2

$$b_{n+1} - b_n \to 0. \tag{4.29}$$

It follows from (4.28) - (4.29) that for every $\beta > 0$ there exists an integer-valued $m = m(\beta,n)$ such that

$$b_{n+m} - b_n \to \beta. \tag{4.30}$$

The subsequence $u_{n+m}(x+b_{n+m})$ can be presented in the form

$$u_{n+m}(x+b_{n+m}) = [p(n)/p(n+m)]^{\frac{1}{2}} u_n(x+b_{n+m}) + \psi_{n,m}(x), \tag{4.31}$$

where

$$\psi_{n,m}(x) = [\sum_{i=n+1}^{n+m} F_i(x+b_{n+m}) + p(n+m) - p(n) - m]/\sqrt{p(n+m)}.$$

By our assumption about the relationship between the sequences $\{b_n\}$ and $\{p_n\}$ we get that $p(n)/p(n+m) \to 1$. Thus by (4.30), when $n \to \infty$, we conclude from (4.31) that

$$u(x) = u(x+\beta) + \psi(x),$$

where $\psi(x)$ is the limit of the sequence $\psi_{n,m}(x)$. Obviously, $\psi(x)$ is a non decreasing function. Thus we proved that the sf $u$ must have the following property: for every $\beta > 0$ the difference $u(x) - u(x+\beta)$ does not decrease in $(\underline{u},\infty)$. Hence, $u$ is concave in $(\underline{u},\infty)$.

## REFERENCES

1. Balkema, A.A. and De Haan, L. (1978). "Limit distributions for order statistics, I and II." Theory Prob. and its Applications V.23, N. 1, pp.80-96 and N.2, pp. 358-375.

2. Chibisov, D.M. (1964). "On limit distributions for order statistics." Theory Prob. and its Applications, V.9, pp.142-148

3. Fréchet, M. (1927). "Sur la loi de probabilité de l'écart maximum." Ann. Soc. Polon. Math. Cracovie, V. 6, pp.93-116.

4. Gnedenko, B.V. (1943). Sur la distribution limit du maximum d'une série aléatoire." Ann. Math. V. 44, pp.423-453.

5. Gnedenko, B.V. and Kolmogorov, A.N. (1954). "Limit distributions for sums of independent random variables." Addison-Wesley, Cambridge, Mass.

6. Juncosa, M.L. (1949). "The asymptotic behaviour of the minimum in a sequence of random variables." Duke Math. J., V. 16, N. 4. pp.609-618.

7. Mejzler, D. (1949). "On a problem of B.V. Gnedenko." Ukrain Mat. Ž., N. 2, pp. 67-84 (Russian).

8. Mejzler, D. (1950). "On the limit distribution of the maximal term of a variational series." Dopovidi Akad. Nauk Ukrain. SSR, N. 1, pp. 3-10 (Ukrainian, Russian Summary).

9. Mejzler, D. (1953). "The study of the limit laws for the variational series." Trudy Inst. Mat. Meh. Akad. Nauk Uzbek. SSR., N. 10, pp. 96-105. (Russian).

10. Mejzler, D. (1956). "On the problem of the limit distributions for the maximal term of a variational series." Lvov. Politechn. Inst. Naučn. Zap., Ser. Fiz-Mat., V. 38, pp. 90-109 (Russian).

11. Mejzler, D. (1978). "Limit distributions for the extreme order statistics." Canad. Math. Bull., V. 21, N. 4, pp. 447-459.

12. Mejzler, D. and Weissman, I. (1969). "On some results of N.V. Smirnov concerning limit distributions for variational series." Ann. Math. Statist., V. 40, N. 2., pp. 480-491.

13. Smirnov, N.V. (1949). "Limit distributions for the terms of a variational series." Trudy Mat. Inst. Steklova, V. 25 (Russian). Amer. Math. Soc. Translations, V. 11(1962), pp. 82-143.

# LIMIT DISTRIBUTION OF THE MINIMUM DISTANCE BETWEEN INDEPENDENT AND IDENTICALLY DISTRIBUTED d-DIMENSIONAL RANDOM VARIABLES

Takuji Onoyama, Masaaki Sibuya and Hiroshi Tanaka

Department of Mathematics
Keio University
3-14-1 Hiyoshi, Kohoku-ku, Yokohama, Japan 223

ABSTRACT. For a sequence $X_1, X_2, \ldots$ of i.i.d. d-dimensional random variables let $Y_n$ denote the minimum Euclidean distance among the first n variables. It is shown that if the probability density function f of $X_i$ belongs to $L^2(R^d)$ then the limit distribution of $n^2 Y_n^d$ is an exponential distribution whose intensity is proportional to the square of the $L^2$-norm $\|f\|$. The limit joint distribution of the smallest k distances and the midpoints of the k nearest pairs of variables is also obtained.

## 1. INTRODUCTION

The theory of statistical extremes was developed first and naturally in the case of independent and identically distributed (i.i.d. for short) random variables. Later it was extended to the case of exchangeable random variables. Galambos (1978) expounded the main streams of research and referred briefly to another line of studies on extremes of dependent variables, that is, extremes among random spacings by i.i.d. random variables.

In the classical case of one-dimensional uniform distribution the smallest and the largest spacings in random partition of the distribution interval are well studied. See Darling (1953) and Feller (1971). The limit distribution of the smallest spacing normalized by $n^2$ is the exponential distribution. It was extended to the case of histogram distributions by Weiss (1959 and 1969). Motivated by studies of the spectrum of random Jacobi matrices which appear in solid physics, Molchanov and Reznikova (1982) obtained the limit joint distribution of the k smallest spacings assuming that the probability density function belongs to $L^2$. They also obtained the limit joint distribution of the k largest spacings under stronger conditions.



J. Tiago de Oliveira (ed.), Statistical Extremes and Applications, 549–562.

In this paper, Molchanov and Reznikova's result is extended to the d-dimensional case. They discussed actually the minimum distance between random variables since contribution from the spacings at the both ends are asymptotically negligible. Let $X_1, X_2, \ldots$ be a sequence of i.i.d. $R^d$-valued random variables having a probability density function $f \in L^2 = L^2(R^d)$. Let $Y_n$ be the minimum Euclidean distance among the first n random variables, namely,

$$Y_n = \min_{1 \le i < j \le n} |X_i - X_j| \tag{1.1}$$

Theorem 1 of Section 3 states that the limit distribution of $n^2 Y_n^d$ is an exponential distribution with intensity parameter proportional to $\|f\|^2$. In other words, the limit distribution of $n^{2/d} Y_n$ is the Weibull distribution with power parameter (shape parameter) 1/d and scale parameter proportional to $\|f\|^{-2/d}$. Further, Theorem 2 of Section 4 gives the limit joint distribution of the k smallest distances and the midpoints of the nearest k pairs of random variables; $Y_n^d$ and the differences of the k smallest distances to the power d and the midpoints are asymptotically independent. In the final Section 5 another proof of Theorem 1 is given. This is more related to the proof of Theorem 2, while the first proof in Section 3 is more elemental.

## 2. PRELIMINARIES

The probability of an event $\{\min|X_i - X_j| < \varepsilon\}$ for some $\varepsilon > 0$ will be evaluated as union of events $\{|X_i - X_j| < \varepsilon\}$ using Bonferroni's inequalities: For any events $E_1, E_2, \ldots, E_n$,

$$P\{\bigcup_{i=1}^{n} E_i\} \begin{matrix} \le \\ = \\ \ge \end{matrix} \sum_{k=1}^{m} (-1)^k \sum_{1 \le i_1 < \ldots < i_k \le n} P\{E_{i_1} \cap E_{i_2} \cap \ldots \cap E_{i_k}\}, \tag{2.1}$$

according as

$$m \begin{cases} <n \text{ and odd}, \\ =n, \\ <n \text{ and even}. \end{cases} \tag{2.2}$$

The equality is the inclusion-exclusion formula.

Let $\varepsilon$ be an arbitrarily fixed positive constant. To describe which of the events

$$A_{ij} = \{|X_i - X_j| < \varepsilon\} \tag{2.3}$$

occur, let $G_n$ denote the set of all ordered pairs $(i,j)$, $1 \le i < j \le n$. To each G of $G_n$, there corresponds a graph $g = \gamma(G)$ which has n vertexes labeled $1,2,\ldots,n$, and undirected edges connecting vertexes i and j if $(i,j)\in G$. The mapping $\gamma$ gives a 1-1 correspondence between $G_n$ and the set of all undirected graphs with labeled n vertexes. Let $V_g$ denote the set of unisolated vertexes in g and define

$$v(g) = \# V_g = \text{the number of unisolated vertexes in } g,$$
$$e(g) = \text{the number of edges of } g.$$

If $v(g) = 2e(g)$, then g consists of isolated vertexes and isolated edges. Such a graph is said to be pairwise connected in this paper. Another type of graph, which has no closed loop, is said to be simple. A simple connected graph is a tree. It is often convenient to take any of its vertexes as a root and regard the connection by an edge as parent-child relation. A single isolated vertex is a tree by this definition, but such a trivial tree is excluded in the following discussions. Thus if g is a tree, then $v(g) = e(g)+1$. If a simple graph g consists of $\alpha$ tree-subgraphs and possibly of some isolated vertexes, then $v(g) = e(g)+\alpha$.

Using (2.1) and (2.3) with $\varepsilon = (n^{-2}t)^{1/d}$,

$$P\{Y_n < t\} = P\{\bigcup_{(i,j)\in G_n} A_{ij}\} \begin{matrix}\le\\=\\\ge\end{matrix} \sum_{k=1}^{m} (-1)^k \sum_{e(g)=k} P\{A_g\} \tag{2.4}$$

according as (2.2), where $A_g$ is defined by

$$A_g = \bigcap_{(i,j)\in\gamma^{-1}(g)} A_{ij} \quad . \tag{2.5}$$

## 3. FIRST THEOREM

Theorem 1. Let $X_1, X_2, \ldots$ be a sequence of i.i.d. $R^d$-valued random variables with a probability density function f belonging to $L^2(R^d)$ and set $Y_n = \min_{1\le i<j\le n} |X_i - X_j|$. Then, for any $t>0$

$$\lim_{n\to\infty} P\{n^2 Y_n^d > t\} = e^{-ct} \quad , \tag{3.1}$$

where

$$c = c_d \| f \|^2/2 = (c_d/2)\int_{R^d} f(x)^2 dx \quad , \tag{3.2}$$

and $c_d = \pi^{d/2}/\Gamma(1+d/2)$ is the volume of the d-dimensional unit ball.

Proof. The events $A_g$'s are classified according as the graph g is pairwise connected or not. Since $X_1, X_2, \ldots$ are i.i.d.,

$$P\{A_g\} = \{P(A_{12})\}^{e(g)}$$

provided that g is pairwise connected. Using (2.4),

$$P\{Y_n < t\} \begin{smallmatrix}\le\\=\\\ge\end{smallmatrix} M_n^{(m)} + R_n^{(m)}$$

according as (2.2), where the first term corresponds to the summation over the pairwise connected graphs g:

$$M_n^{(m)} = \sum_{k=1}^{m} (-1)^{k-1} \sum_{e(g)=k,v(g)=2k} \{P(A_{12})\}^k \quad , \tag{3.3}$$

and the second term is the remainder:

$$R_n^{(m)} = \sum_{k=1}^{m} (-1)^{k-1} \sum_{e(g)=k,v(g)<2k} P\{A_g\} \quad . \tag{3.4}$$

To prove the theorem, it is enough to show that for $\varepsilon = (n^{-2}t)^{1/d}$

$$\lim_{n\to\infty} M_n^{(m)} = \sum_{k=1}^{m} (-1)^{k-1} (ct)^k/k! \quad , \tag{3.5}$$

$$\lim_{n\to\infty} R_n^{(m)} = 0, \quad \text{for } \forall\, m \ge 1 \quad , \tag{3.6}$$

because

$$\lim_{m\to\infty} \sum_{k=1}^{m} (-1)^{k-1} (ct)^k/k! = 1 - e^{-ct}, \quad t > 0 \quad .$$

1: To prove (3.5), first notice that the probability density function of $X_1 - X_2$,

$$\phi(x) = \int_{R^d} f(x+y)f(y)dy \quad ,$$

is continuous and $\phi(0) = \|f\|^2$. Thus $P\{A_{12}\} \sim 2c\varepsilon^d$ as $\varepsilon \downarrow 0$ and hence for $t>0$ and $\varepsilon = (n^{-2}t)^{1/d}$

$$P\{|X_1 - X_2| < \varepsilon\} \sim 2ctn^{-2}, \quad n\to\infty \quad . \tag{3.7}$$

Therefore

$$\lim_{n\to\infty} M_n^{(m)} = \lim_{n\to\infty} \sum_{k=1}^{m} (-1)^{k-1} \frac{1}{k!} \binom{n}{2}\binom{n-2}{2}\cdots\binom{n-2k+2}{2}(2ctn^{-2})^k$$

$$= \sum_{k=1}^{m} (-1)^{k-1}(ct)^k/k! \quad .$$

2: The proof of (3.6) will be given first assuming f to be bounded and this assumption will be removed later.

2.1: Assuming that there exists a positive constant M such that $f(x)\leq M$, we have the following lemma.

Lemma 1. If g is a simple graph, then

$$P\{A_g\} \leq (Mc_d\varepsilon^d)^{e(g)} \quad . \tag{3.8}$$

Proof of Lemma 1. First assume that g is a tree, that is, all vertexes are connected. Fix any vertex of g as a root, and let vertex j be a leaf of the tree. The vertex j is connected only to one vertex (say) k which is closer to the root. Write

$$P\{A_g\} = \int\cdots\int \prod_{i=1}^{n} f(x_i) \prod_{i=1}^{n} dx_i \quad , \tag{3.9}$$

where the integration region is $\bigcap_{(i,j)\in G} \{x\in R^d: |x_i - x_j| < \varepsilon\}$ and G is a subset of $G_n$ such that $g=\gamma(G)$. Integrating (3.9) first with respect to $x_j$ in the range $|x_j - x_k| < \varepsilon$,

$$P\{A_g\} \leq (Mc_d\varepsilon^d)\int\cdots\int \prod_{i\neq j} f(x_i) \prod_{i\neq j} dx_i \quad .$$

The vertex j and the edge connecting vertexes j and k being deleted, the remaining subgraph is still a tree and hence we can proceed the integration similarly with respect to all edges to get the inequality (3.8).

In case when g is a simple graph, it is enough to repeat the above integration process for each component tree-subgraph. The proof of Lemma 1 is completed.

Going back to the proof of Theorem 1, we evaluate $R_n^{(m)}$ of (3.4). Since $X_i$'s are i.i.d., $P\{A_g\}$ does not depend on labeling of vertexes, and $P\{A_g\}$ is the same for the g's which are the same as unlabeled undirected graphs. For a given g the number of such equivalent graphs are at most $n^{v(g)}$. To complete the proof of the theorem in the case $f(x)\leq M$, it is enough to prove

$$\lim_{n\to\infty} n^{v(g)} P\{A_g\} = 0 \quad , \tag{3.10}$$

because (3.6) follows immediately from (3.10) and the expression (3.4). (3.10) holds if g is a simple but not pairwise connected graph. In fact, the number $\alpha$ of component tree-subgraphs is less than the number of edges $e(g)\geq 2$, and hence

$$n^{v(g)} P\{A_g\} \leq n^{e(g)+\alpha} (Mc_\alpha t n^{-2})^{e(g)} \to 0, \quad n\to\infty \quad .$$

If g is a general graph, then opencircuit some edges keeping $V_g$ unchanged to get a simple but not pairwise connected graph g'. The event $A_{g'}$ has less conditions than $A_g$ has, and has larger probability. Thus

$$n^{v(g)} P\{A_g\} \leq n^{v(g')} P\{A_{g'}\} \to 0, \quad n\to\infty \quad .$$

2.2: When f is unbounded, take an arbitrary positive constant M, put

$$\mathcal{M} = \{x \in R^d : f(x) \leq M\} \quad ,$$

and define random variables $V_{ij}$ and $W_{ij}$ by

$$V_{ij} = \begin{cases} |X_i - X_j|^d, & \text{if both } X_i \text{ and } X_j \text{ are in } \mathcal{M}, \\ \infty, & \text{otherwise,} \end{cases}$$

$$W_{ij} = \begin{cases} \infty, & \text{if both } X_i \text{ and } X_j \text{ are in } \mathcal{M}, \\ |X_i - X_j|^d, & \text{otherwise.} \end{cases}$$

Then

$$Y_n^d = \min\{\min_{1\leq i<j\leq n} V_{ij}, \min_{1\leq i<j\leq n} W_{ij}\}$$

and it will be proved that for any t>0

$$\lim_{n\to\infty} P\{n^2 \min_{1\leq i<j\leq n} V_{ij} < t\} = 1 - e^{-c't} \quad , \tag{3.11}$$

where

$$c' = (d_d/2)\int_{\mathcal{M}} f(x)^2 dx \quad ,$$

and that for any t>0

$$\lim_{M\to\infty} \sup_n P\{n^2 \min_{1\leq i<j\leq n} W_{ij} < t\} = 0 \quad . \tag{3.12}$$

In fact, (3.11) is proved just in the same way as the case where f is bounded with f replaced by $f\cdot 1_M$, where $1_M$ is the indicator function of the set $M$. (3.12) is shown as follows:

$$P\{n^2 \min_{1\le i<j\le n} W_{ij} < t\} = \frac{n(n-1)}{2} P\{W_{12} < tn^{-2}\}$$

$$= n(n-1)P\{|X_1 - X_2| < tn^{-2},\ X_1 \notin M\}$$

$$= n(n-1)\int_{M^c} f(x_1)\,dx_1 \int_{|x_2-x_1|<tn^{-2}} f(x_2)\,dx_2$$

$$= n(n-1)\int_{|y|<tn^{-2}} dy \int_{M^c} f(x)f(x+y)\,dx$$

$$\le n(n-1)\,||f||\,\{\int_{M^c} f(x)^2 dx\}^{1/2} \int_{|y|<tn^{-2}} dy$$

$$\le c_d t\,||f||\,\{\int_{M^c} f(x)^2 dx\}^{1/2} \to 0, \quad M\to\infty .$$

The proof of the theorem is now completed.

## 4. SECOND THEOREM

Let $X_1, X_2, \ldots$ be the same sequence as before, and let $Y_n^{(1)}, Y_n^{(2)}, \ldots, Y_n^{(k)}$ be the k smallest distances to the power d among the first n random variables. $Z_n^{(j)}$ is the mean of the pair of variables whose distance to the power d is $Y_n^{(j)}$, that is,

$$Y_n^{(1)} = \min_{1\le i<j\le n} |X_i - X_j|^d = |X_{\alpha(1)} - X_{\beta(1)}|^d,\ 1\le\alpha(1)<\beta(1)\le n ,$$

$$Z_n^{(1)} = \{X_{\alpha(1)} + X_{\beta(1)}\}/2 ,$$

$$Y_n^{(2)} = \min_{\substack{1\le i<j\le n \\ (i,j)\ne(\alpha(1),\beta(1))}} |X_i - X_j|^d = |X_{\alpha(2)} - X_{\beta(2)}|^d,\ 1\le\alpha(2)<\beta(2)\le n ,$$

$$Z_n^{(2)} = \{X_{\alpha(2)} + X_{\beta(2)}\}/2 ,$$

and so on. $Y_n^{(1)}$ is $Y_n^d$ of Section 3. These have the following limit joint distribution.

Theorem 2. Let k be any fixed positive integer. If the probability density f of $X_1$ belongs to $L^2(R^d)$, then for any positive

$t_1, t_2, \ldots, t_k$ and for any Borel sets $B_1, B_2, \ldots, B_k$ of $R^d$,

$$\lim_{n\to\infty} P\{n^2 Y_n^{(1)} > t_1,\ n^2(Y_n^{(2)} - Y_n^{(1)}) > t_2,\ \ldots,\ n^2(Y_n^{(k)} - Y_n^{(k-1)}) > t_k,$$
$$Z_n^{(1)} \in B_1,\ Z_n^{(2)} \in B_2,\ \ldots,\ \text{and } Z_n^{(k)} \in B_k\}$$
$$= \prod_{i=1}^{k} e^{-ct_i} \lambda(B_i) \quad , \tag{4.1}$$

where c is given by (3.2) and

$$\lambda(B) = \|f\|^{-2} \int_B f(x)^2 dx \quad . \tag{4.2}$$

Before giving the proof we prepare two lemmas.

Lemma 2. If $f \in L^2(R^d)$, then

$$P\{|X_1 - X_2| < \varepsilon \text{ and } |X_2 - X_3| < \varepsilon\} = o(\varepsilon^{3d/2}), \quad \varepsilon \downarrow 0 \quad . \tag{4.3}$$

Proof of Lemma 2. It is shown first that there exists positive $c(\varepsilon)$ such that

$$\left.\begin{aligned} & c(\varepsilon) \to 0 \text{ as } \varepsilon \downarrow 0, \text{ and} \\ & \int_{|x_3 - x_2| < \varepsilon} f(x_3) dx_3 \le c(\varepsilon) \varepsilon^{d/2} \quad . \end{aligned}\right\} \tag{4.4}$$

In fact, by the Schwarz inequality the integration in (4.4) is equal to or less than

$$(c_d \varepsilon^d)^{1/2} \{ \int_{|x - x_2| < \varepsilon} f(x)^2 dx \}^{1/2}$$

The integration region in this expression is included in $\{x : f(x) > N\} \cup \{x : |x - x_2| < \varepsilon \text{ and } f(x) \le N\}$, and hence

$$\int_{|x - x_2| < \varepsilon} f(x)^2 dx \le c_d \varepsilon^d N^2 + \int_{\{f(x) > N\}} f(x)^2 dx \quad .$$

Therefore, for any positive N

$$\overline{\lim_{\varepsilon \downarrow 0}} \sup_{x_2} \int_{|x - x_2| < \varepsilon} f(x)^2 dx \le \int_{\{f(x) > N\}} f(x)^2 dx \quad ,$$

and the last integration decreases to zero as $N \uparrow \infty$, and hence (4.4) is proved. Now the probability in (4.3) is expressed by the following integration which can be evaluated by (4.4):

$$\iiint_{|x_2-x_1|<\varepsilon,\,|x_3-x_2|<\varepsilon} f(x_1)f(x_2)f(x_3)\,dx_1dx_2dx_3$$

$$\le c(\varepsilon)\varepsilon^{d/2}\iint_{|x_2-x_1|<\varepsilon} f(x_1)f(x_2)\,dx_1dx_2$$

$$= c(\varepsilon)\varepsilon^{d/2}\int_{|x|<\varepsilon}dx\int_{R^d} f(y)f(y+x)\,dy$$

$$\le c(\varepsilon)c_d\varepsilon^{3d/2}\|f\|^2 = o(\varepsilon^{3d/2}), \quad \varepsilon\downarrow 0 \quad .$$

Lemma 3. For a Borel set B and for $t>0$ define

$$F_B(t) = P\{|X_1 - X_2|^d \le t \text{ and } \tfrac{1}{2}(X_1 + X_2) \in B\} \quad .$$

If $\lambda(B)$ of (4.2) is positive, then

$$F_B(t) \sim 2c\lambda(B)t, \quad t\downarrow 0 \quad .$$

Proof of Lemma 3.

$$F_B(t) = \iint_{|x_1-x_2|\le t^{1/d},\,(x_1+x_2)/2\in B} f(x_1)f(x_2)\,dx_1dx_2$$

$$= \int_{|y|\le t^{1/d}}dy\int_{R^d} f(x)f(x+y)\,1_B(x+\tfrac{y}{2})\,dx \quad .$$

$$\sim c_d t\int_B f(x)^2dx = 2c\lambda(B)t, \quad t\downarrow 0 \quad ,$$

because

$$\lim_{y\to 0}\int_{R^d} f(x)f(x+y)\,1_B(x+\tfrac{y}{2})\,dx = \int_B f(x)^2dx \quad .$$

The proof of the lemma is finished.

Going back to the proof of Theorem 2, we show only the special statement

$$\lim_{n\to\infty} P\{n^2Y_n^{(1)} \le t,\ n^2(Y_n^{(2)} - Y_n^{(1)}) > s \text{ and } Z_n^{(1)} \in B\}$$

$$= (1 - e^{-ct})e^{-cs}\lambda(B), \quad t,s > 0, \quad B\in \mathcal{B}(R^d) \quad , \qquad (4.5)$$

for simplicity. The general statement (4.1) can be proved in quite a similar way.

The event of the probability in (4.5) is a union of the disjoint events $H_{\alpha\beta}$, $1 \le \alpha < \beta \le n$:

$$H_{\alpha\beta} = \{n^2 |X_\alpha - X_\beta|^d \le t,\ \tfrac{1}{2}(X_\alpha + X_\beta) \in B \text{ and}$$
$$n^2 \cdot \min_{1 \le i < j \le n,\,(i,j) \ne (\alpha,\beta)} |X_i - X_j|^d > s + n^2 |X_\alpha - X_\beta|^d\}, \quad (4.6)$$

and $P\{\bigcup_{1 \le \alpha < \beta \le n} H_{\alpha\beta}\} = 2^{-1} n(n-1) P\{H_{12}\}$. The last condition on $\min|X_i - X_j|^d$ in (4.6) is decomposed, when $(\alpha,\beta) = (1,2)$, in such a way that all $\min_{3 \le i < j \le n} |X_i - X_j|^d$, $\min_{3 \le j \le n} |X_1 - X_j|^d$ and $\min_{3 \le j \le n} |X_2 - X_j|^d$ satisfy the condition. Thus, in terms of the events

$$C_n = \{n^2 |X_1 - X_2|^d \le t,\ \tfrac{1}{2}(X_1 + X_2) \in B \text{ and}$$
$$n^2 \min_{3 \le i < j \le n} |X_i - X_j|^d > s + n^2 |X_1 - X_2|^d\},$$

$$D_{nj} = \{n^2 |X_1 - X_j|^d \le s + n^2 |X_1 - X_2|^d\},$$

$$E_{nj} = \{n^2 |X_2 - X_j|^d \le s + n^2 |X_1 - X_2|^d\},$$

we can write

$$H_{12} = C_n - C_n \cap \{\bigcup_j (D_{nj} \cup E_{nj})\}.$$

Since

$$(n(n-1)/2) P[C_n \cap \{\bigcup_{j=3}^{n} (D_{nj} \cup E_{nj})\}]$$
$$\le n(n-1)(n-2) P\{n^2 |X_1 - X_2|^d \le t,\ n^2 |X_1 - X_3|^d \le s + n^2 |X_1 - X_2|^d\}$$
$$\le n(n-1)(n-2) P\left\{|X_1 - X_2| \le \left(\frac{t}{n^2}\right)^{1/d},\ |X_1 - X_3| \le \left(\frac{s+t}{n^2}\right)^{1/d}\right\}$$
$$= o(1), \quad n \to \infty,$$

by Lemma 2, we have

$$P\{\bigcup_{1 \le \alpha < \beta \le n} H_{\alpha\beta}\} = \frac{n(n-1)}{2} P\{C_n\} + o(1).$$

On the other hand, Theorem 1 asserts that a function defined by

$$\phi_n(t) = P\{n^2 Y_n^{(1)} > t\}$$

converges to $\phi(t) = e^{-ct}$, $t>0$, and the convergence is uniform in $t>0$. Therefore

$$\begin{aligned}
\frac{n(n-1)}{2} P\{C_n\} &= \frac{n(n-1)}{2} E\{\phi_{n-2}((\tfrac{n-2}{n})^2 \cdot (s+n^2|X_1-X_2|^d)); \\
&\qquad\qquad n^2|X_1-X_2|^d \le t,\ \tfrac{1}{2}(X_1+X_2) \in B\} \\
&= \frac{n(n-1)}{2} E\{\phi((\tfrac{n-2}{n})^2 \cdot (s+n^2|X_1-X_2|^d)); \\
&\qquad\qquad n^2|X_1-X_2|^d \le t,\ \tfrac{1}{2}(X_1+X_2) \in B\} + o(1) \\
&= \frac{n(n-1)}{2} \int_0^{tn^{-2}} \phi(s+n^2u)\, dF_B(u) + o(1) \quad ,
\end{aligned}$$

where $F_B$ is defined in Lemma 3, since the error term caused by replacing $\phi_{n-2}$ by $\phi$ is dominated, in modulus, by

$$\begin{aligned}
&\frac{n(n-1)}{2} \|\phi_{n-2} - \phi\|_\infty \cdot P\{n^2|X_1 - X_2|^d \le t\} \\
&\sim \frac{n(n-1)}{2} \|\phi_{n-2} - \phi\|_\infty \cdot 2ctn^{-2} \\
&\le t\|\phi_{n-2} - \phi\|_\infty \to 0, \qquad n\to\infty \quad .
\end{aligned}$$

Integration by parts yields

$$\begin{aligned}
&\int_0^{tn^{-2}} \phi(s+n^2u)\, dF_B(u) \\
&= \phi(s+t) F_B(tn^{-2}) - \int_0^{tn^{-2}} F_B(u) n^2 \phi'(s+n^2u)\, du \\
&= \phi(s+t) F_B(tn^{-2}) - \int_0^t F_B(vn^{-2}) \phi'(s+v)\, dv \quad .
\end{aligned}$$

Making use of Lemma 3, we have

$$\lim_{n\to\infty} \frac{n(n-1)}{2} P\{C_n\}$$

$$= \lim_{n\to\infty} \frac{n(n-1)}{2}\{\phi(s+t)2c\lambda(B)tn^{-2} - \int_0^t 2c\lambda(B)vn^{-2}\phi'(s+v)dv\}$$

$$= c\lambda(B)t\phi(s+t) - c\lambda(B)\int_0^t v\phi'(s+v)dv$$

$$= c\lambda(B)\int_0^t \phi(s+u)du = (1-e^{-ct})e^{-cs}\lambda(B) \quad ,$$

proving (4.5).

## 5. ANOTHER PROOF OF THEOREM 1

Theorem 1 was used in the proof of Theorem 2, but the proof suggests an alternative proof of Theorem 1. In the preceding section it was shown that

$$P\{n^2Y_n^{(1)} \le t,\ n^2(Y_n^{(2)} - Y_n^{(1)}) > s\}$$

$$= \frac{n(n-1)}{2} E\{\phi_{n-2}((\tfrac{n-2}{n})^2 \cdot (s+n^2|X_1-X_2|^d));$$

$$n^2|X_1-X_2|^d \le t\} + o(1) \quad . \tag{5.1}$$

Starting from this expression another proof of Theorem 1 is given. To prove both Theorems 1 and 2 this is a shorter way, but the proof in Section 3 makes the essential points clearer.

Lemma 4. The family of functions $\{\phi_n,\ n\ge 1\}$, is equicontinuous.

Proof of Lemma 4. In fact, for $0\le s<t$,

$$\phi_n(s) - \phi_n(t) = P\{s < n^2Y_n^{(1)} \le t\}$$

$$\le \sum_{1\le i<j\le n} P\{s < n^2|X_i-X_j|^d \le t,\ (\alpha(1),\beta(1)) = (i,j)\}$$

$$\le \frac{n(n-1)}{2} P\{s < n^2|X_1-X_2|^d \le t\}$$

$$= \frac{n(n-1)}{2}\iint_{\varepsilon_1<|x_1-x_2|\le\varepsilon_2} f(x_1)f(x_2)dx_1dx_2$$

$$\leq \frac{n(n-1)}{2}\|f\|^2 c_d(\varepsilon_2^d - \varepsilon_1^d) \leq c(t-s) \quad ,$$

where $\varepsilon_1 = (sn^{-2})^{1/d}$ and $\varepsilon_2 = (tn^{-2})^{1/d}$.

Lemma 5. $\phi_n(t) - \phi_{n-1}(t)$ converges to zero as $n\to\infty$ uniformly in $t \in [0,T]$ for any $T>0$.

Proof of Lemma 5. We have

$$\phi_n(t) = P\{(n-1)^2 Y_n^{(1)} > (\frac{n-1}{n})^2 t,\ n^2|X_i - X_n|^d > t,\ 1\leq i\leq n\}$$

$$= P\{(n-1)^2 Y_{n-1}^{(1)} > (\frac{n-1}{n})^2 t\}$$

$$- P\{(n-1)^2 Y_n^{(1)} > (\frac{n-1}{n})^2 t,\ \bigcup_{i=1}^{n-1}(|X_i - X_n| \leq (tn^{-2})^{1/d})\}.$$

The second term on the right hand side of the above is $O(1/n)$ because it is less than or equal to

$$\sum_{i=1}^{n-1} P\{|X_i - X_n| \leq (tn^{-2})^{1/d}\} = (n-1)P\{|X_1 - X_2| \leq (tn^{-2})^{1/d}\}$$

which is $O(1/n)$ by (3.7). Therefore

$$\phi_n(t) = \phi_{n-1}((\frac{n-1}{n})^2 t) + O(\frac{1}{n}) = \phi_{n-1}(t) + o(1), \qquad n\to\infty \quad ,$$

since $\{\phi_n\}$ is equicontinuous.

Proof of Theorem 1. Putting $s=0$ in (5.1), we have

$$1 - \phi_n(t) = \frac{n(n-1)}{2} E\{\phi_{n-2}((n-2)^2|X_1 - X_2|^d);\ n^2|X_1 - X_2|^d \leq t\} + o(1)$$

$$= \frac{n(n-1)}{2} E\{\phi_n(n^2|X_1 - X_2|^d);\ n^2|X_1 - X_2|^d \leq t\} + o(1)$$

by Lemma 4 and Lemma 5, and this is equivalent to

$$1 - \phi_n(t) = \frac{n(n-1)}{2}\int_0^{tn^{-2}} \phi_n(sn^2)dF(s) + o(1) \quad , \tag{5.2}$$

where

$$F(s) = P\{|X_1 - X_2|^d \leq s\} \sim 2cs \quad . \tag{5.3}$$

Let $\phi(t)$ be any limit function of $\phi_n(t)$ as $n\to\infty$ via some sequence $n_1<n_2<\ldots$ Then (5.2) implies

$$1-\phi(t) = \frac{n(n-1)}{2}\int_0^{tn^{-2}} \phi(sn^2)\,dF(s) + o(1)$$

as $n\to\infty$ via $n_1<n_2<\ldots$ Using (5.3) it is shown that the right hand side of the above expression tends to $c\int_0^t \phi(s)ds$, and therefore

$$1-\phi(t) = c\int_0^t \phi(s)ds \quad ,$$

that is, $\phi(t)=e^{-ct}$. Since $\phi(t)$ is any limit function of $\phi_n(t)$, we have $\phi_n(t)\to e^{-ct}$, completing the proof.

## REFERENCES

Darling, D.A. (1953). On a class of problems related to the random division of an interval, *Ann. Math. Stat.* 24, pp. 239-253.

Feller, W. (1971). *An Introduction to Probability Theory and Its Applications*, Vol. 2, Second ed., John Wiley and Sons, New York.

Galambos, J. (1978). *The Asymptotic Theory of Extreme Order Statistics*, John Wiley and Sons, New York, §2.11.

Molchanov, S.A. and A. Ya. Reznikova (1982). Limit theorems for random partitions, *Theory of Probability and Its Applications* 27, pp. 310-323 (in Russian edition pp. 296-307).

Weiss, L. (1959). The limiting joint distribution of the largest and smallest spacings, *Ann. Math. Statist.* 30, pp. 590-592.

Weiss, L. (1969). The joint asymptotic distribution of the k-smallest sample spacings, *J. Appl. Probability* 6, pp. 442-448.

# APPROXIMATE VALUES FOR THE MOMENTS OF EXTREME ORDER STATISTICS IN LARGE SAMPLES

G Ramachandran

Fire Research Station, Borehamwood, Herts

## SUMMARY

For large samples, precise calculations of moments of order statistics by an exact approach are time consuming and impracticable. For such samples a method of approximation is developed in this paper using the values of the failure rate function and its derivatives and the moments of order statistics in an exponential distribution. Application of the method is illustrated with reference to extreme order statistics in exponential, gamma and normal distributions.

## 1 INTRODUCTION

Suppose a distribution $F(x)$ with density $f(x)$ has a location parameter $\mu$ and a scale parameter $\sigma$. Consider the standard variable

$$t = (x - \mu)/\sigma \tag{1}$$

If n observations in a sample from $F(x)$ are arranged in decreasing order of magnitude with $x_{(m)n}$ as the $m^{th}$ observation in that arrangement. Let

$$t_{(m)n} = (x_{(m)n} - \mu)/\sigma \tag{2}$$

This paper is concerned with the moments of $t_{(m)n}$ in repeated samples of size n from $F(x)$.



*J. Tiago de Oliveira (ed.), Statistical Extremes and Applications, 563–578.*

For some distributions exact values of the moments for small samples have been discussed and tabulated - see Sarhan and Greenberg (1962), David (1970) and Owen, Odeh and Davenport (1977). For large samples (large n) the precise calculations of moments of order statistics by the exact approach are time consuming and impracticable. For such samples the following method is generally suggested for obtaining approximate values of the moments - see David (1970, p 65-66). If G(t) is the distribution function of t, the probability integral transform $u = G(t)$ takes the continuous order statistic $t_{(m)n}$ into the $m^{th}$ order statistic $u_{(m)n}$ (from the top) in a sample of n from a rectangular distribution. Then, by considering the inverse function

$$t_{(m)n} = G^{-1}\left[u_{(m)n}\right] = Q\left[u_{(m)n}\right] \tag{3}$$

and expanding $Q\left[u_{(m)n}\right]$ in a Taylor series about

$$E\left[u_{(m)n}\right] = 1 - \{m/(n+1)\} \tag{4}$$

the moments of $t_{(m)n}$ can be obtained from those of $u_{(m)n}$. For small m compared with large n the value given by equation (4) would tend to the mode rather than the expectation of $u_{(m)n}$. For this reason, defining the parameter $B_{(m)n}$ given by

$$G\left(B_{(m)n}\right) = 1 - (m/n) \tag{5}$$

as the mode of the standard variable $t_{(m)n}$, a method of approximation is developed using the values of the failure rate function

$$h(u) = g(u)/\{1 - G(u)\} \tag{6}$$

and its derivatives at $B_{(m)n}$. G(t) and g(t) are distribution and density functions of the standard variable t. This method involves the moments of order statistics in an exponential distribution rather than the rectangular distribution. Application of the method is illustrated with reference to extreme order statistics in exponential, gamma and normal distributions.

## 2 THE METHOD

The distribution function G(t) can be written as

$$G(t) = 1 - \exp(-H(t)) \tag{7}$$

$$H(t) = \int^{t} h(u)\,du$$

The derivative of H(t) is h(t). From equations (5) and (7).

$$H(B_{(m)n}) = \log_e (n/m) \tag{8}$$

Exapnding in a Taylor series

$$y_{(m)n} = H(t_{(m)n}) - H(B_{(m)n}) \tag{9}$$

$$= A_{(m)n} C_{(m)n} + \frac{A^{I}_{(m)n} \cdot C^2_{(m)n}}{2} + \frac{A^{II}_{(m)n} \cdot C^3_{(m)n}}{6}$$

$$+ \frac{A^{III}_{(m)n} \cdot C^4_{(m)n}}{24} + \ldots\ldots = P(C_{(m)n}) \tag{10}$$

$$C_{(m)n} = t_{(m)n} - B_{(m)n} \tag{11}$$

$$A_{(m)n} = h(B_{(m)n}) \tag{12}$$

as given by equation (6) with $A^{I}_{(m)n}$, $A^{II}_{(m)n}$, $A^{III}_{(m)n}$ and so on as successive derivatives of $h(B_{(m)n})$. We may invert the relationship in equation (10) and expand as follows

$$C_{(m)n} = P^{-1}(y_{(m)n}) = Q(y_{(m)n}) \tag{13}$$

$$C_{(m)n} = Q(0) + y_{(m)n}\, Q^{I}(0) + \frac{y^2_{(m)n}}{2}\, Q^{II}(0) + \frac{y^3_{(m)n}}{6}\, Q^{III}(0)$$

$$+ \frac{y^4_{(m)n}}{24}\, Q^{IV}(0) + \ldots\ldots \tag{14}$$

If $y_{(m)n} = 0$, $C_{(m)n} = 0$; hence $Q(0) = P^{-1}(0) = 0$.

The successive derivates of $Q(y)$ in terms of those of $P(C)$ are given by

$$Q^{I}(y) = \frac{d}{dy}\, Q(y) = \frac{dC}{dy} = 1/P^{I}(C)$$

$$Q^{II}(y) = \frac{d}{dC}\, \left\{ \frac{1}{P^{I}(C)} \right\} \frac{dC}{dy} = -P^{II}(C)/\{P^{I}(C)\}^3$$

$$Q^{III}(y) = \frac{1}{\{P^{I}(C)\}^5}\, \left[3\{P^{II}(C)\}^2 - P^{I}(C).\ P^{III}(C)\right]$$

$$Q^{IV}(y) = \frac{1}{\{P^{I}(C)\}^6}\left[5\, P^{II}(C) \,.\, P^{III}(C) - P^{I}(C)\, P^{IV}(C)\right]$$

$$- \frac{5}{\{P^{I}(C)\}^7}\left[3\{P^{II}(C)\}^3 - P^{I}(C) \,.\, P^{II}(C) \,.\, P^{III}(C)\right]$$

From equation (10)

$P^{I}(0) = A_{(m)n}$, $P^{II}(0) = A^{I}_{(m)n}$, $P^{III}(0) = A^{II}_{(m)n}$,
$P^{IV}(0) = A^{III}_{(m)n}$.

so that

$$Q^{I}(0) = 1/A_{(m)n}$$

$$Q^{II}(0) = -\left\{A^{I}_{(m)n}/A^{3}_{(m)n}\right\}$$

$$Q^{III}(0) = \left[3\, A^{I^2}_{(m)n} - A_{(m)n} \,.\, A^{II}_{(m)n}\right]/A^{5}_{(m)n}$$

$$Q^{IV}(0) = \left[10\, A_{(m)n} .\, A^{I}_{(m)n}\, A^{II}_{(m)n} - 15\, A^{I^3} - A^{2}_{(m)n}\, A^{III}_{(m)}\right]/A^{7}_{(m)n}$$

Using equation (7) the density function of $t_{(m)n}$ is

$$\frac{n!}{(n-m)!(m-1)!}\left[1 - \exp(-w_{(m)n})\right]^{n-m} \exp\left(-m\, w_{(m)n}\right)$$

where the variable $w_{(m)n} = H(t_{(m)n})$ is the $m^{th}$ order statistic (from the top) in a sample of size n from the standard exponential distribution

$$G(w) = 1 - \exp(-w)$$

The first four cumulants of $W_{(m)n}$ are known to be

$$K_{1m} = \sum_{v=m}^{n} (1/v),\quad K_{2m} = \sum_{v=m}^{n} (1/v^2) \qquad (15)$$

$$K_{3m} = 2\sum_{v=m}^{n} (1/v^3),\quad K_{4m} = 6\sum_{v=m}^{n} (1/v^4)$$

Hence from equations (8), (9) and (15), taking expectations

$$E(y_{(m)n}) = \bar{y}_{(m)n} = \sum_{v=m}^{n} (1/v) - \log_e (n/m) = K_{1my} \tag{16}$$

$$E(y^2_{(m)n}) = K_{2m} + K^2_{1my}$$

$$E(y^3_{(m)n}) = K_{3m} + 3K_{2m} K_{1my} + K^3_{1my}$$

$$E(y^4_{(m)n}) = K_{4m} + 4K_{3m} K_{1my} + 3K^2_{2m} + 6K_{2m} K^2_{1my} + K^4_{1my}$$

## 3 EXPECTED VALUE

From equations (14) and (11)

$$E(t_{(m)n}) = B_{(m)n} + Q^{I}(0)\, E(y_{(m)n}) + \frac{Q^{II}(0)}{2}\, E(y^2_{(m)n})$$

$$+ \frac{Q^{III}(0)}{6}\, E(y^3_{(m)n}) + \frac{Q^{IV}(0)}{24}\, E(y^4_{(m)n}) \tag{17}$$

approximately. For the (standard) exponential distribution the first two terms on the right-hand side of equation (17) are sufficient since the failure rate function of this distribution has a constant value of unity. Hence $Q^{I}(0) = 1$.
Also $B_{(m)n} = \log_e(n/m)$ so that

$$E(t_{(m)n} = \sum_{v=m}^{n} (1/v) = \sum_{v=1}^{n} (1/v) - \sum_{v=1}^{m-1} (1/v) \tag{18}$$

For 'exponential type' distributions defined by Gumbel (1958, p. 120) the derivatives of the failure rate function have negligible values for small m compared with large n. In such cases an approximate value is provided by

$$E(t_{(m)n}) = B_{(m)n} + (1/A_{(m)n}) \left[ \sum_{v=m}^{n} (1/v) - \log_e(n/m) \right] \tag{19}$$

In Table 1 approximate values based on equation 19 and the inclusion of terms up to the third and fourth are given for the top 5 order statistics in samples of sizes 50 and 100 from the gamma distribution

$$g(t) = \{1/\Gamma(r)\} \exp(-t) t^{r-1}$$

with r = 5 and 10. The following relationships are used.

$$A^{I}_{(m)n} = A_{(m)n}\left[(r-1-B_{(m)n})/B_{(m)n} + A_{(m)n}\right]$$

$$A^{II}_{(m)n} = A_{(m)n}\left[(r-1-B_{(m)n})^2 - r + 1\right]/B^2_{(m)n} + 3\,A^2_{(m)n}\left[(r-1-B_{(m)n})/B_{(m)n}\right] + 2\,A^3_{(m)n}$$

For the gamma distribution, inclusion of the first four terms in the Taylor series expansion of equation (17) appears to be sufficiently accurate at least for samples of sizes greater than 50.

Similar figures for the (standard) normal distribution are shown in Table 2 for samples of sizes 50 and 500. Allowing for rounding-off errors values in in the last column for n = 50 compare well with the exact values (in brackets) tabulated by Tietjen, Kahaner and Beckman (1977). The following relationships have been used in obtaining these results.

$$A^{I}_{(m)n} = A_{(m)n}\left[A_{(m)n} - B_{(m)n}\right]$$

$$A^{II}_{(m)n} = 2A^3_{(m)n} - 3A^2_{(m)n}\cdot B_{(m)n} + A_{(m)n}\,B^2_{(m)n} - A_{(m)n}$$

$$A^{III}_{(m)n} = 6A^4_{(m)n} - 12A^3_{(m)n}\,B_{(m)n} + 7B^2_{(m)n}\,A^2_{(m)n} - 4A^2_{(m)n} + 3A_{(m)n}\,B_{(m)n} - B^3_{(m)n}\,A_{(m)n}$$

## 4 COVARIANCE

Consider the two order statistics from the top

$$w_{(i)n} = H(t_{(i)n}),\; w_{(j)n} = H(t_{(j)n}) \tag{20}$$

If i > j, as discussed in the appendix, the expected value of the product $w^p_{(i)n} \cdot w^q_{(j)n}$ is given by

$$\sum_{r=o}^{q} qCr.\mu^1_{rj}(i-1) \,.\, \mu^1_{(p+q-r)i}(n) \tag{21}$$

where $\mu^1_{rj}(i-1)$ is the $r^{th}$ moment about the origin of the $j^{th}$ order statistic in a sample of size (i-1) from the standard exponential distribution. $\mu^1_{(p+q-r)i}(n)$ has a similar definition.

From equation (21) it may be proved that

$$E(w_{(i)n} \cdot w_{(j)n}) = K_{2i} + K_{1i} \cdot K_{1j} \tag{22}$$

$$E(w^2_{(i)n} \cdot w_{(j)n}) = K_{3i} + 2K_{2i} \cdot K_{1i} + K_{2i} K_{1j} + K_{1i}^2 \cdot K_{1j}$$

$$E(w_{(i)n} \cdot w^2_{(j)n}) = K_{3i} + 2K_{2i} \cdot K_{1j} + K_{1i} K_{2j} + K_{1i} K_{1j}^2$$

$$E(w^2_{(i)n} \cdot w^2_{(j)n}) = K_{4i} + 3K_{3i} K_{1i} + 2K_{2i}^2 + 3K_{2i} K_{1i}^2$$

$$+ K_{1j} \cdot K_{3i} + K_{2i} \cdot K_{1i} \cdot K_{1j} - K_{1j} \cdot K_{1i}^3 + K_{2j} K_{2i}$$

$$+ K_{1j}^2 K_{2i} + K_{1i}^2 K_{2j} + K_{1j}^2 K_{1i}^2 + K_{1i}^4$$

It may be noted that since $i > j$

$$\mu^1_{1j}(i-1) = \mu^1_{1j}(n) - \mu^1_{1i}(n) = K_{1j} - K_{1i}$$

$$\mu^1_{2j}(i-1) = K_{2j} - K_{2i} + (K_{1j} - K_{1i})^2$$

Equations (22) and (9) provide the following results

$$E(y_{(i)n} \cdot y_{(j)n}) - E(y_{(i)n}) \cdot E(y_{(j)n}) = K_{2i} = D_1$$

$$E(y_{(i)n} \cdot y^2_{(j)n}) - E(y_{(i)n})\ E(y^2_{(j)n}) = K_{3i} + 2K_{2i}\ \bar{y}_{(j)n} = D_2$$

$$E(y^2_{(i)n} \cdot y_{(j)n}) - E(y^2_{(i)n}) \cdot E(y_{(j)n}) = K_{3i} + 2K_{2i}\ \bar{y}_{(i)n} = D_3$$

$$E(y^2_{(i)n} \cdot y^2_{(j)n}) - E(y^2_{(i)n})\ E(y^2_{(j)n})$$

$$= K_{4i} + 2K_{3i}\bar{y}_{(i)n} + 2K_{3i}\bar{y}_{(j)n} + 2K_{2i}^2 + 4K_{2i} \cdot \bar{y}_{(i)n}\bar{y}_{(j)n} = D_4$$

Hence, using the first three terms on the right-hand side of equation (14) with equation (11), the covariance of $t_{(i)n}$ and $t_{(j)n}$ with $i > j$ is

$$Q^{I}_{(i)} \cdot Q^{I}_{(j)} \cdot D_1 + \frac{Q^{I}_{(i)} \cdot Q^{II}_{(j)}}{2} D_2 + \frac{Q^{II}_{(i)} \cdot Q^{I}_{(j)}}{2} D_3 + \frac{Q^{II}_{(i)} \cdot Q^{II}_{(j)}}{4} D_4 \quad (23)$$

$$Q^{I}_{(i)} = 1/A_{(i)n}; \; Q^{II}_{(i)} = - \{A^{I}_{(i)n}/A^{3}_{(i(n}\}$$

$$Q^{I}_{(j)} = 1/A_{(j)n}; \; Q^{II}_{(j)} = - \{A^{I}_{(j)n}/A^{3}_{(j)n}\}$$

If, for small i and j compared with large n, the derivatives of the failure rate function A have negligible values, from equation (23)

$$\mathrm{CoV}[t_{(i)n}, t_{(j)n}] = \frac{K_{2i}}{A_{(i)n} \cdot A_{(j)n}} \quad (24)$$

In Table 3 for gamma distribution the first value (a) in each cell refers to equation (23) and the second value (b) to equation (24). Table 4 is for the standard normal distribution. Judging from the values in the tables equation (24) is not a good approximation especially for a small sample. For the normal distribution with sample size of 50, the estimates (a) given by equation (23) are better than those against (b) when compared with the exact values against (c) tabulated by Tietjen, Kahaner and Beckman (1977). Inclusion of more terms from the Taylor expansion of equation (14) could improve the approximation.

## 5 ASYMPTOTIC THEORY

Using equations (7), (8), (9) we may write

$$G(t_{(m)n}) = 1 - (m/n) \exp(-y_{(m)n}) \quad (25)$$

If for small m and large n, the derivatives of $A_{(m)n}$ have negligible values equation (10) reduces to

$$y_{(m)n} = A_{(m)n} (t_{(m)n} - B_{(m)n}) \quad (26)$$

This simplification used by Gumbel (1958) and others leads to the following asymptotic density function

$$\frac{m^m}{(m-1)!} \exp[- m\, y_{(m)n} - m \exp(-y_{(m)n})] \quad (27)$$

for the 'reduced variable' $y_{(m)n}$ whose central moments are

$$\mu_{m1y} = \sum_{v=1}^{\infty} (1/v) + \log_e(m/n) - \sum_{v=1}^{m-1} (1/v) = 0.5772 + \log_e m - \sum_{v=1}^{m-1} (1/v) \qquad (28)$$

$$\mu_{m2y} = \sum_{v=1}^{\infty} (1/v^2) - \sum_{v=1}^{m-1} (1/v^2) = 1.6449 - \sum_{v=1}^{m-1} (1/v^2) \qquad (29)$$

$$\mu_{m3y} = 2\sum_{v=1}^{\infty} (1/v^3) - 2\sum_{v=1}^{m-1} (1/v^3) = 2.4042 - 2\sum_{v=1}^{m-1} (1/v^3) \qquad (30)$$

$$\mu_{m4y} = 6\sum_{v=1}^{\infty} (1/v^4) - 6\sum_{v=1}^{m-1} (1/v^4) = 6.4938 - 6\sum_{v=1}^{m-1} (1/v^4) \qquad (31)$$

These moments are asymptotic (large n) values of those in equation (15). Using equations (26), (28) and (29) and $\mu_{i^2y}$ for $k_{2_i}$ in equation (24), Ramachandran (1975) has tabulated the asymptotic expected values, variances and covariances of extreme order statistics (m = 1 to 10) for samples of selected sizes from normal and gamma distributions. Another approximation would result by using the asymptotic moments given by equations (28) to (31) in equation (14) and including as many terms as necessary.

In some accident studies it will be time consuming and expensive to collect data on all incidents and hence figures for damage caused may be available only for a few large events of economic importance. For assessing the economic value of fire protection measures, for example, loss data are generally available only for large fires. In such cases, the location and scale parameters as mentioned in equation (1) have to be estimated only from large observations assuming a specific form for the parent distribution F(x). For this purpose, in applying the Generalised Least Square Method, estimated values are required for the expected values, variances and covariances of extreme order statistics. The asymptotic theory has been applied by Ramachandran to fire protection problems, assuming a log normal probability distribution for fire loss. A review of his work is contained in a recent paper (1982). Rogers (1977) has applied this method to investigate the economic value of sprinklers in industrial and commercial buildings.

ACKNOWLEDGEMENT

The work described has been carried out as part of the research programme of the Building Research Establishment of the Department of the Environment and this paper is published by permission of the Director.

REFERENCES

1 David, H.A. (1970), Order Statistics. John Wiley & Sons, New York.
2 Gumbel, E J. (1958), Statistics of extremes, Columbia University Press, New York.
3 Owen, D.B., Odeh, R.E. and Davenport, J.M. (Eds). (1977), Selected Tables in Mathematical Statistics, Vol 5, American Mathematical Society, Providence, Rhode Island.
4 Ramachandran, G. (1975), Extreme order statistics from exponential type distributions with applications to fire protection and insurance. Ph.D. thesis, University of London.
5 Ramachandran, G. (1982), Properties of extreme order statistics and their application to fire protection and insurance problems. Fire Safety Journal, 5, pp. 59-76.
6 Rogers, F.E. (1977), Fire losses and the effect of sprinkler protection of buildings in a variety of industries and trades. Building Research Establishment Current Paper CP 9/77.
7 Sarhan, A.E and Greenberg, B.G. (Eds). (1962), Contributions to order statistics, Wiley, New York.
8 Tietjen, G.L., Kahaner, D.K. and Beckman, R.J. (1977), pp. 1-74 in reference 3

## APPENDIX PRODUCT MOMENTS

Omitting the subscript n and writing $u_{(i)} = \exp(-w_{(i)})$ and $u_{(j)} = \exp(-w_{(j)})$ the joint distribution of $u_{(i)}$ and $u_{(j)}$ for $i > j$ is

$$D\{1 - u_{(i)}\}^{n-i} \{u_{(i)} - u_{(j)}\}^{i-j-1} u_{(j)}^{j-1} \qquad (1)$$

$$D = \frac{n!}{(n-i)!(i-j-1)!(j-1)!}$$

The expected value of the product $w_{(i)}^p \cdot w_{(j)}^q$ is given by multiplying the expression (1) by $(-1)^{p+q} \{\log u_{(i)}\}^p \{\log u_{(j)}\}^q$ and integrating over the domain $0 \leqslant u_{(j)} \leqslant u_{(i)} < 1$. Consider first

$$I_1 = \int_0^{u_{(i)}} \{u_{(i)} - u_{(j)}\}^{i-j-1} u_{(j)}^{j-1} \{\log u_{(j)}\}^q$$

$$= u_{(i)}^{i-1} \int_0^1 \{1-\psi_j\}^{i-j-1} \psi_j^{j-1} \{\log u_{(i)} + \log \psi_j\}^q$$

with $\psi_j = u_{(j)}/u_{(i)}$. Hence $I_1 = u_{(i)}^{i-1} \cdot I_2$ where

$$I_2 = \sum_{r=0}^{q} {}_qC_r \{\log u_{(i)}\}^{q-r} \int_0^1 \{1-\psi_j\}^{i-j-1} \psi_j^{j-1} \{\log \psi_j\}^r$$

$$I_3 = \int_0^1 \{1-\psi_j\}^{i-j-1} \psi_j^{j-1} \{\log \psi_j\}^r$$

$$= \sum_{s=0}^{k} (-1)^s \, {}_kC_s \int_0^1 \psi_j^{s+j-1} (\log \psi_j)^r$$

where $k = i-j-1$. Also

$$\int_0^1 \psi_j^{s+j-1} (\log \psi_j)^r = (-1)^r r!/(s+j)^{r+1}$$

Since in a sample of size n from the standard exponential distribution the $p^{th}$ moment about the origin of $w_{(m)}$, the $m^{th}$ order statistic (from the top), is

$$\mu^1_{pm}(n) = \frac{n!\ p!}{(m-1)!} \sum_{r=0}^{n-m} \frac{(-1)^r}{(n-m-r)!\ r\ !(r+m)^{p+1}}$$

$$I_3 = (-1)^r\ k!\ \mu^1_{rj}(i-1)\ \frac{(j-1)!}{(i-1)!}$$

where $\mu^1_{rj}(i-1)$ is the $r^{th}$ moment about the origin of the $j^{th}$ order statistic (from the top) in a sample of size (i-1) from the standard exponential distribution. Hence

$$I = (-1)^{p+q}.D\ \frac{(-1)^r(i-j-1)!(j-1)!}{(i-1)!} \sum_{r=0}^{q} \mu^1_{rj}(i-1).\ q\ C_r.\ I_4$$

Following the evaluation of $I_3$

$$I_4 = \int_0^1 u_{(i)}^{\ i-1}\ (1-u_{(i)})^{n-i}\ \{\log\ (u_i)\}^{\ p+q-r}$$

$$= (-1)^{p+q-r}\ (n-i)!\ \frac{(i-1)!}{n!}\quad \mu^1_{(p+q-r)i}(n)$$

where $\mu^1_{(p+q-r)i}(n)$ is the $(p+q-r)^{th}$ moment about the origin of the $i^{th}$ order statistic (from the top) in a sample of size n from the exponential distribution.

Thus the expected value of the produce $w_{(i)}^{\ p}\ .\ w_{(j)}^{\ q}$ with $i > j$ is

$$\sum_{r=0}^{q} q\ C_r\ .\ \mu^1_{rj}(i-1)\ .\ \mu^1_{(p+q-r)i}(n)$$

TABLE 1 GAMMA DISTRIBUTION - EXPECTED VALUES OF ORDER STATISTICS

| Order (m) | $A_{(m)n}$ | $B_{(m)n}$ | No. of terms in the Taylor Expansion | | |
|---|---|---|---|---|---|
| | | | Two terms | Three terms | Four terms |
| r = 5, n = 100 | | | | | |
| 1 | 0.6894 | 11.6046 | 12.4491 | 12.3783 | 12.3968 |
| 2 | 0.6634 | 10.5804 | 10.9955 | 10.9620 | 10.9670 |
| 3 | 0.6455 | 9.9610 | 10.2411 | 10.2175 | 10.2201 |
| 4 | 0.6313 | 9.5104 | 9.7245 | 9.7055 | 9.7072 |
| 5 | 0.6192 | 9.1535 | 9.3284 | 9.3121 | 9.3134 |
| r = 10, n = 100 | | | | | |
| 1 | 0.5581 | 18.7831 | 19.8263 | 19.7083 | 19.7388 |
| 2 | 0.5301 | 17.5098 | 18.0292 | 17.9735 | 17.9817 |
| 3 | 0.5112 | 16.7312 | 17.0849 | 17.0457 | 17.0500 |
| 4 | 0.4965 | 16.1603 | 16.4325 | 16.4009 | 16.4037 |
| 5 | 0.4841 | 15.7052 | 15.9289 | 15.9017 | 15.9039 |
| r = 5, n = 50 | | | | | |
| 1 | 0.6634 | 10.5804 | 11.4655 | 11.3727 | 11.4011 |
| 2 | 0.6313 | 9.5014 | 9.9545 | 9.9086 | 9.9170 |
| 3 | 0.6086 | 8.8566 | 9.1619 | 9.1286 | 9.1332 |
| 4 | 0.5902 | 8.3767 | 8.6142 | 8.5866 | 8.5898 |
| 5 | 0.5744 | 7.9936 | 8.1908 | 8.1665 | 8.1691 |
| r = 10, n = 50 | | | | | |
| 1 | 0.5301 | 17.5098 | 18.6175 | 18.4629 | 18.5100 |
| 2 | 0.4965 | 16.1603 | 16.7249 | 16.6487 | 16.6626 |
| 3 | 0.4734 | 15.3244 | 15.7169 | 15.6614 | 15.6692 |
| 4 | 0.4550 | 14.7049 | 15.0129 | 14.9670 | 14.9724 |
| 5 | 0.4394 | 14.2060 | 14.4638 | 14.4233 | 14.4276 |

TABLE 2 NORMAL DISTRIBUTION - EXPECTED VALUES OF ORDER STATISTICS

| Order (m) | $A_{(m)n}$ | $B_{(m)n}$ | No. of terms in the Taylor Expansion | | | |
|---|---|---|---|---|---|---|
| | | | Two terms | Three terms | Four terms | Five terms |
| n = 50 | | | | | | |
| 1 | 2.4209 | 2.0537 | 2.2963 | 2.2345 | 2.2590 | 2.2407 (2.2491) |
| 2 | 2.1543 | 1.7507 | 1.8808 | 1.8502 | 1.8575 | 1.8536 (1.8549) |
| 3 | 1.9854 | 1.5548 | 1.6484 | 1.6260 | 1.6301 | 1.6281 (1.6286) |
| 4 | 1.8583 | 1.4051 | 1.4805 | 1.4619 | 1.4648 | 1.4634 (1.4637) |
| 5 | 1.7550 | 1.2816 | 1.3462 | 1.3297 | 1.3319 | 1.3309 (1.3311) |
| 6 | 1.6670 | 1.1750 | 1.2324 | 1.2173 | 1.2192 | 1.2183 (1.2184) |
| 7 | 1.5898 | 1.0803 | 1.1326 | 1.1184 | 1.1201 | 1.1193 (1.1195) |
| 8 | 1.5207 | 0.9945 | 1.0430 | 1.0295 | 1.0310 | 1.0303 (1.0305) |
| 9 | 1.4578 | 0.9154 | 0.9611 | 0.9480 | 0.9494 | 0.9488 (0.9489) |
| 10 | 1.3998 | 0.8416 | 0.8850 | 0.8723 | 0.8737 | 0.8731 (0.8731) |
| n = 500 | | | | | | |
| 1 | 3.1701 | 2.8782 | 3.0606 | 3.0319 | 3.0398 | 3.0356 |
| 2 | 2.9618 | 2.6521 | 2.7437 | 2.7311 | 2.7327 | 2.7322 |
| 3 | 2.8338 | 2.5121 | 2.5745 | 2.5660 | 2.5668 | 2.5666 |
| 4 | 2.7399 | 2.4089 | 2.4568 | 2.4502 | 2.4507 | 2.4506 |
| 5 | 2.6652 | 2.3263 | 2.3654 | 2.3600 | 2.3603 | 2.3602 |
| 6 | 2.6028 | 2.2571 | 2.2904 | 2.2856 | 2.2859 | 2.2858 |
| 7 | 2.5491 | 2.1973 | 2.2264 | 2.2221 | 2.2223 | 2.2223 |
| 8 | 2.5017 | 2.1444 | 2.1703 | 2.1664 | 2.1666 | 2.1666 |
| 9 | 2.4593 | 2.0969 | 2.1203 | 2.1168 | 2.1169 | 2.1169 |
| 10 | 2.4209 | 2.0537 | 2.0751 | 2.0718 | 2.0719 | 2.0719 |

TABLE 3 GAMMA DISTRIBUTION - COVARIANCES OF ORDER STATISTICS

| Order (j) | Order (i) | | r = 5 n = 50 | r = 5 n = 100 | r = 10 n = 50 | r = 10 n = 100 |
|---|---|---|---|---|---|---|
| 1 | 1 | (a) | 3.12395 | 3.0172 | 4.6278 | 4.3963 |
| | | (b) | 3.69265 | 3.4401 | 5.7833 | 5.2492 |
| 1 | 2 | (a) | 1.34107 | 1.2792 | 2.0615 | 1.9226 |
| | | (b) | 1.49266 | 1.3884 | 2.3752 | 2.1463 |
| 1 | 3 | (a) | 0.85028 | 0.8093 | 1.3300 | 1.2341 |
| | | (b) | 0.92913 | 0.8651 | 1.4949 | 1.3494 |
| 1 | 4 | (a) | 0.62248 | 0.5930 | 0.9865 | 0.9130 |
| | | (b) | 0.67432 | 0.6293 | 1.0946 | 0.9884 |
| 1 | 5 | (a) | 0.49060 | 0.4687 | 0.7841 | 0.7270 |
| | | (b) | 0.52885 | 0.4952 | 0.8652 | 0.7824 |
| 2 | 2 | (a) | 1.42116 | 1.3401 | 2.2293 | 2.0485 |
| | | (b) | 1.56856 | 1.4428 | 2.5359 | 2.2597 |
| 2 | 3 | (a) | 0.90297 | 0.8490 | 1.4419 | 1.3172 |
| | | (b) | 0.97638 | 0.8990 | 1.5960 | 1.4207 |
| 2 | 4 | (a) | 0.66169 | 0.6225 | 1.0695 | 0.9752 |
| | | (b) | 0.70861 | 0.6539 | 1.1687 | 1.0406 |
| 2 | 5 | (a) | 0.52179 | 0.4922 | 0.8516 | 0.7768 |
| | | (b) | 0.55574 | 0.5146 | 0.9237 | 0.8237 |
| 3 | 3 | (a) | 0.93821 | 0.8745 | 1.5166 | 1.3705 |
| | | (b) | 1.01279 | 0.9240 | 1.6739 | 1.4732 |
| 3 | 4 | (a) | 0.68809 | 0.6415 | 1.1261 | 1.0153 |
| | | (b) | 0.73504 | 0.6721 | 1.2257 | 1.0790 |
| 3 | 5 | (a) | 0.54287 | 0.5073 | 0.8971 | 0.8091 |
| | | (b) | 0.57647 | 0.5288 | 0.9688 | 0.8541 |
| 4 | 4 | (a) | 0.70970 | 0.6565 | 1.1725 | 1.0468 |
| | | (b) | 0.75795 | 0.6872 | 1.2753 | 1.1110 |
| 4 | 5 | (a) | 0.56017 | 0.5193 | 0.9346 | 0.8345 |
| | | (b) | 0.59444 | 0.5407 | 1.0080 | 0.8794 |
| 5 | 5 | (a) | 0.57540 | 0.5297 | 0.9677 | 0.8565 |
| | | (b) | 0.61079 | 0.5513 | 1.0438 | 0.9019 |

TABLE 4 NORMAL DISTRIBUTION - COVARIANCES OF ORDER STATISTICS

| Order (j) | Order (i) | n = 50 (a) | n = 50 (b) | n = 50 (c) | n = 500 (a) | n = 500 (b) |
|---|---|---|---|---|---|---|
| 1 | 1 | 0.1850 | 0.2773 | 0.2157 | 0.1282 | 0.1635 |
| 1 | 2 | 0.0930 | 0.1199 | 0.1023 | 0.0593 | 0.0685 |
| 1 | 3 | 0.0634 | 0.0780 | 0.0688 | 0.0390 | 0.0437 |
| 1 | 4 | 0.0487 | 0.0587 | 0.0525 | 0.0294 | 0.0324 |
| 1 | 5 | 0.0399 | 0.0474 | 0.0428 | 0.0237 | 0.0260 |
| 2 | 2 | 0.1079 | 0.1347 | 0.1165 | 0.0649 | 0.0733 |
| 2 | 3 | 0.0738 | 0.0877 | 0.0785 | 0.0428 | 0.0468 |
| 2 | 4 | 0.0568 | 0.0660 | 0.0600 | 0.0322 | 0.0347 |
| 2 | 5 | 0.0465 | 0.0533 | 0.0490 | 0.0260 | 0.0278 |
| 3 | 3 | 0.0807 | 0.0952 | 0.0855 | 0.0450 | 0.0489 |
| 3 | 4 | 0.0622 | 0.0716 | 0.0654 | 0.0339 | 0.0363 |
| 3 | 5 | 0.0510 | 0.0578 | 0.0535 | 0.0274 | 0.0290 |
| 4 | 4 | 0.0667 | 0.0765 | 0.0700 | 0.0352 | 0.0375 |
| 4 | 5 | 0.0547 | 0.0618 | 0.0572 | 0.0284 | 0.0300 |
| 5 | 5 | 0.0580 | 0.0654 | 0.0606 | 0.0293 | 0.0309 |

# ESTIMATION OF PARAMETERS OF EXTREME ORDER DISTRIBUTIONS OF EXPONENTIAL TYPE PARENTS

G Ramanchandran

Fire Research Station, UK

## SUMMARY

This paper is concerned with the estimation of parameters of asymptotic distributions of extreme order statistics from exponential type parents. Moment and maximum likelihood methods are discussed; the asymptotic efficiency of the former method relative to the latter is evaluated for the top ten order statistics.

## 1. INTRODUCTION

Consider N samples, each with n observations, from a distribution F(x) with density f(x). If the observations in the jth sample are arranged in decreasing order of magnitude let $x_{(m)nj}$ be the mth observation in that arrangement. It is known that, if F(x) is of the 'exponential type', the asymptotic density function of

$$y_{(m)j} = a_{(m)n} (x_{(m)nj} - b_{(m)n}) \tag{1}$$

is given by

$$\frac{m^m}{(m-1)!} \exp \left[-m\, y_{(m)j} - m \exp (-y_{(m)j})\right] \tag{2}$$



*J. Tiago de Oliveira (ed.), Statistical Extremes and Applications, 579–588.*

The parameters $a_{(m)n}$ and $b_{(m)n}$ are solutions of

$$F(b_{(m)n}) = 1 - (m/n) \tag{3}$$

$$a_{(m)n} = (n/m)\, f(b_{(m)n}) \tag{4}$$

The exponential type distributions, as defined by Gumbel (1958, p. 120) with reference to their limiting behaviour, include the normal, log normal, gamma, chi-square and logistic as well as the exponential distribution. These distributions are also in the 'domain of attraction' of Gumbel's double exponential law $\Lambda(x) = \exp(-\exp(-x))$ as defined by other authors on extreme value theory.

If the value of n and the precise form of F(x) are known, values of $a_{(m)n}$ and $b_{(m)n}$ can be easily obtained from equations (3) and (4). But in many practical situations this is not the case; only values of $x_{(m)nj}$ are usually available for N samples from F(x) which could be assumed to be of the exponential type. In such cases estimation of $a_{(m)n}$ and $b_{(m)n}$ is an important problem. The moment and maximum likelihood methods of estimation are discussed in this paper. The efficiency of the moment method relative to the maximum likelihood is evaluated for the top ten order statistics.

## 2. MOMENT METHOD

For typographical ease, hereafter, we shall drop the subscripts (m) and (m)n and use

$$x_{(m)nj} = x_j \qquad a_{(m)n} = a$$

$$y_{(m)j} = y_j \qquad b_{(m)n} = b$$

$$c_{(m)n} = c \quad \text{etc}$$

From equation (1), c (= 1/a) is first estimated by

$$c = s_x/\sigma_y \tag{5}$$

and then b by

$$\hat{b} = \bar{x} - \hat{c}\,\bar{y} \tag{6}$$

where $\bar{x}$ and $s^2_x$ are the mean and variance of the observations $x_j$ (j = 1,2 ...N) pertaining to the mth order. The asymptotic mean $\bar{y}$ and variance $\sigma^2_y$ of $y_j$ are known to be given by

$$\bar{y} = \nu + \log_e m - \sum_{v=1}^{m-1} (1/v) \tag{7}$$

$$\sigma^2_y = \pi^2/6 - \sum_{v=1}^{m-1} (1/v^2) \tag{8}$$

where

$\nu$ = Euler's constant = 0.5772 ..... and $\pi^2/6 = 1.6449$ ...

The variance of $\hat{c}$ using the result of Cramer (1951, p. 353), is

$$\hat{V}(\hat{c}) = V(s)/\sigma^2_y$$

$$= \frac{1}{\sigma^2_y} \frac{\mu_{m4x} - \mu^2_{m2x}}{4\ N\ \mu_{m2x}}$$

where $\mu_{m4x}$ and $\mu_{m2x}$ are the fourth and second central moments of $x_j$ and $\mu_{m2x} = s^2_x$.

If $\beta_{m^2x} = \mu_{m4x}/\mu^2_{m2x}$ is the asymptotic kurtosis coefficient of $x_j$.

$$V(\hat{c}) = c^2\ (\beta_{m2x} - 1)/4N \tag{9}$$

The variance of $\hat{b}$ is given by

$$V(\hat{b}) = V(\bar{x}) + \bar{y}^2\ V(\hat{c}) - 2\ \bar{y}\ \mathrm{Cov}[\bar{x},\ \hat{c}]$$

$$= (\hat{c}^2/N)[\sigma^2_y + \frac{\bar{y}^2\ (\beta_{m2x} - 1)}{4}]$$

$$- (2\bar{y}/\sigma_y)\ \mathrm{Cov}[\bar{x},\ s_x]$$

Following Menon (1963)

$$\mathrm{Cov}[\bar{x},\ s_x] = \mu_{m3x}/2\ N\ (\mu_{m2x})^{\frac{1}{2}}$$

approximately where $\mu_{m^3x}$ is the third central moment of $x_j$. If $\beta_{m1x} = \mu^2_{m3x}/\mu^3_{m2x}$ is the asymptotic coefficient of skewness of $x_j$

$$V(\hat{b}) = \frac{\hat{c}^2}{N} \left[\sigma^2_y + \frac{\bar{y}^2 (\beta_{m2x} - 1)}{4} - \sigma_y \bar{y} (\beta_{m1x})^{\frac{1}{2}}\right] \tag{10}$$

## 3. MAXIMUM LIKELIHOOD METHOD

From the joint density function

$$\left[\frac{m^m a}{(m-1)!}\right]^N \exp\left[- m \sum_{j=1}^{N} y_j - m \sum_{j=1}^{N} \exp(-y_j)\right]$$

the log likelihood, omitting the absolute constants, is

$$L = N \log a - m \sum_{j=1}^{N} y_j - m \sum_{j=1}^{N} \exp(-y_j)$$

Differentiating L with respect to a and b

$$\frac{\partial L}{\partial a} = \frac{N}{a} - m \sum_{j=1}^{N} (x_j - b) + m \sum_{j=1}^{N} \exp(-y_j)(x_j - b)$$

$$\frac{\partial L}{\partial b} = m\, a \left[N - \sum_{j=1}^{N} \exp(-y_j)\right]$$

Maximum likelihood estimates are therefore given by

$$\sum_{j=1}^{N} \exp(-y_j) = \exp(a \,.\, b) \sum_{j=1}^{N} \exp(-a .\, x_j) = N \tag{11}$$

$$(1/a) = m\,\bar{x} - m \frac{\sum_{j=1}^{N} x_j \exp(-a \,.\, x_j)}{\sum_{j=1}^{N} \exp(-a \,.\, x_j)} \tag{12}$$

Equations (11) and (12) can be solved for a and b only by an iterative process. Estimates provided by the method of moments may be used as initial values. The maximum likelihood estimation in the case of the largest value (m=1) has been discussed by Kimball (1946) and Mann (1968).

Consider now

$$\frac{d^2L}{da^2} = -\frac{N}{a^2} - \frac{m}{a^2} \sum_{j=1}^{N} y^2_j \exp(-y_j)$$

Using equation (2) and taking the expectation

$$E\left[y^2_j \exp(-y_j)\right]$$

$$= \frac{m^m}{(m-1)!} \int_{-\infty}^{\infty} \exp\left[-(m+1)y_j - m \exp(-y_j)\right]y^2_j \, dy \qquad (13)$$

we have

$$\frac{d}{dy}\left[\exp\left\{-my - m\exp(-y)\right\} y^2\right]$$

$$= (2y - my^2)\exp\left[-my - m\exp(-y)\right]$$

$$+ my^2 \exp\left[-(m+1)y - m\exp(-y)\right] \qquad (14)$$

Therefore, the right hand side of equation (13) is equal to

$$0 + \frac{m^m}{(m-1)!} \int_{-\infty}^{\infty} (y^2_j - \frac{2y_j}{m}) \exp\left[-my_j - m\exp(-y_j)\right] dy_j$$

$$= \sigma^2_y + \bar{y}^2 - \frac{2}{m}\bar{y}$$

Hence

$$E\left(\frac{\partial^2 L}{\partial a^2}\right) = -\frac{N}{a^2}\left[1 + m(\sigma^2_y + \bar{y}^2) - 2\bar{y}\right]$$

It may be seen that, since the expected value of $\exp(-y_j)$ is

$$\frac{m^m}{(m-1)!} \int_{-\infty}^{\infty} \exp\left[-(m+1)y_j - m\exp(-y_j)\right]dy_j = 1$$

$$E\left(\frac{\partial^2 L}{\partial b^2}\right) = -m a^2 E \sum_{j=1}^{N} \exp(-y_j) = -N ma^2$$

As before, since

$$\frac{d}{dy} \left[\exp\left\{- my - m \exp(-y)\right\} y\right]$$

$$= (1 - my) \exp\left[- my - m \exp(-y)\right] + my \exp\left[-(m+1)y - m \exp(-y)\right]$$

$$E\left(\frac{\partial^2 L}{\partial b\, \partial a}\right) = m N E \left[\exp(-y_j)\, y_j\right] = N (m\bar{y} - 1)$$

The variances and covariance of $\hat{a}$ and $\hat{b}$ are therefore given by

$$V(\hat{a}) = - E \frac{\partial^2 L}{\partial b^2} \quad /D$$

$$= ma^2 \,/\, \left[N \left\{(m-1) + m^2 \sigma^2{}_y\right\}\right] \tag{15}$$

$$V(\hat{b}) = - E \left(\frac{\partial^2 L}{\partial a^2}\right) \quad /D$$

$$= \left[1 + m(\sigma^2 y + \bar{y}^2) - 2\bar{y}\right] \,/\, \left[Na^2 \left\{(m-1) + m^2 \sigma^2{}_y\right\}\right] \tag{16}$$

$$\mathrm{Cov}|\hat{a}, \hat{b}| = E \left(\frac{\partial\ {}^2 L}{\partial a\ db}\right) /D \tag{17}$$

$$= (m\ \bar{y} - 1)/\left[N\left\{(m-1) + m^2\ \sigma^2{}_y\right\}\right]$$

$$D = E\left(\frac{\partial^2 L}{\partial a^2}\right) \ E \ \left(\frac{\partial^2 L}{\partial b^2}\right) - \left[E\left(\frac{\partial^2 L}{\partial a\ \partial b}\right)^2\right]$$

$$= N^2 \left[(m-1) + m^2\ \sigma^2{}_y\right]$$

For the largest value, m = 1, the following results agree with those of Kimball (1946)

$$V(\hat{a}_{(1)n}) = \hat{a}^2{}_{(1)n} \,/\, \left[N\ \pi^2/6\right]$$

$$V(\hat{b}_{(1)n}) = \left[1 + \left\{(1-\nu)^2/(\pi^2/6)\right\}\right]/Na^2_{(1)n}$$

$$\mathrm{Cov}\left[\hat{a}_{(1)n}, \hat{b}_{(1)n}\right] = - (1-\nu)/N\ \pi^2/6$$

where we have used $y_{(1)} = \nu$ (Euler's constant) and $\sigma^2{}_{(1)y} = \pi^2/6$

## 4. ASYMPTOTIC EFFICIENCY

Using the moment generating function

$$G(t) = m^t \, \Gamma(m-t)/\Gamma(m)$$

it can be seen that the third and fourth central moments of $y_j$ are

$$\mu_{m3y} = 2 \sum_{v=1}^{\infty} (1/v^3) - 2 \sum_{v=1}^{m-1} (1/v^3)$$

$$\mu_{m4y} = 6 \sum_{v=1}^{\infty} (1/v^4) - 6 \sum_{v=1}^{m-1} (1/v^4)$$

The skewness and kurtosis coefficients of $x_j$ are asymptotically equal to those of $y_j$ so that

$$\beta_{m1x} = \beta_{m1y} = \mu^2{}_{m3y}/\mu^3{}_{m2y}$$

$$\beta_{m2x} = \beta_{m2y} = \mu_{m4y}/\mu^2{}_{m2y}$$

where $\mu_{m2y} = \sigma^2{}_y$ as given by equation (8).

It can be shown by using the $\Delta$ method that asymptotically the coefficient of variation of $\hat{c}$ is equal to the coefficient of variation of its reciprocal $\hat{a}$ ($= 1/\hat{c}$) - Ehrenfeld and Littauer (1964, p. 129). Hence in the case of the moment method, from equation (9).

$$V(\hat{a})/\hat{a}^2 = V(\hat{c})/\hat{c}^2 = (\beta_{m2y} - 1)/4N$$

Therefore, for the moment estimator

$$k(\hat{a}) = NV(\hat{a})/\hat{a}^2 = (\beta_{m2y} - 1)/4$$

and for the maximum likelihood estimator, from equation (15),

$$k(\hat{a}) = (m)/\left[(m-1) + m^2 \sigma^2{}_y\right]$$

TABLE 1. RELATIVE EFFICIENCY

| Extreme (m) | $\bar{y}$ | $\sigma^2_y$ | $\beta_{m1y}$ | $\beta_{m2y}$ | k(a) | | | k(b) | | |
|---|---|---|---|---|---|---|---|---|---|---|
| | | | | | Moment estimator | Maximum likelihood estimator | Relative efficiency Col(7)/Col(6) | Moment estimator | Maximum likelihood estimator | Relative efficiency Col(10)/Col(9) |
| (1) | (2) | (3) | (4) | (5) | (6) | (7) | (8) | (9) | (10) | (11) |
| 1 | 0.5772 | 1.6449 | 1.2987 | 5.4000 | 1.1000 | 0.6079 | 0.5526 | 1.1679 | 1.1087 | 0.9493 |
| 2 | 0.2704 | 0.6449 | 0.6091 | 4.1874 | 0.7969 | 0.5587 | 0.7011 | 0.5338 | 0.5294 | 0.9918 |
| 3 | 0.1758 | 0.3949 | 0.3862 | 3.7616 | 0.6904 | 0.5401 | 0.7823 | 0.3476 | 0.3467 | 0.9974 |
| 4 | 0.1302 | 0.2838 | 0.2815 | 3.5584 | 0.6396 | 0.5304 | 0.8293 | 0.2578 | 0.2576 | 0.9992 |
| 5 | 0.1033 | 0.2213 | 0.2196 | 3.4284 | 0.6071 | 0.5245 | 0.8639 | 0.2048 | 0.2048 | 1.0000 |
| 6 | 0.0857 | 0.1813 | 0.1803 | 3.3466 | 0.5867 | 0.5205 | 0.8872 | 0.1702 | 0.1701 | 1.0000 |
| 7 | 0.0731 | 0.1535 | 0.1537 | 3.3061 | 0.5765 | 0.5177 | 0.8980 | 0.1455 | 0.1454 | 1.0000 |
| 8 | 0.0637 | 0.1331 | 0.1344 | 3.2683 | 0.5671 | 0.5155 | 0.9090 | 0.1271 | 0.1270 | 1.0000 |
| 9 | 0.0565 | 0.1175 | 0.1256 | 3.2248 | 0.5562 | 0.5138 | 0.9238 | 0.1126 | 0.1126 | 1.0000 |
| 10 | 0.0508 | 0.1051 | 0.1085 | 3.2138 | 0.5535 | 0.5126 | 0.9261 | 0.1013 | 0.1013 | 1.0000 |

The function k(a) is a scaled coefficient of variation. Results given in Table 1 show that the moment method is less efficient than the maximum likelihood for estimating a; but the loss of efficiency decreases with m.

For the parameter $\hat{b}$, from equation (10)

$$k(\hat{b}) = N\hat{a}^2 V(\hat{b})$$

$$= \sigma^2_y + \frac{\bar{y}^2 (\beta_{m2y^{-1}})}{4} - \sigma_y \bar{y} (\beta_{m1y})^{\frac{1}{2}}$$

for the moment method and from equation (16),

$$k(\hat{b}) = \left[1 + m(\sigma^2_y + \bar{y}^2) - 2\bar{y}\right] / \left[(m-1) + m^2 \sigma^2_y\right]$$

for the maximum likelihood estimator. Results indicate that, in the case of b efficiency of the moment estimator relative to maximum likelihood is as high as 0.95 even for $m = 1$ and almost unity for $m > 5$. The results for $m = 1$ in Tables 1 and 2 agree with those of Menon (1963), for the parameters of Weibull distribution which could be transformed to the extreme (largest) value distribution.

This paper is based on Chapter 2 of the author's Ph.D. thesis (1975).

ACKNOWLEDGEMENT

The work described has been carried out as part of the research programme of the Building Research Establishment of the Department of the Environment and this paper is published by permission of the Director.

REFERENCES

1 Cramer, H. (1951), Mathematical methods of statistics. Princeton University Press, Princeton, N.J.

2 Ehrenfeld, S and Littauer, S.B. (1964), Introduction to statistical Method, McGraw Hill, New York.

3 Gumbel, E.J. (1958), Statistics of extremes, Columbia University Press, New York.

4 Kimball, B.F. (1946), Sufficient statistical estimation functions for the parameters of the distribution of maximum values. Ann. Math. Statist., 17, pp. 299-309.

5 Mann, N.R. (1968), Point and Interval estimation procedures for the two parameter Weibull and extreme value distributions. Technometrics, 10, pp. 231-256.

6 Menon, M.V. (1963), Estimation of the shape and scale parameters of the Weibull distribution. Technometrics, 5, pp. 175-182.

7 Ramachandran, G. (1975), Extreme order statistics from exponential type distributions with applications to fire protection and insurance. Ph.D. thesis, University of London.

ON ORDERED UNIFORM SPACINGS FOR TESTING GOODNESS OF FIT

S. Rao Jammalamadaka

Univ. of California
Santa Barbara
U.S.A.

ABSTRACT. Tests for the goodness of fit problem, based on sample spacings i.e., observed distances between successive order statistics, have been used in the literature. This paper reviews some recent work on tests which make use of ordered spacings, like the largest spacing, sum of the $k$ largest spacings and a test based on counting the number of "small" spacings where "small" is defined so as to optimize the large sample efficiency of the test.

## 1. INTRODUCTION

Many interesting statistical problems can be reduced to the following simple form: given some independent and identically distributed (i.i.d.) observations on $[0,1]$, test if they are uniformly distributed on the unit interval. Such problems include

(i) testing goodness of fit: given $X_1,\ldots,X_{n-1}$ i.i.d. from some cumulative distribution function (cdf) $F$, test if this $F$ is a given (continuous) cdf $F_0$. By making the so called probability integral transformation $U_i=F_0(X_i)$, $i=1,\ldots,(n-1)$ on the data, this reduces to the interval $[0,1]$ and to testing uniformity.

(ii) testing for a Poisson process and/or exponentiality of inter-arrivals: given the times of occurrences of events in a finite interval, one would want to verify if these were generated by a Poisson process. From the property that a homogeneous Poisson process, suitably scaled, behaves like the uniform distribution, this problem is equivalent to testing uniformity on the unit interval.



J. Tiago de Oliveira (ed.), Statistical Extremes and Applications, 589–596.

(iii) testing for no preferred direction in circular data: A novel area of statistics is where the measurements are directions. Such directions in 2-dimensions can be represented as points on the perimeter of a unit circle and are referred to as the circular data. See for instance J.S. Rao [6]. One of the important questions here is whether the data is uniformly distributed (isotropic) or if indeed, there is a preferred direction. By cutting open the circle at anyone of the observations, this problem reduces to testing uniformity on $[0,1]$.

Among the several possible approaches to testing uniformity which include empirical distribution function methods and $\chi^2$ methods, we would like to focus on those which utilize the spacings. Let $U_1,\ldots,U_{n-1}$ be $(n-1)$ i.i.d. random variables (r.v.'s) from a continuous cdf $F(u)$ on $[0,1]$. The null hypothesis of interest is

$$H_0: F(u) = u, \; 0 \leq u \leq 1. \tag{1.1}$$

Define the order statistics

$$0 = U_{0,n} \leq U_{1,n} \leq \ldots \leq U_{n-1,n} \leq U_{n,n} = 1$$

and the spacings

$$Y_{i,n} = U_{i,n} - U_{i-1,n}, \quad i = 1,\ldots,n. \tag{1.2}$$

For notational convenience, we drop the second subscript $n$ on the spacings and refer to them as $\{Y_i, i=1,\ldots,n\}$ which add upto 1. It is easy to see that the joint distribution of $(Y_1,\ldots,Y_{n-1})$ is a Dirichlet distribution with constant density $(n-1)!$ over $\{\underline{y}: y_i \geq 0, \sum_1^{n-1} y_i \leq 1\}$. Recall that a collection of r.v.'s $(Y_1,\ldots,Y_b)$ is said to have a Dirichlet distribution with parameters $(\nu_1,\ldots,\nu_b; \nu_{b+1})$, written as $D(\nu_1,\ldots,\nu_b; \nu_{b+1})$ if they have the joint density

$$\frac{\Gamma(\nu_1+\ldots+\nu_{b+1})}{\Gamma(\nu_1)\ldots\Gamma(\nu_{b+1})}\left(\prod_{i=1}^{b} y_i^{\nu_i-1}\right)(1-y_1-\ldots-y_b)^{\nu_{b+1}-1} \tag{1.3}$$

over the b-dimensional simplex $S_b = \{(y_1,\ldots,y_b): y_i \geq 0,$ $i = 1,\ldots,b,$ $\sum_1^b y_i \leq 1\}$ and zero outside $S_b$. For an elementary exposition on Dirichlet distributions and some properties, see Wilks [9], pp. 177-182 and for a more detailed discussion as well as tables, see Sobel et al [8]. One important property that we shall use later on, is that the marginal distributions of Dirichlet are also Dirichlet i.e., if $a < b$, then $(Y_1,\ldots,Y_a) \frown D(\nu_1,\ldots,\nu_a;$ $\nu_{a+1}+\ldots+\nu_{b+1})$ where "$\frown$" denotes "distributed as".

Coming back to uniform spacings, these are exchangeable with $E(Y_i) = \frac{1}{n}$ for all $i$ under $H_0$. Thus tests of the form $\sum_{i=1}^{n} (Y_i - \frac{1}{n})^2$ and $\sum_{i=1}^{n} |Y_i - \frac{1}{n}|$ have been proposed and used to test $H_0$. See for instance Rao and Sethuraman [4], [5] for a unified treatment of the asymptotic theory and efficiencies. Simpler and more intuitive tests based on ordered spacings have also been used for this problem. Let

$$Y_{(1)} \leq \cdots \leq Y_{(n)} \tag{1.4}$$

be the ordered uniform spacings. Fisher [1] used the largest spacing $Y_{(n)}$ to construct a test of significance of the largest amplitude in harmonic analysis. J.S. Rao [3] defined the complement of the largest gap on the circle as the "circular range" since this is the smallest interval containing all the observations and proposed a test of uniformity on the circle based on it.

## 2. SOME RESULTS ON EXACT DISTRIBUTIONS

For $b \leq (n-1)$, consider the $b$ spacings $(Y_1,\ldots,Y_b)$ which have a $D(1,\ldots,1;\ (n-b))$ distribution. J.S. Rao and Sobel [7] define the following two b-dimensional incomplete Dirichlet integrals:

$$I_p^{(b)}(1,n) = \frac{(n-1)!}{(n-b-1)!} \int_0^p \cdots \int_0^p (1-\sum_1^b y_i)^{n-b-1} \prod_1^b dy_i \tag{2.1}$$

$$\{\sum_1^b y_i \le 1\}$$

for $0 < p < 1$ and

$$J_p^{(b)}(1,n) = \frac{(n-1)!}{(n-b-1)!} \int_p^1 \ldots \int_p^1 (1-\sum_1^b y_i)^{n-b-1} \prod_1^b dy_i \qquad (2.2)$$
$$\{\sum_1^b y_i \le 1\}$$

for $0 < p < \frac{1}{b}$. These integrals represent respectively, $P(Y_i < p, i=1,\ldots,b)$ ie., the maximum of these $b$ spacings is less than $p$ and $P(Y_i > p, i=1,\ldots,b)$ ie., the minimum of these $b$ spacings exceeds $p$. Using recurrence relations on these, one can then obtain the distributions of ordered spacings and statistics based on these. For instance, it can be seen

$$I_p^{(b)}(1,n) = I_p^{(b-1)}(1,n) - (1-p)^{n-1} I_{p/(1-p)}^{(b-1)}(1,n) \qquad (2.3)$$

and by successive iterations, this reduces to

$$I_p^{(b)}(1,n) = \sum_{j=0}^{b} (-1)^j \binom{b}{j} \langle 1 - jp \rangle^{n-1}, \qquad (2.4)$$

where $\langle x \rangle = x$ if $x > 0$ and $=0$ if $x \le 0$. It can be shown (refer Rao and Sobel [7]) that the joint distribution of the $k$ largest spacings viz. $Y_{(n-k+1)}, \ldots, Y_{(n)}$ is given by an I-type integral

$$f_{Y_{(n-k+1)},\ldots,Y_{(n)}}(a_1,\ldots,a_k) = (n-1)_{P_k} \cdot n_{P_k} \cdot A_k^{n-1-k} I_{p_k}^{(n-k)}(1,n-k-1)$$
$$= (n-1)_{P_k} \cdot n_{P_k} \cdot \sum_{j=0}^{\infty} (-1)^j \binom{n-k}{j} \langle A_k - ja_k \rangle^{n-k} \qquad (2.5)$$

for $0 < a_1 < \ldots < a_k < 1$, $\sum_1^k a_i \le 1$, and with the notation

$A_k = (1 - \sum_1^k a_i)$ and $p_k = a_k/A_k$. Starting from (2.5) one can obtain the distribution of the $k^{th}$ largest spacing $Y_{(n-k+1)}$ or the sum of the $k$ largest gaps $\sum_{i=n-k+1}^{n} Y_{(i)}$. The density functions of these two statistics are

$$f_{Y_{(n-k+1)}}(x) = (n-1).k.\binom{n}{k} \sum_{j=0}^{n-k} (-1)^j \binom{n-k}{j} \langle 1-(j+k)x \rangle^{n-2} \quad (2.6)$$

for $0 < x < \frac{1}{k}$ and

$$f_S(s) = n!(n-1) \sum_{j=q}^{n} \frac{(-1)^{k-j+1}}{(j-k)!(j-k)^{k-1}.k!k^{n-k-1}(n-j)!} \langle js-k \rangle^{n-2} \quad (2.7)$$

for $k/q < s \leq k/(q-1)$, $q=k+1,\ldots,n$. Analogous results for the smallest spacings may be obtained by using the J-function in (2.2) or more easily by using the complementary nature of these, namely that the $k^{th}$ largest is the $(n-k+1)^{th}$ smallest and that the sum of the $k$ smallest spacings is equal to the complement of the sum of the $(n-k)$ largest spacings. These and other details including fuller derivations may be found in Rao and Sobel [7] which streamlines and unifies the distribution theory, also derived by other methods in the literature before.

## 3. A TEST BASED ON THE NUMBER OF "SMALL" SPACINGS

To detect certain clustering alternatives one may simply count the number of "small" spacings and reject the hypothesis of uniformity if there are too many "small" spacings. This type of a statistic was investigated in Puri, Rao and Yoon [2]. Recall that an average spacing is of the order of $\frac{1}{n}$ under $H_0$. Therefore one may define for $0 < \delta < 1$, a spacing as "small" if it is smaller than $\delta_n = \delta/n$. Let

$$R_n = R_n(\delta) = \text{number of } Y_i \leq \delta/n . \quad (3.1)$$

The following results are quoted from Puri, Rao and Yoon (Ibid)

Theorem 1 (Exact null distribution). Under the hypothesis of uniformity, the probability mass function of $R_n$ defined in (3.1) is given by

$$P(R_n = k) = \binom{n}{k} \sum_{j=0}^{k} (-1)^j \binom{k}{j} \langle 1-(n-k+j)\delta_n \rangle^{n-1} \tag{3.2}$$

for $k=0,1,\ldots,n-1$ .

Theorem 2 (Asymptotic distribution under close alternatives). When $U_1,\ldots,U_{n-1}$ are i.i.d. from the sequence of close alternatives densities

$$a_n(x) = 1 + \frac{\ell(x)}{n^{1/4}}, \quad 0 \leq x \leq 1 \tag{3.3}$$

then $\sqrt{n}\left(\frac{R_n}{n} - G_n(\delta)\right)$ is asymptotically $N(0,\sigma^2)$ where

$$G_n(x) = (1-e^{-x}) + \frac{e^{-x}}{\sqrt{n}}\left(x-\frac{x^2}{2}\right)\left(\int_0^1 \ell^2(p)dp\right), \quad x \geq 0$$

and

$$\sigma^2 = e^{-\delta}(1-e^{-\delta}-\delta^2 e^{-\delta}) .$$

The asymptotic null distribution of $R_n$ is obtained from Theorem 2 by putting $\ell(x) \equiv 0, 0 \leq x \leq 1$, which affects only the mean. This type of a result allows one to compute the asymptotic relative efficiency (Pitman efficiency), which is the reciprocal of the sample size required for the test to attain a specified power. Because of the nature of the alternatives (3.3), the Pitman efficiency for $R_n(\delta)$ is given by the expression

$$\left(\frac{\mu_\Delta}{\sigma}\right)^4 = \frac{\left(\int_0^1 \ell^2(p)dp\right)^4 \left(\delta-\frac{\delta^2}{2}\right)^4}{(e^{\delta}-1-\delta^2)^2} \tag{3.4}$$

where $\mu_\Delta$ denotes the change in the mean under the alternatives from that under the null hypothesis. By numerical evaluations, it can be seen that among the several possible definitions of "small" ie. choices of $\delta$, the value $\delta = 0.7379$ yields the largest asymptotic efficiency. Thus, an optimal definition of a "small" spacing is when the spacing is about 73.79% smaller than the average value $1/n$. One may use this as the definition of "small" spacing and test the hypothesis (1.1) using Theorem 1 for the null distribution.

An illustrative example. Suppose a fire station received 20 calls on a particular day and we wish to test if these are uniformly distributed over the entire day or they tend to cluster around some particular time of the day. Suppose the calls are received at the following times:

1.10, 4.30, 6.00, 6.10, 7.00, 8.00, 8.30, 8.45, 9.30, 10.05, 13.00, 14.10, 16.00, 17.50, 19.30, 21.15, 22.00, 22.15, 23.00, 23.30.

The optimal definition of "small" spacing is when it is smaller than $\delta_n = (0.7379)\frac{24}{20}$ hrs. $\approx$ 53 mts. The observed $R_n$, the number of small spacings, for this data is, 10. From Theorem 1, it can be checked that this value of $R_n = 10$ is not significant even $\alpha = 0.10$. Therefore, it may be concluded that there is no reason to reject the hypothesis of randomness of these calls in time.

REFERENCES

Fisher, R.A. (1929), "Tests of Significance in Harmonic Analysis". Proc. Roy. Soc. Ser. A. 125, pp. 54-59.

Puri, M.L., Rao, J.S. and Yoon, Y. (1979), "A Simple Test for Goodness of Fit Based on Spacings with some Efficiency Comparisons", in Contributions to Statistics (J. Jurecková, Ed.) Academia (Prague), pp. 197-209.

Rao, J.S. (1969), "Some Contributions to the Analysis of Circular Data". Unpublished Ph. D. thesis, Indian Statistical Inst., Calcutta.

Rao, J.S. and Sethuraman, J. (1970), "Pitman Efficiencies of Tests based on Spacings", in Nonparametric Techniques in Statistical Inference (Ed. M.L. Puri) Cambridge Univ. Press, pp. 405-415.

Rao, J.S. and Sethuraman, J. (1975), "Weak Convergence of Empirical Distribution Functions of Random Variables Subject to Perturbations and Scale Factors". Ann. Statist. 3, pp. 299-313.

Rao, J.S. (1976), "Some Tests based on Arc Lengths for the Circle". Sankhyā, Ser. B. 38, pp. 329-338.

Rao, J.S. and Sobel, M. (1980), "Incomplete Dirichlet Integrals with Applications to Ordered Uniform Spacings". Jour. Multivariate Analysis, 10, pp. 603-610.

Sobel, M., Uppuluri, V.R.R. and Frankowski, K. (1977), "Dirichlet Distributions Type 1", in Selected Tables in Math. Statistics, Vol. IV, Amer. Math. Soc., Providence, R.I.

Wilks, S.S. (1962), "Mathematical Statistics". John Wiley, N.Y., pp. 177-182.

# INEQUALITIES FOR THE RELATIVE SUFFICIENCY BETWEEN SETS OF ORDER STATISTICS

R.-D. Reiss, M. Falk and M. Weller

University of Siegen

By following ideas of Weiss [8] we investigate the approximate relative sufficiency between certain sets of ordered observations. The results enable us to give some bounds for the loss of power of test procedures when some of the observations are omitted from the sample. Moreover, it is shown that statistical procedures which are based on central ordered values or, respectively, on extreme values can approximately be studied within a normal model or an univariate extreme value model.

## 1. INTRODUCTION AND AN OUTLINE OF THE MAIN IDEAS

Let $z_1 \leq \ldots \leq z_n$ be the ordered sample of independent and identically distributed random observations $x_1, \ldots, x_n$ with common distribution function F. It is well known that the joint distribution of any subcollection, say $z_{r_1} \leq \ldots \leq z_{r_k}$, of the ordered sample can explicitly be described by its joint density function however this function has a complex structure. Thus it is desirable to establish asymptotic results to get theoretical insight and to simplify the statistical inference. In order to indicate that the asymptotic results can be meaningful for practical purposes it is necessary to construct error bounds. In connection with ordered samples such work has been done e.g. in [3], [4] and [5].

The present study concerns the amount of information which is lost if some of the observations are omitted from the ordered sample. As an example let us consider the extreme values $z_1 \leq \ldots \leq z_m$ of a sample of size n with the common distribution



J. Tiago de Oliveira (ed.), Statistical Extremes and Applications, 597–610.

function $F_o$. It is intuitively clear that no information is lost if $z_1 \leq \ldots \leq z_m$ is replaced by $y_1 \leq \ldots \leq y_{m-1} \leq z_m$ where, given $z_m$, $y_1, \ldots, y_{m-1}$ is distributed as the ordered sample of m-1 independent random variables which are distributed according to the appropriately defined restriction of $F_o$ to $(0, z_m)$. In the next step this procedure is also applied when the original sample has a common distribution function F which is close - in a sense to be described later - to $F_o$; thus, the values $y_1 \leq \ldots \leq y_{m-1}$ are still generated according to the restriction of the distribution function $F_o$ being known to the statistician (see also Weiss [8], page 796). It can be proved that for every critical function $\psi$ the power function of $\psi(y_1, \ldots, y_{m-1}, z_m)$ is an approximation to the power function of $\psi(z_1, \ldots, z_m)$ if F is close to $F_o$. These ideas can be expressed in a mathematical model by using the conditional distribution $K(z_m, \cdot)$ of $z_1 \leq \ldots \leq z_m$ given $z_m$ where $F_o$ is the actual distribution function. The final outcomes $y_1 \leq \ldots \leq y_{m-1} \leq z_m$ of the random experiment are governed by the distribution which is induced by $z_m$ and the Markov kernel K. By making use of the Fubini theorem for the distributions of Markov kernels it can easily be shown that the test procedure based on $\psi(y_1, \ldots, y_{m-1}, z_m)$ is equivalent - as far as power functions are concerned - to the test procedure based on $\tilde{\psi}(z_m)$ where $\tilde{\psi} = \int \psi(x) K(\cdot, dx)$. Thus, it is not necessary that the statistician consults his random number generator to obtain the values $y_1, \ldots, y_{m-1}$ as described above.

One has to be cautious when applying this method to other cases. As a second example we mention the case where $z_m$ is omitted from $z_1 \leq \ldots \leq z_m$. It can be shown by examples that there are cases where, roughly speaking, all the information which is of interest for us concerning the unknown distribution function F is carried by $z_m$; thus, omitting $z_m$ from $z_1, \ldots, z_m$ can possibly result in the loss of all power within a particular testing problem.

Moreover, we indicate the possibility of making the statistical inference within an approximate, simplified model. For this reason we study again the case that the ordered values $z_1 \leq \ldots \leq z_{m-1}$ are removed from $z_1 \leq \ldots \leq z_m$. If the underlying distribution function belongs to some domain of attraction of an extreme value distribution then it is well known that for m being fixed the standardized distributions of $z_m$ and $(z_1, \ldots, z_m)$ converge in distribution (weakly) to an extreme value variable and, respectively, to an extreme value vector as the sample size goes to infinity. Under appropriate regularity conditions this convergence also holds in a stronger sense (see [5]). These limit theorems show that our statistical investigations which are based on $z_m$ or on $(z_1,\ldots,z_m)$ can asymptotically be made within an univariate or a multivariate extreme value model. Moreover, in view of the results indicated above it is possible to replace the multivariate model by the univariate one.

The paper is organized as follows: In Section 2 we shall state the results, indicate the proofs and make some further discussions. Two highly technical proofs will be postponed until Section 3.

## 2. THE MAIN RESULTS

Hereafter we shall use a notation which describes the general concept as short and transparent as possible; in our opinion this can be done most successfully by using a mathematical language. Given a probability measure (p-measure) P on the Borel-σ-algebra $\mathbb{B}$ on the real line $\mathbb{R}$ let $P^n$ and $\mathbb{B}^n$ denote the n-fold product of P and $\mathbb{B}$. The i-th order statistic $Z_{i:n}:\mathbb{R}^n \to \mathbb{R}$ (of a sample of size n) is defined by $Z_{i:n}(\underline{x}) = z_i$ for $\underline{x} = (x_1,\ldots,x_n) \in \mathbb{R}^n$ where $z_1 \leq \ldots \leq z_n$ are the components of $\underline{x}$ arranged in the nondecreasing order. The p-measure induced by $P^n$ and a measurable map $\psi$ will be denoted by $P^n * \psi$.

For certain families of p-measures P and a vector $\underline{r} = (r_1,\ldots,r_k)$ with $1 \leq r \leq r_1 < \ldots < r_k \leq s \leq n$ we shall give an estimate of

$$|P^n\{(Z_{i:n})_{i=r}^{s} \in B\} - \int K_{\underline{r}}((Z_{r_i:n})_{i=1}^{k},B)dP^n|$$

uniformly over measurable sets B where $K_{\underline{r}}$ is a Markov kernel which does not depend on the special p-measure P. This implies, in particular, that for every critical function of the form $\psi((Z_{i:n})_{i=r}^{s})$ we can construct a critical function $\tilde{\psi}((Z_{r_i:n})_{i=1}^{k})$ where $\tilde{\psi} = \int \psi(\underline{z})K_{\underline{r}}(\cdot,d\underline{z})$ which has the same power function with respect to the p-measure $P^n$ within the given error bound. Our first result will be formulated by means of a Markov kernel $K_{\underline{r}}$ which has the property

$$\int K_{\underline{r}}((Z_{r_i:n})_{i=1}^{k},\cdot)\,dQ^n = Q^n*(Z_{i:n})_{i=1}^{n} \tag{2.1}$$

where Q is the uniform distribution on (0,1). Hereafter, let $r_o = 0$ and $r_{k+1} = n+1$.

Theorem 2.2: Let P be a p-measure with support (0,1) such that the distribution function F has four derivatives on (0,1). Then,

$$\sup_{B\in\mathbb{B}^n} |P^n\{(Z_{i:n})_{i=1}^{n}\in B\} - \int K_{\underline{r}}((Z_{r_i:n})_{i=1}^{k},B)\,dP^n| \tag{2.3}$$

$$\leq [\delta(F)\sum_{i=1}^{k+1}(r_j-r_{j-1}-1)\left(\frac{r_j-r_{j-1}}{n}\right)^2]^{1/2}$$

where $\delta(F) = \dfrac{d}{19c^3} + \dfrac{d^2\,\tilde{c}^2}{2c^6}$ and

$$d = \max_{k=2}^{4} \sup_{x\in(0,1)} |F^{(k)}(x)|,\ \tilde{c} = \sup_{x\in(0,1)} F^{(1)}(x),\ c = \inf_{x\in(0,1)} F^{(1)}(x).$$

The continuity of F implies that $P^n\{0<Z_{r_1:n}<\ldots<Z_{r_k:n}<1\} = 1$ so that the kernel $K_{\underline{r}}$ has to be defined for $\underline{z} = (z_1,\ldots,z_k)$ with $0 = z_o<z_1<\ldots<z_k<z_{k+1} = 1$ only. We define

$$K_{\underline{r}}(\underline{z},\cdot) = \prod_{i=1}^{2k+1} Q_{\underline{z},i} \tag{2.4}$$

where $Q_{\underline{z},2i} = \varepsilon_{z_i}$ is the Dirac-measure at $z_i$ for $i=1,\ldots,k$ and

$$Q_{\underline{z},2i-1} = Q_{(z_{i-1},z_i)}^{r_i-r_{i-1}-1} * (Z_{j:r_i-r_{i-1}-1})^{r_i-r_{i-1}-1} \quad \text{for } i=1,\ldots,k+1$$

with $Q_{(z_{i-1},z_i)}$ denoting the uniform distribution on $(z_{i-1},z_i)$. Moreover, $K_{\underline{r}}(\underline{z},\cdot)$ is understood as a p-measure on $\mathbb{B}^n$. Finally, if $r_1 = 1$ or $r_k = n$ then the factor $Q_{\underline{z},1}$ or $Q_{\underline{z},2k+1}$ has to be omitted. Notice that $\delta(F)$ measure the distance of P from Q in an appropriate way.

Elementary calculations show (see also [8], page 796) that

the p-measure $G_{\underline{r}} = \int K_{\underline{r}}((Z_{r_i:n})_{i=1}^{k}, \cdot)dP^n$ has a Lebesgue-density $g_{\underline{r}}$ which is given by

$$g_{\underline{r}}(\underline{x}) = n!\left(\prod_{i=1}^{k} f(x_{r_i})\right) \prod_{i=1}^{k+1} \left(\frac{F(x_{r_i})-F(x_{r_{i-1}})}{x_{r_i}-x_{r_{i-1}}}\right)^{r_i-r_{i-1}-1} \tag{2.5}$$

for $\underline{x}\in(0,1)^n$ with $x_1<\ldots<x_n$ where $f = F^{(1)}$, $x_{r_0}=0$ and $x_{r_{k+1}}=1$. Let $h_n$ denote the Lebesgue-density of $P^{n*}(Z_{i:n})_{i=1}^{n}$.
In the proof of Theorem 2.2 (see (3.1)) we shall make use of the Kullback-Leibler information number (see [2], page 711). As a first estimate of the left-hand side of (2.3) we get $[1-\exp(\int \log(h_n/g_{\underline{r}})dG_{\underline{r}})]^{1/2}$ which is smaller or equal to $[\int(\log(g_{\underline{r}}/h_n)dG_{\underline{r}}]^{1/2}$. Further computations which lead to the right-hand side of (2.3) will be made in (3.1).

Theorem 2.2 can be regarded as a refinement of the result in Section 2 of [8]. Weiss made some superfluous assumptions which can be understood in view of his intention to include a normal approximation argument (see [8], Section 3 and 4). In [8] and [9] it is assumed that, roughly speaking $r_i-r_{i-1}$ is of order $n^{3/4+\delta}$ and, respectively, $n^{2/3+\delta}$. Weiss [9], page 442, also refers to Reiss [3] that these conditions can be weakened to $r_i-r_{i-1}$ being of order $n^{1/2+\delta}$. To clarify the situation we state the following result which can be proved by starting e.g. from formula (3.6) in [6]:

$$\sup_{B\in\mathbb{B}^k} |P^n\{(Z_{r_i:n})_{i=1}^{k}\in B\} - N(\underline{r},P)(B)| = O\left(\sum_{j=1}^{k+1} \frac{1}{r_j-r_{j-1}}\right)^{1/2} \tag{2.6}$$

uniformly over all p-measures P which fulfill the condition of Theorem 2.2 such that $\delta(F)$ is uniformly bounded. Moreover, $N(\underline{r},P)$ is the k-variate normal distribution with mean vector $(F^{-1}(r_i/n))_{i=1}^{k}$ and covariances $\sigma_{ij} = r_i(1-r_j)/[n^3 f(F^{-1}(r_i/n)) f(F^{-1}(r_j/n))]$ for $1\le i\le j\le k$. Thus, Theorem 2.2 and (2.6) imply that

$$\sup_{B\in\mathbb{B}^n}|P^n\{(Z_{i:n})_{i=1}^{n}\in B\} - \int K_{\underline{r}}(\cdot,B)dN(\underline{r},P)|$$

$$= O[(\delta(F)\sum_{j=1}^{k+1}\frac{(r_j-r_{j-1})^3}{n^2})^{1/2} + (\sum_{j=1}^{k+1}\frac{1}{r_j-r_{j-1}})^{1/2}] \quad (2.7)$$

Since the order statistic is sufficient for a model which contains product measures $P^n$ only it is clear from (2.7) that as an approximation to such a model we can study normal distributions $N(\underline{r},P)$ if the right-hand side of (2.7) is sufficiently small. If all the differences $r_i-r_{i-1}$ are equivalent to $l_n$, say, so that k is equivalent to $n/l_n$ then the right-hand side of (2.7) is of order $O(\delta(F)l_n/n^{1/2}+n^{1/2}/l_n)$. This term converges to zero uniformly over all p-measures P which fulfill the conditions of Theorem 2.2 and for which, moreover, uniformly $\delta(F) = o(n^{1/2}/l_n) = o(1)$ as $n\to\infty$. These considerations show that we obtain a normal approximation to the given model (more precisely, the sequence of models) if this model shrinks towards the uniform distribution on (0,1) as the sample size goes to infinity.

In cases where we are only interested in local properties of the p-measure P then our statistical considerations will be based on certain sets of extreme or central order statistics, say $Z_{r:n}\leq\ldots\leq Z_{s:n}$ where $1\leq r\leq s\leq n$. Let us again reduce the number of order statistics. For this purpose we introduce the marginal p-measures $K_{\underline{r},\tau}(\underline{z},\cdot) = K_{\underline{r}}(\underline{z},\cdot)*\tau$ where $\tau$ is the projection $(x_1,\ldots,x_n)\to(x_r,\ldots,x_s)$. With the help of Theorem 2.2 we can prove

<u>Corollary 2.8:</u> Under the conditions of Theorem 2.2 we have

$$\sup_{B\in\mathbb{B}^{s-r+1}}|P^n\{(Z_{i:n})_{i=r}^{s}\in B\} - \int K_{\underline{r},\tau}((Z_{r_i:n})_{i=1}^{k},B)dP^n|$$

$$\leq [\,\delta(F)\sum_{j=2}^{k}(r_j-r_{j-1}-1)(\frac{r_j-r_{j-1}}{n})^2\,]^{1/2} \quad (2.9)$$

if $r = r_1<\ldots<r_k = s$.

Corollary 2.8 is an immediate consequence of Theorem 2.2. Put $\underline{s} = (s_i)_{i=1}^{r_1+k+n-r_k-1} = (1,2,\ldots,r_1-1,r_1,r_2,\ldots,r_k,r_k+1,\ldots,n)$ and apply Theorem 2.2 to $\{(Z_{i:n})_{i=1}^{n}\in\{\tau\in B\}\} = \{(Z_{i:n})_{i=r}^{s}\in B\}$

and $K_{\underline{s}}((Z_{s_i:n})^{r_1+k+n-r_k-1}, \{\tau \in B\}) = K_{\underline{r},\tau}((Z_{r_i:n})_{i=1}^{k}, B)$.

The same argument yields that Corollary 2.8 also holds with $\sum_{j=2}^{k}$ replaced by $\sum_{j=1}^{k}$ or $\sum_{j=2}^{k+1}$ if $r = 1$ and $1<r_1 \leq \ldots < r_k = s$ or $s = n$ and $r = r_1 < \ldots < r_k < n$.

Theorem 2.2 and Corollary 2.8 are concerned with p-measures P which are - in a certain sense - close to the uniform distribution Q. With the help of Lemma 2.11 one can generalize these results to the case that Q and P are replaced by some arbitrary but fixed p-measure $P_o$ which has a continuous distribution function $F_o$ and, respectively, a p-measure $\tilde{P}$ with distribution function $\tilde{F} = F \circ F_o$ where F fulfills the conditions of Theorem 2.2. The distance of $\tilde{P}$ from $P_o$ will again be given by $\delta(F)$; it is obvious that $F = \tilde{F} \circ F_o^{-1}$.

We shall construct a Markov kernel $\tilde{K}_{\underline{r}}$ which has the property

$$\int \tilde{K}_{\underline{r}}((Z_{r_i:n})_{i=1}^{k}, \cdot) dP_o^n = P_o^n * (Z_{i:n})_{i=1}^{n}. \tag{2.10}$$

More precisely, for every $\underline{z} \in \mathbb{R}^k$ with $0<F_o(z_{i-1})<F_o(z_i)<1, i=2,\ldots,k$, the kernel $\tilde{K}_r$ is defined as $K_{\underline{r}}$ with $Q_{(z_{i-1},z_i)}$, $z_o = 0$ and $z_{k+1} = 1$ replaced by the restriction $P_{(z_{i-1},z_i)}$ of $P_o$ to $(z_{i-1},z_i)$, $z_o = \inf\{x:F_o(x)>0\}$ and $z_{k+1} = \sup\{x:F_o(x)<1\}$ for $i=1,\ldots,k+1$. Thus we have $P_{(z_{i-1},z_i)}(B) = P_o(B \cap (z_{i-1},z_i))/P_o(z_{i-1},z_i)$ for $B \in \mathcal{B}$. In the proof of Lemma 2.11 (see (3.2)) we shall only assume that F is a continuous distribution function with $F(0) = 0$ and $F(1) = 1$.

<u>Lemma 2.11:</u> Let $1 \leq r \leq r_1 < \ldots < r_k \leq s \leq n$. Denote by $\tau$ again the projection $(x_1,\ldots,x_n) \to (x_r,\ldots,x_s)$. We have

$$\sup_{B \in \mathcal{B}^{s-r+1}} |P^n\{(Z_{i:n})_{i=r}^{s} \in B\} - \int \tilde{K}_{\underline{r},\tau}((Z_{r_i:n})_{i=1}^{k}, B) d\tilde{P}^n|$$

$$\leq \sup_{B\in\mathcal{B}^{s-r+1}} |P^n\{(Z_{i:n})_{i=r}^s \in B\} - \int K_{\underline{r},\tau}((Z_{r_i:n})_{i=1}^k, B)dP^n|,$$

where P is the p-measure with distribution function F.

Below we discuss several consequences of Lemma 2.11.

Remark 2.12: Let $\tilde{P}$ be a p-measure with distribution function $\tilde{F} = F \circ F_o$ where $F_o$ is a continuous distribution function and F fulfills the conditions of Theorem 2.2. Then Theorem 2.2 holds for $\tilde{P}$ and $\tilde{K}_{\underline{r}}$ in place of P and $K_{\underline{r}}$. The error bound is again determined by $\delta(F)$ and $\underline{r}$.

Remark 2.13: When Corollary 2.8 is applied together with Lemma 2.11 then the kernel $\tilde{K}_{\underline{r},\tau} = K_{\underline{r},\tau}$ is independent of a particular uniform distribution $P_o$. We have

$$P^n\{(Z_{i:n})_{i=r}^s \in B\} = \int K_{\underline{r},\tau}((Z_{r_i:n})_{i=1}^k, B)dP_o^n$$

for every uniform distribution $P_o$. Moreover, for every p-measure $\tilde{P}_{\mu,\sigma}$ with distribution function $x \mapsto F(\frac{x-\mu}{\sigma}), \mu \leq x \leq \mu+\sigma, \mu \in \mathbb{R}, \sigma > 0$, the estimate of

$$\sup_{B\in\mathcal{B}^{s-r+1}} |P^n_{\mu,\sigma}\{(Z_{i:n})_{i=r}^s \in B\} - \int K_{\underline{r},\tau}((Z_{r_i:n})_{i=1}^k, B)dP^n_{\mu,\sigma}|$$

only depends on F and $\underline{r}$. Further results concerning location and scale parameter families can be found in [9] and [10].

Remark 2.14: If $\underline{r} = (r,s)$ then the error bound as given in Corollary 2.8 is of order $(s-r)^{3/2}/n$ uniformly for p-measures P with $\delta(F)$ being bounded. Let $C_n$ be a critical region which is based on a kernel type statistic of the form $T = \int k_n(\frac{q-\xi}{\alpha_n})F_n^{-1}(\xi)d\xi, q \in (0,1)$, where $F_n^{-1}$ is the empirical quantile function, $\alpha_n = o(n^{-1/3})$ and $k_n$ has the support $[-1,1]$ for every n. We have $T = \tilde{T}(Z_{r:n}, \ldots, Z_{s:n})$ with $s-r = O(n\alpha_n)$. Thus we can find a critical function $\psi = \tilde{\psi}(Z_{r:n}, Z_{s:n})$ which has asymptotically (within the indicated error bound) the same power function as the critical region $C_n$. This context is evident when testing e.g. the q-quantile (see [7]) or the density function at the q-quantile. Continuing these considerations we can find an argument that the error bound in Theorem 2.2 is sharp in the sense that it cannot be improved in general even if a different method of rebuilding the initial sample is used. If the method of replacing $Z_{r:n} \leq \ldots \leq Z_{s:n}$ by $(Z_{r:n}, Z_{s:n})$ is still successful for $n = o((s-r)^{3/2})$ then it would be possible to construct consistent tests $\tilde{\psi}(Z_{r:n}, Z_{s:n})$ of the 1st derivative of the density function in contrast to any intuition.

Remark 2.15: In the case that $\delta(F)$ is bounded away from infinity over the given model and the sample size n is sufficiently large then it is clear from Theorem 2.2 and Lemma 2.11 that extreme observations do not heavily influence the statistical inference. The situation changes drastically when the statistical procedure is based on ordered observations $z_r, \ldots, z_n$ with $r > \alpha n$ and $\alpha > 0$ (see e.g. [11] where it is assumed that $z_{n-k+1}, \ldots, z_n$ are only given). If $z_r$ is replaced by the largest observation of a sample of size r which is governed by the restriction of $P_o$ to $(-\infty, z_{r+1})$ then, as examples show, the loss of power can be bounded away from zero when n goes to infinity.

## 3. PROOFS

In this section we complete the proofs of Theorem 2.2 and Lemma 2.11.

(3.1) Proof of Theorem 2.2: It suffices to prove that

$$\int \log\left(\frac{h_n}{g_{\underline{r}}}\right) dG_{\underline{r}} \geq -\left(\frac{5d}{96c^3} + \frac{961\, d^2\, \tilde{c}^2}{2304\, c^6}\right) \sum_{j=1}^{k+1} (r_j - r_{j-1} - 1)\left(\frac{r_j - r_{j-1}}{n}\right)^2 . (1)$$

It is straightforward to see that

$$\int \log\left(\frac{h_n}{g_{\underline{r}}}\right) dG_{\underline{r}} = \int \log\left\{ \prod_{j=1}^{k+1} \prod_{i=1}^{r_j - r_{j-1} - 1} \left(f\left(x_{r_{j-1}} + i \frac{x_{r_j} - x_{r_{j-1}}}{F(x_{r_j}) - F(x_{r_{j-1}})}\right)\right)\right\} G_{\underline{r}}(d\underline{x}).$$

Setting

$\Delta_j(\underline{x}) := x_{r_j} - x_{r_{j-1}}, \bar{x}_j := (x_{r_j} + x_{r_{j-1}})/2,$

$u_{ij}(\underline{x}) := (x_{r_{j-1+i}} - \bar{x}_j)/\Delta_j(\underline{x}), j = 1, \ldots, k+1, i = 1, \ldots, r_j - r_{j-1} - 1,$

we can write

$$\int \log(h_n/g_{\underline{r}})dG_{\underline{r}} \tag{2}$$

$$= \int \sum_{j=1}^{k+1} \sum_{i=1}^{r_j - r_{j-1}-1} \log\{\frac{\Delta_j f(\bar{x}_j + \Delta_j u_{ij})}{F(x_{r_j}) - F(x_{r_{j-1}})}\} G_{\underline{r}}(d\underline{x})$$

$$= \sum_{j=1}^{k+1} \sum_{i=1}^{r_j - r_{j-1}-1} \left\langle \int \log\{1 + (\frac{f(\bar{x}_j + \Delta_j u_{ij})}{f(\bar{x}_j)} - 1)\}\right.$$

$$\left. - \log\{1 + (\frac{F(x_{r_j}) - F(x_{r_{j-1}})}{\Delta_j f(\bar{x}_j)} - 1)\} G_{\underline{r}}(dx) \right\rangle$$

$$= \sum_{j=1}^{k+1} \sum_{i=1}^{r_j - r_{j-1}-1} \left\langle \int \frac{f(\bar{x}_j + \Delta_j u_{ij})}{f(\bar{x}_j)} - \frac{1}{2}\left[\frac{f(\bar{x}_j + \Delta_j u_{ij}) - f(\bar{x}_j)}{f(\bar{x}_j)(1+\xi_{ij})}\right]^2\right.$$

$$\left. - \frac{F(x_{r_j}) - F(x_{r_{j-1}})}{\Delta_j f(\bar{x}_j)} + \frac{1}{2}\left[\frac{F(x_{r_j}) - F(x_{r_{j-1}}) - \Delta_j f(\bar{x}_j)}{\Delta_j f(\bar{x}_j)(1+\xi_j)}\right] G_{\underline{r}}(d\underline{x}) \right\rangle$$

by using the expansion $\log(1+x) = x - x^2/2(1+\xi)^2$ for $x > -1$ and $\xi \in (0,x)$ where $(1+\xi_{ij})^{-2} \leq \tilde{c}^2/c^2$. Taylor's formula implies

$$f(\bar{x}_j + \Delta_j u_{ij}) = \sum_{l=0}^{2} f^{(l)}(\bar{x}_j)(\Delta_j u_{ij})^l/l! + f^{(3)}(\theta_{ij})(\Delta_j u_{ij})^3/6$$

where $\theta_{ij}$ is between $\bar{x}_j$ and $\bar{x}_j + \Delta_j u_{ij}$,

$$F(x_{r_j}) = \sum_{l=0}^{3} F^{(l)}(\bar{x}_j)(x_{r_j} - \bar{x}_j)^l/l! + f^{(3)}(\theta_j)(x_{r_j} - \bar{x}_j)^4/24,$$

$$F(x_{r_{j-1}}) = \sum_{l=0}^{3} F^{(l)}(\bar{x}_j)(x_{r_{j-1}} - \bar{x}_j)^l/l! + f^{(3)}(\theta_j)(x_{r_{j-1}} - \bar{x}_j)^4/24$$

where $\theta_j$ lies between $\bar{x}_j$ and $x_{r_j}$ and $\bar{\theta}_j$ between $\bar{x}_j$ and $x_{r_{j-1}}$.

Thus,

$$f(\bar{x}_j+\Delta_j u_{ij}) = f(\bar{x}_j)\{1+\sum_{l=1}^{2}\frac{f^{(l)}(\bar{x}_j)(\Delta_j u_{ij})^l}{f(\bar{x}_j)l!}+\eta_{ij}\} \tag{3}$$

where $|\eta_{ij}|\leq\frac{d}{48c}\Delta_j^3$ since $|u_{ij}|\leq\frac{1}{2}$, and

$$F(x_{r_j})-F(x_{r_{j-1}}) = \Delta_j f(\bar{x}_j)\ \{1+\frac{f^{(2)}(\bar{x}_j)}{24f(\bar{x}_j)}\Delta_j^2+\eta_j\} \tag{4}$$

where $|\eta_j|\leq\frac{d}{192c}\Delta_j^3$. Now (2) - (4) imply

$$\int\log(\frac{h_n}{g_{\underline{r}}})dG_{\underline{r}} = \sum_{j=1}^{k+1}\frac{f^{(1)}(\bar{x}_j)\Delta_j}{f(\bar{x}_j)}\left\langle\int\sum_{i=1}^{r_j-r_{j-1}-1}u_{ij}dG_{\underline{r}}\right\rangle \tag{5}$$

$$+\sum_{j=1}^{k+1}\frac{f^{(2)}(\bar{x}_j)\Delta_j^2}{2f(\bar{x}_j)}\left\langle\int\sum_{i=1}^{r_j-r_{j-1}-1}(u_{ij}^2-\frac{1}{12})dG_{\underline{r}}\right\rangle$$

$$+\sum_{j=1}^{k+1}\sum_{i=1}^{r_j-r_{j-1}-1}\int(\eta_{ij}-\eta_j)dG_{\underline{r}}$$

$$-\frac{1}{2}\sum_{j=1}^{k+1}\sum_{i=1}^{r_j-r_{j-1}-1}\int\{\sum_{l=1}^{2}\frac{f^{(l)}(\bar{x}_j)}{f(\bar{x}_j)l!}(\Delta_j u_{ij})^l+\eta_{ij}\}(1+\xi_{ij})^{-2}\,G_{\underline{r}}(d\underline{x})$$

$$+\frac{1}{2}\sum_{j=1}^{k+1}\sum_{i=1}^{r_j-r_{j-1}-1}\int\{\frac{f^{(2)}(\bar{x}_j)\Delta_j^2}{24f(\bar{x}_j)}+\eta_j\}^2(1+\xi_j)^{-2}\,G_{\underline{r}}(dx)$$

$$=:\ A_n+B_n+C_n+D_n+E_n.$$

Since

$$\int\sum_{i=1}^{r_j-r_{j-1}-1}u_{ij}\ dG_{\underline{r}} = (r_j-r_{j-1}-1)\int_{-1/2}^{1/2}x\ dx = 0$$

and

$$\int\sum_{i=1}^{r_j-r_{j-1}-1}(u_{ij}^2-\frac{1}{12})dG_{\underline{r}} = (r_j-r_{j-1}-1)\int_{-1/2}^{1/2}(x^2-\frac{1}{12})\ dx = 0,$$

we have $A_n = 0$ and $B_n = 0$. (6)

Since $|\eta_{ij}-\eta_j| \leq \frac{5d}{192c}\Delta_j^3$ Fubini's Theorem implies

$$C_n \geq \frac{5d}{192c} \sum_{j=1}^{k+1} (r_j-r_{j-1}-1)\int (Z_{r_j:n}-Z_{r_{j-1}:n})^2 \, dP^n \tag{7}$$

$$\geq - \frac{5d}{192c^3} \sum_{j=1}^{k+1} (r_j-r_{j-1}-1)\int (Z_{r_j:n}-Z_{r_{j-1}:n})^2 \, dQ^n$$

$$\geq - \frac{5d}{96c^3} \sum_{j=1}^{k+1} (r_j-r_{j-1}-1)\left(\frac{r_j-r_{j-1}}{n}\right)^2$$

where the second and third inequality is immediate from the probability transformation theorem and formula (3.1.6) in [1].

Finally,

$$D_n \geq - \frac{\tilde{c}^2}{2c^2} \sum_{j=1}^{k+1} \sum_{i=1}^{r_j-r_{j-1}-1} \int \Big\{ \sum_{l=1}^{2} \frac{f^{(l)}(\bar{x}_j)u^l_{ij}}{f(\bar{x}_j)l!} \Delta^l_j + \eta_{ij} \Big\}^2 \, G_{\underline{r}}(dx) \tag{8}$$

$$\geq - \frac{d^2\tilde{c}^2}{2c^4} \sum_{j=1}^{k+1} \sum_{i=1}^{r_j-r_{j-1}-1} \int \left(\frac{\Delta_j^2}{4} + \frac{\Delta_j^3}{8} + \frac{\Delta_j^4}{64} + \frac{\Delta_j^4}{48} + \frac{\Delta_j^5}{192} + \frac{\Delta_j^6}{2304}\right) dG_{\underline{r}}$$

$$\geq - \frac{961\, d^2\tilde{c}^2}{4608\, c^6} \sum_{j=1}^{k+1} (r_j-r_{j-1}-1)\int (Z_{r_{j:n}}-Z_{r_{j-1:n}})^2 \, dQ^n$$

$$\geq - \frac{961\, d^2\tilde{c}^2}{2304\, c^6} \sum_{j=1}^{k+1} (r_j-r_{j-1}-1)\left(\frac{r_j-r_{j-1}}{n}\right)^2 .$$

Since $E_n > 0$, (1) is immediate from (5) - (8).

<u>(3.2) Proof of Lemma 2.11:</u> Let $F_o^{-1}$ be the generalized inverse of $F_o$. Since $F_o^{-1}(q) \leq x$ iff $q \leq F_o(x)$ for $q \in (0,1)$ and $x \in \mathbb{R}$ we obtain $P{*}F_o^{-1} = \tilde{P}$ whence

$$\tilde{P}^n * (Z_{i:n})_{i=r}^{s} = P^n * (F_o^{-1} \circ Z_{i:n})_{i=r}^{s}. \tag{1}$$

This implies

$$\int \tilde{K}_{\underline{r},\tau}((Z_{r_i:n})_{i=1}^{k}, B)dP^n = \int \tilde{K}_{\underline{r},\tau}((F_o^{-1} \circ Z_{r_i:n})_{i=1}^{k}, B)dP^n. \tag{2}$$

Let $0 = u_o < u_1 < \ldots < u_k < u_{k+1} = 1$. Since $F_o$ is continuous we know that $F_o^{-1}$ is strictly increasing whence $\inf\{x:F_o(x)>0\}=: v_o < v_1 < \ldots < v_k < v_{k+1} := \sup\{x:F_o(x)<1\}$, where $v_i = F_o^{-1}(u_i)$. Since $\varepsilon_{v_i} = \varepsilon_{u_i} * F_o^{-1}$, $i=1,\ldots,k$, and

$$P_{(v_{i-1},v_i)} = Q_{(u_{i-1},u_i)} * F_o^{-1}, \quad i=1,\ldots,k+1,$$

we have

$$P_{(v_{i-1},v_i)}^{r_i-r_{i-1}-1} * (Z_{j:r_i-r_{i-1}-1})_{r=1}^{r_i-r_{i-1}-1}$$

$$= Q_{(u_{i-1},u_i)}^{r_i-r_{i-1}-1} * (F_o^{-1} \circ Z_{j:r_i-r_{i-1}-1})_{j=1}^{r_i-r_{i-1}-1}$$

and $\tilde{K}_{\underline{r},\tau}((v_i)_{i=1}^{k}, B) = K_{\underline{r},\tau}((u_i)_{i=1}^{k}, \tilde{B})$ for every $B \in \mathbb{B}^{s-r+1}$ (3)

where $\tilde{B} = \{\underline{x}:(F_o^{-1}(x_i))_{i=1}^{n} \in B\}$. Thus, it is immediate from (2) and (3) that

$$\int \tilde{K}_{\underline{r},\tau}((Z_{r_i:n})_{i=1}^{k}, B)d\tilde{P}^n = \int K_{\underline{r},\tau}((Z_{r_i:n})_{i=1}^{k}, \tilde{B})dP^n. \tag{4}$$

From (1) we also conclude that $\tilde{P}^n\{(Z_{i:n})_{i=r}^{s} \in B\} = P^n\{(Z_{i:n})_{i=r}^{s} \in \tilde{B}\}$. This together with (4) implies the inequality of Lemma 2.11.

## REFERENCES

[ 1] David, H.A. 1981, *Order statistics*, 2nd edition, Wiley, New York.

[ 2] Hoeffding, W. and Wolfowitz, J. 1958, *Distinguishability of sets of distributions,* Ann. Math. Statist. 29, pp. 700-718

[ 3] Reiss, R.-D. 1975, *The asymptotic normality and asymptotic expansions for the joint distribution of several order statistics,* in: Limit Theorems of Prob. Theory (P. Révész, ed.), Keszthely 1974, North Holland, Amsterdam 297-340.

[ 4] Reiss, R.-D. 1981 a, *Asymptotic independence of distribution of normalized order statistics of the underlying probability measure,* J. Mult. Analysis 11, pp. 386-399.

[ 5] Reiss, R.-D. 1981 b, *Uniform approximation to distributions of extreme order statistics,* Adv. Appl. Probab. 13, pp. 533-547.

[ 6] Reiss, R.-D. 1981 c, *Approximation of product measures with an application to order statistics,* Ann. Probab. 9, pp. 335-341.

[ 7] Reiss, R.-D. 1982, *One-sided tests for quantiles in certain nonparametric models,* in: Nonparametric statistical inference (B.V. Gnedenko et al.) Budapest 1980, North Holland, Amsterdam, pp. 759-772.

[ 8] Weiss, L. 1974, *The asymptotic sufficiency of a relatively small number of order statistics in tests of fit,* Ann. Statist. 2, pp. 795-802.

[ 9] Weiss, L. 1979, *The asymptotic distribution of order statistics,* Nav. Res. Logist. Q. 26, pp. 437-445.

[10] Weiss, L. 1980, *The asymptotic sufficiency of sparse order statistics in tests of fit with nuisance parameter,* Nav. Res. Logist. Q. 27, pp. 397-406.

[11] Weissman, I. 1978, *Estimation of parameters and large quantiles based on the k largest observations.* J. Amer. Statist. Assoc. 73, pp. 812-815.

# POT-ESTIMATION OF EXTREME SEA STATES AND THE BENEFIT OF USING WIND DATA

Dan Rosbjerg*) and Jesper Knudsen**)

*) Technical University of Denmark

**) Danish Hydraulic Institute

## ABSTRACT

The distribution function for the maximum peak during T years of the significant wave height is derived by means of a POT-method. Accordingly, the T-year wave peak (i.e. the level which - on the average - will be exceeded once per T years) and the sample variance of the T-year estimate can be determined. Based on a period with concurrent measurements of wave heights and wind speeds a regression analysis can be applied which allows wave peaks to be calculated during periods with only wind observations. Thus an updated estimate of the T-year event as well as the variance of the estimate can be developed, and the obtained variance reduction used as a measure of the benefit of using wind data. An example is included in order to illustrate the presented theory.

## SEVERITY OF THE SEA STATE

A common measure of the natural sea state is the current value of the so-called significant wave height, $H_s$, defined as the mean value of individual heights greater than the 67% quantile. For sea states with a narrow-band spectrum $H_s$ is close to $H_{m_0} = 4\sqrt{m_0}$ where $m_0$ denotes the integral of the spectral density function (the variance). However, $H_{m_0}$, is also used as a measure of the severity of the sea state independently of the bandwidth of the spectrum. Fig. 1 shows an example of the variation of $H_{m_0}$, where the values of $H_{m_0}$ have been calculated every 3 hours based on continuous observations of the sea for a period of 20 minutes.



*J. Tiago de Oliveira (ed.), Statistical Extremes and Applications, 611–620.*

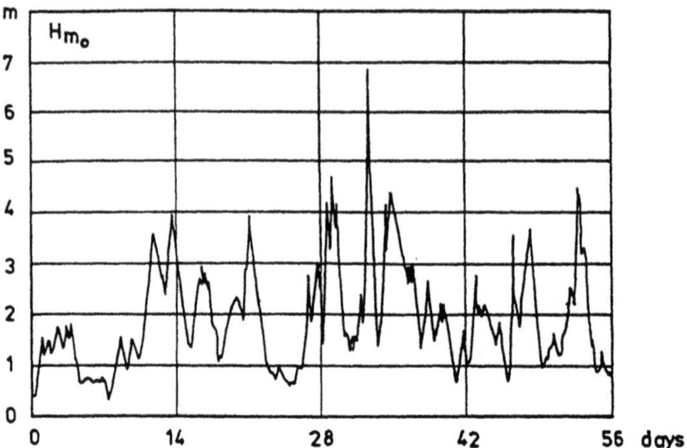


Figure 1. Variation of $H_{m_0}$ during 56 days. (Values calculated every 3 hours.)

## EXTREME VALUE DISTRIBUTION AND T-YEAR EVENT

For the statistical analysis of extreme events a peak-over-threshold (POT) method can be applied. If an appropriate threshold is introduced and the recorded wave heights below this level are deleted, the remaining data constitute a partial duration series. This series is likely to represent independent storm events provided that the threshold is reasonably defined. Selecting the peak value within each storm for further analysis, the POT-method attempts to focus on the relevant statistical variable for extreme value analysis, namely the maximum value within independent storm events. During this process it is crucial to verify that the basic requirements of the further analysis is fulfilled, i.e. that selected peaks are independent and identically distributed.

Denote either the storm peak itself or some suitable transformation by $H_s'$ and the threshold level by $h_*$. On the condition that $H_s'$ is greater than $h_*$, the distribution function for $H_s'$ is then introduced as

$$P\ \{H_s' \le h \mid H_s' > h_*\} = F(h) \tag{1}$$

Further assume that the occurrence of wave peaks takes place according to a Poisson process with a seasonally varying intensity. This implies that the number of storms per year follows a Poisson distribution with the parameter equal to the integral of the intensity over a year. If this quantity is denoted by $\lambda$, the expected number of storms in t years with wave peaks greater than an arbitrary level $h(> h_*)$ becomes

$$\nu_h = \lambda\, t\, [1 - F(h)] \tag{2}$$

The T-year event, $h_T$, is now defined as the level which, on the average, will be exceeded once per T years. Since this situation corresponds to $\nu_h = 1$ for $t = T$, the T-year event is obtained by solving Eq. (2) under these conditions leading to

$$h_T = F^{-1}\left(1 - \frac{1}{\lambda T}\right) \tag{3}$$

By use of the above mentioned assumptions implying that the number of storms with wave peaks greater than h in t years follows a Poisson distribution with $\nu_h$ as parameter, the distribution for the maximum value of $H'_s$ in t years, $H'_{s_{max,t}}$ becomes

$$P\{H'_{s_{max,t}} \leq h\} = e^{-\nu_h} = e^{-\lambda\, t\, [1-F(h)]} \tag{4}$$

Now denote by R the probability (the risk) that the level $h_{L,R}$ will be exceeded once or more during a time period of L years (the lifetime). Accordingly, the left hand side of Eq. (4) equals 1-R for t = L, and a relation between R, L and the corresponding value of T can then be obtained by solving the equation for $h_{L,R}$ and equalizing the expression with Eq. (3). Hereby the following relationship is obtained

$$R = 1 - e^{-L/T} \tag{5}$$

which is shown in Fig. 2. Note that risk values less than 10% correspond to design return periods much greater than the length of the assumed lifetime.

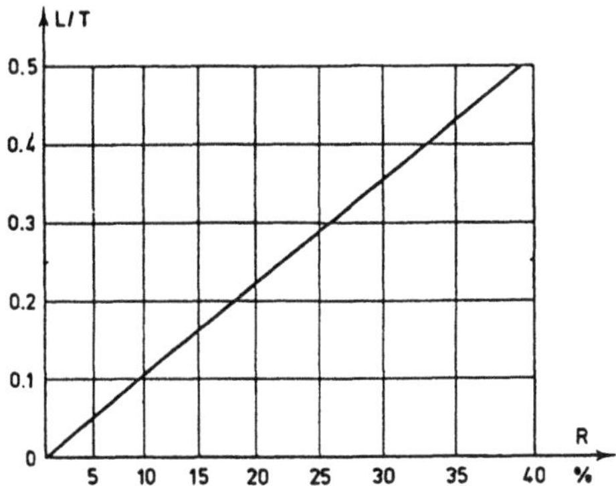


Figure 2. Relation between R and L/T.

In case of exponentially distributed wave peak exceedances i.e.

$$F(h) = 1 - e^{-\frac{h-h_*}{\alpha}} \tag{6}$$

where the parameter $\alpha$ is equal to the expected value of the exceedances above the threshold value $h_*$, the expression for the T-year event becomes particularly simple. Combining Eqs. (3) and (6) results in

$$h_T = h_* + \alpha \ln \lambda T \tag{7}$$

which will be used in the next section in order to exemplify an approximate method for calculating the standard error of the T-year estimate.

## STANDARD ERROR OF T-YEAR ESTIMATE

Assume than $n_1$ storm events are observed during a time period of $t_1$ years. Then the maximum likelihood estimates of the parameters are

$$\hat{\alpha} = \frac{1}{n_1} \sum_{i=1}^{n_1} (h_i - h_*) = \bar{h} - h_* \tag{8}$$

$$\hat{\lambda} = \frac{n_1}{t_1} \tag{9}$$

One of the advantages of the POT-estimation is the possibility of establishing confidence limits for the T-year estimate. Due to a limited sample the estimated distribution parameters $\hat{\alpha}$, $\hat{\lambda}$ are inaccurate which in turn makes the estimate of $h_T$ as obtained by Eq. (10) uncertain:

$$\hat{h}_T = h_* + \hat{\alpha} \ln \hat{\lambda} T = g(\hat{\alpha}, \hat{\lambda}; \ T) \tag{10}$$

The sampling variance of $\hat{h}_T$ can be computed approximately by

$$\mathrm{Var}\{h_T\} \simeq \left(\frac{\partial g}{\partial \hat{\alpha}}\right)_m^2 \mathrm{Var}\{\hat{\alpha}\} + \left(\frac{\partial g}{\partial \hat{\lambda}}\right)_m^2 \mathrm{Var}\{\hat{\lambda}\}$$

$$+ 2\left(\frac{\partial g}{\partial \hat{\alpha}}\right)_m \left(\frac{\partial g}{\partial \hat{\lambda}}\right)_m \mathrm{Cov}\{\hat{\alpha}, \hat{\lambda}\} \tag{11}$$

where the index m denotes that the partial derivatives are to be evaluated at the mean values of $\hat{\alpha}$ and $\hat{\lambda}$. Inserting

$$\left(\frac{\partial g}{\partial \hat{\alpha}}\right)_m = \ln \lambda T; \quad \left(\frac{\partial g}{\partial \hat{\lambda}}\right)_m = \frac{\alpha}{\lambda} \tag{12}$$

and

$$\mathrm{Var}\{\hat{\alpha}\} = \frac{\alpha^2}{\lambda t_1}; \quad \mathrm{Var}\{\hat{\lambda}\} = \frac{\lambda}{t_1}; \quad \mathrm{Cov}\{\hat{\alpha}, \hat{\lambda}\} = 0 \tag{13}$$

into Eq. (11) leads to

$$\mathrm{Var}\{\hat{h}_T\} = \frac{\alpha^2}{\lambda t_1}\left[1 + (\ln \lambda T)^2\right] \tag{14}$$

The standard error of $h_T$ is therefore given by

$$\hat{s}_T = \frac{\alpha}{\sqrt{\hat{\lambda}\, t_1}} \sqrt{1 + (\ln \hat{\lambda}\, T)^2} \tag{15}$$

If the T-year estimated is normally distributed, 68% and 95% confidence limits are accordingly determined as $\hat{h}_T \pm \hat{s}_T$ and $\hat{h}_T \pm 2\hat{s}_T$, respectively.

As a concluding remark to this section it should be stressed that the POT-approach is not limited to the exponential distribution assumption applied above. If another type of distribution, (e.g. lognormal or gamma) is justified, a similar procedure can be applied leading to revised expressions for the standard error of the T-year estimate.

## BENEFIT OF USING WIND DATA

Unfortunately, wave measurements usually cover time periods much shorter ($\sim$ 1-2 years) than the long return period selected for design conditions. An inevitable consequence of this fact is that the standard error of $h_T$ becomes relatively large, and wide confidence bands are therefore obtained.

In many cases, however, the existing data basis may be improved by extending the measured series by use of hindcast techniques. For hindcasts, a model is used with past meteorological conditions as input to compute extreme wave conditions. Such a model may either be a physically based wind-wave model or a statistically based regression model obtained on the basis of simultaneous wind and wave measurements. In both cases a deterministic model is applied to extend the length of the measured wave series. Since more information is included, the uncertainty of $h_T$ is immediately expected to be reduced. However, it is necessary also to take into consideration that the generated values besides the relevant information induce some further uncertainty (noise) due to the imperfection of the model.

The following theory is based on the assumption that simultaneous wind and wave measurements are available, and that it is possible to select some characteristic wind speeds within each storm event to be correlated with the wave peak.

In this case a statistical regression approach is applicable. When some suitable transformation of the wind speed is denoted by u, the general linear regression equation of u on h reads

$$h(i) = \bar{h}_1 + r \frac{s_{h_1}}{s_{u_1}} (u(i) - \bar{u}_1) \tag{16}$$

The index 1 refers to a time period of length $t_1$ in which $n_1$ concurrent observations of h and u are available. $s_{h_1}$ and $s_{u_1}$ denote the standard deviations of the wave and wind observations, respectively, and r denotes the obtained correlation coefficient whose theoretical value is $\rho$.

The length of the period with only wind observations is denoted $t_2$, and it is assumed that the use of the regression equation (16) results in $n_2$ generated values of h greater than $h_*$. The wave peak data now comprise $n_1$ primary observations and $n_2$ secondary calculated events, and the question therefore arises to what extent the secondary data will improve the extreme wave peak estimate.

The estimate of the T-year event based only on observed peak values is given by Eq. (10). An updated estimate including the secondary data is obtained as

$$\tilde{h}_T = h_* + \tilde{\alpha} \ln \tilde{\lambda} T \tag{17}$$

in which $\tilde{\alpha}$ and $\tilde{\lambda}$ are determined as

$$\tilde{\alpha} = \bar{h}_{1+2} - h_* \tag{18}$$

and

$$\tilde{\lambda} = \frac{\hat{\lambda}_1 t_1 + \hat{\lambda}_2 t_2}{t_1 + t_2} \tag{19}$$

where

$$\bar{h}_{1+2} = \frac{1}{n_1+n_2} \sum_{i=1}^{n_1+n_2} h(i) = \bar{h}_1 + \frac{n_2}{n_1+n_2} \frac{s_{h_1}}{s_{u_1}} r (\bar{u}_2 - \bar{u}_1) \tag{20}$$

The benefit of using wind data is now evaluated by comparing the variance of this estimate by the variance of the estimate based only on the primary observations as given by Eq. (14).

The approximate method previously applied to determine the sampling variance $Var\{\hat{h}_T\}$, cf. Eq. (11), is also used to develop $Var\{\tilde{h}_T\}$. For this reason $Var\{\tilde{\alpha}\}$ and $Var\{\tilde{\lambda}\}$ must be worked out. By use of the exponential assumption together with the results of Matalas and Jacobs (1964) it is found from Eqs. (18) and (20) that

$$Var\{\tilde{\alpha}\} = \frac{\alpha^2}{\lambda\, t_1}\left[1 - \frac{t_2}{t_1+t_2}\left(\rho^2 - \frac{1-\rho^2}{\lambda t_1 - 3}\right)\right] \tag{21}$$

From Eq. (19) together with the Poisson assumption follows that

$$Var\{\tilde{\lambda}\} = \frac{\lambda}{t_1+t_2} \tag{22}$$

An approximate expression for the variance of the updated T-year estimate is then

$$Var\{\tilde{h}_T\} = \frac{\alpha^2}{\lambda\, t_1}\left\{\frac{t_1}{t_1+t_2} + (\ln \lambda T)^2\left[1 - \frac{t_2}{t_1+t_2}\left(\rho^2 - \frac{1-\rho^2}{\lambda\, t_1 - 3}\right)\right]\right\} \tag{23}$$

As a control $\rho = 1$ is inserted in Eq. (23) resulting in

$$Var\{\tilde{h}_T\} = \frac{\alpha^2}{\lambda(t_1+t_2)}\left[1 + (\ln \lambda\, T)^2\right] \tag{24}$$

which exactly corresponds to the case with primary observation during the whole period $t_1 + t_2$. For $\rho < 1$ Eq. (23) leads to a larger value than Eq. (24) reflecting the fact that the information value of secondary data is smaller than the value of the real observations.

By comparing Eqs. (23) and (14) it is found that a critical value $\rho = \rho_c$ exists so that $Var\{\tilde{h}_T\} < Var\{\hat{h}_T\}$ for $\rho > \rho_c$ and the opposite for $\rho < \rho_c$. The critical value is

$$\rho_c^{\,2} = \frac{1 - \dfrac{\lambda t_1 - 3}{(\ln \lambda T)^2}}{\lambda t_1 - 2} \tag{25}$$

As expected, it is seen that $\rho_c$ is decreasing for increasing values of $t_1$ and $\lambda$, while $\rho_c$ is increasing for increasing values of T. In applications the value of r obtained by the regression analysis of the concurrent wind and wave data should be compared with $\rho_c$, and only in case of $r > \rho_c$ should the secondary data be generated and introduced in the analysis. The relative variance reduction is found to be

$$\varepsilon = \frac{\mathrm{Var}\{\hat{h}_T\} - \mathrm{Var}\{\tilde{h}_T\}}{\mathrm{Var}\{\hat{h}_T\}} = \frac{t_2}{t_1 + t_2} \frac{1 + (\ln \lambda T)^2 \left(\rho^2 - \frac{1-\rho^2}{\lambda t_1 - 3}\right)}{1 + (\ln \lambda T)^2} \tag{26}$$

In terms of $\varepsilon$ the relation between the standard errors becomes

$$\tilde{s}_T = \hat{s}_T \sqrt{1 - \varepsilon} \tag{27}$$

## ILLUSTRATIVE EXAMPLE

In the following example the data comprise concurrent wind and wave measurements for a certain site in the North Sea during a period of 9 months. In addition, 4½ years of wind measurements are available at another site located some distance from the wave field. Because the example is included only for *illustrative* purposes, no attempt is made to correct the bias due to either the use of two different observation locations or to the fact that the primary observation period is less than one year.

The wave data to be analysed were 3-hour values of $H_{m_0}$ during the above mentioned period. Using a threshold of 2.8 m combined with some independence criteria implied that 34 wave peaks were selected for further analysis. On the basis of some simple tests it was accepted that the exceedances of the squared peak values above the value $h_* = (2.8\ \mathrm{m})^2$ could be considered exponentially distributed. The results of a graphical test is shown in Fig. 3.

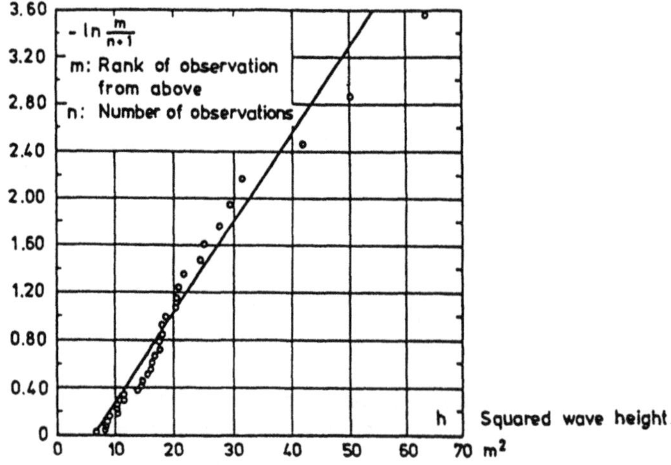


Figure 3. Graphical test of exponential distribution of exceedances of squared wave peaks above the threshold value.

The application of Eqs. (10) and (14) for these data showed a 50-year wave peak estimate of 10.45 m with a corresponding standard error of 0.85 m.

The wind observations to be considered together with the wave observations comprised 10 minutes average wind speeds with 3-hourly time intervals. A preliminary multiple regression analysis indicated that it was sufficient to select the maximum wind peak during the storm as the representative wind speed. A linear regression of the fourth power of this wind speed on the squared wave peak was then performed, See Fig. 4. The correlation coefficient was determined to $r = 0.82$.

When the estimation of the 50-year wave peak was assumed as the purpose of the analysis, $\rho_c$ was by means of Eq. (25) found to be 0.12, i.e. $r > \rho_c$. Generation of wave peak values by means of Eq. (16) resulted in further 73 wave peaks to be taken into account. By us of Eqs. (17) and (23) the revised estimate of the 50-year wave peak was then determined to be 9.65 m with a standard error of 0.5 m. A comparison of this standard error with the original one showed that the use of available wind data in this case reduced the sampling variance by 66%. Fig. 5 summarizes the benefit of using wind data in the present calculation example.

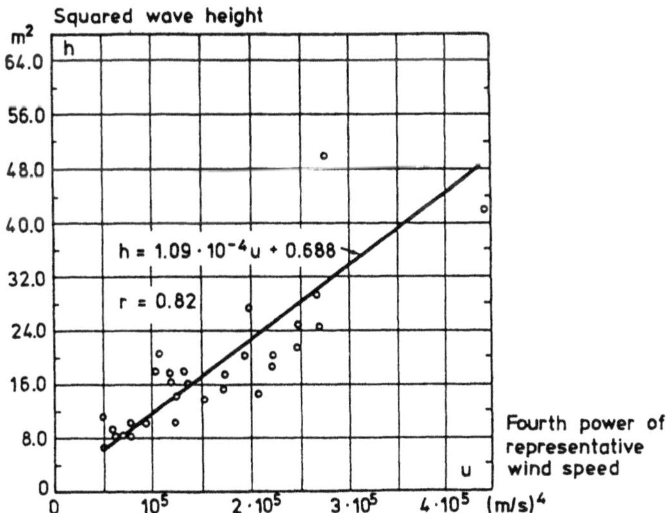


Figure 4. Linear regression of transformed wind peak on transformed wave peak.

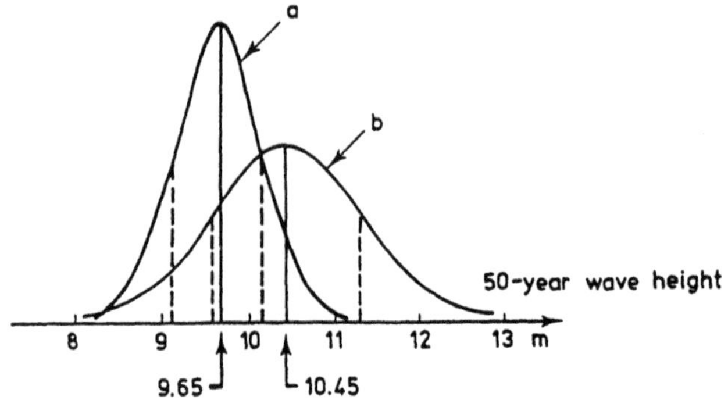


Figure 5. Distribution of the 50-year wave peak estimate showing 68% confidence intervals a) using supplementary wind data, b) without use of wind data.

## CONCLUSION

The use of POT-estimation of extreme sea states is found to have some advantages. The method is flexible with only few really fundamental assumptions, and it allows calculation of the reliability of the T-year wave estimate by means of confidence intervals. Furthermore, the benefit of using wind data to extend the observed wave peak series by synthetic events can be explicitly evaluated. The worked example illustrates that this benefit in some cases can be rather substantial.

## REFERENCE

Matalas, N.C. and Jacobs, B. 1964, "A correlation procedure for augmenting hydrologic data". U.S. Geological Survey, Professional Paper 434-E.

# THRESHOLD METHODS FOR SAMPLE EXTREMES

Richard L. Smith

Department of Mathematics, Imperial College, London SW7

The aim of this paper is to bring together two previously unrelated areas of the statistical analysis of extremes, and to suggest how they might form the basis of a general technique of extreme value analysis of time series.

The first is the class of Peaks Over Threshold methods developed by hydrologists. These methods are based on seemingly ad hoc statistical models, but they are flexible enough to cope with seasonality and serial dependence.

The second area is the analysis of the tail of a distribution based on extreme order statistics. Such techniques have been developed only for i.i.d. observations, and there remain open questions about the sampling properties of the estimators.

These two areas of work may be regarded as contributions to the same general problem, i.e. the modelling of the extreme characteristics of a series in terms of its exceedances over a high threshold level. We discuss here a general method based on (a) modelling the exceedance times as a point process and (b) use of the Generalised Pareto Distribution for the exceedance values. As an example, the method is applied to an analysis of wave heights off the coast of Britain.

## 1. INTRODUCTION

This paper is about methods for estimating return values and related quantities from series of observations. For some of the discussion we shall assume that the observations are independent



*J. Tiago de Oliveira (ed.), Statistical Extremes and Applications, 621–638.*

and identically distributed (i.i.d.), though we are ultimately interested in techniques which will also handle seasonality and serial dependence which are present in many enviromental series.

We shall start with a brief outline of the Peaks Over Threshold (POT) method developed by hydrologists. The method is described in some detail in the English Flood Studies Report [18]; subsequent developments are due to Todorovic [27,28], North [19] and Revfeim [22,23], amongst others. The analysis is based on peaks over a high threshold level. The models which have been adopted are somewhat arbitrary, and the justification for them is mainly empirical.

A parallel, but entirely distinct, development has been the use of methods based on extreme order statistics. Weiss [30] appears to have been the first to propose such a method; Pickands [21], Hill [13] and Weissman [31-33] have been among the subsequent contributors. These procedures have been developed from a more rigorous mathematical standpoint than the POT procedures, but they have been restricted entirely to the case of i.i.d. observations and there remain some open questions even in that case.

In Section 3 we shall propose a slight modification of the extreme order statistics method which leads naturally into an extension of the POT method based on all exceedances (rather than just the peaks) over the threshold. The most general model of this form has not been developed but in Section 4 we shall present a particular model which combines some of the main features of the POT and extreme order statistics methods. In Section 5 we discuss the connections with Leadbetter's theory of extremes in stationary sequences. Finally in Section 6 we present an example of these techniques and propose some areas for further investigation.

## 2. THE PEAKS OVER THRESHOLD METHOD

The analysis is based on the sequence of peaks of a series over a specified high threshold level. A peak is defined as the maximum value achieved during any sequence of consecutive observations above the threshold. The times at which peaks occur are called peak times, and the corresponding excesses over the threshold are called peak values. The data may be summarised as $\{(T_i, Y_i), i=1,2,\ldots\}$ where $T_i$ is the i'th peak time and $Y_i$ the i'th peak value. The most flexible model in the literature appears to be that of [19]:

1. The peak times $\{T_i\}$ are taken to be a nonhomogeneous Poisson process with intensity function

$$\lambda_p(t) = \lambda_p \exp[\sum_{m=1}^{r} \beta_{pm} \sin(\omega_0 mt + \vartheta_{pm})]$$

where $\omega_0$ is the fundamental frequency (corresponding to a period of one year in most environmental applications) and r the number of significant cycles.

2. Conditionally on $\{T_i\}$, the peak values $\{Y_i\}$ are independent and exponentially distributed

$$P\{ Y_j > y \mid \{T_i\} \} = \exp[-\lambda_c(T_j)\, y],$$

where

$$\lambda_c(t) = \lambda_c \exp\left[ \sum_{m=1}^{r} \beta_{cm} \sin(\omega_0 mt + \vartheta_{cm}) \right].$$

The Poisson nature of the peak times and the conditional independence of peak values are believed to be valid assumptions provided the threshold is sufficiently high; this is supported by empirical evidence and by asymptotic arguments as in [27]. Note, however, that it is essential to restrict attention to peaks for this assumption to be reasonable. It would not be appropriate to analyse the full sequence of exceedances in this manner because of the clustering of high values that arises in practice. The assumption that peak values are exponentially distributed appears to be entirely *ad hoc*. The form of the functions $\lambda_p$ and $\lambda_c$ is, of course, intended to allow for seasonal variation.

In subsequent sections we shall focus on two aspects of the model which seem capable of improvement. The first is the arbitrary assumption of exponential peak values. It will be argued, following Pickands [21], that the Generalised Pareto Distribution is a more natural assumption. The exponential distribution is a special case of this. The second aspect is the restriction to peak times. It seems better to consider the point process of all exceedances over the threshold, rather than just the peaks. To analyse such a process, however, we shall need to employ models which allow for clustering.

## 3. METHODS BASED ON EXTREME ORDER STATISTICS

Let $X_1,\ldots,X_n$ denote i.i.d. observations with unknown distribution function F. Inference about the upper (or lower) tail of F may be based on the largest (or smallest) r order statistics, for some value of r. Several authors have treated procedures of this form, starting with Weiss [30], Pickands [21] and Hill [13]. A good exposition of the methods was given by Weissman [31]. Weissman treated r as a fixed constant, and showed how to construct statsitical procedures based on the asymptotic distribution of the r largest order statistics. This distribution, when it exists, is determined by the asymptotic distribution of the sample maximum.

The case of two unknown parameters was dealt with in [31]; more complex three-parameter problems have been considered in [33], [12] and [25]. In the three-parameter case the classical asymptotic results of maximum likelihood theory (as $r \to \infty$) fail; a substantial contribution to this problem was made by Hall [12] but many questions remain unanswered. We now propose an alternative approach to these procedures. Let some threshold level, which we denote by u, be fixed. The procedure is based on all exceedances of u; the number r of such exceedances is random and has, obviously, a binomial distribution with parameters n and $1-F(u)$. Such a procedure has been considered by Weissman [32]. Our proposal is that subsequent inference should be based on the conditional distribution of the exceedance values given r. The theoretical justification for this is that r is an ancillary statistic for the distribution of the exceedance values, i.e. the distribution of r depends only on $F(u)$ and not otherwise on $\{F(x), x > u\}$. There is also a practical advantage in that, conditionally on r, the exceedance values are independent.

The conditional distribution of $X-u$ given $X \geqslant u$ is given by the distribution function

$$F_u(y) = \frac{F(u+y) - F(u)}{1 - F(u)}, \qquad 0 \leqslant y \leqslant x_0 - u,$$

where $x_0 \leqslant \infty$ is the upper boundary point of the distribution.

At this point we introduce the Generalised Pareto Distribution (henceforth referred to as GPD) given by the distribution function

$$G(y \mid \sigma,k) = \begin{cases} 1 - (1 - ky/\sigma)^{1/k}, & k \neq 0, \\ 1 - \exp(-y/\sigma), & k = 0, \end{cases}$$

where $\sigma>0$ and $-\infty<k<\infty$; in the case $k>0$ the range of y is $0 \leqslant y \leqslant \sigma/k$, otherwise the range is $0 \leqslant y < \infty$. The importance of the GPD is given by Pickands' [21] result: for fixed k, we have

$$\lim_{u \to x_0} \inf_{0<\sigma<\infty} \sup_{0 \leqslant y \leqslant \infty} |F_u(y) - G(y \mid \sigma,k)| = 0$$

if and only if there exists $a_n>0$ and $b_n$ such that

$$F^n(a_n y + b_n) \to \begin{cases} \exp[-(1-ky)^{-1/k}], & k \neq 0, \\ \exp[-e^{-y}], & k = 0, \end{cases}$$

for all y with $ky<1$. Essentially, $F_u$ is well approximated by the GPD (with some fixed k) as $u \to x_0$ if and only if F lies in the maximum domain of attraction of an extreme value distribution (with the same k).

This result motivates our next step: as an approximation, we assume that $F_u$ has the GPD with unknown parameters $\sigma$ and k. We shall not consider here how this assumption might be checked in practice; it is an important question, however, and we refer to Davison's [7] paper in this Proceddings for discussion of it.

The density function of the GPD is

$$g(y \mid \sigma,k) = \begin{cases} \sigma^{-1} (1 - ky/\sigma)^{1/k}, & k \neq 0, \\ \sigma^{-1} \exp(-y/\sigma), & k = 0, \end{cases}$$

defined when $y>0$, $ky/\sigma < 1$. The elements of the Fisher information matrix are given by

$$E\{(\frac{\partial}{\partial\sigma} \ln g(Y \mid \sigma,k))^2\} = 1/\{\sigma^2(1-2k)\},$$

$$E\{(\frac{\partial}{\partial\sigma} \ln g(Y \mid \sigma,k))(\frac{\partial}{\partial k} \ln g(Y \mid \sigma,k))\} = -1/\{\sigma(1-k)(1-2k)\},$$

$$E\{(\frac{\partial}{\partial k} \ln g(Y \mid \sigma,k))^2\} = 2/\{(1-k)(1-2k)\},$$

provided $k<1/2$; for $k \geq 1/2$ these expectations are infinite.

For $k<1/2$, it can be shown that the classical asymptotic results for maximum likelihood estimation are valid: for a sample of size r, there exists a solution $(\hat{\sigma},\hat{k})$ of the likelihood equations which as $r\to\infty$ is consistent, efficient and normal with covariance matrix $r^{-1}M$, where

$$M = \sigma^2 (1 - k)^2 \begin{bmatrix} 2/(1-k) & 1/\{\sigma(1-k)\} \\ 1/\{\sigma(1-k)\} & 1/\sigma^2 \end{bmatrix}.$$

is the inverse of the Fisher information matrix.

For $k \geq 1/2$, this result is false and the correct result is very complicated. For $k=1/2$ we have

$$\hat{k} - k = 0_p(r^{-1/2}), \quad \hat{\sigma} - \sigma = 0_p((r \ln r)^{-1/2})$$

with the terms asymptotically independent and normal. For $1/2<k<1$ we have

$$\hat{k} - k = 0_p(r^{-1/2}), \quad \hat{\sigma} - \sigma = 0_p(r^{-k}),$$

and the asymptotic distribution of $\hat{\sigma}$ is the same as that obtained by Woodroofe [34], which is not normal. For $k \geq 1$ and r sufficiently large, the likelihood has no local maximum but has a singularity at the endpoint of the distribution, i.e. when $\max(Y_i) = \sigma/k$. Proofs of these are in [24]. Fortunately, in practice it usually appears to be the case that $k<1/2$, so that classical maximum likelihood is applicable. Davison [7] considers extensions to cases where $\sigma$ and $k$ are not fixed parameters but depend on covariates.

We may summarise the method as follows. First, a high threshold $u$ is chosen and all exceedances above the threshold noted. The number $r$ of such exceedances has a binomial distribution from which inferences about the exceedance probability $p=1-F(u)$ may be made. Conditionally on $r$, the exceedance values are independent with a distribution which we take to be GPD, and whose parameters $\sigma$ and $k$ may be estimated by maximum likelihood, at least so long as $k<1/2$. From these assumptions, the return level $Q_N$ defined by the equation $F(Q_N) = 1-1/N$ is given by

$$Q_N = u + \sigma \frac{1-(Np)^{-k}}{k}, \quad k \neq 0, \tag{3.1}$$

provided $Q_N > u$. The m.l.e. $Q_N$ may therefore be constructed by substituting the estimated values $p,k$ and $\sigma$ in (3.1), and its variance may be estimated approximately as $V^T\Sigma V$, where $V = \left(\frac{\partial Q_N}{\partial p}, \frac{\partial Q_N}{\partial \sigma}, \frac{\partial Q_N}{\partial k}\right)^T \Big|_{p,\sigma,k}$ is the gradient of $Q_N$ at the m.l.e., and $\Sigma$ is the estimated covariance matrix of $(\hat{p},\hat{\sigma},\hat{k})$.

## 4. THRESHOLD METHODS FOR GENERAL TIME SERIES

Our intention now is to suggest ways in which the method of Section 3 might be adapted to general time series sampled at regular intervals. Examples are series of river levels, wave heigths and wind speeds, which are typically sampled at hourly or three-hourly intervals.

The POT method is based just on peaks, rather than all exceedances, over the threshold. This is open to the objection that it may not just be the peaks which are important; in some applications (e.g. assessing the effects of high winds on buildings) the duration of high-level activity may be more important than the peak value attained. We therefore propose to analyse the series consisting of all exceedances over the threshold and the associated exceedance times. To distinguish this from the POT method, we shall refer to it as the Exceedances Over Threshold or EOT method. However, this idea will not be carried to its logical conclusion here: although we model the exceedances times by a clustered point process, the exceedance values will be analysed as in the POT method, by considering cluster peaks.

For the remainder of the paper we shall model the exceedance times $\{T_i\}$ as a very simple doubly stochastic process [6,11] whose supplementary process is a two-state markov chain. Explicitly, $\Lambda(t)$ is assumed to be a Markov chain with states 0 and 1 and jump rates $\lambda$ (from 0 to 1) and $\mu$; conditionally on $\Lambda(t)$ (unobserved), the exceedance time form a Poisson process with rate $\varphi\Lambda(t)$. The unknown parameters are $\varphi, \lambda$ and $\mu$.

The inter-event times $Z_1 = T_1$, $Z_i = T_i - T_{i-1}$ $(i>1)$ are independent with a mixed exponential distribution with p.d.f.

$$f(z) = \frac{\varphi}{\alpha-\beta}\left[(\mu+\varphi-\beta)e^{-\alpha z} - (\mu+\varphi-\alpha)e^{-\beta z}\right], \qquad z>0 \tag{4.1}$$

where $\alpha, \beta$ are the roots (positive and distinct) of the quadratic equation

$$x^2 - (\lambda+\mu+\varphi)x + \lambda\varphi = 0 . \tag{4.2}$$

This follows from formulae in [11], pages 40-44. Maximum likelihood estimation of $\varphi, \lambda$ and $\mu$ is easily performed (numerically) using these equations.

We also need some means of identifying clusters (so that we can pick out cluster maxima). Intuitively, $T_i$ and $T_{i-1}$ belong to the same cluster if $\Lambda(t) = 1$ for $T_{i-1} \leqslant t \leqslant T_i$. Under this definition, the mean number of clusters per unit time is $\lambda\mu\varphi/\{(\lambda+\mu)(\mu+\varphi)\}$. This cannot be applied directly becuase $\Lambda$ is unobserved, so we propose the following empirical rule: assign $T_{i-1}$ and $T_i$ to the same cluster if and only if $Z_i = T_i - T_{i-1} < z^*$. Under this rule, the mean number of clusters per unit time is $\int_{z^*}^{\infty} f(z)dz \,/\, \int_0^{\infty} zf(z)dz$. (These results for mean numbers of clusters follow from elementary renewal theory.) The proposed procedure is to choose $z^*$ so that the mean numbers of clusters under the two rules are the same; this is then determined by setting

$$\int_{z^*}^{\infty} f(z)dz = \left\{\int_0^{\infty} zf(z)dz\right\} \lambda\mu\varphi/\{(\lambda+\mu)(\mu+\varphi)\} = \mu/(\mu+\varphi). \tag{4.3}$$

The proposed method is therefore as follows:

1. From the exceedance times $\{T_i\}$, the intervals $\{Z_i\}$ are obtained and the parameters $\varphi, \lambda, \mu$ estimated by maximum likelihood.
2. The cutoff point $z^*$ is determined by (4.3) and the peak values, i.e. maximum values of $Y_i$ within each cluster, obtained.
3. The GPD is fitted to the peak values and estimates of $k, \sigma$ derived.

Note that there is no allowance for seasonality in this model. Ways of handling that are discussed in Section 6. Also the clustering model is very simple, e.g. it may well be unreasonable to assume the $Z_i$'s independent, in which case a more complicated model would be needed.

Suppose now we want to calculate the return period associated with some high level $q>u$. We define this to be the mean time between successive peaks above the level q. The mean number of clusters per unit time, as already noted, is $\lambda\mu\varphi/\{(\lambda+\mu)(\mu+\varphi)\}$. Therefore, if T is large and $\lambda << \mu$, the mean number of clusters in time [0,T] is approximately $T\lambda\varphi/(\mu+\varphi)$. Within each cluster, the probability of a peak value above q is $1 - G(q-u \mid \sigma,k)$. Thus the mean number of exceedances of q in [0,T] is approximately $T\lambda\varphi(1-k(q-u)/\sigma)^{1/k}/(\mu+\varphi)$, form which it follows that the return period is $(\mu+\varphi)(\lambda\varphi)^{-1}(1-k(q-u)/\sigma)^{-1/k}$. This argument has deliberately been given in a loose and heuristic manner but is easily made rigorous using renewal theory.

The return level $q_T$ associated with a given return period T is given by

$$q_T = u + \frac{\sigma}{k}\left(1 - \left(\frac{T\lambda\varphi}{\mu+\varphi}\right)^{-k}\right) . \tag{4.4}$$

For k=0, the last expression simplifies to

$$q_T = u + \sigma \ln \frac{T\lambda\varphi}{\mu+\varphi} . \tag{4.5}$$

Finally, since maximum likelihood estimators of the five unknown parameters, and the approximate variances and covariances of the estimators, are available, it is possible to estimate $q_T$ and to obtain approximate confidence limits.

## 5. EXTREMES OF STATIONARY SEQUENCES

The asymptotic theory of extremes in stationary sequences was introduced by Watson [29], Berman [2] and Loynes [17], and extensively developed in a series of papers by Leadbetter; see [16]. As the subject has been surveyed elsewhere in this Proceedings [15], we give only a brief review here.

For a stationary sequence $\{\xi_n\}$ and an increasing sequence of threshold levels $\{u_n\}$, Leadbetter introduced two general conditions, $D(u_n)$, and $D'(u_n)$. Condition $D(u_n)$ is a "tail mixing" which states that extreme values far apart in the sequence are approximately independent. Condition $D'(u_n)$ is a "short range independence" condition asserting that the probability of two extreme observations occuring close to each other is asymptotically negligible. If D and D' are satisfied for suitable $\{u_n\}$ then the asymptotic distribution of extreme values from the stationary sequences is the same as that from an i.i.d. sequence with the same marginal distribution. Moreover, the two-dimensional point process of exceedances times and exceedance values converges weakly, under the appropriate renormalisation, to a Poisson process in this case.

Unfortunately, D' is not a reasonable condition for most environmental time series. Indeed, the clustering of high values is a major practical problem. In the case where D is satisfied but not D', Leadbetter [14] has introduced the extremal index $\vartheta$, one definition being that $\vartheta^{-1}$ is the mean number of exceedances per cluster ($0 \leqslant \vartheta \leqslant 1$). The idea, though not the name, first appeared in [20].

These ideas provide some justification for considering the sequence of exceedance times as a clustered Poisson process. The doubly stochastic model we have examined is approximately of this form if $\lambda \ll \mu$. For this model, the mean number of exceedances per unit time is $\lambda\varphi/(\lambda+\mu) \sim \lambda\varphi/\mu$. The asymptotic mean number of peaks per unit time was calculated in Section 4 as $\lambda\varphi/(\mu+\varphi)$. The extremal index is the ratio of these, i.e.

$$\vartheta = \mu/(\mu + \varphi). \qquad (5.1)$$

Our procedure therefore provides an estimator of the extremal index for this model.

## 6. AN APPLICATION

The data were recorded by the Seven Stones light vessel in the English Channel. They consist of recordings of significant wave heights, in metres, taken every three hours by a Shipborne wave recorder. They were collected between 1968 and 1977 but, because of several gaps in the data, the total period spanned by the observations is only seven years. Previous analyses of these data have been made in [3,4,5,8]. The analysis here is an extension of an earlier threshold analysis by Turner [28].

The raw data displayed an obvious annual cyle. To study the nature of the seasonal variation, the data were split into blocks of ten days and within-block means and variances calculated. From the application of standard time-series techniques to these series Turner [28] concluded that the data $\{X(t), t=1,2,...\}$ followed the model

$$\ln X(t) = M + A \cos (\omega t+\vartheta) + Z(t) \qquad (6.1)$$

with $\{Z(t)\}$ a stationary series. The constants M and A (recomputed for the present study) are found to be 0.710 and 0.402, and $\omega = 2\pi/2922$ is the frequency corresponding to a cycle of period one year (2922 observations). The analysis which follows is based on the assumptions that the model (6.1) is correct. We therefore fit a threshold model to the series Z.

The next problem is to choose a threshold. This should be high enough to justify the assumptions of the model but low enough to capture a reasonable number of peaks. In the Flood Studies Report [18], a number of thresholds are considered so that the number of peaks varies between one and five a year. Our analysis was repeated for several thresholds so as to examine the sensitivity of the conclusions to choice of threshold.

For four thresholds u, the model of Section 4 was fitted to the exceedances of Z over u. Estimated parameter values and their standard errors are given in Table 1.

TABLE 1

ESTIMATED PARAMETER VALUES FOR FOUR THRESHOLDS

| Parameter | Threshold u | | | |
|---|---|---|---|---|
| | 1.0 | 1.1 | 1.2 | 1.3 |
| $N_e^{(i)}$ | 295 | 140 | 74 | 36 |
| $N_p^{(ii)}$ | 81 | 43 | 27 | 11 |
| $\varphi$ | 0.46 | 0.52 | 0.47 | 0.50 |
| | (0.04) | (0.06) | (0.08) | (0.11) |
| $\lambda$ | 0.0067 | 0.0036 | 0.0025 | 0.0009 |
| | (0.0008) | (0.0006) | (0.0005) | (0.0003) |
| $\mu$ | 0.18 | 0.23 | 0.27 | 0.22 |
| | (0.02) | (0.04) | (0.06) | (0.07) |
| $\vartheta^{(iii)}$ | 0.28 | 0.31 | 0.37 | 0.31 |
| | (0.03) | (0.04) | (0.06) | (0.08) |
| k | 0.12 | 0.16 | -0.02 | -0.05 |
| | (0.10) | (0.13) | (0.20) | (0.32) |
| $\sigma^{(iv)}$ | 0.17 | 0.17 | 0.12 | 0.12 |
| | (0.02) | (0.03) | (0.03) | (0.05) |
| $\sigma^{(v)}$ | 0.15 | 0.15 | 0.12 | 0.12 |
| | (0.02) | (0.02) | (0.02) | (0.04) |

Standard errors are shown in parentheses beneath the estimates. The units of $\varphi^{-1}$, $\lambda^{-1}$ and $\mu^{-1}$ are the time interval between observations, i.e. three hours.

(i) $N_e$ = Number of exceedances
(ii) $N_p$ = Number of peaks
(iii) Extremal index $\vartheta = \mu/(\mu+\varphi)$
(iv) Assuming k unknown
(v) Assuming k known and equal to zero

It can be seen that $\hat{k}$ is close to zero in each case, and the question arises as to whether the exponential distribution of exceedance values (k=0) is appropriate in place of the GPD. This was tested by a likelihood ratio test. For u=1.1, twice the difference in log likelihoods under the two models k=0, k unrestricted, is 0.92. Since this statistic has a limiting $\chi_1^2$ distribution when k=0, this value is not significant. Similar results were obtained for the other thresholds. Therefore it seems reasonable to assume k=0 for this data, though this is a point to which we return later. The estimates and standard errors of $\sigma$, under the assumption k=0, are also shown in Table 1.

We now turn to the computation of return values from the fitted model. Since we are really interested in the extremes of the series X rather than Z, the method given in Section 4 is not adequate.

Suppose that in a single year (of length $2\pi/\omega$) the exceedance times are denoted $T_1,\ldots,T_N$ (where N is random). We calculate the probability

$$H(q) = P\{\ln X(T_i) \leqslant q \text{ for } i=1,\ldots,N\} \tag{6.2}$$

for q>M+A+u assuming (6.1) and that the exceedances of Z over u are given by the model of Section 4. Thus H is approximately the d.f. of the annual maximum of the series ln X.

Assuming the GPD for excess values we have for q>M+A+u,

$$P\{\ln X(T_i) \leqslant q,\ 1\leqslant i\leqslant N \mid T_1,\ldots,T_N\} = \prod_{i=1}^{N} [1-\{1-k(q-M-u-A\cos(\omega T_i+\vartheta))/\sigma\}^{1/k}]$$

But the process of <u>peak</u> times is approximately Poisson with intensity $\lambda\varphi/(\mu+\varphi)$. Therefore, the conditional distribution of $T_1,\ldots,T_N$ given N is approximately the distribution of uniform order statistics over $[0,2\pi/\omega]$, so that

$$P\{\ln X(T_i) \leqslant q,\ 1\leqslant i\leqslant N \mid N\} \simeq [1-\int_0^{2\pi/\omega}\{1-k(q-M-u-A\cos(\omega t))/\sigma\}^{1/k}(\omega/2\pi)dt]^N. \tag{6.3}$$

Taking expectations in (6.3) assuming the Poisson distribution for N gives

$$H(q) \simeq \exp[-(2\pi/\omega)(\lambda\varphi/(\mu+\varphi)).\int_0^{\pi}\{1-k(q-M-u-A\cos t)/\sigma\}^{1/k}\,dt/\pi]. \tag{6.4}$$

For k=0, the integral in (6.4) simplifies to

$\int_0^{\pi} \exp(-(q-M-u-A\cos t)/\sigma)\, dt/\pi = \exp(-(q-M-u)/\sigma)\, I_0(A/\sigma)$ where $I_0$ is the modified Bessel function of order zero [1], (equation 9.6.16). In this case, the expression for H(q) reduces to the d.f. of the Gumbel (Type 1 Extreme Value) distribution. If $k \neq 0$, the corresponding expression is not one of the classical extreme value distributions (except when A=0, corresponding to the case of no seasonal variation) but a mixture of them.

In the case k=0, the n-year return level $q_n$, which we take to be defined by $H(q_n) = 1/n$, is given explicitly by

$$q_n \simeq M + u + \sigma \ln \{n(2\pi/\omega)(\lambda\varphi/(\mu+\varphi))\, I_0(A/\sigma) \; . \qquad (6.5)$$

This expression may be used to construct an estimator $\hat{q}_n$ of $q_n$. Its standard error may be estimated (approximately) using the procedure described at the end of Sections 3 and 4. In obtaining the derivatives of $q_n$ with respect to the unknown parameters, we use the relation $I_o'(x) = I_1(x)$ ([1], equation 9.6.27).

For $k \neq 0$, there is no closed form expression for $q_n$, which must therefore be found by numerical interpolation in (6.4).

Estimates of $q_{50}$ are shown in Table 2 for each of the threshold levels.

TABLE 2

VARIOUS ESTIMATES OF $q_{50}$

| Source | Estimate | Standard Error |
|---|---|---|
| This paper, u=1.0 | 2.92 | 0.05 |
| This paper, u=1.1 | 2.89 | 0.07 |
| This paper, u=1.2 | 2.77 | 0.08 |
| This paper, u=1.3 | 2.78 | 0.12 |
| [28], u=1.1 (i) | 2.74 | 0.12 |
| [28], u=1.2 (i) | 2.73 | 0.16 |
| [3], (ii) | 2.65 | - |
| [3], (iii) | 2.52 | - |
| [3], (iv) | 2.69 | - |
| [5], (v) | 2.73 | - |
| [8], (vi) | 2.56 | - |

Estimates are for the natural logarithm of the wave height in metres.
(i) Based on 2.5 years of data
(ii) Gumbel method applied to monthly maximum for March
(iii) Gumbel method applied to annual maxima
(iv) Based on product of estimated c.d.f.'s for monthly maxima
(v) Monthly maxima with sinusoidal location parameter
(vi) Fitted distribution to all values
(Partly adapted from [28])

In estimating the standard errors, M and A were treated as known constants, since their sampling errors are negligible compared with those of the other parameters. Also shown for comparison are the point estimates obtained in previous studies of this data set. Turner's [28] analysis differed from the present one in three respects: (i) he used only two and a half years of the data, (ii) he used only the exponential distribution and not the GPD, (iii) the peak times were modelled by a simple Poisson process.

Table 1 includes estimates of the extremal index, which appears to be around 0.3, i.e. the mean cluster size is between 3 and 4.

The GPD did not play a very important role in this analysis as we took k=0 in the final calculations. It is, however, worth examining the sensitivity of the final conclusions to this assumption. Using u=1.2, we recomputed $q_{50}$ for several values of k within two estimated standard errors of zero, keeping the other parameters fixed (Table 3). It can be seen that statistically insignificant variations in k produce considerable variations in $q_{50}$, especially when k is negative.

TABLE 3

POINT ESTIMATES OF $q_{50}$ FOR VARIOUS VALUES OF k

(Other parameters held fixed)

| k | $q_{50}$ |
|---|---|
| 0.4 | 2.52 |
| 0.3 | 2.56 |
| 0.2 | 2.60 |
| 0.1 | 2.67 |
| 0.0 | 2.77 |
| -0.1 | 2.92 |
| -0.2 | 3.18 |
| -0.3 | 3.59 |
| -0.4 | 4.24 |

This highlights what may well be the principal difficulty of this kind of analysis, namely the sensitivity of the conclusions to the assumptions of the model, especially when return periods much longer than the length of the data are involved. It is difficult to say what should be done about this, but we do suggest that it is important to analyse different models and to compare the conclusions they lead to in any analysis of this sort.

Another important question is whether the very simple model (6.1) is really appropriate or whether some more complicated procedure for handling seasonality is needed. First, let us remark that additional sinusoidal terms in (6.1) may be included without any change in the methodology. The analysis is unchanged up to (6.4) in which the additional sinusoidal terms must be included. In that case, when k=0 the integral no longer reduces to a Bessel function but must be integrated numerically. A more fundamental problem is whether the residual process Z can really be assumed stationary - the standard time series tests based on the autocovariances are only tests of second-order stationarity. The sensitivity of the conclusions to the stationary assumption can be examined by testing for seasonality in the series of exceedance times and exceedance values. As a useful diagnostic tool, we propose the "total excess statistic" $\Sigma(Z(t)-u)_+$, calculated for each month of the data. That is, for each month we add up all the exceedance values over the threshold. For u=1.0, this was calculated and the spectral density of the resulting sequence plotted (Figure 1). It is consistent with a low-order stationary model, and in particular there is no evidence of residual seasonality, as might be indicated by a peak near the frequency 1/12 or one of its multiples.

Another important issue is the assumed independence of peak values. A correlogram of peak values showed no evidence of correlation, though this is one aspect which could be taken further.

In conclusion, if we accept k=0 then, allowing for the different estimates in Table 2 and their standard errors, we might conservatively conclude that $q_{50}$ lies between 2.5 and 3.0. This is a very wide range, probably too wide to be of much practical use. Taking the variation of k into account would lead to an even wider range. It is noticeable that the estimated return values are well above those of Carter and Challenor. Challenor (personal communication) has suggested that this may be an artefact of the logarithmic transformation.

Although we have discussed a number of aspects of threshold models it is clear that there are other aspects needing further study. Two of these which we have touched on are the choice of threshold and the handling of seasonal series. Other questions are the handling of multivariate series and the incorporation of subsidiary information as might be contained in covariates.

Finally, it may often be the case that calculated return levels are very sensitive to the precise model adopted, especially where significant extrapolation is involved. In view of this, it may be necessary to repeat the analysis under different model assumptions in order to obtain a realistic assessment of the accuracy of the conclusions.

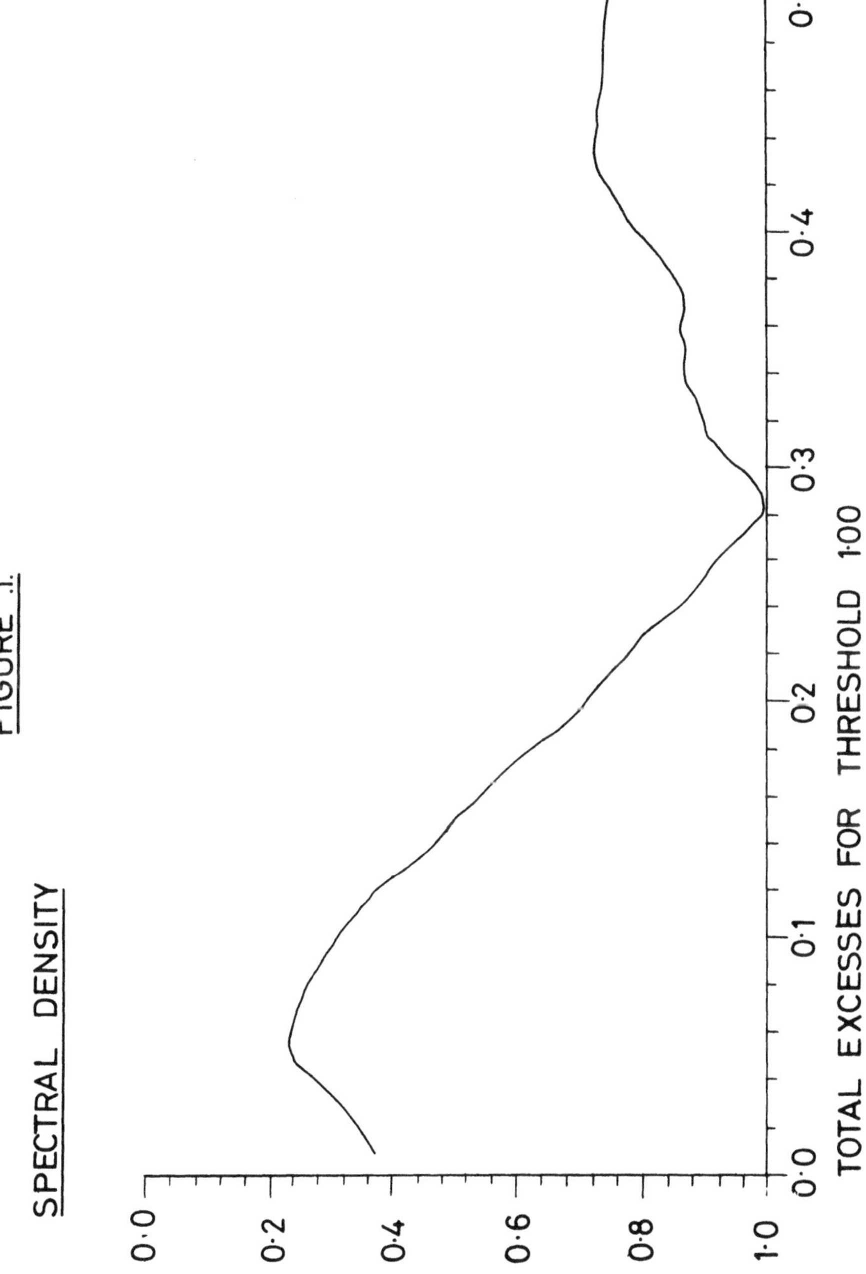
FIGURE 1
SPECTRAL DENSITY
0·0
0·2
0·4
0·6
0·8
1·0
0·0
0·1
0·2
0·3
0·4
0·5
TOTAL EXCESSES FOR THRESHOLD 1·00

Acknowledgements The wave height data were produced by the Institute of Oceanographic Sciences. I am grateful to Peter Challenor for supplying me with the data, and to Anthony Davison and Martyn Turner for many suggestions. The work was carried out during the tenure of an SERC award.

## REFERENCES

1. Abramowitz, M. and Stegun, I. 1964. Handbook of Mathematical Functions. National Bureau of Standards, U.S.A.

2. Berman, S.M. 1962. Limiting distributions of the maximum term in a sequence of dependent random variables. Ann. Math. Statist. 33, pp. 894-908.

3. Carter, D.J.T. and Challenor, P.G. 1981. Estimating return values of environmental parameters. Quart. J. Roy. Met. Soc. 107, pp. 259-266.

4. Carter, D.J.T. 1981. Analysis of Seven Stones wave height data 1968-1978. Institute of Oceanographic Sciences, personal communication.

5. Challenor, P.G. 1982. A new distribution for annual extremes of environmental variables. Quart. J. Roy. Met. Soc. 108, pp. 975-980.

6. Cox, D.R. 1955. Some statistical methods connected with series of events. J. Roy. Statist. Soc. B 27, pp. 129-164.

7. Davison, A.C. 1984. Modelling excesses over high thresholds, with an application, in Statistical Extremes and Applications, this volume.

8. Fortnum, B.C.H. and Tann, H.M. 1977. Waves at Seven Stones Light Vessel. Institute of Oceanographic Sciences, personal communication.

9. Gnedenko, B.V. 1943. Sur la distribution limite du terme maximum d'une serie aleatoire. Ann. Math. 44, 423-453.

10. Grandell, J. 1972. Statistical inference for doubly stochastic Poisson processes. In Stochastic Point Processes, Statistical Analysis, Theory and Applications, ed. P.A.W. Lewis, Wiley New York.

11. Grandell, J. 1976. Doubly Stochastic Poisson Processes. Lecture Notes in Mathematics No. 529, Springer, Berlin.

12. Hall, P. 1982. On estimating the endpoint of a distribution. Ann. Statist. 10, pp. 556-568.

13. Hill, B.M. 1975. A simple general approach to inference about the tail of a distribution. Ann. Statist. 3, pp. 1163-1174.

14. Leadbetter, M.R. 1982. Extremes and local dependence in stationary sequences. Preprint 1982 no.5, University of Copenhagen.

15. Leadbetter, M.R. 1984. Statistical Extremes and Applications, this volume.

16. Leadbetter, M.R., Lindgren, G. and Rootzen, H. 1983. Extremal and Related Properties of Random Sequences and Processes. Springer, New York.

17. Loynes, R.M. 1965. Extreme values in uniformly mixing stationary stochastic processes. Ann. Math. Statist. 36, pp. 993-999.

18. NERC 1975. Flood Studies Report, Vol.1. Natural Environment Research Council, London.

19. North, M. 1980. Time-dependent stochastic models of floods. J. Hydraulics Division, A.S.C.E., pp. 649-655.

20. O'Brien, G.L. 1974. The maximum term of uniformly mixing stationary stochastic processes. Z. Wahr. verv. Geb. 30, pp. 57-63.

21. Pickands, J. 1975. Statistical inference using extreme order statistics. Ann. Statist. 3, pp. 119-131.

22. Revfeim, K.J.A. 1982. Seasonal patterns in extreme 1-hr rainfalls. Wat. Res. Res.

23. Revfeim, K.J.A. 1983. On the analysis of extreme rainfalls. J. Hydrology.

24. Smith, R.L. 1983. Maximum likelihood estimation in a class of non-regular cases. Paper in preparation.

25. Smith, R.L. and Weissman, I. 1983. Maximum likelihood estimation of the lower tail of a probability distribution. Submitted for publication.

26. Todorovic, P. (1978). Stochastic models of floods. Wat. Res. Res. 14, pp. 345-356.

27. Todorovic, P. 1979. A probabilistic approach to analysis and prediction of floods. I.S.I. Proceedings.Vol.1, pp 113-124.

28. Turner, M.J. 1982. Estimation of the fifty year return wave height at Seven Stones. M.Sc. and D.I.C. Report, Department of Mathematics, Imperial College.

29. Watson, G.S. 1954. Extreme values from samples in m-dependent stationary processes. Ann. Math. Statist. 25, pp. 798-800.

30. Weiss, L. 1971. Asymptotic inference about a density function at the end of its range. Nav. Res. Log. Quart. 18, pp. 111-114.

31. Weissman, I. 1978. Estimation of parameters and large quantiles, based on the k largest observations. J. Amer. Statist. Soc. 73, 812-815.

32. Weissman, I. 1980. Estimation of tail parameters under Type I censoring. Commun. Statist. Theor. Meth. A9 (11), pp. 1165-1175.

33. Weissman, I. 1982. Confidence intervals for the threshold parameter II: Unknown shape parameter. Commun. Statist. Theor. Meth. All. pp. 2451-2474

34. Woodroofe, M. 1974. Maximum likelihood estimation of translation parameter of a truncated distribution II. Ann. Statist. 3, 474-488.

ON SUCCESSIVE RECORD VALUES IN A SEQUENCE OF INDEPENDENT IDENTICALLY DISTRIBUTED RANDOM VARIABLES

Jozef L. Teugels

Katholieke Universiteit Leuven

ABSTRACT : We investigate the probability that a certain function of the record values exceeds a fixed quantity. Originating from an application in insurance mathematics, the results can be applied to other instances.

## 1. INTRODUCTION AND MOTIVATION

Let $Y_1, Y_2, \ldots$ be a sequence of independent random variables with common continuous distribution as the variable Y, say F, and such that $F(o-)=o$. Let $X_1, X_2, \ldots$ be the resulting sequence of record values of the sequence $\{Y_k\}_1^\infty$

In insurance mathematics one thinks of the sequence as the claims in a specific portfolio. In this context one almost always assumes that Y has moments of all order; or even that for positive constants c and k, $1-F(x) < k \exp - cx$. In practice however the mere existence of reinsurance evidently points out a possibility for large claims, so that an exponential bound on F is no longer acceptable. For information on insurance, see [1,2]. As explained in [4] the simplest model for such a situation is to assume that the claim size distribution is *Pareto*, i.e.

$$1 - F(x) = x^{-\alpha} \quad , \; x > 1 \qquad (1)$$

where the *index* $\alpha$ is positive. The smaller $\alpha$ the more probable are large values of the claims.

In practice the insurer uses i.a. data from previous insurance periods to estimate unknown quantities like the mean claim



*J. Tiago de Oliveira (ed.), Statistical Extremes and Applications, 639–650.*

size EY or the variance of the claim size, say Var Y. These values are then used in the formulas for the premium calculation. However if (1) applies, then EY will only be finite if $\alpha > 1$ while $\text{Var } Y < \infty$ iff $\alpha > 2$. Statistical tests usually are not designed to investigate expressions like $EY = \infty$.

In the Pareto-case however the hypothesis $EY < \infty$ leads to a composite hypothesis on the parameter $\alpha$ and as such a set of classical large sample tests is available. From the managerial point of view it is time-consuming to follow-up the data while checking any of the hypotheses resulting from these large sample techniques. A decision based solely on the consecutive record values looks appealing.

Let us then assume that the investigator is interested in knowing whether or not $\alpha > \alpha_o$ where $\alpha_o$ is a preassigned quantity. He is allowed to look at the consecutive record values $X_1, X_2, \ldots$ coming from the batch of claims. For him these values are *warning values*. As a single estimator of $\alpha^{-1}$ based on $X_k$ he should use

$$Z_k = (\log X_k)/k \ . \tag{2}$$

Indeed $Z_k$ has some desirable properties. Every time however $Z_k \geq \alpha_o^{-1}$ he fears that the hypothesis $H(\alpha > \alpha_o)$ will be violated; the warning value becomes an *alarming value*. On the basis of the number and of the sizes of the alarmvalues the investigator hopes to decide on accepting or rejecting the hypothesis.

In §2 we recall some properties of records; record values have a number of distribution free properties. Using the exponential in §3 we find the probabilities that the first n warning values are actually alarming.

## 2. ON RECORD VALUES

Let $Y_1, Y_2, \ldots$ be i.i.d. positive r.v. with d.f. F. Define the *record times* by

$$N_1 = 1$$

$$N_{k+1} = \inf \{m: Y_m > Y_{N_k}\} \quad ,$$

and the *record values*

$$X_k = Y_{N_k} \ .$$

Shorrock [3] has shown that the process $\{(X_k, N_k);\ 1 \leq k\}$ is a bivariate Markov process with transition function

$$P\{X_{k+1} > a \, , \, N_{k+1} = m | X_k, N_k\}$$
$$= \{1-F(a)\}\{F(X_k)\}^{m-N_k-1} \, , \quad 1 \leqslant N_k \leqslant m \, , \; X_k \leqslant a \, .$$

Turning to new variables

$$T_k = -\log\{1 - F(X_k)\} \tag{3}$$

this Markov character can be transformed into the basic property: the sequence $\{T_k; 1 \leqslant k\}$ acts as the arrival process of a homogeneous Poisson process with $P(T_1 > a) = e^{-a}$. But then

$$f_{T_1,T_2,\ldots,T_n}(t_1,t_2,\ldots,t_n) = e^{-t_n} \quad 0 < t_1 < t_2 < \ldots < t_n$$
$$= 0 \quad \text{elsewhere.} \tag{4}$$

In particular

$$f_{T_k}(t) = \frac{1}{(k-1)!} e^{-t} t^{k-1} \qquad 0 < t \, ,$$

and for $m < k$ and $x < y$

$$f_{T_m,T_k}(x,y) = \{(m-1)!(k-m-1)!\}^{-1} \, e^{-y}(y-x)^{k-m-1} x^{m-k} .$$

It follows that $E(T_k^r) = (r+k-1)!/(k-1)!$; hence $E(T_k) = \text{Var}T_k = k$; also $\text{cov}(T_m,T_k) = m$ so that the correlation coefficient between $T_m$ and $T_k$ equals $(m/k)^{1/2}$.

Note that if $F^i$ is the inverse of F then

$$[T_k > a] = [X_k \geqslant F^i(1 - e^{-a})] \tag{5}$$

and for appropriate functions g

$$E\{g(X_k)\} = \frac{1}{(k-1)!} \int_0^\infty g\{F^i(1-e^{-t})\} e^{-t} t^{k-1} dt \, . \tag{6}$$

## 3. AN AUXILIARY PROBLEM

In this section we would like to evaluate the probability that the first n warning values are alarming. In terms of the $T_k$ defined above we have the following theorem.

THEOREM I : For $c \geqslant 0$ and $n \geqslant 1$,

$$p_n(c) \equiv P\{\bigcap_{k=1}^{n} (T_k > ck)\} = \frac{e^{-nc}}{n} \sum_{m=o}^{n-1} \frac{(n-m)}{m!} (nc)^m . \qquad (7)$$

Proof : Using the density of the sequence $T_1, T_2, \ldots, T_n$ we find

$$p_n(c) = \int_c^\infty \int_{2c}^\infty \int_{nc}^\infty f_{T_1,T_2,\ldots,T_n}(t_1,t_2,\ldots,t_n)dt_n \ldots dt_2\, dt_1.$$

Using $a \vee b = \max(a,b)$ and (4) this can be rewritten as

$$p_n(c) = \int_c^\infty dt \int_{2c\vee t_1}^\infty dt_2 \ldots \int_{nc\vee t_{n-1}}^\infty e^{-t_n}\, dt_n .$$

Reverse the order of integration to find

$$p_n(c) = \int_{nc}^\infty \exp(-t_n)dt_n \int_{(n-1)c}^{t_n} dt_{n-1} \ldots \int_{2c}^{t_3} dt_2 \int_c^{t_2} dt_1.$$

Now make the substitutions

$$t_k = nc + y_k$$

to find

$$p_n(c) = e^{-nc} \int_o^\infty e^{-y_n} F_{n-1}(c,y_n)dy_n \qquad (8)$$

where for $n \geqslant 1$

$$F_n(c,y) = \int_o^{c+y} dy_n \int_o^{c+y_n} dy_{n-1} \ldots \int_o^{c+y_2} dy_1 \qquad (9)$$

and where we define $F_o(c,y) = 1$ for convenience.
Note that

$$F_{n+1}(c,y) = \int_o^{c+y} F_n(c,u)du \qquad (10)$$

so that we can set up a recursion. For example $F_1(c,y) = c+y$, $F_2(c,y) = 1/2(y^2 + 4cy + 3c^2)$.
Let us put $F_n$ into a series expansion :

$$F_n(c,y) = \sum_{k=o}^{n} b_{n,k}\, c^{n-k}\, y^k . \qquad (11)$$

The relation (10) is transformed into

$$b_{n+1,k} = \sum_{m=(k-1)^+}^{n} \frac{1}{m+1} \binom{m+1}{k} b_{n,m} \qquad o \leqslant k \leqslant n+1 \tag{12}$$

where $a^+ = a \vee o$.

Using (11) into (8) yields an expression of the form (7). Therefore it remains to be proved that $\{b_{n,k} \; ; \; o \leqslant k \leqslant n\}$ is given by the formula

$$b_{n,k} = \frac{1}{k!} (k+1) \frac{(n+1)^{n-k-1}}{(n-k)!} \qquad o \leqslant k \leqslant n \; . \tag{13}$$

We use mathematical induction. The case n=1 is obvious. So let $n > 1$ but take $k \geqslant 1$ first. Using (13) in the right hand side of (12) leads to

$$b_{n+1,k} = \frac{1}{k!} \sum_{m=k-1}^{n} \frac{m+1}{(m-k+1)!} \frac{1}{(n-m)!} (n+1)^{n-m-1} \; .$$

Change m into n-m to obtain :

$$b_{n+1,k} = \frac{1}{k!} \frac{1}{(n-k+1)!} \sum_{m=o}^{n-k+1} \binom{n-k+1}{m} (n+1-m)(n+1)^{m-1} \; .$$

Using the binomium expansion for $(1+t)^r$ and its derivative with respect to t we get after easy algebra :

$$b_{n+1,k} = \frac{1}{k!} (k+1) \frac{(n+2)^{n-k}}{(n+1-k)!}$$

which is (13) for n replaced by n+1.

Finally let k=o; then a similar calculation can be performed to prove that (13) also holds in this case. This finishes the proof of theorem 1.

Since the expression (7) is vital to the applications we include a short table in the appendix.

Recall also that with the explicit form of the coefficients $b_{n,k}$ the quantities $F_n$ can be put in closed form as well. For later record we include two useful expressions here :

$$F_n(c,y) = \frac{1}{n!} (y+c)\{y+c(n+1)\}^{n-1} \tag{14}$$

$$\int_a^b F_n(c,y)dy = \frac{1}{(n+1)!}\{b[b+(n+1)c]^n - a[a+(n+1)c]^n\},\ a < b\ . \qquad (15)$$

It is by no means necessary to include all of the n warning values in the statement on the left of (7). If one leaves one value free then we obtain the next result :

THEOREM 2 : For $c \geqslant o$ and $1 \leqslant j < n$,

$$P\{\bigcap_{k\neq j}^{n} [T_k > kc]\} = p_n(c) + e^{-jc}p_{n-j}(c)\frac{1}{j!}(j-1)^{j-1}c^j\ . \qquad (16)$$

Proof : Proceeding like in the case where the j-th event was included we can write that the left hand side equals the integral

$$\int_{nc}^{\infty} e^{-t_n}dt_n \dots \int_{(j+1)c}^{t_{j+2}} dt_{j+1} \int_{(j-1)c}^{t_{j+1}} dt_j \dots \int_c^{t_2} dt_1$$

$$= e^{-nc}\int_o^{\infty} e^{-y_n}\, dy_n\, F_{n-1}(c,y_n)$$

$$+ e^{-nc}\int_o^{\infty} e^{-y_n}\, dy_n \dots \int_o^{c+y_{j+2}} dy_{j+2}\, g_j$$

where

$$g_j = \int_{-c}^{o} F_{j-1}(c,y_j)dy_j = \frac{1}{j!}c^j(j-1)^{j-1}\ .$$

Compare the first integral with (8) and the second with (8) but where n has been replaced by (n-j). The value of $g_j$ follows from (15). This finishes the proof of (16).

If one leaves out two of the events then an entirely similar argument yields

$$P\{\bigcap_{\substack{1\leqslant k\leqslant n\\ \neq j,\ \neq m}} [T_k > kc]$$

$$= p_n(c) + e^{-jc}\, p_{n-j}g_j \qquad (17a)$$

$$+ e^{-mc}\, p_{n-m}(g_m + g_jg_{m-j})\ ,\ 2\leqslant j+1 < m < n$$

$$= p_n(c) + e^{-jc} p_{n-j} g_j \qquad (17b)$$

$$+ e^{-(j+1)c} p_{n-j-1} \frac{4jc}{j+1} g_j , \quad 2 \leq j+1 = m < n.$$

We can continue this way although the expressions become more and more complicated. One finally could have none of the events; as alternative :

THEOREM 3 : For $c \geq o$ and $1 \leq n$,

$$P\{\bigcap_{k=1}^{n} [T_k \leq kc]\} = 1 - \sum_{o}^{n-1} \exp\{-(k+1)c\}c^k \frac{(k+1)^k}{(k+1)!} .$$

Proof : Denote the required probability by $q_n(c)$; we have

$$q_n(c) = \int_o^c dx_1 \int_{x_1}^{2c} dx_2 \ldots \int_{x_{n-2}}^{(n-1)c} dx_{n-1} \int_{x_{n-1}}^{nc} e^{-x_n} dx_n .$$

Work out the last integral to set up a recursion. Setting aside $q_{n-1}$ the remaining integral is denoted by $r_{n-1}(c)$. Hence

$$q_n(c) = q_{n-1}(c) - e^{-nc} r_{n-1}(c) .$$

It easily follows that $r_{n-1}(c) = r_{n-1}(1)c^{n-1} \equiv r_{n-1}c^{n-1}$. It remains to evaluate the integral

$$r_{n-1} = \int_o^1 dx_1 \int_{x_1}^{2} dx_2 \ldots \int_{x_{n-1}}^{n-1} dx_{n-1}$$

$$= \int_o^1 dy_{n-1} \int_o^{1+y_{n-1}} dy_{n-2} \ldots \int_o^{1+y_3} dy_2 \int_o^{1+y_2} dy_1$$

$$= F_{n-1}(1,o) .$$

Solving the resulting recursion yields the statement of the theorem since $q_1(c) = 1-e^{-c}$. This ends the proof.

## 4. APPLICATIONS

We now give a number of applications of the above results.

1. *Pareto-case*. Adapting the notation of §2 to the case of a Pareto-law with index $\alpha$, (3) reads : $T_k = \alpha \log X_k$. But then $Z_k = k^{-1}\log X_k$ is an unbiased estimator for $(1/\alpha)$. By using the fact that $T_k$ has a gamma-density one also easily checks that the Fisher-information contained in $Z_k$ equals $k\alpha^2 = (\text{Var } Z_k)^{-1}$. As an estimator of $(1/\alpha)$ $Z_k$ has then some desirable properties. Exact confidence intervals can be constructed since $2\alpha k Z_k$ has a chi-square distribution with 2k degrees of freedom. Combining the successive intervals we arrive at events like those treated in the theorems of the previous section. Indeed fixing a value of $\alpha$, say $\alpha_o$, we see that

$$[X_k \text{ is alarming}] = [Z_k > \alpha_o^{-1}];$$

Under the hypothesis that the original sample comes from a Pareto-law with index $\alpha$, we have

$$P_\alpha(Z_k > \alpha_o\} = P\{T_k > k(\alpha/\alpha_o)\}$$

and hence

$$P_\alpha\{\bigcap_{k=1}^{n} [Z_k > \alpha_o^{-1}] = p_n(\alpha/\alpha_o) .$$

To be even more specific, let us assume that we would not like EY to be infinite. If this were so probably some kind of reinsurance might be pondered. So we put the alarm value at $\alpha_o = 1$. We look for the probability that the first three record or warning values give estimates for $\alpha$ below the alarm level. In the table for n=3 we find that any $\alpha$ larger than 2 is highly improbable. The variance is "likely" to be infinite.

As another application let us fix a significance level, say q. For a given n what is the smallest common value, say $c_{n,q}$, that we can accept from the consecutive records ? If the alarm is fixed at $\alpha_o$

$$P_\alpha\{[\bigcap_{k=1}^{n} [Z_k > c_{n,q}]\} = P\{\bigcap_{k=1}^{n} [T_k > k\alpha c_{n,q}]\}$$

$$< P\{\bigcap_{k=1}^{n} [T_k > k\alpha_o c_{n,q}]\} = p_n(\alpha_o c_{n,q}) = q .$$

From the table we can find $c_{n,q}$. If for example n=4, $\alpha_o$=2 and

$q = .05$ then $2c_{4,.05} = 1.56$. So for any $\alpha > 2$ the probability that the first four records give a value $\log X_k > k(1.56)$ is less than .05.

2. *Exponential case.* Let the underlying distribution be exponential with unknown parameter $\lambda$ : $1-F(x) = \exp(-\lambda x)$. Take $Z_k = X_k/k$. Then as before we get

$$P_\lambda\{\bigcap_{k=1}^{n} [Z_k > c]\} = P\{\bigcap_{k=1}^{n} [T_k > \lambda kc]\} = p_n(\lambda c).$$

The theorems of section 2 allow a variety of applications concerning hypotheses on $\lambda$.

3. *Distributions with power-exponential tails.* As a third example take $1-F(x) = \exp(-x^\alpha)$ where now $\alpha$ is unknown. Take $Z_k = \log X_k - (1/\alpha)\log k$. Then (3) and a little algebra show that

$$P_\alpha\{\bigcap_{k=1}^{n} [Z_k > c]\} = p_n(e^{\alpha c}) .$$

In this case $Z_k$ is not unbiased. Indeed by (6) we find that

$$\begin{aligned} EZ_k &= E(\log X_k) - (1/\alpha)\log k \\ &= (1/\alpha)\left\{\int_0^\infty \log t\, e^{-t}\, t^{k-1} \frac{1}{(k-1)!}\, dt - \log k\right\} \\ &= (1/\alpha)\{\psi(k) - \log k\} . \end{aligned}$$

## 5. CONCLUDING REMARKS

1. It is clear how the formulas (16) and (17) can be used in case some of the warning values are not alarming. Theorem 3 gives some idea of the probability that no alarming value occurs.

2. The table in the appendix is given for illustrative purposes; the explicit expressions for $p_n(c)$ make it possible to evaluate these quantities to whatever precision one desires. The reason why the table allows n to vary between 1 and 10 only lies in the fact that one should not expect too many records in a set of data. Referring again to [3] one knows that

$$P(N_k = n+1) = \frac{|S_n^{k-1}|}{(n+1)!}$$

where $S_n^k$ are the Stirling numbers of the first kind. Using properties of the Stirling numbers one finds that the record times have fat-tailed distributions. As a result the number of such values will be rather limited.

3. If one is only interested in the applications to insurance, then an interesting extension of the theory has to be worked out. In insurance mathematics one assumes that the claims arrive according to a Poisson process. To have a more realistic application one ought to consider the sequence of record values of a compound Poisson process.

## ACKNOWLEDGEMENT

The author thanks G. Lata of the Naval Postgraduate School, Monterey, for his help with the numerical aspects of this paper.

## REFERENCES

[1] BUHLMAN, H. Mathematical Methods in Risk Theory. New York : Springer-Verlag, 1970.

[2] GERBER, H.U. An Introduction to Mathematical Risk Theory. S.S. Huebner Foundation for Insurance Education, Wharton School, Univ. Pennsylvania, Philadelphia, 1979.

[3] SHORROCK, R.W. On record values and record times. J. Appl. Probability, 9, 1972, 316-326.

[4] TEUGELS, J.L. Large claims in insurance mathematics. Astin Bulletin, 15, 1983, 81-88.

APPENDIX

| c ↓ / n → | 1 | 2 | 3 | 4 | 5 |
|---|---|---|---|---|---|
| .1 | .904837 | 900094 | 900016 | 900003 | 900001 |
| .2 | .818731 | 804384 | 801265 | 800405 | 800138 |
| .3 | .740818 | 713455 | 705398 | 702385 | 701117 |
| .4 | .670320 | 629061 | 614436 | 607843 | 604498 |
| .5 | .606531 | 551819 | 529934 | 518785 | 512390 |
| .6 | .548812 | 481911 | 452919 | 436898 | 426924 |
| .7 | .496585 | 419215 | 383901 | 363320 | 349802 |
| .8 | .449329 | 363414 | 322956 | 298597 | 282061 |
| .9 | .406570 | 314068 | 269830 | 242744 | 224037 |
| 1.0 | .367879 | 270671 | 224042 | 195367 | 175467 |
| 1.1 | .332871 | 232687 | 184969 | 155791 | 135649 |
| 1.2 | .301194 | 199579 | 151920 | 123183 | 103612 |
| 1.3 | .272512 | 170829 | 124184 | 096643 | 078267 |
| 1.4 | .246597 | 145944 | 101070 | 075528 | 058520 |
| 1.5 | .223130 | 124468 | 081929 | 058251 | 043344 |
| 1.6 | .201896 | 105982 | 066167 | 044800 | 031824 |
| 1.7 | .182683 | 090108 | 053255 | 034261 | 023178 |
| 1.8 | .165299 | 076506 | 042767 | 026065 | 016756 |
| 1.9 | .149586 | 064875 | 034179 | 019733 | 012029 |
| 2.0 | .135335 | 054970 | 027266 | 014872 | 008581 |
| 2.1 | .122456 | 046486 | 021696 | 011161 | 006084 |
| 2.2 | .110803 | 039287 | 017222 | 008344 | 004290 |
| 2.3 | .100259 | 033171 | 013640 | 006214 | 003009 |
| 2.4 | .090718 | 027981 | 010781 | 004613 | 002100 |
| 2.5 | .082085 | 023583 | 008504 | 003413 | 001459 |
| 2.6 | .074276 | 019860 | 006695 | 002517 | 001009 |
| 2.7 | .067205 | 016711 | 005262 | 001851 | 000695 |
| 2.8 | .060810 | 014052 | 004129 | 001358 | 000477 |
| 2.9 | .055023 | 011807 | 003234 | 000993 | 000326 |
| 3.0 | .049787 | 009915 | 002530 | 000725 | 000222 |
| 3.1 | .045049 | 008321 | 001976 | 000528 | 000151 |
| 3.2 | .040762 | 006978 | 001541 | 000384 | 000102 |
| 3.3 | .036883 | 005850 | 001201 | 000278 | 000069 |
| 3.4 | .033373 | 004901 | 000934 | 000201 | 000047 |
| 3.5 | .030197 | 004103 | 000726 | 000145 | 000031 |
| 3.6 | .027324 | 003434 | 000564 | 000105 | 000021 |
| 3.7 | .024723 | 002873 | 000437 | 000075 | 000014 |
| 3.8 | .022371 | 002402 | 000339 | 000054 | 000009 |
| 3.9 | .020242 | 002008 | 000262 | 000039 | 000006 |
| 4.0 | .018316 | 001677 | 000203 | 000028 | 000004 |

Values of $p_n(c)$ as defined by (7).

| c ↓ / n → | 6 | 7 | 8 | 9 | 10 |
|---|---|---|---|---|---|
| .1 | 900000 | 900000 | 900000 | 900000 | 899999 |
| .2 | 800049 | 800018 | 800007 | 800003 | 800001 |
| .3 | 700544 | 700273 | 700140 | 700073 | 700038 |
| .4 | 602675 | 601633 | 601018 | 600644 | 600413 |
| .5 | 508450 | 505905 | 504203 | 503036 | 502219 |
| .6 | 420255 | 415578 | 412183 | 409655 | 407733 |
| .7 | 340264 | 333204 | 327795 | 323543 | 320132 |
| .8 | 269999 | 260769 | 253458 | 247515 | 242586 |
| .9 | 210152 | 199339 | 190621 | 183409 | 177321 |
| 1.0 | 160623 | 149003 | 139586 | 131756 | 125110 |
| 1.1 | 120699 | 109054 | 099663 | 091891 | 085327 |
| 1.2 | 089279 | 078260 | 069492 | 062329 | 056359 |
| 1.3 | 065080 | 055146 | 047398 | 041195 | 036127 |
| 1.4 | 046805 | 038208 | 031674 | 026579 | 022522 |
| 1.5 | 033245 | 026062 | 020771 | 016771 | 013684 |
| 1.6 | 023343 | 017522 | 013385 | 010367 | 008119 |
| 1.7 | 016217 | 011625 | 008488 | 006287 | 004712 |
| 1.8 | 011156 | 007618 | 005303 | 003747 | 002680 |
| 1.9 | 007605 | 004935 | 003267 | 002197 | 001496 |
| 2.0 | 005140 | 003164 | 001988 | 001269 | 000821 |
| 2.1 | 003447 | 002008 | 001195 | 000723 | 000443 |
| 2.2 | 002295 | 001263 | 000710 | 000406 | 000236 |
| 2.3 | 001517 | 000788 | 000418 | 000226 | 000124 |
| 2.4 | 000997 | 000487 | 000244 | 000124 | 000064 |
| 2.5 | 000651 | 000299 | 000141 | 000067 | 000033 |
| 2.6 | 000423 | 000182 | 000081 | 000036 | 000017 |
| 2.7 | 000273 | 000110 | 000046 | 000019 | 000008 |
| 2.8 | 000175 | 000067 | 000026 | 000010 | 000004 |
| 2.9 | 000112 | 000040 | 000015 | 000005 | 000002 |
| 3.0 | 000071 | 000024 | 000008 | 000003 | 000001 |
| 3.1 | 000045 | 000014 | 000004 | 000001 | |
| 3.2 | 000029 | 000008 | 000002 | 000001 | |
| 3.3 | 000018 | 000005 | 000001 | | |
| 3.4 | 000011 | 000003 | 000001 | | |
| 3.5 | 000007 | 000002 | | | |
| 3.6 | 000004 | 000001 | | | |
| 3.7 | 000003 | 000001 | | | |
| 3.8 | 000002 | | | | |
| 3.9 | 000001 | | | | |
| 4.0 | 000001 | | | | |

Values of $p_n(c)$ as defined by (7).

# TWO TEST STATISTICS FOR CHOICE OF UNIVARIATE EXTREME MODELS

J. Tiago de Oliveira and M. Ivette Gomes

Departamento de Estatística e
Centro de Estatística e Aplicações
da Universidade de Lisboa

ABSTRACT. Asymptotic properties of the statistic proposed by Gumbel, $W_n = \{\max(X_i) - \operatorname{med}(X_i)\}/\{\operatorname{med}(X_i)-\min(X_i)\}$, are obtained for testing the shape parameter $k = 0$, in von Mises-Jenkinson form. Similar results are obtained for a ratio of variances test statistic modified from one suggested by Jenkinson. Comparison is made; as could be expected Gumbel statistic turns out to be the better one.

## 1. INTRODUCTION AND PRELIMINARIES

Let $\{X_i\}_{i \geqq 1}$ be a sequence of independent, identically distributed (i.i.d.) random variables (r.v.'s) and let us denote by $\{X_{n:n}\}_{n \geqq 1}$, the sequence of associated maximum values, i.e.

$$X_{n:n} = \max_{1 \leqq i \leqq n} X_i.$$

It is well known that if the distribution function (d.f.) of $X_{n:n}$, suitably normalized, converges to a proper d.f. $L(z)$, than $L(z)$ must be of one of the following three types — the so-called <u>extreme value stable types</u>:

<u>Type I</u> (Gumbel) $\quad \Lambda(z) = \exp(-\exp(-z)),\ z \in \mathbb{R}$



*J. Tiago de Oliveira (ed.), Statistical Extremes and Applications, 651–668.*

$$\underline{\text{Type II}} \text{ (Frechet)} \quad \Phi_\alpha(z) = \exp(-z^{-\alpha}),\ z \geq 0\ (\alpha>0) \tag{1.1}$$

$$\underline{\text{Type III}} \text{ (Weibull)} \quad \Psi_\alpha(z) = \exp(-(-z)^{\alpha}),\ z<0\ (\alpha>0)$$

More precisely, if for each n, there exist attraction coefficients $\{a_n\}_{n\geq 1}$ $(a_n>0)$ and $\{b_n\}_{n\geq 1}$, such that $F^n(a_n z+b_n)\to G(z)$, as $n\to\infty$, $G(.)$ a proper d.f., then $G(z) = L(\beta z+\upsilon)$, $\beta>0$ and $\upsilon\in \mathbb{R}$, with $L(.)$ as in (1.1) [Gnedenko, 1943]. We then say that F is in the domain of attraction of L and denote this fact by $F\in \mathcal{D}_{max}(L)$. Asymptotic results for minimum values are easily derived from the ones for maxima, taking into account the fact that $\min_{1\leq i\leq n} X_i = -\max_{1\leq i\leq n} (-X_i)$ whenever F belongs to the domain of attraction of a stable d.f. for minimum, $L^*$, we shall denote the fact by $F\in \mathcal{D}_{min}(L^*)$, $L^*(z) = 1-L(-z)$, $z\in \mathbb{R}$, $L(.)$ as in (1.1).

The three types of extreme value stable d.f.'s (1.1) are the only members of the one parameter von Mises-Jenkinson family:

$$L(z) = L(z;k) = \exp(-(1+kz)^{-1/k},\ 1+kz>0,\ k\in \mathbb{R} \tag{1.2}$$

(with the usual continuation for k=0). Remark that for $k>0$, $L(z;k) = \Phi_{1/k}(kz+1)$, that for $k<0$, $L(z;k) = \Psi_{1/|k|}(|k|z-1)$, and that $\lim_{k\uparrow 0} L(z;k) = \lim_{k\downarrow 0} L(z;k) = \Lambda(z)$.

The Gumbel type d.f. $\Lambda(z)$ is a favourite one in statistical theory of extremes, essentially because of the greater simplicity of inference associated to Gumbel populations. Thus, separation among extreme value models, with $\Lambda$ playing a central and proeminent position, turns out to be an important statistical problem. Empirical tests of the hypothesis $H_0$:k=0 versus a sensible one--sided or two-sided alternative were suggested by Jenkinson (1955) and Gumbel (1965). Other papers on the subject are those of van Montfort (1970, 1973), Galambos (1980) and also Tiago de Oliveira (1981), where a locally most powerful statistical choice statistic was developed. In Gomes (1982), a statistic used by Gumbel (1965) in connection with estimation in Frechet populations was put forward as a test statistic of the null hypothesis $H_0$:k=0. The sampling distribution of such a statistic for values of sample size $n \leqq 50$ and under the null hypothesis, was therein obtained by simulation, as well as the power function of a test based on it. The results there obtained lead us to go further in the subject. We shall here be mainly interested in obtaining asymptotic properties of a test based on this statistic as well as of a test based on a suitable version of von Mises-Jenkinson (1955) statistic.

## 2. GUMBEL STATISTIC AND ASSOCIATED TEST OF SIGNIFICANCE

Let $\underset{\sim}{X}= (X_1, X_2,\ldots,X_n)$ be a random independent sample of maximum values.

The statistic used by Gumbel (1965) to estimate the unknown parameters in Frechet model is

$$W_n(\underset{\sim}{X}) = \{X_{n:n}-X_{[n/2]+1:n}\}/\{X_{[n/2]+1:n}-X_{1:n}\} \qquad (2.1)$$

where $X_{j:n}$ denotes the j-th ascending order statistic (o.s) corresponding to the random sample $\underset{\sim}{X}$. Notice that such a statistic, independent of location and scale parameters, may be regarded as the sample characteristic corresponding to the characteristic, dependent on n,

$$\omega_n = \{\text{med}(\max_{1\leqq j\leqq n} X_j) - \text{med}(X_1)\}/\{\text{med}(X_1) - \text{med}(\min_{1\leqq j\leqq n} X_j)\}$$

where med(X) stands for the median of the r.v.X. If we assume that the d.f. of X is $L((x-\lambda)/\delta;k)$, von Mises-Jenkinson d.f., with unknown parameters $\lambda$ and $\delta$ respectively, we get

$$\left.\begin{aligned} \omega_n = \omega_{n,k} &= \{n^k-1\}/\{1-(-\log(2)/\log(1-.5^{1/n}))^k\} \\ \underset{k\to 0}{\to}\ d_n &= -\log(n)/\log(-\log(2)/\log(1-.5^{1/n})). \end{aligned}\right\} \qquad (2.2)$$

### 2.1. Asymptotic Properties of $W_{n,k}(\underset{\sim}{X})$

We are here interested in the asymptotic behaviour of $W_{n,k}(\underset{\sim}{X}) = W_{n,k}(\underset{\sim}{Z})$, $\underset{\sim}{Z}= (\underset{\sim}{X}-\underset{\sim}{\lambda})/\delta$, $\underset{\sim}{X}$ a random sample of size n from a von Mises-Jenkinson population with location parameter $\lambda$ and scale parameter $\delta>0$, $\underset{\sim}{\lambda}$ being a column vector of dimension n, with all its components equal to $\lambda$. Notice first of all that

$L(z;k) \in \mathcal{D}_{min}(\Lambda^*)$, $k \in \mathbb{R}$, with attraction coefficients $a_{n,k} =$

$(\log(n))^{-k-1} \underset{k\to 0}{\to} 1/\log(n)$ and $b_{n,k} = \{(\log(n))^{-k}-1\}/k \underset{k\to 0}{\to} -\log(\log(n))$,

and this obviously means that , whenever Z has a d.f. $L(z;k)$,

$$(\log(n))^{k+1}\{Z_{1:n}-((\log(n))^{-k}-1)/k\} \underset{n\to\infty}{\overset{w}{\to}} Z^*$$

such that $P(Z^* \leqq z) = \Lambda^*(z)$, and consequently, $\{Z_{1:n}-((\log(n))^{-k}-1)/$ $/k\} \underset{n\to\infty}{\to} 0$, whenever $k > -1$.

Let us put, for simplicity of notation, $W_{n,k} = \dfrac{T_{n,k}-U_{n,k}}{U_{n,k}-V_{n,k}}$.

If we assume $k = 0$, we have that $T^*_{n,0} = \{T_{n,0}-\log(n)\}$ is a r.v. with d.f. $\Lambda(z)$, $U^*_{n,0} = \sqrt{n}\,\log(2)\,\{U_{n,0}+\log(\log(2))\}$ is asymptotically standard normal, as $n\to\infty$, and $V^*_{n,0} = \log(n)\{V_{n,0}+\log(n))\}$ $\underset{n\to\infty}{\overset{w}{\to}} Z^*$, with d.f. $\Lambda^*(z)$. Consequently, $U_{n,0}+\log(\log(2)) \overset{p}{\to} 0$ and $V_{n,0}+\log(\log(n)) \underset{n\to\infty}{\overset{p}{\to}} 0$. The use of the $\delta$-method (Tiago de Oliveira, 1982) leads us to the following result:

$$W_{n,k}(\underset{\sim}{X}) - \frac{\log(n)+\log(\log(2))}{\log(\log(n)) - \log(\log(2))} = \frac{T^*_{n,0}}{\log(\log(n))-\log(\log(2))} +$$

$$U^*_{n,0}o(1/\log(\log(n)))+V^*_{n,0}o(1/\log(\log(n)))+ R_{n,0},$$

where $R_{n,0} = o_p(1/\log(\log(n)))$ [ $o_p(u_n)$ denotes as usual (and in analogy with the symbol $o(.)$) a r.v. of smaller order in probability than $u_n$, i.e., $o_p(u_n)/u_n \overset{p}{\to} 0$, as $n\to\infty$]. Noticing that the coefficien of $T^*_{n,0}$ is of the order of $1/\log(\log(n))$, we have that, with

$$\left.\begin{aligned} &\beta_{n,0} = \{\log(n)+\log(\log(2))\}/\{\log(\log(n))-\log(\log(2))\} \\ &\text{and} \quad \alpha_{n,0} = 1/\log(\log(n)) \\ &\frac{W_{n,0}(\underset{\sim}{X}) - \beta_{n,0}}{\alpha_{n,0}} \underset{n\to\infty}{\overset{w}{\to}} Z, \text{ with d.f. } \Lambda(z) . \end{aligned}\right\} \quad (2.3)$$

Notice also that $\beta_{n,0}/d_n \underset{n\to\infty}{\to} 1$, $d_n$ as in (2.2).

Let us assume now that $k \neq 0$.

With the same notation for $W_{n,k}$, we have now that $T^*_{n,k} =$

$\{T_{n,k}-(n^k-1)/k\}/n^k$ is a r.v. with d.f. $L(z;k)$, $U^*_{n,k}=\sqrt{n}(\log(2))^{k+1}$

$\{U_{n,k}-((\log(2))^{-k}-1)/k\}$ is asymptotically standard normal, as

$n\to\infty$, and $V^*_{n,k}=(\log(n))^{k+1}\ \{V_{n,k}-((\log(n))^{-k}-1)/k\} \underset{n\to\infty}{\overset{w}{\to}} Z^*$, with

d.f.$\Lambda^*(z)$. Let us state three results used separately in the sequel: $T_{n,k}-(n^k-1)/k \underset{n\to\infty}{\overset{p}{\to}} 0$, whenever $k < 0$, $U_{n,k}-((\log(2))^{-k}-1)/k$ $\underset{n\to\infty}{\overset{p}{\to}} 0$, for every $k\in \mathbb{R}$ and $V_{n,k}-((\log(n))^{-k}-1)/k \underset{n\to\infty}{\overset{p}{\to}} 0$, whenever $k > -1$.

It seems natural, if possible, to choose an attraction location coefficient $\beta_{n,k}$ for $W_{n,k}(\underset{\sim}{X})$, which goes to $\beta_{n,0}$ as $k$ approaches 0, and the $\beta_{n,k}$ which is immediately suggested is

$$\beta_{n,k}=\{n^k-(\log(2))^{-k}\}/\{(\log(2))^{-k}-(\log(n))^{-k}\}.$$

We then have

$$W_{n,k}(\underset{\sim}{X}) - \beta_{n,k} = T^*_{n,k} \frac{k\, n^k}{(\log(2))^{-k}-(\log(n))^{-k}}$$

$$+ U^*_{n,k} \frac{k((\log(n))^{-k}-n^k)}{(\log(2))^{k+1}\sqrt{n}\{(\log(2))^{-k}-(\log(n))^{-k}\}^2}$$

$$+V^*_{n,k} \frac{k(n^k-(\log(2))^{-k})}{(\log(n))^{k+1}\{(\log(2))^{-k}-(\log(n))^{-k}\}^2} +R_{n,k}.$$

Whenever $k > 0$, the coefficient of $T^*_{n,k}$ is of the order of $n^k$, the coefficient of $U^*_{n,k}$ is of the order of $n^{k-1/2}=o(n^k)$ and the coefficient of $V^*_{n,k}$ is of the order of $n^k/(\log(n))^{k+1}=o(n^k)$. We also have $R_{n,k}=o_p(n^k)$. Consequently if we choose

$\alpha_{n,k} = k(n \log(2))^k$, we have that

$\dfrac{W_{n,k}(\underset{\sim}{X}) -\beta_{n,k}}{\alpha_{n,k}}$ converges weakly, as $n\to\infty$, to a r.v. with d.f.

$L(z,k)$.

We could obviously have chosen other equivalent attraction coefficients for $W_{n,k}(\underset{\sim}{X})$. The simplest ones turn out to be

$\beta'_{n,k}=(n \log(2))^k$ and $\alpha'_{n,k}=k\beta'_{n,k}$ . If we want to have the scale attraction coefficient approaching also $\alpha_{n,0}=1/\log(\log(n))$, as $k \to 0$, and this seems sensible to us, we then need to choose

$$\alpha^+_{n,k} = \frac{k(n \log(2))^k}{1-(\log(n))^{-k}} , \; \beta^+_{n,k}=\beta_{n,k} . \tag{2.4}$$

Whenever $k < 0$, the dominant coefficient is the one of

$V^*_{n,k}$, which is of the order of $1/(\log(n))^{1-k}$. The coefficient of $T^*_{n,k}$ is of the order of $(n\log(n))^k = o(1/(\log(n))^{1-k}$ and the one of $U^*_{n,k}$ is of the order of $(\log(n))^k/\sqrt{n} = o(1/(\log(n))^{1-k})$. Moreover $R_{n,k} = o_p(1/(\log(n))^{1-k})$. Consequently, if we choose $\alpha^*_{n,k} = (\log(2))^{-k}(\log(n))^{k-1}$, we have

$$\frac{W_{n,k}(\underset{\sim}{X})-\beta_{n,k}}{\alpha^*_{n,k}}$$ converges weakly, as $n\to\infty$, to a r.v. with d.f. $\Lambda^*(z)$.

Notice here that the attraction location coefficient $\beta_{n,k}$ may be replaced by $\beta'_{n,k} = (\log(2)/\log(n))^{-k}$ if $k < -1$.

It is here impossible to achieve continuity of the attraction coefficients, as $k \to 0$. We may however choose

$$\beta^-_{n,k}=\beta_{n,k} \text{ and } \alpha^-_{n,k} = k(\log(2))^{-k}/\{\log(n)(1-(\log(n))^{-k})\}. \quad (2.5)$$

## 2.2 Asymptotic Power Function of a Test Based on Gumbel Statistic

In the sequel we will deal with one-sided tests. Similar results were obtained for two-sided tests.

If we desire an asymptotic test of $H_0{:}k=0$ versus $H_1{:}k>0$ based on the statistic $W_n(\underset{\sim}{X})$ and if we assume $\alpha$ to be the significance level of the test, the rejection region $R^+$ will be, asymptotically

$$R^+ : W_n(\underset{\sim}{X}) \geq \beta_{n,0} - \alpha_{n,0} \log(-\log(1-\alpha)) = X^+_{n,\alpha},$$

with $\alpha_{n,0}$ and $\beta_{n,0}$ as in (2.3).

Consequently,

$$P_n(k|k>0) = P[\text{rej}.H_0|k>0] = P[W_n(\underset{\sim}{X}) \geqq \chi^+_{n,\alpha}|k>0]$$

$$= P\left[\frac{W_{n,k}(\underset{\sim}{X})-\beta^+_{n,k}}{\alpha^+_{n,k}} \geqq \frac{\chi^+_{n,\alpha}-\beta^+_{n,k}}{\alpha^+_{n,k}}\right], \text{ with}$$

$\alpha^+_{n,k}$, $\beta^+_{n,k}$ as in (2.4). This power function is well approximated, for large values of n, by

$$P_{n,a}(k|k>0) = 1-\exp\{-\left[1+k\,\frac{\beta_{n,0}-\beta^+_{n,k}}{\alpha^+_{n,k}} - k\,\frac{\alpha_{n,0}}{\alpha^+_{n,k}}\log(-\log(1-\alpha))\right]^{-1/k}$$

and since $\alpha_{n,0}/\alpha^+_{n,k}=O(1/(n^k\log(\log(n)))) \underset{n\to\infty}{\to} 0$ and $(\beta_{n,o}-\beta^+_{n,k})/\alpha^+_{n,k}$ $\underset{n\to\infty}{\to}$ $-1/k$, we have obviously a consistent test, i.e. $P_n(k|k>0) \underset{n\to\infty}{\to} 1$, for every $k>0$.

For the case $H_1$:$k<0$, and at the significance level $\alpha$, the rejection region $R^-$ will be, asymptotically

$$R^- : W_n(\underset{\sim}{X}) \leqq \beta_{n,0}-\alpha_{n,0}\log(-\log(\alpha)) = \chi^-_{n,\alpha},$$ with $\alpha_{n,0}$ and $\beta_{n,0}$ as before.

Consequently,

$$P_n(k|k<0) = P[\text{rej}.H_0|k<0]=P\left[\frac{W_{n,k}(\underset{\sim}{X})-\beta^-_{n,k}}{\alpha^-_{n,k}} \leqq \frac{\chi^-_{n,\alpha}-\beta^-_{n,k}}{\alpha^-_{n,k}}\right],$$

with $\alpha^-_{n,k}$ and $\beta^-_{n,k}$ as in (2.5). This power function is well approximated, for large values of n, by

$$P_{n,a}(k|k<0)=1-\exp\{-\exp\left[-\frac{\beta_{n,0}-\beta_{n,k}}{\alpha^-_{n,k}} - \frac{\alpha_{n,0}}{\alpha^-_{n,k}}\log(-\log(\alpha))\right]\}.$$

We have obviously $P_n(k|k<0) \underset{n\to\infty}{\to} 1$, for every $k < 0$.

Although aware of the asymptotic validity of the expressions previously obtained for the power function of the test here developed, we present in table 1., for values of n=20 and 50 and for $\alpha$=.05, the values of $P_{n,a}(k|k\ 0)$ and of $P_{n,a}(k|k\ 0)$. The results here obtained for small sample sizes agree quite well with the results, concerning the exact power function, obtained by simulation in Gomes (1982).

| | k | n=20 | n=50 |
|---|---|---|---|
| $P_{n,a}(k\|k<0)$ | -.7 | 1.000 | 1.000 |
| | -.6 | .978 | 1.000 |
| | -.5 | .845 | 1.000 |
| | -.4 | .595 | 1.000 |
| | -.3 | .351 | 1.000 |
| | -.2 | .182 | .672 |
| | -.1 | .085 | .128 |
| $P_{n,a}(k\|k>0)$ | .1 | .170 | .239 |
| | .2 | .323 | .490 |
| | .3 | .473 | .699 |
| | .4 | .598 | .833 |
| | .5 | .697 | .909 |
| | .6 | .770 | .949 |
| | .7 | .824 | .970 |
| | .8 | .863 | .982 |
| | .9 | .891 | .988 |
| | 1.0 | .912 | .991 |

Table 1. Asymptotic power function of a test based on $W_n(\underset{\sim}{X})$, for n=20 and 50 and $\alpha$=0.05.

## 3. JENKINSON'S MODIFIED ASYMPTOTIC TEST

Let us assume now that we have a sample $(X_1, Y_1, \ldots, X_n, Y_n)$ of size 2n, from a population with d.f. $L((x-\lambda)/\delta;k)$, von Mises-Jenkinson d.f. Our objective is to select a statistic, of the type of of Jenkinson's statistic, which is invariant under location and scale transformations, the sampling distribution of which is, at least asymptotically, easy to obtain.

Consider the pairs (S,M), with S=X+Y and M=max(X,Y) and the new bivariate sample of size n, $(S_i, M_i)$, $S_i=X_i+Y_i$, $M_i=\max(X_i,Y_i)$, $1\leq i\leq n$.

Let $\mu_k$ and $\sigma_k^2$ be the mean value and variance associated to $L(z;k)$, $z=(x-\lambda)/\delta$, i.e.

$\mu_k = \{\Gamma\,(1-k)-1\}/k$ if $k < 1$, which goes to $\mu_0=\gamma$, Euler's constant, as $k\to 0$, and

$\sigma_k^2 = \{\Gamma(1-2k)-\Gamma^2(1-k)\}/k^2$ if $k<1/2$, which goes to $\sigma_0^2=\pi^2/6$, as $k\to 0$, where $\Gamma(\cdot)$ stands for the complete gamma function, $\Gamma(\alpha) = \int_0^{+\infty} t^{\alpha-1}e^{-t}dt$.

Putting $S^* = (S-2\lambda)/\delta$, $M^*=(M-\lambda)/\delta$, we thus have $E(S^*)=2\mu_k$ and $Var(S^*)=2\sigma_k^2$, and the fact that $\{M^*-(2^k-1)/k\}\overset{d}{=}Z$ implies that $E(M^*)=\{2^k\Gamma(1-k)-1\}/k$ and $Var(M^*)=2^{2k}\,\sigma_k^2$. Consequently,

$$Var(S)/Var(M) = 2^{1-2k} \begin{cases} >2 \text{ if } k < 0, \\ \to 2 \text{ if } k \to 0, \\ <2 \text{ if } k > 0. \end{cases}$$

Considering the statistic $Q_{2n} = S_n^2(S)/S_n^2(M)$, where $S_n^2(Z) = \sum_{i=1}^{n} (Z_i-\bar{Z})^2/n$, $\bar{Z} = \sum_{i=1}^{n} Z_i/n$, stands for the empirical variance associated to a sample $Z_i$, $1\leq i\leq n$, we can thus avoid eventual location and dispersion parameters in $L(.;k)$.

Cramer's results (1946) immediately suggest the asymptotic normality of $Q_{2n}$, under certain regularity conditions.

Let us deal into detail with the sampling distribution of $Q_{2n}=Q_{2n,k}$ as $n\to\infty$.

First of all, for large values of n, $Var(S_n^2(Z))$ is asymptotically equivalent to $\{\mu_4(Z)-\mu_2^2(Z)\}/n$, where $\mu_4(Z)=E((Z-E(Z))^4)$, $\mu_2(Z)=E((Z-E(Z))^2)$. As we are using $\mu_4(Z)$ which introduces $\Gamma(1-4k)$ put, for $k<1/4$,

$$A_k^2 = \mu_4(S^*)-\mu_2^2(S^*) = 2(\beta_k+1)\sigma_k^4$$

and $B_k^2 = \mu_4(M^*)-\mu_2^2(M^*) = 2^{4k}(\beta_k-1)\ \sigma_k^4$ ,

with $\beta_k = E((Z-\mu_k)^4)/\sigma_k^4 = \{\Gamma(1-4k)-4\Gamma(1-3k)\Gamma(1-k)+6\Gamma(1-2k)\Gamma^2(1-k)$
$-3\Gamma^4(1-k)\}/\{\Gamma(1-2k)-\Gamma^2(1-k)\}^2$.

We then have that $(S_n^*,M_n^*) = \left(\sqrt{n}\ \dfrac{S_n^2(S^*)-2\sigma_k^2}{A_k}\ ,\ \sqrt{n}\ \dfrac{S_n^2(M^*)-2^{2k}\sigma_k^2}{B_k}\right)$

is asymptotically standard binormal, with correlation coefficient $\rho_k$, whose computation is irrelevant at the moment.

Consequently

$$Q_{2n,k} = \frac{S_n^2\ (S)}{S_n^2\ (M)} = \frac{S_n^*A_k/\sqrt{n}+2\sigma_k^2}{M_n^*B_k/\sqrt{n}+2^{2k}\sigma_k^2} \ \underset{n\to\infty}{\overset{p}{\to}}\ 2^{1-2k}$$

and since $\sqrt{n}\{Q_{2n,k}-2^{1-2k}\}$ and $\{A_kS_n^*-2^{1-2k}B_kM_n^*\}/(2^{2k}\sigma_k^2)$ converge

weakly to the same limit law, we have that $\sqrt{n}\{Q_{2n,k}-2^{1-2k}\}$ is

asymptotically normal, with variance $c_k^2$, to be computed.

Noticing that

$$\lim_{n\to\infty} n\ \mathrm{Cov}\left[S_n^2(S^*),\ S_n^2(M^*)\right] = \mathrm{Cov}\left[(S^*-E(S^*))^2,\ (M^*-E(M^*))^2\right] ,$$

we have

$$c_k^2 = \left[\ A_k^2+2^{2-4k}B_k^2-2^{2-2k}\ \mathrm{Cov}((S^*-E(S^*))^2,\ (M^*-E(M^*))^2)\right]/(2^{4k}\sigma_k^4)$$

$$=\{2(\beta_k+1)+4(\beta_k-1)-2^{2-2k}\ \phi_k+8\}/2^{4k}$$

$$=\{6\beta_k+6-2^{2-2k}\phi_k\}/2^{4k},$$

where $\phi_k=E((S^*-E(S^*))^2(M^*-E(M^*))^2)/\sigma_k^4$. When $k\to 0$, $\beta_k\to 3+6\zeta(4)/\zeta^2(2)=5.4$, the Riemann zeta function, and $c_0^2 = 38.4 - 4\phi_0$.

Consequently, we have that

$$\sqrt{n}\ \frac{Q_{2n,k}-2^{1-2k}}{c_k} \text{ is asymptotically standard normal if } k<1/4.$$

A significance test of the hypothesis $H_0$:k=0 versus a suitable one-sided or two-sided alternative, depends heavily on the computation of this function $\phi_k$, $k\in\mathbb{R}$, which requires cumbersome computations. It depends essentially on integrals of the type:

$$I(\alpha,\beta) = \iint_{u>v>0} u^{\alpha}\, v^{\beta} \exp\{-(u+v)\}dudv \ .$$

We have obviously $I(\alpha,\beta)+I(\beta,\alpha)=\Gamma(\alpha+1)\Gamma(\beta+1)$, $I(\alpha,\alpha) = \Gamma^2(1+\alpha)/2$. Moreover

$$I(\alpha,\beta) = \iint_{\substack{0<v<1\\ u>0}} u^{\alpha+\beta+1}\, v^{\beta}\exp\{-(1+v)u\}dudv$$

$$= \Gamma(\alpha+1)\Gamma(\beta+1)\ I_{1/2}\ (\beta+1,\alpha+1),$$

where $I_x(a,b) = (\Gamma(a+b)/(\Gamma(a)\Gamma(b))) \int_0^x t^{a-1}(1-t)^{b-1}dt$ denotes, as usual, the incomplete beta function.

Consequently, for values of $k \neq 0$

$$\sigma_k^4\ \phi_k = 2\{\ \Gamma^2(1-2k)/2+2^k\Gamma^4(1-k)\left[4-5.2^k\right] + 2^{4k-1}\Gamma(1-4k)$$

$$-2^{3k}\Gamma(1-k)\Gamma(1-3k)\left[2+2^k\right] + 2^2\Gamma^2(1-k)\Gamma(1-2k)\left[3.2^k+2^{2k+2}-2\right]$$

$$+2\Gamma(1-k)\Gamma(1-3k)\ I_{1/2}(1-3k,\ 1-k)$$

$$-\Gamma^2(1-k)\Gamma(1-2k)\ I_{1/2}(1-2k,\ 1-k)\left[4+2^{k+1}\right]\}/k^2$$

Notice that, even with a great accuracy in the gamma function, the rounding errors in the neighbourhood of $k = 0$ turn out to be significant .

For $k = 0$, we have to compute integrals of the form

$$II(m,n) = \iint_{u>v>0} (\log(u))^m (\log(v))^n \exp\{-(u+v)\}\, du\,dv.$$

We have $II(m,n)+II(n,m) = \Gamma^{(m)}(1)\,\Gamma^{(n)}(1)$ and $II(n,n)=(\Gamma^{(n)}(1))^2/2$, where $\Gamma^{(n)}(.)$ stands for the n-th derivative of the gamma function. For the values of $II(2,1)$ and $II(3,1)$, needed in the computation of $\phi_0$, we had to sum the series $\sum_{r\geq 1} \frac{(-1)^r}{r} \sum\sum_{1\leq i<j\leq r} (1/(ij))$ and $\sum_{r\geq 1} \frac{(-1)^r}{r} \sum\sum\sum_{1\leq i<j<l\leq r} (1/(ijl))$; this was done numerically. Finally,we get

$$\sigma_0^4\,\phi_0 = \pi^4/45 - 4\psi''(1)\log(2) + (\pi\log(2))^2/2 - \log^4(2)/2$$
$$-2\log(2) \sum_{r\geq 1} (-1)^r a_r - 2 \sum_{r\geq 1} (-1)^r c_r,$$

where $\psi''(1)$, $\psi''(.)$ the second derivative of the digamma function $\psi(.) = \Gamma'(.)/\Gamma(.)$, is equal to $-2\zeta(3)$, $\zeta(.)$ the Riemann zeta function, $a_r = 2 \sum_{i<j} (1/(ij))/r$ and $b_r = 6 \sum_{i<j<l} (1/(ijl))/r$.

The values of $\phi_k$ and $c_k$, $k=-1\ (.1)\ 0$ and $k=.05(.05).2$, are shown in table 2.

Analagously to what we did in 2.2., we compute here the asymptotic power function of tests of the hypothesis $H_0{:}k=0$ versus one--sided alternatives, based on the sequence of statistics $Q_{2n}$.

Under the validity of the null hypothesis, we have the asymptotic normality of $\sqrt{n}(Q_{2n}-2)/c_0$, as $n\to\infty$.

If we are thinking on an alternative hypothesis $H_1 : k>0$, and since $Q_{2n} \xrightarrow[n\to\infty]{p} 2^{1-2k} < 2$ if $k>0$, we should choose a rejection region $R^+$, at the significance level $\alpha$, asymptotically given by

$R^+$: $Q_{2n} < 2+\chi_\alpha c_0/\sqrt{n}$, where $\chi_\alpha$ is the $\alpha$-quantile of a standard normal d.f.

| $k$ | $\phi_k$ | $c_k$ |
|---|---|---|
| -1.0 | 2.500 | 17.889 |
| -.9 | 2.503 | 13.226 |
| -.8 | 2.506 | 9.791 |
| -.7 | 2.514 | 7.298 |
| -.6 | 2.540 | 5.532 |
| -.5 | 2.601 | 4.319 |
| -.4 | 2.736 | 3.519 |
| -.3 | 3.021 | 3.011 |
| -.2 | 3.619 | 2.697 |
| -.1 | 4.929 | 2.509 |
| 0 | 8.125 | 2.434 |
| .05 | 11.601 | 2.432 |
| .10 | 18.230 | 2.521 |
| .15 | 33.864 | 2.769 |
| .20 | 89.931 | 3.549 |

Table 2. Values of $\phi_k = E_k[(S-E(S))^2 (M-E(M))^2]/\sigma_k^4$ and of $c_k$, for $k=-1.(.1)\ 0$ and $k=.05(.05).2$

Consequently

$$P_{2n}(k|k>0) = P\left[\sqrt{n}\ \frac{Q_{2n,k}-2^{1-2k}}{c_k} < \sqrt{n}\left[2+\chi_\alpha c_0/\sqrt{n}-2^{1-2k}\right]\right]$$ is well

approximated for large values of n, by

$$P_{2n,a}(k|k>0) = N((2+\chi_\alpha c_0/\sqrt{n}-2^{1-2k})\sqrt{n}/c_k),$$

where N(.) stands for the standard normal d.f.

Analagously for an alternative hypothesis $H_1$:k<0, we have

$$R^- : Q_{2n} > 2+\chi_{1-\alpha}c_0/\sqrt{n}$$

and consequently

$$P_{2n,a}(k|k<0) = 1-N((2+\chi_{1-\alpha}c_0/\sqrt{n}-2^{1-2k})\sqrt{n}/c_k).$$

Both tests are obviously consistent.

In table 3 we present for $\alpha$=0.05, and for sample sizes n'=2n=20 and 50 the values of $P_{n',a}(k|k<0)$ and of $P_{n',a}(k|k>0)$.

| | k | n'=20 | n'=50 |
|---|---|---|---|
| $P_{n',a}(k\|k<0)$ | -1.0 | .900 | .984 |
| | -.9 | .916 | .991 |
| | -.8 | .927 | .994 |
| | -.7 | .928 | .996 |
| | -.6 | .916 | .995 |
| | -.5 | .875 | .991 |
| | -.4 | .773 | .968 |
| | -.3 | .583 | .864 |
| | -.2 | .337 | .579 |
| | -.1 | .145 | .227 |
| $P_{n',a}(k\|k>0)$ | .05 | .082 | .106 |
| | .10 | .131 | .198 |
| | .15 | .203 | .316 |
| | .20 | .305 | .436 |

Table 3. Asymptotic power function of a test based on $Q_{2n}$, for sample sizes 20, 50 and $\alpha$=.05

## 4. JENKINSON'S MODIFIED TEST — EXACT RESULTS.

We are now mainly interested in obtaining, by simulation, the d.f. of $Q^*_{2n,0} = \sqrt{n}\{Q_{2n,0}-2\}/c_0$ for small sample sizes. Replication was used and convergence was accelerated by the use of antithetic variables. In table 4 we present for values of n' = 2n =10(10)50, estimates of a few quantiles of $Q^*_{2n,0}$ as well as estimates of its mean value, variance, skewness $\beta_1$ and kurtosis $\beta_2$. The associated standard error is placed between brackets below the corresponding estimate. Since the results are based on 50 independent replications, each replication with 500 runs and $t_{49,.975}=2.0$, $t_{n,\xi}$ the $\xi$-quantile of a t distribution with n degrees of freedom, a 95% confidence interval for a characteristic $\theta$, to whose estimate $\theta^*$ is associated an error $e_{\theta^*}$ is $(\theta^*-2e_{\theta^*},\theta^*+2e_{\theta^*})$.

| n' | E(.) | Var(.) | $\beta_1$(.) | $\beta_2$(.) | $X_{.025}$ | $X_{.05}$ | $X_{.5}$ | $X_{.95}$ | $X_{.975}$ |
|---|---|---|---|---|---|---|---|---|---|
| 10 | .45<br>(.01) | 2.82<br>(.33) | 3.85<br>(.43) | 39.14<br>(8.30) | -1.21<br>(.01) | -1.05<br>(.01) | .12<br>(.01) | 2.88<br>(.04) | 3.93<br>(.07) |
| 20 | .29<br>(.01) | 1.20<br>(.02) | 1.19<br>(.03) | 5.49<br>(.21) | -1.25<br>(.01) | -1.10<br>(.01) | .10<br>(.01) | 2.33<br>(.01) | 2.88<br>(.02) |
| 30 | .24<br>(.01) | 1.08<br>(.01) | .89<br>(.04) | 4.35<br>(.23) | -1.31<br>(.01) | -1.14<br>(.01) | .09<br>(.01) | 2.13<br>(.02) | 2.66<br>(.03) |
| 40 | .22<br>(.01) | 1.03<br>(.01) | .77<br>(.02) | 3.86<br>(.13) | -1.35<br>(.01) | -1.17<br>(.01) | .08<br>(.01) | 2.07<br>(.03) | 2.52<br>(.02) |
| 50 | .20<br>(.01) | 1.00<br>(.01) | .67<br>(.02) | 3.48<br>(.05) | -1.40<br>(.01) | -1.21<br>(.01) | .08<br>(.01) | 2.02<br>(.02) | 2.47<br>(.03) |

Table 4. Estimates of population characteristics of $Q^*_{n',0}$

Another simulation was carried out , to evaluate for different values of n', the power function of a test of the hypothesis $H_0$:k=0 versus sensible one-sided or two-sided alternatives. In table 5, we present for values of n'=20,40 and 50, 95% confidence intervals for $P_{Q^*}(k) = P[$ rejecting $H_0$ | the sample comes from a population with d.f. $L((x-\lambda)/\delta;k)]$, when we are using the statistic $Q^*_{n';0}$ and working at a significance level $\alpha$=.05.

## 5. FINAL COMMENTS

Tha results obtained in section 2. and 3. lead us to the conclusion that, asymptotically the test based on von Mises-Jenkinson statisti

| k | n'=20 | | n'=40 | | n'=50 | |
|---|---|---|---|---|---|---|
| | one sided | two sided | one sided | two sided | one sided | two sided |
| -2.0 | (1.00,1.00) | ( .99, .99) | (1.00,1.00) | (1.00,1.00) | (1.00,1.00) | (1.00,1.00) |
| -1.5 | ( .99, .99) | ( .97, .98) | (1.00,1.00) | (1.00,1.00) | (1.00,1.00) | (1.00,1.00) |
| -1.0 | ( .92, .94) | ( .88, .90) | (1.00,1.00) | ( .99,1.00) | (1.00,1.00) | (1.00,1.00) |
| -.75 | ( .80, .83) | ( .71, .75) | ( .98, .99) | ( .96, .98) | ( .99,1.00) | ( .99, .99) |
| -.5 | ( .53, .57) | ( .42, .46) | ( .83, .86) | ( .73, .77) | ( .89, .92) | ( .83, .87) |
| -.4 | ( .38, .41) | ( .27, .31) | ( .66, .70) | ( .54, .59) | ( .75, .79) | ( .63, .69) |
| -.3 | ( .24, .27) | ( .16, .20) | ( .42, .48) | ( .29, .35) | ( .52, .58) | ( .38, .45) |
| -.2 | ( .14, .17) | ( .10, .12) | ( .23, .27) | ( .14, .18) | ( .27, .30) | ( .16, .21) |
| -.1 | ( .07, .10) | ( .05, .07) | ( .10, .13) | ( .07, .09) | ( .11, .14) | ( .06, .09) |
| .1 | ( .08, .10) | ( .05, .06) | ( .11, .14) | ( .09, .10) | ( .13, .17) | ( .08, .11) |
| .2 | ( .11, .15) | ( .05, .08) | ( .24, .27) | ( .15, .19) | ( .27, .31) | ( .17, .22) |
| .3 | ( .17, .21) | ( .07, .11) | ( .35, .40) | ( .24, .29) | ( .41, .47) | ( .30, .37) |
| .4 | ( .23, .28) | ( .09, .14) | ( .47, .52) | ( .36, .42) | ( .56, .60) | ( .44, .50) |
| .5 | ( .30, .36) | ( .12, .18) | ( .59, .63) | ( .48, .53) | ( .67, .70) | ( .56, .62) |
| .75 | ( .47, .54) | ( .23, .32) | ( .77, .80) | ( .68, .74) | ( .83, .86) | ( .76, .80) |
| 1.0 | ( .59, .66) | ( .33, .42) | ( .85, .88) | ( .79, .83) | ( .90, .92) | ( .86, .89) |
| 1.5 | ( .75, .80) | ( .53, .64) | ( .93, .95) | ( .90, .93) | ( .95, .96) | ( .93, .95) |
| 2.0 | ( .84, .88) | ( .67, .76) | ( .96, .98) | ( .94, .97) | ( .97, .98) | ( .96, .97) |

Table 5. 95% confidence interval for $Q^*_{n',0}(k)$, for different values of n' and k.

$Q_n$ is slightly better than the one based on Gumbel statistic $W_n$ only for values of k quite close to zero, when the alternative is $H_1: k<0$; in all other situations this last statistic turns out to be the better one.

On the other side, in what concerns the exact power function of both tests, obtained in Gomes (1982) and in the previous paragraph, we come to the conclusion that for small values of n the test based on Gumbel statistic is much more powerful than the one based on von Mises-Jenkinson statistic, whatever the alternative.

## REFERENCES

Galambos, J. (1980), Colloq. Math. Soc. Bolyai 32,"Nonparametric Methods of Statistical Inference", Budapest, pp. 221-230.
Gnedenko, B.V. (1943), Ann. Math. 44, pp.423-453.
Gomes, M.I. (1982), IX Jornadas Matemáticas Hispano-Lusas, Salamanca.
Gumbel, E.J. (1965), Rev. Inst. Intern. Statist. 33, pp. 349-363.
Jenkinson, A.F. (1955), Quart. J. Royal Meteor. Soc. 87, pp. 158--171.
Tiago de Oliveira, J. (1981), In C.Taillie et al. (eds.), Statistical Distributions in Scientific Work, vol. 6, pp. 367-387, D. Reidel.
Tiago de Oliveira, J. (1982), Publ. Inst. Statist. Univ. Paris, vol. XXVII, pp. 49-70.
van Montfort, M.A.J. (1970), J. Hydrology 11, pp. 421-427.
van Montfort, M.A.J. (1973), Mededelingen Landbouwhogeschool 73, pp. 1-15.

# ON THE ASYMPTOTIC UPCROSSINGS OF A CLASS OF NON-STATIONARY SEQUENCES

K. F. Turkman

Faculty of Sciences of Lisbon
Center of Statistics and Applications
(I.N.I.C.)

ABSTRACT. A class of non-stationary sequences will be introduced and asymptotic Poisson character of the upcrossings will be shown. Some examples will be given to demonstrate that this class is particularly capable of describing sequences with seasonal components.

## 1. INTRODUCTION

Many stochastic processes arising in various physical sciences such as hydrology, meteorology, oceanography exhibit non-stationary due to the variations in their mean functions. However, they retain some stationary dependence structure. In this paper, a class of sequences which is capable of describing such processes will be defined and some results on the upcrossings and related topics of the class will be given.

### 1.1 Definition of the Class

Consider a class $\ell$ of sequences of random variables $\{X_n, n=1,\dots\}$ having the following properties.

(a) There exists a bivariate sequence of random variables $(S_n, X_n), n=1,\dots$ such that, for each $n$,

$$P[S_n=k]=\delta_{n,k}, \quad k=1,\dots m, \quad \sum_{k=1}^{m} \delta_{n,k}=1 . \tag{1.1}$$

$$P[X_n \leq x | S_1=s_1,\dots,S_n=s_n]=P[X_n \leq x | S_n=s_n]=F_{s_n}(x), s_n=1,\dots m, \tag{1.2}$$



J. Tiago de Oliveira (ed.), Statistical Extremes and Applications, 669–678.

thus depending only on the value of $S_n$.

(b) Conditional on $\{S_1=s_1,\ldots,S_n=s_n\}$, the set of random variables $(X_1,X_2,\ldots,X_n)$ is independent of $(S_{n+1},S_{n+2},\ldots)$, for each $n$; thus for any integer $k>0$, we can write

$$P[X_1\leqq x_1,\ldots,X_n\leqq x_n|S_1=s_1,\ldots,S_n=s_n]=P[X_1\leqq x_1,\ldots,X_n\leqq x_n|S_1=s_1,\ldots,S_{nk}=s_{nk}] \tag{1.3}$$

(c) The sequence $\{X_n\}$ conditional on the sequence $\{S_n\}$ satisfies the $D(u_n)$ and $D^*(u_n)$ conditions for a sequence $u_n$ specified in (e) below.

We will say that $\{X_n\}$ conditional on $\{S_n\}$ satisfies the $D(u_n)$ condition if, for any two sets of integers $(i_1,\ldots,i_p)$ and $(j_1,\ldots,j_q)$ such that $1\leqq i_1<\ldots<i_p<j_1\ldots<j_q\leqq n$ and $j_1-i_p<k$, we have, writing $\tilde{s}_n=(s_1,\ldots,s_n)$,

$$|P[X_{i_1}\leqq u_n,\ldots,X_{i_p}\leqq u_n,X_{j_1}\leqq u_n,\ldots,X_{j_q}\leqq u_n|\tilde{s}_n]$$

$$-P[X_{i_1}\leqq u_n,\ldots,X_{i_p}\leqq u_n|\tilde{s}_n]P[X_{j_q}\leqq u_n,\ldots,X_{j_q}\leqq u_n|\tilde{s}_n]| \leqq \alpha_{\tilde{s}_n,k}, \tag{1.4}$$

where,

$$\lim_{k\to\infty}\lim_{n\to\infty}\alpha_{\tilde{s}_n,k}=0\ , \quad \text{uniformly in } \tilde{s}_n=(s_1,\ldots,s_n). \tag{1.5}$$

Also $\{X_n\}$ conditional on $\{S_n\}$ satisfies the $D^*(u_n)$ condition if, for any set of consecutive integers $(i_1,\ldots,i_n)$ and some integer $k>0$,

$$\limsup_{n\to\infty}\sum_{i_1\leqq j_1<j_2\leqq i_n} P[X_{j_1}>u_{nk},X_{j_2}>u_{nk}|S_{i_1}=s_{i_1},\ldots,S_{i_n}=s_{i_n}]=o\left(\frac{1}{k}\right), \tag{1.6}$$

uniformly in $(s_{i_1},\ldots,s_{i_n})$.

Note that the $D(u_n)$ and $D^*(u_n)$ conditions are the natural extensions of the conditions given by Leadbetter (1974) for stationary sequences.

(d) Let $I_j=(j_1,\ldots,j_n)$ be a set of $n$ consecutive integers. If $n_i(I_j)$ denotes the number of times $\{S_r=i\}$ occurs among $(S_{j_1},\ldots,S_{j_n})$, then $\frac{n_i(I_j)}{n} \to p_i$ in probability, where $0<p_i<1$, $i=1,\ldots,m$. Clearly $\sum_{i=1}^{m} p_i=1$.

(e) There exists a sequence of real numbers $\{u_n\}$ and real numbers $z_i$, $i=1,\ldots,m$, such that for every $F_i$, $i=1,\ldots,m$, defined by (1.2),

$$1 - F_i(u_n) = \frac{z_i}{n} + o(\tfrac{1}{n}), \text{ as } n\to\infty . \qquad (1.7)$$

Let $M_n = \max(X_1,\ldots,X_n)$, where $\{X_n\}\varepsilon\ell$. The following theorem and the corollary are proved in Turkman and Walker (1983).

<u>Theorem 1</u>

Let $\{u_n\}$ be a sequence of real numbers satisfying conditions (c) and (c). Then

$$\lim_{n\to\infty} P\left[M_n \leq u_n\right] = \exp\left[-\sum_{i=1}^{m} p_i z_i\right] \qquad (1.8)$$

<u>Corollary 1</u>

Let $x_o^i = \sup\{x: F_i(x)<1\}$, $i=1,\ldots,m$. Let $x_o=x_o^1 \geq x_o^j$, $j=2,\ldots,m$ and assume that

$$\lim_{x\to x_o} \frac{1-F_1(x)}{1-F_j(x)} = \alpha_j , \quad 1 \leq \alpha_j \leq \infty . \qquad (1.9)$$

Also assume that there exists a real sequence $\{u_n\}$, where

$u_n = a_n x+b_n$, $a_n>0$, such that as $n\to\infty$,

$$\lim_{n\to\infty} F_1^n(u_n) = G(x),\ G(x) \text{ non-degenerate.}$$

Then

$$\lim_{n\to\infty} P|M_n \leqq u_n| = G(Ax+b)\ , \tag{1.10}$$

where, (defining $\alpha_1=1$)

(1) when $G(x) = \exp(-e^{-x})$, $-\infty \leqq x \leqq \infty$ ,

then

$$A = 1,\ B = -\log\left[\sum_{i=1}^{m} \frac{p_i}{\alpha_i}\right] \tag{1.11}$$

(2) when

$$G(x) = \Phi_\alpha(x) = \begin{cases} \exp(-x^{-\alpha}) , & x \geqq 0 \\ 0 , & x \geqq 0 \end{cases} ,\ \alpha>0$$

then

$$A = \left[\sum_{i=1}^{m} \frac{p_i}{\alpha_i}\right]^{-\frac{1}{\alpha}} ,\quad B=0 \tag{1.12}$$

(3) when

$$G(x) = \Psi_\alpha(x) = \begin{cases} \exp(-(-x)^{\alpha}) , & x<0 \\ 1 & x \geqq 0 \end{cases} ,\alpha>0$$

then

$$A = \left[\sum_{i=1}^{m} \frac{p_i}{\alpha_i}\right]^{\frac{1}{\alpha}} ,\quad B=0\ . \tag{1.13}$$

The proof of the theorem 1 is an generalization of the proof given for stationary sequences. Here we give an outline of the proof.

Let k be a positive integer and divide the set of integers $(1,2,\ldots,N)$, where $N=nk$, into $2k$ consecutive intervals $I_j$, $I^*_j$ such that $I_j = [(j-1)n+1,\ldots,jn-t]$, $I^*_j = [jn-t+1,\ldots,jn]$, $j=1,\ldots,$ where $t$ is fixed and strictly less then $n$. Let $M(I_j) = \max_{i\varepsilon I_j} (X_i)$.

(1) Let $\tilde{s}_N = (s_1,\ldots,s_N)$ then

$$\left|P\left[\bigcap_{j=1}^{k} (M(I_j)\leqq u_N|\tilde{s}_N\right] - \prod_{j=1}^{k} P\left[M(I_j)\leqq u_N|\tilde{s}_N\right]\right| \leqq (k-1)\,\alpha_{\tilde{s}_N,t} \quad (1.14)$$

(2) Let $\tilde{S}_N = (S_1,\ldots,S_N)$ then

$$\limsup_{n\to\infty} E\left|\prod_{j=1}^{k} P\left[M(I_j)\leqq u_N|\tilde{S}_N\right] - \left[1-\frac{1}{k}\sum_{i=1}^{m} p_i z_i\right]^k\right| \leqq k\,o\left(\frac{1}{k}\right) \quad (1.15)$$

(3) $$\limsup_{n\to\infty} E\left|P\left[\bigcap_{j=1}^{k} (M(I_j)\leqq u_N)|\tilde{S}_N\right]-P\left[M_N\leqq u_N|\tilde{S}_N\right]\right|=0 \quad (1.16)$$

Observing the fact that

$$\left|P\left[M_N\leqq u_N\right]-\left(1-\frac{1}{k}\sum_{i=1}^{m} p_i z_i\right)^k\right| \leqq E\left|P\left[M_N\leqq u_N|\tilde{S}_N\right]-\left(1-\frac{1}{k}\sum_{i=1}^{m} p_i z_i\right)^k\right|$$

$$\leqq E\left|P\left[M_N\leqq u_N|\tilde{S}_N\right]- P\left[\bigcap_{j=1}^{k} (M(I_j)\leqq u_N)|\tilde{S}_N\right]\right|$$

$$+E\left|P\left[\bigcap_{j=1}^{m} (M(I_j)\leqq u_N)|\tilde{S}_N\right] - \prod_{j=1}^{k} P\left[M(I_j)\leqq u_N|\tilde{S}_N\right]\right|$$

$$+E\left|\prod_{j=1}^{k} P\left[M(I_j)\leqq u_N|\tilde{S}_N\right]-\left(1-\frac{1}{k}\sum_{i=1}^{m} p_i z_i\right)^k\right|$$

we obtain, from (1.14), (1.15), (1.16),

$$\limsup_{n\to\infty} \left|P\left[M_N\leqq u_N\right]-\left(1-\frac{1}{k}\sum_{i=1}^{m} p_i z_i\right)^k\right| \leqq k\,o\left(\frac{1}{k}\right).$$

(1.8) then follows once we show that $N$ can be replaced by $n$ and letting $k\to\infty$.

## 1.2 Some Examples Illustrating the Class

Example 1

Resnick and Neuts (1970) have considered sequences $\{X_n\}$ which form a subclass of $\ell$. In this class $\{S_n\}$ are restricted to be a finite irreducible aperiodic Markov chain and replacing (d) by the stronger assumption that conditional on $\{S_n\}$, the $\{X_n\}$ are independent.

Example 2

Let $X_n = Y_n + \Theta_n$, where $\{\Theta_n\}$ is a sequence of random variables with state space $(\theta_1, \theta_2, \ldots, \theta_m)$, satisfying (d) and $\{Y_n\}$ is a normal, weakly stationary sequence with mean zero and variance unity, which is independent of $\{\Theta_n\}$. Let $r_n$ be the covariance function of $\{Y_n\}$, and assume that either

$$r_n \log n \to 0 \text{ as } n \to \infty,$$

or

$$\sum_{n=1}^{\infty} r_n^2 < \infty .$$

Then it can be shown that $\{X_n\}$ belongs to the class $\ell$.

Let

$$a_n = (2 \log n)^{-\frac{1}{2}}$$

$$b_n = (2 \log n)^{\frac{1}{2}} - \frac{1}{2} (\log n)^{-\frac{1}{2}} (\log \log n + \log 4\pi)$$

and

$$\theta_1 = \max(\theta_1, \ldots, \theta_m) .$$

Also let

$$u_n = a_n x + b_n + \theta_1 .$$

Then from corollary 1, it follows that (observing that $\alpha_1 = 1$, $\alpha_i = \infty$, $i = 2, \ldots, m$.)

$$\lim_{n \to \infty} P[M_n \leq u_n] = \Lambda(x - \log p_1)$$

Example 3

Let

$$X_n = Y_n + \Psi_n,$$

where $\{Y_n\}$ is sequence of i.i.d. random variables and $\{\Psi_n\}$ is a deterministic periodic sequence of real numbers with integer period $m$. Assume that the distribution function $F$ of $Y_n$ satisfies the condition (e) for some real sequence $\{u_n\}$. Then $\{X_n\}$ belongs to the class $\ell$. (Compare with Tiago de Oliveira (1976)).

In fact this example can be given a more general form where $\{Y_n\}$ is a stationary sequence satisfying the condition $D'(u_n)$ and the condition in the following definition.

Definition 1.1

Given $m$ real sequences $\{u_{n,r},\ r=1,\dots,m\}$, we will say that $\{X_n\}$ satisfies the $D(u_{n,r})$ condition, if for any two sets of integers $(i_1,\dots,i_p)$ and $(j_1,\dots,j_q)$ such that,

$$1 \leqq i_1 < \dots < i_p < j_1 < \dots < j_q \leqq n \quad \text{and} \quad j_1 - i_p > k > 0\ ,$$

we have

$$\Big| P\big[X_{i_1} \leqq u_{n,r_{i_1}},\dots,X_{i_p} \leqq u_{n,r_{i_p}}, X_{j_1} \leqq u_{n,r_{j_1}},\dots,X_{j_q} \leqq u_{n,r_{j_q}}\big]$$

$$- P\big[X_{i_1} \leqq u_{n,r_{i_1}},\dots,X_{i_p} \leqq u_{n,r_{i_p}}\big] P\big[X_{j_1} \leqq u_{n,r_{j_1}},\dots,X_{j_q} \leqq u_{n,r_{j_q}}\big]\Big| \leqq \alpha_{n,k},$$

where

$$\lim_{k\to\infty} \limsup_{n\to\infty} \alpha_{n,k} = 0\ .$$

Let $F$ be the distribution function of $Y_n$ and let $u_n = a_n x + b_n$ be such that $\lim_{n\to\infty} F^n(u_n) = G(x)$, where $G(x)$ is non-degenerate. Let $\Psi_1 = \max(\Psi_1,\dots,\Psi_m)$ and let $u^*_n = u_n + \Psi_1$. Since

$$P[X_n \leq x] = F(x-\Psi_n) = F(x-\Psi_i),$$

for some $\Psi_i$, $i=1,\ldots,m$, when $n=Pm+i$, for some integer $P$, we see that $\{X_n\}$ has $m$ different one dimensional distributions functions $F_1,\ldots,F_m$, such that $F_i(x) = F(x-\Psi_i)$.

Now assume that $\lim\limits_{x \to x_0} \dfrac{1-F(x-\Psi_1)}{1-F(x-\Psi_i)}$ exists and equal to $\beta_i$.

Then from theorem 1, it can be shown that

$$\lim_{n\to\infty} P[M_n \leq u_n^*] = \exp\Big[-\frac{1}{m}\sum_{i=1}^{m}\frac{\log(G(x))}{\beta_i}\Big] .$$

The possible forms of $\beta_i$ are investigated by Tiago de Oliveira (1976) and the following theorem is due to him.

Theorem 1.2

(i) If $x_0<\infty$ then $\beta_1=1$ and $\beta_i=\infty$, $i=2,\ldots,m$.

(ii) If $x_0=\infty$ and $F$ is in the domain of attraction of $\Phi_\alpha(x)$ then $\beta_i=1$, $i=1,\ldots,m$.

(iii) If $x_0=\infty$ and $F$ is in the domain of attraction of $\Lambda$ then there is no general result.

## 2. POISSON CHARACTER OF THE UPCROSSINGS

Let $\{u_n\}$ be a sequence of real numbers satisfying (c) and (e). Denote by $C_n$ the number of upcrossings of the level $u_n$ by a sequence $\{X_n\}\varepsilon\ell$. Then we have the following theorem.

Theorem 2.1

$$\lim_{n\to\infty} P[C_n = r] = \frac{e^{-\lambda}\lambda^r}{r!}, \quad r=0,1,2,\ldots \tag{2.1}$$

where

$$\lambda = \exp(-\sum_{i=1}^{m} p_i z_i)$$

Here, we give a rough outline of the proof, the details can be filled without difficulty. (See Leadbetter (1974)).

Let $k$ and $t$ be fixed positive integers and let $I_1, I_1^*, \ldots, I_k, I_k^*$ be $2k$ consecutive intervals from $(1,\ldots,N)$ as defined in section I. Denote by $C_n(I)$ the number of times $X_r > u_n$, $r \in I$, where $I$ is any set of integers. Similarly denote by $C_N$ the number of times $X_r > u_N$, $r \in (1,\ldots,N)$

(i) $$\limsup_{n\to\infty} \left|P\left[C_N = r|\tilde{S}_N\right] - P\left[B_r(N)|\tilde{S}_N\right]\right| \leqq g(k), \qquad (2.2)$$

where, $g(k)\to 0$ as $k\to\infty$ and $B_r(N)$ is the event that for exactly $r$ of the $I_j$, $C_N(I_j) \geqq 1$.

(ii) $$\lim_{n\to\infty} E\left|P\left[M(I_j) \leqq u_N|\tilde{S}_N\right] - \exp(-\frac{1}{k}\sum_{i=1}^{m} p_i z_i)\right| = 0 \quad (2.3)$$

(iii) $$\limsup_{n\to\infty} E\left|P\left[B_r(N)|\tilde{S}_N\right) - \binom{k}{r}(1-p)^r p^{k-r}\right| = 0 \qquad (2.4)$$

where, $p = \exp(-\frac{1}{k}\sum_{i=1}^{m} p_i z_i)$.

Now, observing that

$$\left|P\left[C_N = r\right] - \binom{k}{r} p^{k-r}(1-p)^r\right| \leqq E\left|P\left[C_N = r|\tilde{S}_N\right) - P\left[B_r(N)|\tilde{S}_N\right)\right|$$

$$+ E\left|P\left[B_r(N)|\tilde{S}_N\right) - \binom{k}{r} p^{k-r}(1-p)^r\right|$$

(2.1) follows from (2.2) and (2.4), once we show that $N$ can be replaced by $n$.

REFERENCES

Leadbetter, M.R. (1974), "On Extreme Values in Stationary Sequences". Zeitschrift fur warhschein. Werw. Geb. 28, pp. 289--303.

Resnick, S.I. and Neuts, M.F. (1970),"Limit Laws for Maxima of a Sequence of Random Variables Defined on a Markov Chain". Adv. Appl. Prob. 2, pp. 323-343.

Tiago de Oliveira, J. (1976), "Asymptotic Behaviour of Maxima with Periodic Disturbances". Ann. Inst. Statist. Math. 28, pp. 19--23.

Turkman, K.F. and Walker, A.M. (1983), "Limit Laws for the Maxima of a Class of Quasi-Stationary Sequences". J. Appl. Probab. To be printed.

# AUTHOR INDEX

SUBJECT INDEX

GPSR Compliance
The European Union's (EU) General Product Safety Regulation (GPSR) is a set of rules that requires consumer products to be safe and our obligations to ensure this.

If you have any concerns about our products, you can contact us on

ProductSafety@springernature.com

In case Publisher is established outside the EU, the EU authorized representative is:

Springer Nature Customer Service Center GmbH
Europaplatz 3
69115 Heidelberg, Germany

www.ingramcontent.com/pod-product-compliance
Ingram Content Group UK Ltd.
Pitfield, Milton Keynes, MK11 3LW, UK
UKHW021903190726
13853UKWH00003B/1391